COARSE-GRAINED DELTAS

Coarse-Grained Deltas

EDITED BY
ALBINA COLELLA AND
DAVID B. PRIOR

SPECIAL PUBLICATION NUMBER 10 OF THE
INTERNATIONAL ASSOCIATION OF SEDIMENTOLOGISTS
PUBLISHED BY BLACKWELL SCIENTIFIC PUBLICATIONS
OXFORD LONDON EDINBURGH BOSTON
MELBOURNE PARIS BERLIN VIENNA

Editorial offices:
Osney Mead, Oxford OX2 0EL
25 John Street, London WC1N 2BL
23 Ainslie Place, Edinburgh EH3 6AJ
3 Cambridge Center, Cambridge
Massachusetts 02142, USA
54 University Street, Carlton
Victoria 3053, Australia

First published 1990

Set by Setrite Typesetters Ltd, Hong Kong
Printed and bound in Great Britain
by The Alden Press, Oxford

DISTRIBUTORS

Marston Book Services Ltd
PO Box 87
Oxford OX2 0DT
(*Orders*: Tel: 0865 791155
Fax: 0865 791927
Telex: 837515)

USA
Blackwell Scientific Publications
3 Cambridge Center, Cambridge
Massachusetts 02142, USA
(*Orders*: Tel: (800) 749–6102)

Canada
Oxford University Press
70 Wynford Drive
Don Mills
Ontario M3C 1J9
(*Orders*: Tel: (416) 441–2941)

Australia
Blackwell Scientific Publications
(Australia) Pty Ltd
54 University Street
Carlton, Victoria 3053
(*Orders*: Tel: (03) 347–0300)

British Library
Cataloging in Publication Data

Coarse-grained deltas.
1. Deltes
I. Colella, Albina II. Prior, David III. International Association of Sedimentologists IV. Series
551.456

ISBN 0–632–02894–7

Library of Congress
Cataloguing in Publication Data

Coarse-grained deltas/edited by Albina Colella and David Prior.
p. cm. — (Special publication number 10 of the International Association of Sedimentologists)
Includes bibliographical references and index.

ISBN 0–632–02894–7

1. Deltas. I. Colella, Albina. II. Prior, David B.
III. Series: Special publication ... of the International Association of Sedimentologists; no. 10.
GB591.C57 1990
551.4′56—dc20 90–614
CIP

Contents

Ancient Alluvial Deltas — Effects of Tectonics

Ancient Alluvial Deltas — Effects of Varying Climate and Water Level

Non-alluvial Deltas

Preface

Coarse-grained deltas, including both river and fan deltas, have become a focus of research in the last few years. Such basin-margin systems are sensitive recorders of tectonic, climatic and base-level conditions. The deposits of these systems are excellent potential reservoirs for hydrocarbons, and contain other mineral resources such as coals and ore placers. Coarse-grained deltas are thus of great interest to academic and industry researchers alike.

The International Symposium on Fan Deltas (1987), held in Norway, was a major forum for review of the state of knowledge concerning coarse-grained deltas. The symposium, and the subsequent volume (Nemec & Steel, 1988), identified a number of significant issues:

1 the concept and original definition of 'fan delta' are controversial and require further consideration;

2 there is a great variety in coarse-grained delta form and structure, and present understanding is, as yet, insufficient to construct adequate stratigraphic and facies models;

3 there is a need to combine the evidence from both ancient and modern examples of coarse-grained deltas in order to better understand sediment transport and depositional processes over underwater slopes;

4 case studies, in a wide range of depositional settings, are necessary in order to provide an adequate data base for a synthesis of coarse-grained delta development.

Accordingly, a major recommendation of the International Symposium was that a series of field-oriented workshops be held, one every 2 or 3 years, focusing on specific research problems and considering different deltaic types. The region of Calabria in southern Italy, with its beautifully exposed Pleistocene coarse-grained deltas, was chosen for the first workshop, to provide examples of Gilbert-type delta systems. The workshop, organized by A. Colella, was held near Cosenza in September 1988, and was sponsored by the University of Calabria, International Association of Sedimentologists, Consiglio Nazionale delle Ricerche, Cassa di Risparmio di Calabria e Lucania, Camera di Commercio di Cosenza, AGIP S.p.A. Milano, Comunitá del Pollino, Comune di Cassano, Assessorato al Turismo Regione Calabria, Amministrazione Provinciale di Cosenza, BP Milano. The workshop was attended by 50 participants from academic and industrial research centres in 15 countries.

Gilbert-type deltas, characterized by steep subaqueous slopes, have a rather unusual history of research. Sedimentological studies of ancient deltas actually commenced with Gilbert's (1885, 1890) classical descriptions of the Pleistocene deltaic sediments in Lake Bonneville. Influenced by these early concepts, researchers have sometimes followed the mistaken notion that deltaic deposits must contain a tripartite structure comprising topset, foreset and bottomset strata. As data from other deltas became available, the Gilbert-type model somehow lost prominence, perhaps also partly due to the mistaken assumption that this type of delta develops almost exclusively in lacustrine basins.

Gilbert-type and other 'steep-face' deltas returned into focus only recently, when their common occurrence in marine settings was realized and how specifically important and instructive such delta systems are in themselves. The renewed interest in Gilbert-type deltas stems partly from the recognition that the internal architecture of such deltas often provides an excellent record of synsedimentary tectonic activity at the basin margin, and of the variations in basinal regime, particularly water level, wave energy and sediment supply. Moreover, sedimentologists have also returned to Gilbert-type deltas in search of diagnostic criteria for identification of such deltaic deposits in the stratigraphic record. Specifically, there is a need to be able to discriminate between these deposits and those which also contain 'foresets', but which have quite different origins, such as steep-face progradational beaches, barrier spits, fluvial bars or large, straight-crested megaripples. Further, Gilbert-type deltas force reconsideration of the mechanisms and effects of a range of high-energy sediment transport processes, including underwater landsliding, avalanching, debris flows, high- and low-density turbidity flows.

The 1st International Workshop on Fan Deltas addressed many of these themes. The richly varied programme comprised field excursions, lecture and poster sessions, and evening discussion sessions. Special emphasis was placed on:

1 selected case histories of modern and ancient coarse-grained deltas;

2 physical processes governing coarse-grained delta geometry, facies and stratigraphic evolution;

3 facies and stratigraphic modelling of coarse-grained deltas;

4 effects of eustatic and tectonic factors on delta architecture;

5 contrasts and similarities between Gilbert-type deltas and progradational gravelly beaches;

6 temporal and geometrical relationships between the development of coarse-grained deltas and deep sea fans.

The discussions identified the need for three-dimensional rather than simple two-dimensional modelling of coarse-grained deltas, with integration of morphological, hydrodynamical and facies information from modern systems. Only through detailed, well-verified facies and related process models, will it be possible to approach stratigraphic and evolutionary modelling of such systems. 'Time-slice' studies of modern coarse-grained deltas are thus of critical importance. Specific results and conclusions of the workshop are featured in this volume.

This publication contains most of the contributions presented at the 1st International Workshop on Fan Deltas (1988), but also several additional papers, so the content of the volume does not strictly correspond to the workshop's programme. The selection of contributions deals with coarse-grained deltas, ranging from coarse sandy to gravelly, and the main focus is on steep-face systems, whose steep subaqueous slopes are dominated by high-energy processes. Both fan deltas and river deltas are represented, and examples of gravelly beaches have been included, for closer comparison and contrast with Gilbert-type deltas. Throughout the volume, there is emphasis on the subaqueous realm of the delta face, its sedimentary processes and facies associations.

The contributions have been arranged into five sections of the book as follows.

'General Considerations' deals with some introductory issues and general sedimentological aspects of coarse-grained deltas. The delta concept and classifications are discussed (Nemec), and a new, original classification for alluvial deltas is proposed (Postma). Gravity-driven sedimentary processes operative on steep, coarse-grained delta slopes are reviewed (Nemec), and varied patterns of sediment dispersal on such slopes are described (Prior and Bornhold). The signatures of varying tectonic and environmental conditions in the facies associations of fan deltas are explored (Dabrio), and the role of tectonics in the formation and evolution of coarse-grained deltas in rift basins is discussed (Gawthorpe and Colella). The interacting roles of faulting and eustatic sea-level changes in the development of Pleistocene fan deltas are considered, and a comparison is made between the deposits of fan deltas and those of coarse-grained linear coasts with no localized alluvial input (Bardaji *et al.*).

'Modern Alluvial Deltas' contains case studies that further illustrate some of the major aspects discussed in the previous section. The morphology and associated facies pattern of a marine Gilbert-type delta are described in detail (Corner *et al.*). The sedimentary processes on the subaqueous slopes of fjord fan deltas are then discussed, with emphasis on the role of density underflow processes (Bornhold and Prior).

'Ancient Alluvial Deltas — Effects of Tectonics' presents case studies of ancient lacustrine and marine coarse-grained deltas, where tectonics are recognized to have played a major role in controlling the delta development and architecture. The effects of graben-margin tectonics on the progradation style and sediment dispersal pattern in a sub-Recent fan-delta system are discussed (Kazancı). The architectural signatures resulting from syndepositional extensional faulting in a marine pull-apart basin are deciphered in deltaic deposits (van der Strateen). An example from the lacustrine North Park Basin (Flores) illustrates the different facies and architecture of transverse Gilbert-type systems deposited along a shallow, rapidly subsiding basin-margin, and longitudinal Gilbert-type systems developed in the deep, slowly subsiding basin centre. The sedimentary facies of a marine fan delta in a rapidly subsiding basin are described (Hwang and Chough), and sequence stratigraphic analysis of a Gilbert-type deltaic complex is presented (García-Mondéjar).

'Ancient Alluvial Deltas — Effects of Varying Climate and Water Level' presents case histories from lacustrine and marine settings. The principal control of climate on the development of diverse coarse-grained deltas in a lacustrine basin is discussed, with emphasis on the palaeoclimatic and palaeohydrologic information recorded in Gilbert-type deltaic sequences (Bowman). Examples of steep-face deltas fed by a variety of glacial outwash systems, both proglacial and englacial, are presented (Martini). The effects of diurnal and seasonal hydrological variations are deciphered from glacial lacustrine delta foreset deposits (Mastalerz). The depositional conditions and facies of wave-dominated Gilbert-type deltas in drowned palaeovalleys are discussed and a comparison is made with fluvial-dominated analogous systems (Massari and Parea), pointing out that a continuum exists between gravelly beach sequences of increasing 'ramp' height and wave-dominated Gilbert-type deltaic sequences.

'Non-Alluvial Deltas' presents an illustrative example of 'lava delta' that shows distinct foreset bedding and points to the striking analogy between conventional, alluvial Gilbert-type systems and those fed by volcanic sources (Porębski and Gradzínski).

A. Colella, *Universita della Calabria, Italy*
D.B. Prior, *Atlantic Geoscience Center, Nova Scotia*

REFERENCES

Gilbert, G.K. (1885) The topographic features of lake shores. *Ann. Rept. U.S. geol. Surv.* **5**, 69–123.

Gilbert, G.K. (1890) Lake Bonneville. *Monogr. U.S. geol. Surv.* **1**, 438 pp.

Nemec, W. & Steel, R.J. (eds.) (1988) *Fan Deltas, Sedimentology and Tectonic Settings.* Blackie and Son, London, 444 pp.

Acknowledgements

We would like to thank the following group of international sedimentologists who assisted us in reviewing of contributions: B.D. Bornhold, L.B. Clemmensen, G.D. Corner, C.J. Dabrio, F.G. Ethridge, R.L. Gawthorpe, R. Higgs, R. Hiscott, R.A. Kostaschuk, M.R. Leeder, P.I. Martini, M. Marzo, F. Massari, A.M. McCabe, S.B. McCann, J.G. McPherson, W. Nemec, L.H. Nielsen, T. Nilsen, G.G. Ori, S.J. Porębski, G. Postma, I. Reid, M. Sagri, R.J. Steel, F. Surlyk.

Finally, we would like to give a special thanks to W. Nemec for his valuable suggestions and assistance with editorial work for this volume.

General Considerations

Spec. Publs int. Ass. Sediment. (1990) **10**, 3–12

Deltas — remarks on terminology and classification

W. NEMEC

Geological Institute (A), University of Bergen, 5007 Bergen, Norway

ABSTRACT

The concept of a 'delta' and some of the major terminological problems associated with recent progress in delta research are discussed. The existing and newly suggested classifications of alluvial deltas are reviewed, with emphasis on their advantages and weaknesses. It is suggested that several classification schemes can be used in delta research, but in a hierarchical way — with a descriptive and relatively detailed classification on the basic level of field studies.

INTRODUCTION

The recent progress in delta research has broadened the concept of a 'delta' as such, and the terminology adopted by researchers for deltas becomes increasingly more elaborate. As the emphasis has shifted towards the feeder systems of the deltas, the researchers find themselves embroiled in terminological controversies and discussions on definitions and actual field criteria. The situation appears to be rather alarming. Until many of the existing definitions are clarified and made applicable to the rock record, the feeder-based terminology for deltas has to be kept simple and rather general in order to be functional.

Another result of recent research progress is that the classification schemes for alluvial deltas have proliferated. There is a wide range of approaches and a great diversity of criteria that have been employed. When applying these schemes, researchers should be more aware of their advantages and shortcomings.

DELTAS

The concept of a delta is one of the oldest in earth sciences, dating back to *c.* 450 B.C., when the ancient voyager and historian Herodotus observed that the alluvial plain at the mouth of the Nile was similar, in plain-view shape, to the Greek letter Δ. Accordingly, the term 'delta' in geomorphology and sedimentary geology has traditionally been associated with rivers, to denote the coastal prism of land-derived sediment built by a river into a lake or a sea (e.g. Barrell, 1912; Johnston, 1921; Holmes, 1965; Coleman & Wright, 1975; Elliott, 1978, 1986; Coleman, 1981; Miall, 1984).

As descriptions of modern deltas increased in number, so it became evident that such river-formed coastal protuberances varied enormously in their characteristics and that very many deltas were not 'delta-shaped' at all. The term has thus lost its original geometric meaning and become essentially a genetic one. Furthermore, researchers observed coastal protuberances formed by terrestrial systems other than simple rivers, and terms like fan delta (Holmes, 1965), braidplain delta (Orton, 1988), slope-apron delta (Busby-Spera, 1988) and lava delta (Holmes, 1965) were introduced into the literature. The connotation of the term delta has thus become even more general, so that the traditional definition apparently requires some reformulation.

A *delta* can be defined as a deposit built by a terrestrial feeder system, typically alluvial, into or against a body of standing water, either a lake or a sea. The result is a localized, often irregular progradation of the shoreline controlled directly by the terrestrial feeder system, with possible modification by basinal processes, such as the action of waves or tides. It does not follow, of course, that deltas are necessarily associated with an overall marine or lacustrine regression; many deltas have formed as elements of retreating shorelines, where the episodes of delta progradation accompanied an overall, longer-term transgression of the sea or lake. Moreover, there are deltas that are totally subaqueous, lacking (as yet) the typical subaerial expression in the form of a prograding 'delta plain'.

Alluvial deltas

The increased recent interest in 'fan deltas', the variety of deltas built by alluvial fans, and the accompanying wave of discussions around the fan-delta concept (McPherson *et al.*, 1987, 1988a,b; Nemec & Steel, 1987, 1988; Dunne, 1988; Orton, 1988; Fernández *et al.*, 1988) have further shifted emphasis towards the delta's feeder system itself. In order to distinguish between a river (common) delta and an alluvial-fan delta, it is obviously crucial that the researcher recognizes the direct influence of either a river or an alluvial fan as the principal supplier of sediment. Although sedimentologists have been recognizing alluvial rivers and fans in the stratigraphic record for many decades, the discussion around the fan-delta concept has made it plainly clear that the distinction, in reality, may not be an easy task. It has suddenly become evident that we simply lack a widely acceptable definition of an alluvial fan, one that would be both general and readily applicable to the rock record.

This is not an isolated case in geology where an earlier definition, accepted for decades on a textbook level, suddenly proves to be inadequate in the light of intensified research. A similar painful realization has been reached by the editors of a recent excellent volume on submarine fans (Bouma *et al.*, 1985). The research on modern submarine fans in the last decade has made it clearly evident that we lack an acceptable set of terms, definitions and practical criteria for such depositional systems. Submarine fans have essentially been rediscovered. The 'real' submarine fans hardly match the simplistic models established in our minds, those 'imaginary' idealized fans that we have derived from textbook definitions. Submarine fans appear to vary enormously. Moreover, very few such fans are 'fan-shaped', and there is also the controversy as to where to place the boundary between 'lower fan' and 'basinal (abyssal) plain'. The problems are quite familiar when one thinks of alluvial fans.

In their concluding remarks, the editors of the aforementioned volume (*op. cit.*, p. 342) quoted a phrase from an American cartoonist: 'We have met the enemy and he is us'. The same, indeed, might be said of our recent attempts to categorize alluvial deltas according to their feeder systems, when we suddenly realize that some of the definitions we have adopted and used for years appear to be inadequate or difficult to apply.

Although there are numerous studies proving that systems like fan deltas, braidplain deltas and river deltas *can* be distinguished in the stratigraphic record (see Koster & Steel, 1984; volume edited by Nemec & Steel, 1988; Van der Meulen, 1989; Nemec, 1990), we will inevitably entangle ourselves in boundless controversies and discussions as long as the definitions that we apply are vague and unaccompanied by well-specified, geologically applicable criteria. The researchers have to decide what is the exact meaning of terms like 'alluvial fan', 'braidplain' or '(braided) river', and more importantly — how such depositional systems, especially their coastal parts, are to be distinguished from one another in the stratigraphic record.

It is beyond the scope of this short note to attempt to answer these questions, and the author does not even claim to have any satisfactory answers at hand. The intention of the present remarks is merely to argue for the readoption of a convenient general term, like *alluvial delta*, which obviates the need to identify the exact nature of the alluvial feeder system of a delta. Such a general term is certainly useful in rock-record studies, where the nature of the alluvial feeder system of an ancient delta is often difficult or virtually impossible to specify, particularly where the exposures are poor or limited. Whenever a more exact recognition can be made, the researcher should be free to adopt a more specific term for a particular alluvial delta (Fig. 1). The fact is that various types of alluvial system have somehow *been* recognized and distinguished from one another in the rock record for many decades, and there is no reason why researchers should not continue such attempts with respect to delta feeder systems.

The specific qualifying term should thus reflect, as closely as possible, our interpretation of the subaerial feeder system of a particular alluvial delta. Some possible terms are listed in Figure 1 (see also Nemec & Steel, 1988, fig. 1). If the subaerial system is dominated by rockfall avalanches and has a conical or broader, apron-like geometry, as observed along some cliffed rocky coasts, terms like 'scree-cone delta' or 'scree-apron delta' would be appropriate

(Fig. 1). Whatever specific terms are adopted for such a purpose, they should be informative and correspond to the terminology that we otherwise use for terrestrial depositional systems.

It is further suggested that the term 'braid delta', as defined by McPherson *et al.* (1987, 1988b), should be abandoned, even though it has already been applied by some authors and was initially adopted by the writer himself (see Nemec & Steel, 1988). The literal meaning of the term is odd, and it inadvertently implies that a term like 'meander delta' might be introduced as well, as a parallel. The traditional terms 'delta with braided distributaries' and 'delta with meandering distributaries' are more informative and far more appropriate.

Non-alluvial deltas

If we accept, after Holmes (1965, 1978), that deltas are not necessarily formed by rivers, or alluvial systems in general, then our terminological scheme must, of course, provide room for non-alluvial deltas. These would include *pyroclastic deltas* and *lava deltas* (see Holmes, 1965, fig. 637, or 1978, fig. 23.61), which are the localized outbuildings of coasts due to pyroclastic flows and lava flows, respectively (Fig. 1). Some of the deltas may be of 'mixed' type, comprising both kinds of volcanogenic deposits, and others may include simple volcaniclastic debris-flows (lahars) or even stream-derived alluvium.

Advancing volcanic coasts are characteristic of many island-arcs or isolated island volcanoes that have established their craters above sea level. Among the more spectacular modern examples are the lobe-shaped extensions of the coasts in Hawaii and Iceland, where voluminous outpourings of lava have advanced over the seafloor. Many such 'lava deltas' involve significant amounts of lava breccia and other hyaloclastic deposits, and may contain intervening marine clastics of purely detrital provenance. In some cases, the subaerial volcanic eruptions appear to have culminated in the seaward progradation of an accumulation of lava flows and breccias that has distinct 'foreset' bedding and resembles a Gilbert-type delta (see Porębski & Gradziński, this volume).

Pyroclastic deltas would normally involve a volcaniclastic fan or steep cone, or a broader apron, prograding into the sea or lake. Depending on the composition of the magma and the physical character of the eruptions (whether phreatic or purely magmatic), the principal depositional mechanisms may range from various pyroclastic flows, with or without internal welding, to strongly turbulent pyroclastic surges and air-borne suspension falls (see

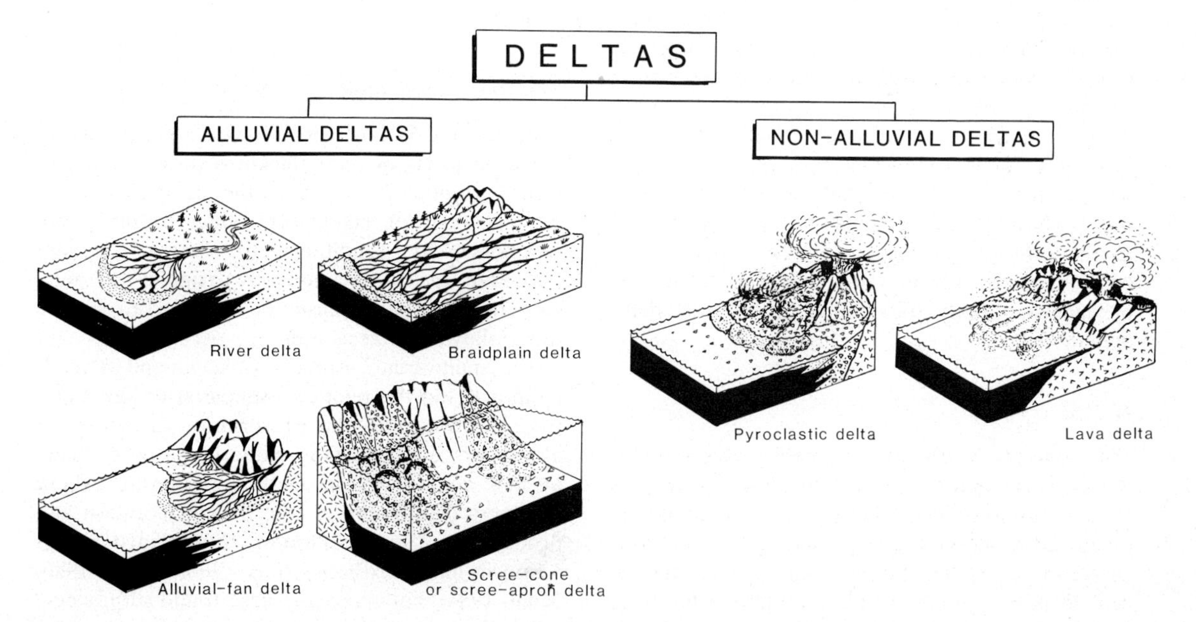

Fig. 1. Suggested broad division of deltas, into *alluvial* and *non-alluvial* varieties, with specific examples of both delta types.

Fisher & Schmincke, 1984; Cas & Wright, 1987).

There are also cases where a localized coastal outbuilding appears to be related to a prograding aeolian dune system. Among spectacular modern examples are the large, tongue-like 'aeolian deltas' that have prograded into the saline lake Ounianga Serir, Sahara (see Beadle, 1981, fig. 11.5).

CLASSIFICATION OF ALLUVIAL DELTAS

A variety of approaches, adopting different criteria, have been proposed in the literature to classify alluvial deltas. These classification schemes are generally well known to delta students and hence are only briefly reviewed here, with some critical comments and new suggestions.

Existing classifications

Five of the existing major concepts of delta classification are reviewed below. These are the classification schemes which pertain, or might potentially pertain, to alluvial deltas in general. Schemes of more limited scope or design, like Colella's (1988) classification of Gilbert-type deltas, are not reviewed here.

A drawback with these classification schemes is that they often employ criteria which are more genetic than descriptive, and hence not readily applicable to the rock record. A classification scheme, in order to be practical and reasonably universal, should rely on some easily recognizable, descriptive attributes, rather than on vaguely specified process interpretations or more complex genetic inferences. Moreover, it is somewhat paradoxical that the schemes tend to emphasize delta aspects other than those which a field sedimentologist might regard as the principal or most essential characteristics of a prism of 'alluvium that has prograded into or against a body of standing water' (cf. delta definitions).

Feeder system

The concept of categorizing deltas according to the gross character of their alluvial feeder systems (Fig. 1 and preceding discussion) can essentially be attributed to Holmes (1965). Among his numerous successors are sedimentologists entangled in the recent dispute around the topic of fan deltas (e.g. McPherson *et al.*, 1987, 1988a,b; Nemec & Steel, 1987, 1988; Dunne, 1988; Orton, 1988; Nilsen, 1989). The concept itself is relatively simple and attractive, and the nature of an alluvial system is, indeed, very important to the character of the resulting delta (e.g. see considerations by Elliott, 1986; Prior & Bornhold, 1988; Orton, 1988; Colmenero *et al.*, 1988).

The classification scheme, however, appears to be poorly defined and has several shortcomings. First, it inadvertently ignores the characteristics of the delta as such, by focusing on the subaerial, landward part of the system. Second, we apparently have problems with categorizing alluvial systems, and even more serious difficulties with applying our ambiguous categories to the rock record (see preceding discussion). Third, the classification is genetic, rather than descriptive, and hence relies heavily on sheer inference when applied to the rock record.

Therefore, one may challenge the wisdom of using this particular criterion as the basis for what should be an objective, verifiable and readily applicable classification. It does not mean that deltas, in principle, should not be categorized according to their feeder systems. However, such broad and genetic categories are not the most useful or most essential, and should rather be considered as higher-order, secondary ones (see Postma, next chapter). Suffice it to realize that the same river, or alluvial fan, might produce quite different deltas if prograding into different basins or even into different parts of the same basin.

Thickness distribution

Coleman and Wright (1975) have classified modern deltas in terms of their thickness distribution patterns, considered to reflect the combined effect of a number of major controlling factors, from sediment-yield conditions to basinal regime. Although the classification was designed for sandy river deltas, it might easily be extended to include other alluvial deltas as well.

The 'multivariate' aspect of Coleman and Wright's approach and the relatively simple criterion it employs, both render the classification attractive. It also bears an important utilitarian aspect; many ancient deltas are hydrocarbon reservoirs, where thickness distribution is crucial to exploration and production. The classification thus gained recognition in many textbooks (e.g. Elliott, 1978, 1986; Miall, 1984), but appears to have found surprisingly few followers.

Although the classification favourably focuses on the delta as such and has many advantages, it is not free of shortcomings — both practical and conceptual:

1 It requires quite detailed thickness data, which — apart from some densely drilled and untectonized areas — are seldom available for ancient deltas.

2 Many ancient deltas were long-lived and appear to be far more complicated, especially in terms of the spatial thickness patterns, than their modern, relatively young and simple counterparts.

3 The classification relies on the empirical assumption that a unique combination of physical factors, or depositional conditions, determines a particular pattern of delta thickness distribution. However, it is uncertain to what extent the limited number of actualistic factor combinations (see the six delta 'types' distinguished by Coleman & Wright, 1975) apply to the geological past and ancient deltas.

4 It is also uncertain to what extent a given thickness pattern is truly unique to a particular combination of factors. Arguably, different combinations of factors may lead to similar results.

5 Last but not least, the assumption that the thickness distribution should faithfully reflect the morphology of a delta may not necessarily be correct. As we have learned from some deltas and other modern systems, the spatial thickness patterns may be heavily affected by differential subsidence or pre-existing basin-floor topography (e.g. Bull, 1977, figs 11–13; Bouma *et al.*, 1985).

In summary, the concept of Coleman and Wright (1975) is generally sound and attractive, but its application to ancient deltas is hindered by practical difficulties and theoretical uncertainties.

Tectono-physiographic setting

Ethridge and Wescott (1984) distinguished three delta categories, to reflect different tectono-physiographic coastal settings: 'shelf-type', 'slope-type' (shelf-margin), and 'Gilbert-type' deltas, the latter ascribed to 'lacustrine and intracratonic' coastal settings. Although the classification was introduced for alluvial-fan deltas, it might equally pertain to any alluvial deltas; in fact, the term 'Gilbert-type delta' derives from a variety of river delta. The concept has found many followers among the fan-delta students (e.g. Colella, 1988; Massari & Colella, 1988; and other examples in same volume), not least because it was the first classification scheme ever suggested for such systems.

The categories are broad and have some special appeal to basin researchers concerned with tectonic settings, but the classification itself has serious shortcomings:

1 The scheme focuses on tectonic/geomorphic 'settings', rather than actual delta characteristics, thus leading to paradoxes. The delta as such becomes essentially irrelevant, and observation of its characteristics virtually unnecessary, if the researcher happens to know the basinal setting beforehand.

2 When applied to the rock record, the classification relies on complex basinal inferences, rather than on objectively observable or measurable features. This renders the approach poorly verifiable. (Gilbert-type deltas are an exception, but in turn — their uniqueness to the assumed settings is questionable.)

3 The terminology adopted for the scheme is unfortunate and misleading. Suffice it to note that Gilbert-type deltas are known also from shelves; 'shelf-type' deltas (i.e. deltas lacking both Gilbertian foresets and deep-sea extensions) are common in intracratonic seas or even inland lakes; 'slope-type' deltas, in their upper, shelf-hosted segments, would normally involve one of the other two delta varieties; and finally, slope-type deltas are known also from settings other than the shelf breaks of continental margins. In addition, Wood and Ethridge (1988) have recently adopted the term 'mouth bar-type' delta (after Dunne & Hempton, 1984), apparently to denote some shelf-type deltas in a non-shelf setting. However, it is unclear whether this additional term, denoting non-Gilbertian delta characteristics and shallow-water settings, should actually replace the shelf-type category altogether.

In conclusion, the idea to categorize deltas according to their tectono-physiographic basinal settings is certainly interesting and has a broader geological appeal, but the classification, in its present form, requires reconsideration and modification. The categories are too broad and pay too little attention to the actual delta sediments (e.g. shelf-type deltas alone vary enormously as sedimentary deposits). Even when improved, the scheme would hardly play the role of a practical, basic classification for use in the field.

Delta-front regime

Galloway (1975), in his well-known ternary diagram, has classified river deltas as 'fluvial-dominated', 'wave-dominated' and 'tide-dominated', with a possible full range of intermediate, 'mixed-type' varieties. The classification refers to the delta-front regime, and apparently stems from the earlier concept of 'high-constructive' deltas, dominated by fluvial processes, and 'high-destructive' deltas, dominated by the basinal processes (Fisher *et al.*, 1969). Galloway's scheme is the one most commonly used by sedimentologists. It has gained much popularity among students of river deltas (see Elliott, 1978, 1986; Miall, 1984), and has also been adopted for fan deltas (Kleinspehn *et al.*, 1984; Orton, 1988).

The scheme is quite comprehensive and, importantly, pertains to the delta as such, but is not free of some inherent weaknesses:

1 The scheme is genetic, rather than descriptive, and relies heavily on the researcher's ability to make the appropriate genetic inferences and semi-quantitative estimates. Even with good-quality exposures, such precise interpretations are often difficult to reach and are necessarily subjective.

2 The concept, though referring to delta-front regime, actually focuses on the *degree of reworking* of the delta front by waves and tides. Suffice it to note that a delta can only be classified as 'fluvial-dominated' if the researcher has recognized that the degree of reworking was negligible. The concept thus puts emphasis on basinal processes, rather than on the prograding coastal alluvium itself. As a consequence, the regime of the delta plain is virtually ignored, although it may not necessarily correspond to the regime of the delta front (see also comments by Elliott, 1978, p. 100; 1986, p. 116).

3 The classification diagram requires that the positions of individual deltas are plotted in a semi-quantitative or, ideally, quantitative way (see also Elliott, 1986, p. 116). However, it is unclear how the degree of reworking is to be quantified, even in approximate terms. Although many researchers perceive it as the relative volume, or thickness, of the preserved wave- or tide-worked sediments, the adequacy of this criterion is by no means obvious. Suffice it to say that a very powerful storm or tsunami wave is capable of removing a large portion of a delta, especially if relatively small and fine-grained, while the actual record of such a destructive event may be little more than a solitary erosional surface. Moreover, the delta-plain distributaries often remove shoreline deposits, thus considerably reducing the actual preservation potential of delta-front facies. Resedimentation processes on steep-face deltas further complicate the problem, and the slopes of such deltas commonly extend far below wave base.

Despite these shortcomings, Galloway's approach is sound and certainly worth pursuing. However demanding the criterion itself may appear to be, the remarkable popularity of this classification scheme clearly testifies to its usefulness in sedimentary geology.

Delta-front regime and grain size

Orton (1988) has recently extended the ternary classification diagram of Galloway (1975) in a fourth dimension — to account for the dominant *grain size* of sediment delivered to the delta front. The reasoning behind this modification is that the degree of reworking of a delta front is not independent of the grain size of the delta-front material. For example, the same basinal wave regime will have a different impact on a fine-grained sandy delta than on a coarse gravelly delta. Thus, it is more reasonable that only deltas of similar grain sizes are compared.

This modification makes Galloway's scheme somewhat more elaborate and more appealing, essentially without complicating its practical aspects. Grain sizes are readily determinable in the field, and are part of routine sedimentological data. Despite its advantages, Orton's approach inherits all of the weaknesses of Galloway's scheme (see preceding section) and has also its own weakness: since the degree of reworking depends on the grain size, one may question the wisdom of building a classification scheme on the basis of two non-independent variables. Furthermore, the criterion of 'dominant grain size' itself may beg the question as to which delta-front facies are actually to be compared or given priority; many delta fronts comprise highly heterogeneous deposits, where the distributary channel-fills, mouth bars, interdistributary bay-fills and related facies differ considerably in their relative abundance and grain-size characteristics.

New classifications

Two of the most recent suggestions will now be reviewed briefly. The two classification schemes are somewhat different from the previous ones, not least because they are more descriptive and place more emphasis on internal delta characteristics.

Grain size and delta-face slope

Corner (this volume, and pers. comm., 1989) suggests that alluvial deltas can be classified quite satisfactorily on the basis of the dominant grain size of sediment delivered to the delta front and the gradient of the delta-face slope. The two variables are among the descriptive parameters readily determinable in a delta outcrop and routinely included in sedimentological descriptions of deltas. The existence of an avalanching slope, whether Gilbert-type (general) or merely related to individual mouth bars, is easily recognizable in ancient deltas by the presence of 'megaforeset' cross-bedding or sets of large-scale, 'microdelta-type' cross-stratification, respectively. The approach is thus practical, elegantly simple and fully descriptive.

Corner's bivariate diagram (Fig. 2) provides room for virtually all alluvial deltas: from steep-face Gilbert-type and related, totally subaqueous deltas (see Nemec, later chapter in this volume), to gentle-face shoal-water deltas (e.g. see Nemec & Steel, 1987, fig. 2b,c); and from muddy lacustrine deltas (some of which are Gilbert-type), to very coarse-grained gravelly deltas. The scheme has no major weaknesses and the criteria adopted are readily verifiable, rendering the classification potentially very useful in field studies. (However, see earlier comment on the dominant grain-size criterion.)

Main physiographic attributes

Postma (this volume, next chapter) has established an elaborate classification that employs the following three main criteria:

1 The physiography of delta-front shoreline, categorized as either *irregular*, with recognizably isolated mouth-bar lobes, or relatively *uniform* — with welded/amalgamated mouth bars and/or smooth outline due to wave action.

2 The gradient of the delta-face slope (as in Corner's scheme above).

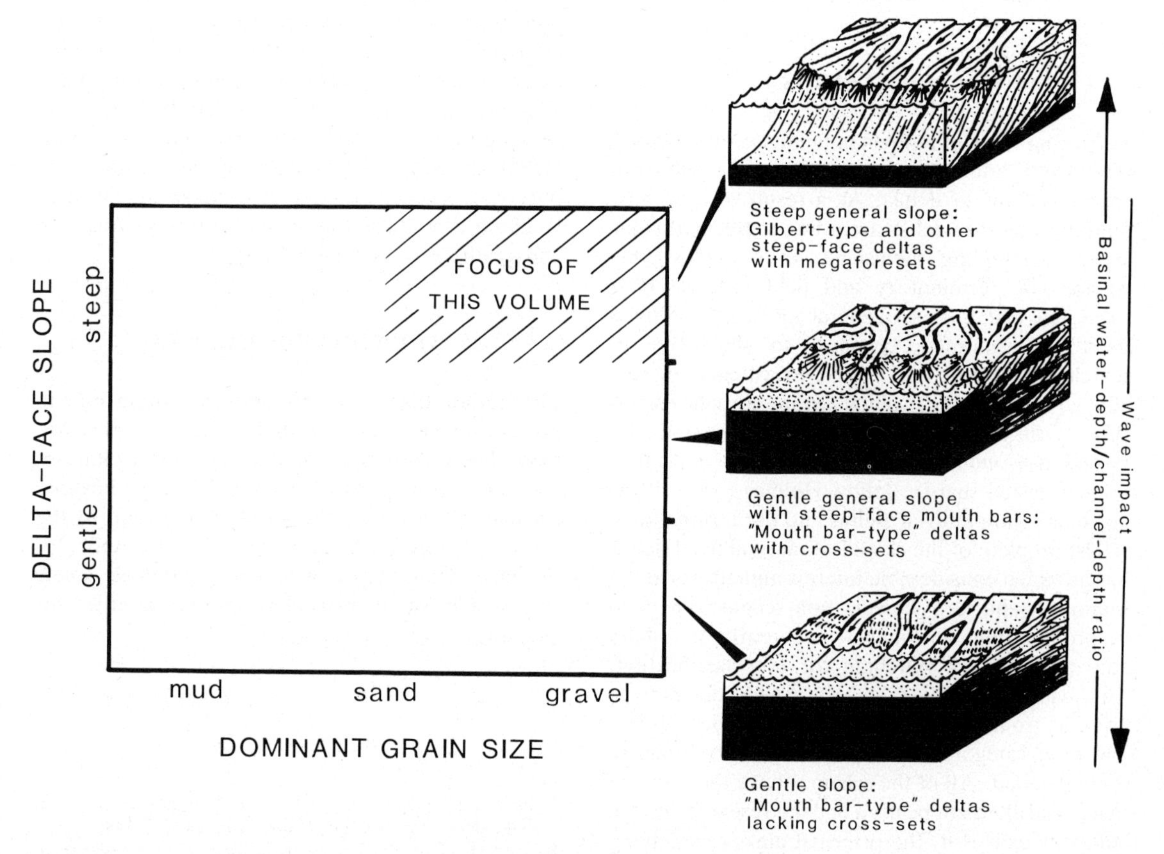

Fig. 2. Corner's classification scheme, in a graphical form, supplemented with explanatory comments and some genetic remarks.

3 The depth of the delta toe, whether *above* or *below* storm-wave base.

In terms of these three basic criteria, Postma has distinguished 12 delta 'prototypes' as some kind of conceptual end-members for comparative purposes and actual categorizing of alluvial deltas. He has further demonstrated that a great number of modern and ancient deltas can readily be accommodated in this classification scheme.

This sophisticated approach clearly focuses on several important physiographic characteristics of the delta as a whole, with emphasis on the delta front and subaqueous slope. In its applications to ancient deltas, the classification relies, of course, on facies analysis and related palaeophysiographic inferences. However, Postma seems to have carefully selected those major attributes of a delta that are almost routinely inferred by the sedimentologists in delta studies. The scheme, once applied to ancient deltas, will thus further stimulate facies-based palaeophysiographic analysis of these systems.

CONCLUSIONS

The recent wave of interest in fan deltas has focused researchers' attention on the nature of the subaerial feeder systems of deltas. As a result, people now tend to categorize deltas according to this particular aspect, entangling themselves in boundless discussions on terminology and field criteria. It is suggested that a useful general term, like 'alluvial delta', is readopted, primarily to allow the researchers to escape the difficulty of categorizing the alluvial feeder of a delta. More specific terms (Fig. 1) can be used once the character of the feeder system has unambiguously been recognized. It is also suggested that we follow Holmes (1965, 1978) and admit 'non-alluvial' deltas into our terminology.

The problem of the classification of alluvial deltas has attracted considerable interest and efforts in the literature, and new classification schemes seem to be proliferating. This means, apparently, that delta students are still searching for *the* classification. The available schemes employ a wide diversity of criteria, from purely genetic to purely descriptive, and offer categories that range from very broad to quite detailed. All of the schemes have their advantages, and all might be used in delta research. As the author perceives it, the principal aim of classifying deltas is to recognize and conceptually systematize their natural variability, and to summarize our knowledge of the essential aspects of this variability. The choice of scheme would then depend on what kind of basic information about a particular delta is to be used for specific comparative purposes.

The schemes, however, should preferably be applied in a hierarchical manner, with a descriptive and relatively detailed scheme on the basic level of field research and the broadest genetic schemes on the highest level of more general considerations. For example: Corner's scheme might serve as a descriptive, practical and reasonably informative 'first-order approximation' of the delta type, with several important genetic implications. Postma's scheme might serve for a more precise descriptive classification, once sufficiently detailed facies analysis has been completed, and its genetic implications would then be quite detailed and satisfactory. The Galloway/Orton scheme would be most useful when the actual impact of basinal processes on the delta front is considered, and the feeder-based scheme for considerations focused on the subaerial component of the deltaic system. The Coleman–Wright approach would be particularly useful for making reservoir geometry predictions, and the Ethridge–Wescott approach, in turn, would be crucial in relating deltas to plate-tectonic settings. The individual schemes would thus serve specific synthesizing purposes, and their combined usage might then facilitate a multilevel approach to the classification and synthesis of alluvial deltas.

ACKNOWLEDGEMENTS

The author thanks the editors for inviting him to present his views and 'second thoughts' in this volume. The manuscript was read by John Collinson, Geoffrey Corner, Huw Davies and George Postma; Michael Talbot drew the author's attention to the aeolian 'deltas' of Saharan lakes. The author appreciates their help and support, but feels solely responsible for the critical views presented in this chapter.

REFERENCES

Barrell, J. (1912) Criteria for the recognition of ancient delta deposits. *Bull. geol. Soc. Am.* **23**, 377–446.

Beadle, L.C. (1981) *The Inland Waters of Tropical Africa: An Introduction to Tropical Limnology*, 2nd edn. Longman, London, 475 pp.

Bouma, A.H., Normark, W.R. & Barnes, N.E. (Eds.) (1985) *Submarine Fans and Related Turbidite Systems*. Springer-Verlag, New York, 351 pp.

Bull, W.B. (1977) The alluvial-fan environment. *Prog. phys. Geogr.* **1**, 222–270.

Busby-Spera, C.J. (1988) Development of fan-deltoid slope aprons in a convergent-margin tectonic setting: Mesozoic, Baja California, Mexico. In: *Fan Deltas: Sedimentology and Tectonic Settings* (Ed. by W. Nemec and R.J. Steel), pp. 419–429. Blackie and Son, London.

Cas, R.A.F. & Wright, J.V. (1987) *Volcanic Successions: Modern and Ancient*. Allen and Unwin, London, 530 pp.

Colella, A. (1988) Pliocene–Holocene fan deltas and braid deltas in the Crati Basin, southern Italy: a consequence of varying tectonic conditions. In: *Fan Deltas: Sedimentology and Tectonic Settings* (Ed. by W. Nemec and R.J. Steel), pp. 50–74. Blackie and Son, London.

Coleman, J.M. (1981) *Deltas: Processes and Models for Exploration*. 2nd edn. Burgess Publ. Co., Minneapolis, 124 pp.

Coleman, J.M. & Wright, L.D. (1975) Modern river deltas: variability of processes and sand bodies. In: *Deltas: Models for Exploration* (Ed. by M.L. Broussard), pp. 99–149. Houston Geological Society, Houston.

Colmenero, J.R., Agueda, J.A., Fernández, L.P., Salvador, C.I., Bahamonde, J.R. & Barba, P. (1988) Fan-delta systems related to the Carboniferous evolution of the Cantabrian Zone, northwestern Spain. In: *Fan Deltas: Sedimentology and Tectonic Settings* (Ed. by W. Nemec and R.J. Steel), pp. 267–285. Blackie and Son, London.

Dunne, L.A. (1988) Fan-deltas and braid deltas: varieties of coarse-grained deltas — Discussion. *Bull. geol. Soc. Am.* **100**, 1308–1309.

Dunne, L.A. & Hempton, M.R. (1984) Deltaic sedimentation in the Lake Hazar pull-apart basin, south-eastern Turkey. *Sedimentology* **31**, 401–412.

Elliott, T. (1978) Deltas. In: *Sedimentary Environments and Facies*, 1st edn. (Ed. by H.G. Reading), pp. 97–142. Blackwell Scientific Publications, Oxford.

Elliott, T. (1986) Deltas. In: *Sedimentary Environments and Facies*, 2nd edn. (Ed. by H.G. Reading), pp. 113–154. Blackwell Scientific Publications, Oxford.

Ethridge, F.G. & Wescott, W.A. (1984) Tectonic setting, recognition and hydrocarbon reservoir potential of fan-delta deposits. In: *Sedimentology of Gravels and Conglomerates* (Ed. by E.H. Koster and R.J. Steel). Mem. Can. Soc. Petrol. Geol. 10, 217–235.

Fernández, L.P., Agueda, J.A., Colmenero, J.R., Salvador, C.I. & Barba, P. (1988) A coal-bearing fan-delta complex in the Westphalian D of the Central Coal Basin, Cantabrian Mountains, northwestern Spain: implications for the recognition of humid-type fan deltas. In: *Fan Deltas: Sedimentology and Tectonic Settings* (Ed. by W. Nemec and R.J. Steel), pp. 286–302. Blackie and Son, London.

Fisher, R.V. & Schmincke, H.-U. (1984) *Pyroclastic Rocks*. Springer-Verlag, New York, 528 pp.

Fisher, W.L., Brown, L.F., Scott, A.J. & McGowen, J.H. (1969) Delta systems in the exploration for oil and gas. *Trans. Gulf-Cst. Ass. geol. Soc.* **17**, 105–125.

Galloway, W.E. (1975) Process framework for describing the morphology and stratigraphic evolution of the deltaic depositional systems. In: *Deltas: Models for Exploration* (Ed. by M.L. Broussard), pp. 87–98. Houston Geological Society, Houston.

Holmes, A. (1965) *Principles of Physical Geology*, 2nd edn. Thomas Nelson, London, 1288 pp.

Holmes, A. (1978) *Principles of Physical Geology*, 3rd edn. Thomas Nelson, London, 730 pp.

Johnston, W.A. (1921) Sedimentation of the Fraser River delta. *Mem. Can. geol. Surv.* **125**, 1–46.

Kleinspehn, K.L., Steel, R.J., Johannessen, E. & Netland, A. (1984) Conglomeratic fan-delta sequences, late Carboniferous–early Permian, western Spitsbergen. In: *Sedimentology of Gravels and Conglomerates* (Ed. by E.H. Koster and R.J. Steel). Mem. Can. Soc. Petrol. Geol. 10, pp. 279–294.

Koster, E.H. & Steel, R.J. (Eds.) (1984) *Sedimentology of Gravels and Conglomerates*. Mem. Can. Soc. Petrol. Geol. 10, 441 pp.

Massari, F. & Colella, A. (1988) Evolution and types of fan-delta systems in some major tectonic settings. In: *Fan Deltas: Sedimentology and Tectonic Settings* (Ed. by W. Nemec and R.J. Steel), pp. 103–122. Blackie and Son, London.

McPherson, J.G., Shanmugam, G. & Moiola, R.J. (1987) Fan-deltas and braid deltas: varieties of coarse-grained deltas. *Bull. geol. Soc. Am.* **99**, 331–340.

McPherson, J.G., Shanmugam, G. & Moiola, R.J. (1988a) Fan-deltas and braid deltas: varieties of coarse-grained deltas — Reply. *Bull. geol. Soc. Am.* **100**, 1309–1310.

McPherson, J.G., Shanmugam, G. & Moiola, R.J. (1988b) Fan deltas and braid deltas: conceptual problems. In: *Fan Deltas: Sedimentology and Tectonic Settings* (Ed. by W. Nemec and R.J. Steel), pp. 14–22. Blackie and Son, London.

Miall, A.D. (1984) Deltas. In: *Facies Models*, 2nd edn (Ed. by R.G. Walker). Geosci. Can. Reprint Ser. 1, pp. 105–118.

Nemec, W. (1990) Depositional controls on plant growth and peat accumulation in a braidplain delta environment: Helvetiafjellet Formation (Barremian–Aptian), Svalbard. In: *Controls on the Distribution and Quality of Cretaceous Coals* (Ed. by P.J. McCabe and J. Totman Parrish). Spec. Pap. geol. Soc. Am. (in press).

Nemec, W. & Steel, R.J. (1987) Convenors' address: What is a fan delta and how do we recognise it? *Abstr. Int. Symp. Fan Deltas: Sedimentology and Tectonic Settings*, pp. 11–17. Bergen.

Nemec, W. & Steel, R.J. (1988) What is a fan delta and how do we recognize it? In: *Fan Deltas: Sedimentology and Tectonic Settings* (Ed. by W. Nemec and R.J. Steel), pp. 3–13. Blackie and Son, London.

Nilsen, T.H. (1989) Book review: 'Fan deltas: sedimentology and tectonic settings'. *Sedimentology* **36**, 509–510.

Orton, G.J. (1988) A spectrum of Middle Ordovician fan deltas and braidplain deltas, North Wales: a consequence of varying fluvial clastic input. In: *Fan Deltas: Sedimentology and Tectonic Settings* (Ed. by W. Nemec and R.J. Steel), pp. 23–49. Blackie and Son, London.

Prior, D.B. & Bornhold, B.D. (1988) Submarine morphology and processes of fjord fan deltas and related high-gradient systems: modern examples from British Columbia. In: *Fan Deltas: Sedimentology and Tectonic Settings* (Ed. by W. Nemec and R.J. Steel), pp. 125–143. Blackie and Son, London.

Van der Meulen, S. (1989) The distribution of Pyrenean erosion material, deposited by Eocene sheet-flood systems and associated fan-deltas. *Geol. Ultraiec.* **59**, 125 pp.

Wood, M.L. & Ethridge, F.G. (1988) Sedimentology and architecture of Gilbert- and mouth bar-type fan deltas, Paradox Basin, Colorado. In: *Fan Deltas: Sedimentology and Tectonic Settings* (Ed. by W. Nemec and R.J. Steel), pp. 251–263. Blackie and Son, London.

Spec. Publs int. Ass. Sediment. (1990) **10**, 13–27

Depositional architecture and facies of river and fan deltas: a synthesis

G. POSTMA

Sedimentology Group, Institute for Earth Sciences, Postbus 80.021, 3508 TA Utrecht, The Netherlands

ABSTRACT

New ways to integrate studies of deltas are discussed with emphasis on the actual delta characteristics (architecture and sedimentary facies) rather than the modifying role of basinal processes or the nature of the alluvial feeder system. It is suggested that the first-order level of delta recognition and classification should pertain to the physiography and facies of the actual delta plain, delta front, delta slope (if present) and prodelta, whereas the nature of the feeder system (whether alluvial fan, braidplain or solitary river) should constitute a second-order level of recognition.

Assuming a 'steady-state' development of a delta, the character of the delta varies with the character of the feeder (distributary) system, the basinal water depth and the different ways of sediment diffusion at the delta front. Analysis of these variables has resulted in 12 prototype deltas which are postulated as sedimentological norms to facilitate comparisons of both modern and ancient delta systems.

INTRODUCTION

Deltas are fed either by a single river which may develop a braided distributary plain, or by an alluvial-fan system. Deltas form in oceanic basins, gulfs, inland seas, bays and lakes, and vary greatly in size, morphology, geometry and facies. The world's largest modern river deltas have attracted considerable study (e.g. Van Straaten, 1960; Coleman & Wright, 1975; Coleman, 1982; Prior & Coleman, 1982; see also Elliott, 1986, and references therein), and the existing delta models and classifications are largely based on these studies (e.g. Fisher *et al.*, 1969; Coleman & Wright, 1975; Galloway, 1975). Deltas fed by alluvial fans (fan deltas) are often much smaller than the huge deltas of large rivers, like the modern Mississippi. The interest in fan-delta systems is relatively young and the results of studies are seldom integrated with those of the large river deltas, partly because their sedimentary facies appear often to be so different. As a result, an aritificial barrier has arisen between river-delta and fan-delta concepts, an impression which is substantiated by, for example, a separate classification for fan deltas (Ethridge & Wescott, 1984).

McPherson *et al.* (1987) made a first attempt to cross this barrier by distinguishing three classes of deltas based mainly on the character of the alluvial feeding system: fan deltas, braid deltas and 'common' deltas. The term 'fan delta' is suggested to denote gravel-rich deltas formed where steep gradient, debris-flow dominated fans (best represented by semi-arid alluvial fans) have prograded directly into a standing body of water. 'Braid deltas' would denote gravel-rich deltas that form where a braided fluvial system (such as an outwash fan), or a braided distributary plain created by a single river progrades into a standing body of water. 'Common deltas' would then mean the finer-grained deltas created by straight or meandering, mixed-load or suspended-load rivers.

McPherson's *et al.* (1987) intention to reserve the term fan delta for mass-flow dominated, tectonically controlled alluvial fans prograding into standing water precludes deltas of wet-type (stream-dominated) alluvial fans, as critically discussed by Nemec and Steel (1988) and Dunne (1988). Nemec and Steel (1988) have correctly pointed out, with reference to

Holmes's (1965) original definitions of an alluvial fan and a fan delta, that the term 'fan delta' should make allowance for even some very large, low-gradient, sandy-to-muddy alluvial fans, dominated by stream processes, as well as for fans not necessarily related to tectonic escarpments. The application of Holmes's original definition, as well as Nemec and Steel's (1988) revised, broader definition of fan delta, however, requires reliable recognition of the type of the alluvial feeder system of a given delta (fan or single river system), which is often difficult to accomplish in the stratigraphic record (e.g. Fernández *et al.*, 1988, p. 299).

Orton (1988), in his attempt to synthesize river and fan deltas, proposed to extend the well-known ternary diagram of Galloway (1975) in a fourth dimension to account for the grain-size variation in delta fronts, so that deltas with similar delta-plain and front gradients, similar responses to nearshore wave energy and similar catchment areas and rates of sediment supply, could be compared. However, as will be discussed in this chapter, the grain size of a delta system may, but may also not be representative of the delta-front gradient and the size of catchment area (drainage basin). For example, both sandy (e.g. Stanley & Surdam, 1978) and gravelly Gilbert-type deltas (e.g. Gilbert, 1885, 1890; Clemmensen & Houmark-Nielsen, 1981; Porębski & Gradziński, 1987) with similar achitectural style have been described.

Instead, it is suggested here that the general, first-order level of delta classification (including both river and fan deltas) should pertain to the actual characteristics of the delta-plain, delta-front, delta-slope and prodelta environments, whose facies and depositional architecture are more readily recognizable in the stratigraphic record. The characteristics of delta-plain, delta-front, delta-slope and prodelta of deltas vary significantly with differences in the alluvial feeder or distributary system (Syvitski & Farrow, 1983; Dunne & Hempton, 1984; McPherson *et al.*, 1987; Orton, 1988; Nemec & Steel, 1988), with differences in basinal water depth (Ethridge & Wescott, 1984), and with differences in sediment diffusion within the deltaic environment (e.g. Prior & Bornhold, 1988; Syvitski *et al.*, 1988) against the background of different climatic, tectonic and physiographic setting (Leeder *et al.*, 1988). The terms 'fan delta', 'braidplain delta' and 'common delta' would then constitute a second-order level of delta classification, where the actual nature of the alluvial feeder system would be expected to be recognized and named.

The objective of this chapter is to discuss some major variations in delta systems, in order to explore the possibilities for a delta classification which might facilitate comparative studies of modern and ancient deltas, and of different delta successions.

CONCEPTUAL FRAMEWORK OF DELTAS

In order to analyse the variability of modern and ancient deltas, Elliott (1986) adopted a conceptual framework (Fig. 1) in which he defined the 'delta regime' as being dependent on the interaction between fluvial regime and sediment load, on the one hand, and basinal regime defined by shape, size, bathymetry and dynamics of that basin, on the other. It can be argued that causative links exist between delta regime, delta morphology and delta facies.

It is noted that in the conceptual framework of Fig. 1, significance has been given to the rate of change in any given delta system. Small deltas, developed in close proximity to small drainage basins are likely to experience much higher rates of changes (induced by climatic, tectonic and sea-level changes) than large deltas fed by large drainage basins that often cover various climatic and tectonic zones (Table 1). Hence, even if the steady-state development of two modern deltas are comparable, their longer-term evolution in time and space may not be.

MAJOR VARIATIONS IN DELTA SYSTEMS

The sedimentary characteristics of delta-plain, delta-front, delta-slope and prodelta segments of deltas prograding in a low-energy basin vary significantly with differences in the alluvial feeder and delta-plain distributary system (Syvitski & Farrow, 1983; Dunne & Hempton, 1984; McPherson *et al.*, 1987; Orton, 1988; Nemec & Steel, 1988), with differences in the basin depth (Ethridge & Wescott, 1984), and differences in the mode of sediment diffusion within the deltaic basin (Syvitski *et al.*, 1988). The origin of these variations is discussed below.

The subaerial delta segment

As pointed out by McPherson *et al.* (1987), deltas

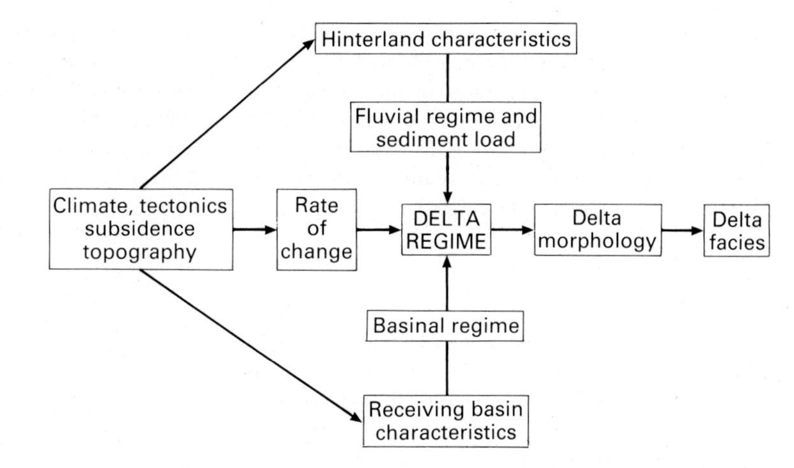

Fig. 1. Conceptual framework for a comparative study of both modern and ancient delta systems (modified from Elliott, 1986).

Table 1. Comparison of qualities which define the rate of change in small and large deltas

	Small delta systems	Large delta systems
Area drainage basin	small	large
Surface-area delta	few kms	10s to 100s kms
Climatic zones covered by drainage basin	one	few
Feeder system	short, few to 10s kms	very long, 100s kms
Gradient delta plain	variable	low
Discharge	highly variable, often ephemeral	relatively constant perennial
Sediment flux over a c. 10 000 yr period	highly variable	relatively constant
Grain-size distribution	boulder–clay	sand–clay
Response to climatic and tectonic changes	fast	slow

vary with the character of their feeder system, which is readily recognizable from the alluvial and upper delta-plain facies. The distributary pattern of alluvial clastic influx to the lower delta plain and front may be classified roughly into four empirical categories, which are further referred to as types A–D feeder systems.

Type-A feeder system

Alluvial systems, generally dominated by gravel, with a very steep gradient (more then a few degrees, see data presented by McPherson *et al.*, 1987; their fig. 5), which would include alluvial cones possibly as steep as 20–30° in extreme cases. These systems are commonly characterized by ephemeral, unconfined streams involving mass flows, are relatively small in radius and form along basin-margin fault-scarps and fjord margins.

Type-B feeder system

Steep-gradient (± 0·4°), often gravelly alluvial systems comprising closely spaced, highly mobile (unstable) bedload channels feeding the delta front essentially as a line source (a multitude of distributary channels whose effluents merge to provide a more-or-less uniform supply of sediment along the delta front). Modern examples include proglacial outwash plains and some other, braided allluvial systems in the head parts of fjords and lakes (including most of the 'braid delta' examples of McPherson *et al.*, 1987; their figs 3A, C,D).

Type-C feeder system

Moderate-gradient, gravelly and sandy alluvial systems of closely spaced, but relatively stable channels acting as a line source, but characterized by well-defined points of sediment supply. Modern examples include the feeder systems in the head parts of fjords and lakes. Stabilization of the outlets is often enhanced by vegetation on mid-channel and mouth bars (e.g. Axelsson, 1967).

Type-D feeder system

Low-gradient alluvial systems of wider-spaced, highly stable channels generally characterized by a low-bedload/total-load ratio (excess of fine-suspension load). These systems develop at the coastal margins of extensive lowland areas, whereby the channels have a tendency to prograde in isolation ('birdfoot' pattern, with leveed distributaries) and act as a number of 'point' sources.

It is to be emphasized that: (1) the distinction between feeder systems A–D is somewhat arbitrary, as a change in discharge, for example, may result in the transformation of one type into another; (2) gradations exist between the four types of alluvial feeder system; (3) although leveed channel systems can be dominated by bedload transport (Chudzikiewicz *et al.*, 1979), in general the bedload/total-load ratio will decrease from type-A to type-D feeder system, as noted by McPherson *et al.* (1987). Importantly, each type of alluvial feeder system may characterize both fan and river-fed deltas (see also discussion by Nemec & Steel, 1988).

The subaqueous delta segment

Variations in delta-front, delta-slope and prodelta architecture and facies will depend not only on the alluvial regime itself, but also on the interaction between the alluvial and the basinal regime. The differences in basinal regime depend on the shape, size (surface area), bathymetry and dynamics of the receiving basin. The three-dimensional accommodation space determines the growth rate of the delta system, while different bathymetries define the following two broad delta categories (see also Ethridge & Wescott, 1984; Elliott, 1986; Fraser & Suttner, 1986).

i Shallow-water deltas (basin depth in the order of some tens of metres): these would include 'shelf-type' deltas of Ethridge and Wescott (1984), but also lacustrine and other deltas unrelated to these shelves. Shallow-water deltas are characterized either by a gently inclined delta front (for which the term 'shoal-water' profile is used) or by a steeply inclined delta front (Gilbert-type profile, see Fig. 2), depending on the ratio channel depth/basin depth (depth ratio, see Jopling, 1965), the gradient of the basin floor immediately basinwards of the stream outlet and the delta front regime (further discussed below);

ii Deep-water deltas: these would include the shelf-edge deltas and the 'slope-type' deltas of Ethridge & Wescott (1984), but also other systems, not necessarily related to true shelf breaks, as in fjords, fault-controlled deep lakes or small marine grabens, slump scars on inclined shelves and other basinal or local depressions.

The interaction of alluvial and basinal dynamics determines the sedimentation rate on the subaqueous delta segment, which controls its growth-style and thus its profile (e.g. Bogen, 1983; Syvitski *et al.*, 1988). The 'steady state' growth style, recognizable from the delta-front facies and depositional architecture, is governed by the relative contribution of the following four modes of sediment-particle transport and deposition (cf. Syvitski *et al.*, 1988) beyond the channel mouths:

1 Bedload sedimentation in the lower delta plain and delta front governed by inertial and frictional forces, on one hand (Wright, 1977), and wave and tidal forces, on the other (e.g. Wright & Coleman, 1973).

2 Hemipelagic sedimentation of fine silt and clay sometimes carried far into the basin by buoyant forces in plumes and settling on the delta slope and prodelta. Domination of hemipelagic sedimentation may cause shallowing near the outlet and will influence the style of delta-front sedimentation.

3 Sediment-bypassing processes that are the result of delta-front failure and/or hyperpycnal effluent (cf. Prior *et al.* 1987; Prior & Bornhold, 1989).

4 Diffusion processes due to waves, tides and gravity, the latter in the form of creep, slide and sedi-

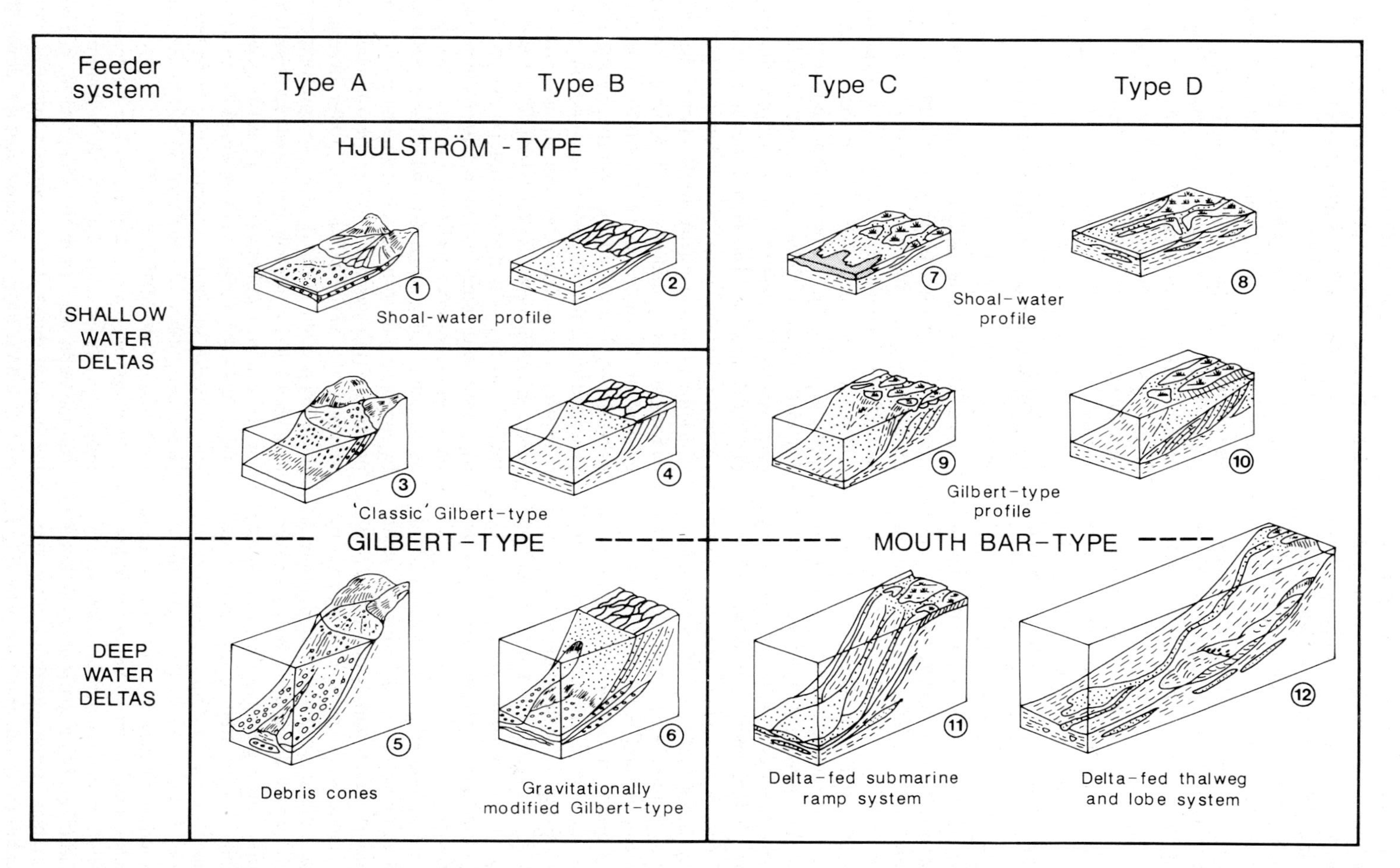

Fig. 2. Twelve major prototype deltas, pictured for simplicity as being dominated by fluvial processes. The prototypes are distinguished on the basis of a unique combination of four different types of distributary systems and two ranges of depth ratios (see text), and also take into account the variation due to inertia-, friction- and buoyancy-dominated effluent.

ment flows that redistribute the sediment of the delta front in a longshore or basinward direction.

Summarizing, the basis for a universal delta classification should take into account (1) feeder system, (2) depth ratio, (3) river-mouth processes, and (4) diffusion processes due to waves, tides and gravity. Similar to wave and tidal reworking, gravitational reworking may control the growth style and architecture of many deltas, though is less important in others.

For deltas prograding in a low-energy basin, at least eight prototypes can be defined on the basis of a combination of the four different feeder systems and the two ranges of basin depth. Consideration of the variation in effluent dynamics (inertial, frictional and buoyant) might increase the number of prototype deltas theoretically to 24. However, the character of particular distributary systems precludes some of these theoretical possibilities (e.g. buoyant effluent will not dominate the delta fronts fed by feeder systems A and B). In total, 12 prototype deltas can be considered essential; they are pictured, for simplicity, as fluvial-dominated systems in Fig. 2, and thus pertain directly to deltas in low-energy basins. For deltas in high-energy basins, the relative role of basinal pocesses can be expressed by using the ternary diagram of Galloway (1975), or its extended version by Orton (1988) for each prototype. Because deltas can be modified by wave, tide *and* gravitational reworking, it may be desirable to use a diagram similar to that of Orton, but where the relative role of gravitational reworking is taken into account, together with wave and tidal influences (Postma, 1990). Gravitational reworking means the redistribution of sediment solely due to slope-instability processes.

Hence, the broad physiographic basis assumed for this classification implies that every prototype delta represents an almost infinite variety of similar deltas, which can be modified to a variable degree by basinal and gravitational processes. For example, a shoal-water delta with a distributary system of type D would comprise, among many others, the wave-dominated Rhône delta, the mixed wave/tide-dominated Niger delta, and the tide-dominated Schelde estuary. The prototype deltas are, therefore, not regarded as a means to refine the existing models for these shallow-water deltas with low-gradient distributary plains, but to improve our understanding of other, often coarse-grained deltas, which the researchers usually found difficult to consider in a systematic and comprehensive way.

DEPOSITIONAL ARCHITECTURE AND SEDIMENTARY FACIES VARIATIONS

A brief account is given here of the physical processes which govern the depositional architecture and sedimentary facies of the various prototype deltas. Table 2 summarizes the general characteristics of the 12 prototype deltas of Fig. 2, based on examples from the literature (Table 4). Table 3 is an explanatory list of the letter symbols used in Table 2.

Distinguishing between shallow- and deep-water deltas

Deltas prograding in *shallow* waters are generally characterized by three physiographic zones; (1) delta plain, comprising the alluvial feeder system, (2) delta front dominated by coarse-load deposition and possibly influenced directly by waves, and (3) prodelta, dominated by hemipelagic sedimentation. The delta front of deltas prograding in relatively *deep* waters is separated from the prodelta by a distinct delta slope. The slope is beyond the direct influence of waves and variously dominated by suspension settling and gravity-driven mass transport of coarser-grained sediment supplied by the stream effluent or derived from the delta front. The delta slope thus constitutes another, fourth physiographic zone and the adjacent deep-water prodelta is then the fifth zone which is often akin to submarine fan environment. The deep-water, axial, delta system of Bella Coola fjord (BC), for example, is characterized by a friction-dominated, lower delta plain at high tides, an inertia-dominated delta front at low tides, and a delta slope and prodelta which are dominated by suspension deposition and sediment gravity-flow processes (Kostaschuk, 1985; Kostaschuk & McCann, 1987; Prior & Bornhold, 1988). In some of these deep-water delta systems (see prototypes 11 and 12 in Fig. 2), the boundary between delta front and delta slope can be drawn on basis of grain size and sedimentary facies, as exemplified by studies of ancient delta successions, particularly those in the Namurian basins of England (Walker, 1966a,b; Collinson, 1969; McCabe, 1978).

In deep-water delta systems with a Gilbert-type profile (prototypes 5 and 6), the delta front merges with the delta slope (is of the same facies), because the bedload dumped at the stream outlet is transported almost instantaneously further down the

Table 2. Generalized characteristics of prototype deltas in low-energy basins

	Feeder (distributary) system											
	A			B			C			D		
	Shallow-water delta			Shallow-water delta			Shallow-water delta with			Shallow-water delta with		
	Shoal-water-type delta	Gilbert-type delta	Deep-water delta	Shoal-water-type delta	Gilbert-type delta	Deep-water delta	shoal-water profile	Gilbert-type profile	Deep-water delta	shoal-water profile	Gilbert-type profile	Deep-water delta
Physiographic setting alluvial system	Faultblock, mountain fronts, fjord margins, volcanic highlands. Very steep gradient enhanced by ephemeral discharges			Fluvio-glacial outwash, steep gradient braidplains, and braided rivers. Also common near mountain fronts, etc. in wet climate			Moderate gradient, commonly vegetated braidplains and braided rivers			Low gradient, commonly vegetated coastal plains and meandering rivers with well-developed levees		
Delta plain processes	Landslides, mass flows, unconfined stream flow			Poorly confined and unconfined stream flow; bulk sediment transport during floods			Mainly confined stream flow; relatively stable channels; channel splitting at the delta front through small mouth bars			Confined stream flow; stable channels which shift by avulsion upstream; large mouth-bar systems		
Lithofacies	Gms, Gm, Gh, Ss (minor Gt, Gp, Sh)			Gh, Gt, Gp, Sh, S1, Se, Ss (minor Gms, Gm)			St, Sp, Sr, Sh, Ss, Fl, Fsc, Fcf, Fr, C, P (Gt, Gp)			St, Sr, Fl, Fsc, Fm, Fcf, Fr, C, P (Sh, Sp)		
Delta–front processes	Land-derived mass flow; hyperpycnal flows bypassing delta front; bedload deposition.			Bedload deposition; hyperpycnal flow during river floods; rare land-derived mass flow			Bedload deposition at the outlet; mud in interdistributary embayments;			Bedload deposition at the outlet; mud in interdistributary bays;		
Lithofacies	Gms, Gmm, Gh, Shl, Sh (Sr)	Gms, Gg, Ss (Sg)	Gmm, Gms, Gg, Sg, Ss (Fl, Fb)	Gm, Gms, Gh, Gt, Gp, Shl, Sh, Sr	Gms, Gg, Sg, Ss	Gmm, Gms, Gg, Sg, Ss, Fl, Fb	Gh, Sh, Shc, Sr, Sg, St, Sp, Fl, Fr, C	Gg, Sg, Gl	Gms, Sg (Gg) Sm, Shl, Sh, Ss, Fb	St, Sr, Sp, Sh, Shl, Fl, Fsc, Fr, Fl	St, Sr, Sg, Fb, Fl, Fr	Sm, St, Sg, Sh, Fl, Fb, Shl
Prodelta processes	Land- and slope-derived mass flow; density currents; hemipelagic sedimentation			Density currents: hemipelagic sedimentation; rare land-derived and common slope-derived mass flow			Hemipelagic sedimentation density currents		Land/slope derived turbidity currents	Hemipelagic sedimentation; density currents		Land/slope derived turbidity currents
Lithofacies	Shl, Sh, Sr	Shl, Sh, Sr, Fl, (Fb)	GMS, Gh, Shl, Fl, Fb, (Sh)	Sr, Fl, Fb	Shl, Sh, Sr, Fl, (Fb)	Gms, Gh(Sh) Ta-e	Sr, Fl, Fb	Shl, Sh, Sr, Fl, (Fb)	Gms, Gh, Sm, Shl, Sh, Ta-e, Fl, Fb	Fb, Fl	–	Sm, Shl, Sh, Ta-e, Fl, Fb
Delta geometry and size	Few km^2, lenticular to wedge-shaped			Few to tens of km^2, planar to wedge-shaped			Tens to hundreds km^2 Planar to wedge shaped			Hundreds to thousands km^2 large lens-shaped sandbodies		
Grain size distribution	Poorly sorted; boulders and cobbles common			Moderate to poorly sorted; pebbles common; rare cobbles and boulders			Moderately sorted, sand and pebbles; rare cobbles			Moderately to well sorted sand and mud; rare pebbles		

Table 3. List of letter symbols used in Table 2

	Sedimentary structure	Depositional process/facies
Gm	Massive clast-supported gravel	Longitudinal bars, lag deposits, sieve deposits
Gg	Graded and massive cross-strata	Slipface processes (including grain flow)
Gmm	Massive matrix–supported gravel	Cohesive debris flow
Gms	Massive clast–supported gravel	Cohesionless debris flow
Gh	Horizontally stratified gravel	Traction carpets driven by stream flow (comparable to sheet flow) or high density turbulent flow
Gt	Trough crossbeds	Traction
Gp	Planar crossbeds	Traction
Sm	Massive (structureless) sand, commonly with water-escape structures	Rapid suspension fall-out
St	Solitary or grouped trough crossbeds in sand	Dunes (lower flow regime)
Sp	Solitary or grouped planar crossbeds in sand	Linguoid, transverse bars, sand waves (lower flow regime)
Sr	Ripple marks of all types	Ripples (lower flow regime)
Sh	Horizontal lamination	Planar bed flow (lower and upper flow regime)
Sg	Graded and massive cross-strata	Slipface processes
Shl	Horizontally layered sand	Traction carpets
Shc	Low-angle and tangential foreset strata	Mouth bars
Se	Crude crossbedding	Scour fills
Ss	Broad shallow scours	Scour fills
Fb	Homogeneous commonly bioturbated muds	Suspension settling and density currents
Fl	Fine lamination, very small ripples	1. Overbank or waning flood 2. Density currents
Fsc	Laminated to massive	Backswamp deposits
Fcf	Massive, with fresh water molluscs	Backswamp pond deposits
Fm	Massive, dessication cracks	Overbank or drape deposits
Fr	Rootlets	Seat-earth
C	Plants and mud films	Swamp deposits
P	Pedogenic structures	Soil
Ta-e	Bouma sequence	Turbidite
Ch	Channel	
UDS	Upper Delta Slope	
LDS	Lower Delta Slope	
BS	Base of Slope	

entire slope by slipface processes. The delta slope then constitutes that part of the delta not influenced by wave processes (Colella, 1988). Very young, immature deep-water systems may be characterized by a 'bypass' slope and represent debris cones (prototype 5 in Fig. 2; Prior & Bornhold, 1988).

The boundary of the prodelta realm of a deep-water delta is a matter of discussion. It can be

Table 4. Examples from the literature of prototype deltas dominated by fluvial processes

Modern deltas	Ancient deltas	Modern deltas	Ancient deltas
Hjulström-type deltas		*Mouth-bar deltas*	
1. Laguna Salada, NE California (In: McPherson *et al.*, 1987)	Domba Conglomerate, Devonian, Hornelen basin, Norway (Nemec *et al.*, 1984)	7 Burdekin delta (weak wave and tidal influence), Australia (Coleman & Wright, 1975)	Helvetiafjellet Formation, Svalbard, Cretaceous (Nemec *et al.*, 1988; Nemec, 1990)
2. Supra-aquatic delta fed by sandurs (Hjulström, 1952)	Late Wisconsin Kame delta, Massachusetts (Jopling & Walker, 1968); Subaquous outwash deltas (Rust & Romanelli, 1975)	8. Shoal-water lobes of Mississippi delta (e.g. Coleman, 1982, 1988)	Yoredale Series, Carboniferous, UK (Elliott, 1975); Pleistocene/Holocene Mississippi delta (Frazier, 1967; Fisher & McGowen, 1969).
Gilbert-type deltas		9 Axial lake system Laitaure Delta, Sweden (Axelsson, 1967)	Athabasca delta, Pleistocene, Alberta (Rhine & Smith, 1988); Lake Burdur, Turkey, Pleistocene (Kazancı, 1988a).
3. Margins of Lake Hazar (Dunne & Hempton, 1984)	Some of the fault-bounded deltas of the Crati Basin, Plio-Pleistocene (Colella, 1988); Pleistocene of Burdur Lake, Turkey (Kazanci, 1988b)	10. Lake Plociczno, Poland (Chudzikiewicz *et al.*, 1979)	No example found
4. Glacial lakes Norway; Losna and Atnsjo deltas (Bogen, 1983)	Lake Bonneville (Gilbert, 1885); Pleistocene glacial lake, Denmark (Clemmensen & Houmark-Nielsen, 1981); Polonoez Cove Formation (volcanic breccia, Miocene) (Porębski & Gradzínski, 1987); Pleistocene Lake Washakie, Wyoming (Stanley & Surdam, 1978); Meteora Conglomerate, Greece (Ori & Roveri, 1987)	11. Axial fjord systems; Bella Coola delta, British Columbia (Kostaschuk & McCann, 1987; Prior & Bornhold, 1988; Prior *et al.*, 1981; Kostaschuk, 1985)	Axial graben systems: Mam Tor–Nether Tor facies, Kinderscout Grit, Carboniferous, UK (Allen, 1960; Walker, 1966a, 1966b; McCabe, 1978)
		Shelf-edge; Crati fan, Ionian Sea (Ricci Lucchi *et al.*, 1984; Colella & Normark, 1984)	Tabernas Fan system: Miocene, SE Spain (Kleverlaan, 1987, 1989); Tyee Formation, Eocene, Oregon (e.g. Heller & Dickinson, 1985)
5. Fjords British Columbia (Prior and Bornhold, 1988, 1989)	Debris aprons: Hunghae Formation, SE Korea (Choe & Chough, 1988); Footwall fan deltas: Gulf of Suez, Miocene (Gawthorpe *et al.*, 1988); Crati Basin, Plio-Pleistocene (Colella, 1988)	12. Axial fjord system; Bute Inlet (Prior *et al.*, 1987)	Some parts of the Kinderscout Grit (Walker, 1966a, Alport castle channel system)
		Shelf-edge: Mississippi delta and fan (e.g. Coleman, 1988; Lindsay *et al.*, 1984)	Deadman Stream, Miocene, New Zealand (Lewis *et al.*, 1980)
6. Fjords British Columbia (Prior and Bornhold, 1986, 1988); Fjords Spitsbergen (Evans, 1987)	Espiritu Santo Formation, Pliocene, SE Spain (Postma & Roep, 1985); Late Pleistocene/Holocene fjord delta, Varangerfjord, Norway (Postma & Cruickshank, 1988); Promina Formation, Eocene, Yugoslavia (Postma *et al.*, 1988)		
Shelf-edge: Yallahs delta, (island-arc slope) Jamaica (Wescott & Ethridge, 1980)	Crati Basin, Plio-Pleistocene (Colella *et al.*, 1987; Colella, 1988)		

argued that those parts of the basin influenced directly by delta-controlled sedimentation processes, albeit undoubtedly combined with other processes (Coriolis force, tidal currents, contour currents, etc.) belongs to the delta system. This means that even some very large, deep-sea-clastic systems extending hundreds of kilometres away from the delta front (e.g. Bengal fan, Mississippi fan) can be considered as a 'prodelta'. However, an ensuing transgression will normally force the delta front to retreat landwards and cause temporary abandonment of both the delta slope and 'prodelta'. This would change not only the name of the deep-sea system from 'prodelta' into 'deep-sea fan', but also the way it is fed until the delta system would be reactivated and reach the shelf edge. The Mississippi delta and its deep-sea fan is an example of such a coupled system. The nomenclature dilemma lies in the fact that the physiographical criteria used to distinguish between the two components of such systems in modern environments can scarcely be applied to ancient settings. The problem seems to be less critical for comparatively small, deep-water delta systems, as in fjords and small graben systems, because the delta system is prograding over a narrow shelf which shortens the period of delta-slope abandonment.

Shallow-water delta profiles

For a river to be able to form a delta or mouth bar with a steeply inclined, Gilbert-type profile where slipface processes dominate, the most important conditions are (see Axelsson, 1967): (1) transport of sufficient bedload as far as the river mouth (required bedload/total-load ratio may also be depth-dependent, see Bogen, 1983); (2) sufficiently large water depths immediately seaward of the mouth (sufficiently low depth ratio); and (3) spreading (expansion) of the effluent as an axial turbulent jet (inertia-dominated effluent diffusion). Homopycnal conditions at the river mouth as suggested by Bates (1953) are less important, evidenced by the occurrence of deltas with Gilbert-type profiles in both fresh- and salt—water basins.

If water depths seaward of the mouth are shallow or shallowing, turbulent diffusion becomes restricted to the horizontal, and bottom friction plays a major role (friction-dominated effluent diffusion; Wright, 1977). This will normally result in a gently inclined, shoal-water profile of commonly less than a few degrees.

Deep-water delta profiles

If the ratio bedload/total-load is high, the front of a deep-water delta progrades faster than the lower part of the delta slope. This results in progressive oversteepening of the delta front and ensuing delta-front failure (cf. Postma, 1984). The resedimentation of the coarse (delta-front) material transported by sediment gravity-flow processes gives the delta a tangential shape (e.g. Postma & Roep, 1985; Kenyon & Turcotte, 1985). Deep-water deltas with type-A and type-B feeder systems are of this type and comprise the gravitationally modified, Gilbert-type deltas (Postma & Roep, 1985; Postma *et al.*, 1988).

At the other extreme, if the ratio bedload/total-load is very low, the progradation of a delta front may lag behind the aggradation and progradation rates of the delta slope. This will produce a convex-upward delta profile due to excessive deposition of fine suspension material. Deep-water deltas with type-D feeder system can be of this type, and the large, suspension-dominated Mississippi shelf-edge delta system may be an example. Excessive mud deposition also promotes slope instability processes and the resulting delta profile may become gradually tangential downslope.

The delta profile and morphology may be modified locally by underflows (hyperpycnal flow). Hyperpycnal flow is known to be important where the density of the effluent is in excess of the density of the basin water (e.g. Gustavson, 1975). Hence, high discharges carrying large suspension loads can be expected to enhance underflow activity, even in saline water. The effects of heavily sediment-laden underflow on delta morphology are not well known, but a reduction of the gradient near the subaqueous 'apex' of a delta, thought to be dominated by these flows, was noted (e.g. Prior & Bornhold, 1989), possibly the result of friction-dominated processes. The gradient of the delta face may steepen again at a point where gravitational forces start to dominate the mechanics of sediment transport.

Mouth-bar deltas

Dunne and Hempton (1984) found that deltas (fed by feeder-system type-A) at the steep margins of the shallow-water Lake Hazar were of the Gilbert-type and that the axial delta (fed by feeder-system type-C) was of a mouth-bar-type. Both delta systems were formed under similar conditions and prograding in similar water depths, with the grain size of the mouth-bar system being slightly finer and the grains better rounded. Therefore, Dunne and Hempton (1984) maintained that the differing achitecture of the two delta systems was the result of the structurally different basin margins and different feeder systems, and concluded that a basin floor steeper than 3° allows the formation of Gilbert-type deltas, whereas lower angles favour the development of mouth-bar-type deltas.

While agreeing with Dunne and Hempton that the feeder system determines whether a mouth-bar-type or a Gilbert-type delta will form in a shallow-water basin, I believe it is not the basin-floor gradient which should be emphasized, but the stability, spacing, discharge and width of the distributary channels. These variables (which are functions of basin relief, climate, tectonics, sediment supply, grain size, vegetation, etc.) determine the rate of lateral sediment dispersal along the delta front in a low-energy basin, which varies from very high in feeder systems of type-A to very low in systems of type-D.

The idealized depositional pattern at the outlet of a single and relatively stable, fluvial distributary in low-energy, shallow water will be in the form of a bar, which may differ in sedimentary facies and architecture according to the varied dynamics of the effluent and the water density of the receiving basin. The width of the bar depends on the spreading angle of the effluent which varies from about 12° for an inertia-dominated effluent to *c.* 16–17° for a friction-dominated effluent (Wright, 1977; Kostaschuk, 1985), with intermediate values for plumes. Hence, the lateral dispersion of sediments is confined to a narrow zone, at least in the region immediately basinward of the outlet. If the outlets are sufficiently widely spaced, then a series of bars will form, shallowing the region immediately basinward of the outlet. As the bars grow in size they will deflect and split the flow and cause channels to bifurcate. A detailed study on the processes of bar formation and channel splitting at the delta front has been carried out by Axelsson (1967) based on the modern axial delta system of Laitaure Lake (Sweden). This process of bar building will eventually lead to a lobate pattern of the delta plain (prototypes 7–10).

Bars will coalesce to create a uniform delta front if the outlets are sufficiently closely spaced and/or shift their positions frequently, as is the case for unconfined, outwash-type and alluvial-cone distributary systems (systems of types A and B). In such cases the delta plain will be smooth and gently sloping and, dependent on the effluent dynamics, the classic Gilbert-type delta (prototypes 3 and 4 in Fig. 2) or the shoal-water delta of Hjulström-type (cited by Church & Gilbert, 1975; prototypes 1 and 2 in Fig. 2) will develop.

Very low depth ratios immediately basinward of the outlet prohibit bar formation, so that the frontal delta profile will remain steep being controlled by hyperpycnal and sediment gravity-flow processes (prototypes 5 and 6). It is to be mentioned that the distributary outlets of deep-water deltas (e.g. prototypes 11 and 12) can be characterized by high depth ratios due to tidal differences (submergence of the lower delta plain during high tides, as in some fjords; e.g. Kostaschuk, 1985), or/and by low bedload/total-load ratios resulting in a sigmoidal delta profile (e.g. Mississippi delta).

Summarizing, the formation of mouth-bar-type deltas is normally related to feeder systems type-C and type-D and shallow-water depths immediately basinward of the outlet zone, whereas the formation of classic Gilbert-type and Hjulström-type deltas is related to the feeder systems of type-A and type-B.

Delta-front reworking and abandonment

Although several studies have been devoted to wave- and tidally influenced modern deltas, relatively few of these studies were concerned with coarse-grained, gravelly or sand-gravelly delta systems. As in the fine-grained delta systems, the relative roles of fluvial and basinal regimes in coarse-grained delta systems is governed essentially by sediment supply and basin dynamics (e.g. Wright & Coleman, 1973). Bartolini and Pranzini (1984) demonstrated the effect of a periodical reduction of sediment supply under similar basinal conditions by repeated survey of a coarse-grained delta in western Italy. During the 'starving' period, the delta mouth retreated, a wave-cut platform developed, and the delta slope prograded due to the deposition of wave-winnowed material.

A classic study of a modern, shallow-water, gravelly delta system dominated by basinal processes is from the Copper River delta, Alaska, by Galloway (1975). This delta, with a feeder system of type-B, has a tidally influenced, lower delta plain and a completely wave/tide-reworked delta front characterized by barrier islands and lagoons. The delta could be classified as a mixed wave/tide-modified variety of prototype 2 (Fig. 2).

Ancient examples of coarse-grained deltas are

found and have been modified by basinal processes, of which the reworking by waves is most readily recognizable. Examples of wave-reworked, shallow-water Gilbert-type deltas have been described by Ori and Roveri (1987), and examples of shallow-water wave-worked, mouth-bar-type deltas are described by Kleinspehn *et al.* (1984), Marzo and Anadón (1988), Orton (1988), Colmenero *et al.* (1988), Robles *et al.* (1988), Wescott (1988) and Dabrio and Polo (1988). An example of a mixed, wave/tide system with fluvial-dominated mouth bars is described by Fernández *et al.* (1988).

Wescott and Ethridge (1982) studied sediment dispersal along the front of the modern, deep-water Yallash delta (Jamaica). Wave action there is sufficient to erode a wave-cut platform up to approximately 8–10 m water depth in the actively prograding, gravelly delta front. They called this wave-worked segment the 'transition zone'.

In several ancient, deep-water, coarse-grained delta systems, a similar zone influenced by waves and tides has been recognized. This zone erosively overlies delta-slope facies and is erosively overlain by delta-plain facies. From Gilbert-type, deep-water delta systems these transition zones have been described by Postma (1984) and Colella (1988). It is to be mentioned, however, that the erosive boundaries above and below the transition zone make it difficult to prove that their superposition is the result of simple delta progradation. The wave/tide-worked zone may be of a much later period due to a relative water-level fall, as exemplified by terraced delta bodies, where regressive beach deposits truncate delta-slope deposits (Postma & Cruickshank, 1988).

Wave and tidal influences have also been recognized in the mouth-bar-type, deep-water deltas from the Kinderscout Grit by Collinson (1969) and McCabe (1978). Phases of delta abandonment in deep-water delta systems may be recorded in the delta-slope sequence. Wave-winnowed, well-sorted fine sands have been found as intercalations within gravelly and muddy delta slopes (Postma, 1984; Colella, 1988; Postma & Cruickshank, 1988).

CONCLUSION

The discussion has been focused on the problem of how to compare the large variety of delta systems, small to large, encountered in nature. I have considered the principal variables which are thought to be responsible for the final delta product over a geologically small time period — small enough to assume the delta progradation to have occurred in a 'steady state' fashion. A logical combination of these variables result in 2 prototype deltas (Fig. 2), which facilitate (Table 4) the comparison of both modern and ancient delta systems.

Yet, much remains to be learned about the coarse-grained delta systems, particularly as regards their evolution in time and space. The rate of evolutionary changes in these, often small, systems may be high (Table 1) and quite different from the changes observed in large, often fine-grained, delta systems. Hence, time slices from ancient deltas may be comparable, but not necessarily their longer-term evolution. The prototypes considered here (Fig. 2) are thus to be considered as purely sedimentological models, rather than stratigraphical norms.

ACKNOWLEDGEMENTS

Trevor Elliott (University of Liverpool, UK), Wojtek Nemec (University of Bergen, Norway), Kick Kleverlaan (BP London, UK) Brian Bornhold (Pacific Geoscience Centre, Canada), and Elizabeth Koster (Utrecht State University, The Netherlands) are thanked for reading an earlier draft of this chapter and for their stimulating discussions. Special thanks go to Wojtek Nemec for his concise editing of this manuscript. Part of the research was completed during my lectureship at the University of East Anglia (UK).

REFERENCES

Allen, J.R.L. (1960) The Mam Tor Sandstone; a turbidite facies of the Namurian deltas of Derbyshire, England. *J. sedim. Petrol.* **30**, 193–208.

Axelsson, V. (1967) The Laitaure Delta, a study of deltaic morphology and processes. *Geogr. Ann.* **49A**, 1–127.

Bartolini, C. & Pranzini, E. (1985) Fan-delta erosion in southern Tuscany as evaluated from hydrographic surveys of 1883 and the late 1970s. *Mar. Geol.* **62**, 181–187.

Bates, C.C. (1953) Rational theory of delta formation. *Bull. Am. Assoc. Petrol. Geol.* **37**, 2119–2162.

Bogen, J. (1983) Morphology and sedimentology of deltas in fjord and fjord valley lakes. *Sedim. Geol.* **36**, 245–267.

Choe, M.Y. & Chough, S.K. (1988) The Hunghae Formation, SE Korea: Miocene debris aprons in a back-arc intraslope basin. *Sedimentology* **35**, 239–255.

Chudzikiewicz, L., Doktor, M., Gradziński, R., Haczewski, G., Leszczyński, S., Laptaś, A., Pawelczyk, J., Porębski, S., Rachocki, A. & Turnau,

E. (1979) Sedimentation of modern sandy delta in Lake Plociczno (West Pomerania). *Studia Geol. Polonica* **62**, 1–61.

CHURCH, M. & GILBERT, R. (1975) Proglacial fluvial and lacustrine environments. In: *Glaciofluvial and Glaciolacustrine Sedimentation* (Ed. by A.V. Jopling and B.C. McDonald). Spec. Publ. Soc. econ. Paleont. Minerol., Tulsa 23, pp. 22–100.

CLEMMENSEN, L.B. & HOUMARK-NIELSEN, M. (1981) Sedimentary features of a Weichselian glaciolacustrine delta. *Boreas* **10**, 229–245.

COLELLA, A. (1988) Pliocene–Holocene fan deltas and braid deltas in the Crati Basin, southern Italy: a consequence of varying tectonic conditions. In: *Fan Deltas: Sedimentology and Tectonic Settings* (Ed. by W. Nemec & R.J. Steel), pp. 50-74. Blackie and Son, London.

COLELLA, A. & NORMARK, W. (1984) High-resolution sidescanning sonar survey of delta slope and inner fan channels of Crati submarine fan (Ionian Sea). *Mem. Soc. Geol. It.* **27**, 381–390.

COLELLA, A., DE BOER, P.L. & NIO, S.D. (1987) Sedimentology of a marine intermontane Pleistocene Gilbert-type fan-delta complex in the Crati Basin, Calabria, Italy. *Sedimentology* **34**, 721–736,

COLEMAN, J.M. (1982) *Deltas, Processes of Deposition and Models for Exploration*. International Human Resources Development Corporation, Boston, 124 pp.

COLEMAN, J.M. (1988) Dynamic changes and processes in the Mississippi River delta. *Bull. geol. Soc. Am.* **100**, 999–1015.

COLEMAN, J.M. & WRIGHT, L.D. (1975) Modern river deltas: variability of processes and sand bodies. In: *Deltas, Models for Exploration* (Ed. by M.L. Broussard), pp. 99–149. Houston Geological Society, Houston.

COLLINSON, J.D. (1969) The sedimentology of the Grindslow Shales and the Kinderscout Grit: a deltaic complex in the Namurian of northern England. *J. sedim. Petrol.* **39**, 194–221.

COLMENERO, J.R., AGUEDA, J.A., FERNANDEZ, L.P., SALVADOR, C.I., BAHAMONDE, J.R. & BARBA, P. (1988) Fan-delta systems related to the Carboniferous evolution of the Cantabrian Zone, northwestern Spain. In: *Fan Deltas: Sedimentology and Tectonic Settings* (Ed. by W. Nemec and R.J. Steel), pp. 267–285. Blackie and Son, London.

DABRIO, C.J. & POLO, M.D. (1988) Late Neogene fan deltas and associated coral reefs in the Almanzora Basin, Almeria Province, southeastern Spain. In: *Fan Deltas: Sedimentology and Tectonic Settings* (Ed. by W. Nemec & R.J. Steel), pp. 354–367. Blackie and Son, London.

DUNNE, L.A. (1988) Fan deltas and braid deltas: varieties of coarse-grained deltas: Discussion. *Bull. geol. Soc. Am.* **100**, 1308–1310.

DUNNE, L.A. & HEMPTON, M.R. (1984) Deltaic sedimentation in the Lake Hazar pull-apart basin, southeastern Turkey. *Sedimentology* **31**, 401–412.

ELLIOTT, T. (1975) The sedimentary history of a delta lobe from a Yoredale (Carboniferous) cyclothem. *Proc. Yorkshire geol. Soc.* **40**, 505–536.

ELLIOTT, T. (1986) Deltas. In: *Sedimentary Environments and Facies* (Ed. by H.G. Reading). Blackwell, Oxford, 615 pp.

ETHRIDGE, F.G. & WESCOTT, W.A. (1984) Tectonic setting, recognition and hydrocarbon reservoir potential of fan-delta deposits. In: *Sedimentology of Gravels and Conglomerates* (Ed. by E.H. Kosters and R.J. Steel). Mem. Can. Soc. Petrol. Geol. 10, pp. 217–235.

EVANS, R.E. (1987) Fan-delta sedimentation in a Spitzbergen fjord. Unpublished PhD dissertation, University of East Anglia, UK, 652 pp.

FERNÁNDEZ, L.P., AGUEDA, J.A., COLMENERO, J.R., SALVADOR, C.I. & BARBA, P. (1988) A coal-bearing fan-delta complex in the Westphalian of the Central Coal Basin, Cantabrian Mountains, northwestern Spain: implications for the recognition of humid type fan deltas. In: *Fan Deltas: Sedimentology and Tectonic Settings* (Ed. by W. Nemec and R.J. Steel), pp. 286–302. Blackie and Son, London.

FISHER, W.L., BROWN, L.F., SCOTT, A.J. & MCGOWEN, J.H. (1969) *Delta systems in the exploration for oil and gas*. Bureau for Economic Geology, University of Texas, Austin, 78 pp.

FISHER, W.L., & MCGOWEN, J.H. (1969) Depositional systems in the Wicox Group (Eocene) of Texas and their relationship to occurrence of oil and gas. *Bull. Am. Assoc. Petrol. Geol.* **53**, 30–54.

FRASER, G.S. & SUTTNER, L. (1986) *Alluvial Fans and Fan Deltas. A Guide to Exploration for Oil and Gas*. International Human Resources Development Corporation, Boston, 199 pp.

FRAZIER, D.E. (1967) Recent deltaic deposits of the Mississippi delta: their development and chronology. *Trans. Gulf-Coast Assoc. Geol. Soc.* **17**, 287–315.

GALLOWAY, W.E. (1975) Process framework for describing the morphologic and stratigraphic evolution of the deltaic depositional systems. In: *Deltas, Models for Exploration* (Ed. by M.L. Broussard), pp. 87–98. Houston Geological Society, Houston.

GALLOWAY, W.E. (1976) Sediments and Stratigraphic framework of the Copper river fan-delta, Alaska. *J. sedim. Petrol.* **46**, 726–737.

GAWTHORPE, R.L., HURST, J.M. & SLADEN, C.P. (1988) Evolution of Miocene footwall-derived fan-deltas, Gulf of Suez (East): Implications for exploration. Abstr., *Int. Works. Fan Deltas, Calabria*, 26–27.

GILBERT, G.K. (1885) The topographic features of lake shores. *Ann. Rep. U.S. geol. Survey* **5**, 69–123.

GILBERT, G.K. (1890) Lake Bonneville. *Monogr. Surv. U.S. geol.* **1**, 438 pp.

GUSTAVSON, T.C. (1975) Sedimentation and physical limnology in proglacial Malaspina Lake, southeastern Alaska, In: *Glaciofluvial and glaciolacustrine sedimentation* (Ed. by A.V. Jopling and B.C. McDonald). Spec. Publ. Soc. econ. Paleont. Minerol., Tulsa, 23, 249–263.

HELLER, P.L. & DICKINSON, W.R. (1985) Submarine Ramp facies model for delta fed, sand-rich turbidite systems. *Bull. Am. Assoc. Petrol. Geol.* **69**, 960–976.

HOLMES, A. (1965) *Principles of Physical Geology*, 2nd ed. The Roland Press Co., New York, 1288 pp.

HJULSTRÖM, F. (1952) *The geomorphology of the alluvial outwash plains (sandurs) of Iceland, and the mechanics of braided rivers*. International Geographical Union, 17th Congress Proceedings, Washington, pp. 337–342.

JOPLING, A.V. (1965) Hydraulic factors controlling the shape of laminae in laboratory deltas. *J. sedim. Petrol.*

35, 777–791.

Jopling, A.V. & Walker, R.G. (1968) Morphology and origin of ripple drift cross-lamination, with examples from the Pleistocene of Massachusetts. *J. sedim. Petrol.* **38**, 971–984.

Kazanci, N. (1988a) Repetitive deposition of alluvial fan and fan-delta wedges at a fault-controlled margin of the Pleistocene–Holocene Burdur Lake graben, southwestern Anatolia, Turkey. In: *Fan Deltas: Sedimentology and Tectonic Settings* (Ed. by W. Nemec and R.J. Steel) pp. 186–196. Blackie and Son, London.

Kazanci, N. (1988b) A bouldery fan-delta succession in the Pleistocene–Holocene Burdur Basin, Turkey: the role of basin margin configuration in sediment entrapment and differential facies development. *Abstr. Int. Works. Fan Deltas, Calabria*, 30–32.

Kenyon, P.M. & Turcotte, D.L. (1985) Morphology of a delta prograding by bulk sediment transport. *Bull. geol. Soc. Am.* **96**, 1457–1465.

Kleinspehn, K.L., Steel, R.J., Johanssen, E. & Netland, A. (1984) Conglomeratic fan-delta sequences, late Carboniferous–early Permian, Western Spitsbergen. In: *Sedimentology of Gravels and Conglomerates* (Ed. by E.H. Koster and R.J. Steel). Mem. Can. Soc. Petrol. Geol. 10, pp. 279–294.

Kleverlaan, K. (1987) Gordo megabed: a possible seismite in a Tortonian submarine fan, Tabernas Basin, Province Almeria, southeast Spain. *Sedim. Geol.* **51**, 165–180.

Kleverlaan, K. (1989) Three distinctive feeder-lobe systems within one time slice of the Tortonian Tabernas fan, SE Spain. *Sedimentology* **36**, 25–46.

Kostaschuk, R.A. (1985) River mouth processes in a fjord delta, British Columbia, Canada. *Mar. Geol.* **69**, 1–23.

Kostaschuk, R.A. & McCann, S.B. (1987) Subaqueous morphology and slope processes in a fjord delta, Bella Coola, British Columbia. *Can. J. Earth Sci.* **24**, 52–59.

Leeder, M.R., Ord, D.M. & Collier, R. (1988) Development of alluvial fans and fan deltas in neotectonic extensional settings: implications for the interpretation of basin fills. In: *Fan Deltas: Sedimentology and Tectonic Settings* (Ed. by W. Nemec and R.J. Steel), pp. 173–185. Blackie and Son, London.

Lewis, D.W., Laird, M.G. & Powell, R.D. (1980) Debris-flow deposits of early Miocene age, Deadman stream, Marlborough, New Zealand. *Sedim. Geol.* **27**, 83–113.

Lindsay, J.F., Prior, D.B. & Coleman, J.M. (1984) Distributary-mouth bar development and role of submarine landslides in delta growth, South Pass, Mississippi Delta. *Bull. Am. Assoc. Petrol. Geol.* **68**, 1732–1743.

Marzo, M. & Anadon, P. (1988) Anatomy of a conglomeratic fan-delta complex: the Eocene Montserrat Conglomerate, Ebro Basin, northeastern Spain. In: *Fan Deltas: Sedimentology and Tectonic Settings* (Ed. by W. Nemec and R.J. Steel), pp. 318–341. Blackie and Son, London.

McCabe, P.J. (1978) The Kinderscoutian delta (Carboniferous) of northern England: a slope influenced by density currents. In: *Sedimentation in Submarine Canyons, Fans and Trenches* (Ed. by D.J. Stanley and G. Kelling), pp. 116–126. Stroudsburg: Dowden, Hutchinson and Ross.

McPherson, J.G., Shanmugam, G. & Moiola, R.J. (1987) Fan deltas and braid deltas; varieties of coarse-grained deltas. *Bull. Am. Assoc. Petrol. Geol.* **99**, 331–340.

Nemec, W. (1990) Depositional controls on plant growth and peat accumulation in a braidplain environment: Helvetiafjellet Formation (Barremian-Aptian), Svalvard. In: *Controls on the Distribution and Quality of Cretaceous Coals* (Ed. by P.J. McCabe & J. Totman Parrish), *Spec. Publ. Geol. Soc. Am.* (in press).

Nemec, W. & Steel, R.J. (1988) What is a fan-delta and how do we recognize it? In: *Fan Deltas: Sedimentology and Tectonic Settings* (Ed. by W. Nemec & R.J. Steel), pp. 3–13. Blackie and Son, London.

Nemec, W., Steel, R.J., Porębski, S.J. & Spinnangr, Å. (1984) Domba Conglomerate, Devonian, Norway: process and lateral variability in a mass flow-dominated, lacustrine fan delta. In: *Sedimentology of Gravels and Conglomerates* (Ed. by E.H. Koster and R.J. Steel). Mem. Can. Soc. Petrol. Geol. 10, pp. 295–320.

Nemec, W., Steel, R.J., Gjelberg, J., Collinson, J.D., Prestholm, E. & Øxnevad, I.E. (1988) Anatomy of collapsed and re-established delta front in Lower Cretaceous of eastern Spitsbergen: gravitational sliding and sedimentation processes. *Bull. Am. Assoc. Petrol. Geol.* **72**, 454–476.

Nøttvedt, A. (1985) Askeladden Delta Sequence (Palaeocene) on Spitsbergen — sedimentation and controls on delta formation. *Polar Research* **3**, 21–48.

Ori, G.G. & Roveri, M. (1987) Geometries of Gilbert-type deltas and large channels in the Meteora Conglomerate, Meso-Hellenic basin (Oligo–Miocene), Central Greece. *Sedimentology* **34**, 845–859.

Orton, G.J. (1988) A spectrum of middle Ordovician fan deltas and braidplain deltas, North Wales: a consequence of varying fluvial clastic input. In: *Fan Deltas: Sedimentology and Tectonic Settings* (Ed. by W. Nemec and R.J. Steel), pp. 23–49. Blackie and Son, London.

Porębski, S.J. & Gradzínski, R. (1987) Depositional history of the Polonez Cove Formation (Oligocene), King George Island, West Antarctica: a record of continental glaciation, shallow-marine sedimentation and contemporaneous volcanism. In: *Geological Results of the Polish Antarctic Expeditions* (Ed. by K. Birkemajer). *Studia Geol. Polonica* **93**, 7–62.

Postma, G. (1984) Mass-flow conglomerates in a submarine canyon: Abrioja Fan-delta, Pliocene, SE Spain. In: *Sedimentology of Gravels and Conglomerates* (Ed. by E.H. Koster and R.J. Steel). Mem. Can. Soc. Petrol. Geol. 10, pp. 237–258.

Postma, G. (1990) An analysis of the variation in delta architecture. *Terra Nova* **2**, 124–130.

Postma, G., Babić, LJ., Zupanič, J. & Røe, S.L. (1988) Delta-front failure and associated bottomset deformation in a marine, gravelly Gilbert-type (fan) delta. In: *Fan Deltas: Sedimentology and Tectonic Settings* (Ed. by W. Nemec and R.J. Steel), pp. 91–102. Blackie and Son, London.

Postma, G. & Cruickshank, C. (1988) Sedimentology of a terraced Gilbert-type delta. In: *Fan Deltas: Sedimentology and Tectonic Settings* (Ed. by W. Nemec and R.J. Steel), pp. 144–157. Blackie and Son, London.

Postma, G. & Roep, T.B (1985) Resedimented conglomerates in the bottomset of a Gilbert-type gravel delta. *J. sedim. Petrol.* **55**, 874–885.

PRIOR, D.B. & BORNHOLD, B.D. (1986) Sediment transport on subaqueous fan delta slopes, Britannia Beach, British Columbia. *Geo-Mar. Lett.*, **5**, 217–224.

PRIOR, D.B. & BORNHOLD, B.D. (1988) Submarine morphology and processes of fjord fan deltas and related high-gradient systems: modern examples from British Columbia. In: *Fan Deltas: Sedimentology and Tectonic Settings* (Ed. by W. Nemec and R.J. Steel), pp. 125–143. Blackie and Son, Glasgow.

PRIOR, D.B. & BORNHOLD, B.D. (1989) Submarine sedimentation on a developing Holocene fan delta. *Sedimentology* **36**, 1053–1076.

PRIOR, D.B., BORNHOLD, B.D., WISEMAN W.J. & LOWE, D.R. (1987) Turbidity current activity in a British Columbia fjord. *Science* **237**, 1330–1333.

PRIOR, D.B. & COLEMAN, J.M. (1978) Disintegrating retrogressive landslides on very-low-angle subaqueous slopes, Mississippi Delta. *Mar. Geotechn.* **3**, 37–60.

PRIOR, D.B. & COLEMAN, J.M. (1982) Active slides and flows in underconsolidated marine sediments on the slopes of the Mississippi Delta. In: *Marine Slides and Other Mass Movements* (Ed. by S. Saxov and J.K. Nieuwenhuis), pp. 21–49. Plenum Press, New York.

PRIOR, D.B., WISEMAN, W.J. & BRYANT, W.R. (1981) Submarine chutes on the slopes of fjord deltas. *Nature* **290**, 326–328.

RHINE, J.L. & SMITH, D.G. (1988) The late Pleistocene Athabasca braid delta of northeastern Alberta, Canada: a paraglacial drainage system affected by aeolian sand supply. In: *Fan Deltas: Sedimentology and Tectonic Settings* (Ed. by W. Nemec and R.J. Steel), pp. 158–172. Blackie and Son, London.

RICCI LUCCHI, F., COLELLA, A., CABBIANELLI, G., ROSSI, S. & NORMARK, W.R. (1984) The Crati submarine fan. *Geo-Mar. Lett.* **3**, 71–77.

ROBLES, S., GARCIÁ-MONDÉJER, J. & PUJALTE, J. (1988) A retreoting fan-delta system in the Albiau of Biscay, northern Spain: facies analysis and palaeotectonic implications. In: *Fan Deltas: Sedimentology and Tectonic Settings* (Ed. by W. Nemec and R.J. Steel), pp. 173–185. Blackie and Son, London.

RUST, B.R. & ROMANELLI, R. (1975) Late Quaternary subaqueous outwash deposits near Ottawa, Canada. In: *Glaciofluvial and glaciolacustrine sedimentation* (Ed. by A.V. Jopling and B.C. McDonald). Spec. Publ. Soc. econ. Paleont. Mineral., Tulsa, **23**, pp. 177–192.

STANLEY, K.O. & SURDAM, R.C. (1978) Sedimentation of the front of Eocene Gilbert-type deltas, Washakie Basin, Wyoming. *J. sedim. Petrol.* **48**, 557–573.

SYVITSKI, J.P.M. & FARROW, G.E. (1983) Structures and processes in bayhead deltas: Knight and Bute inlet, British Columbia. *Sedim. Geol.* **36**, 217–244.

SYVITSKI, J.P.M., SMITH, J.N., CALABRESE, E.A. & BOUDREA, B.P. (1988) Basin sedimentation and the growth of prograding deltas. *J. geophys. Res.* **93**, 6895–6908.

VAN STRAATEN, L.M.J.U. (1960) Some recent advances in the study of deltaic sedimentation. *Liverpool Manchester geol. Journ.* **2**, 411–442.

WALKER, R.G. (1966a) Shale Grit and Grindslow Shales: transition from turbidite to shallow-water sediments in the Upper Carboniferous of northern England. *J. sedim. Petrol.* **36**, 90–114.

WALKER, R.G. (1966b) Deep channels in turbidite-bearing formations. *Bull. Geol. Soc. Am.* **50**, 1899–1917.

WESCOTT, W.A. (1988) A late Permian fan-delta system in the southern Morondava Basin, Madagascar. In: *Fan Deltas: Sedimentology and Tectonic Settings* (Ed. by W. Nemec and R.J. Steel), pp. 226–238. Blackie and Son, London.

WESCOTT, W.A. & ETHRIDGE, G.G. (1980) Fan-delta sedimentology and tectonic setting—Yallahs fan-delta, southeast Jamaica. *Bull. Am. Assoc. Petrol. Geol.* **64**, 374–399.

WESCOTT, W.A. & ETHRIDGE, F.G. (1982) Bathymetry and sediment dispersal dynamics along the Yallahs fan delta front, Jamaica. *Mar. Geol.* **46**, 245–260.

WRIGHT, L.D. (1977) Sediment transport and deposition at rivermouths: a synthesis. *Bull. geol. Soc. Am.* **88**, 857–868.

WRIGHT, L.D. (1978) River Deltas. In: *Coastal Sedimentary Environments* (Ed. by R.A. Davies), pp. 5–68. Springer Verlag, Heidelberg.

WRIGHT, L.D. & COLEMAN, J.M. (1973) Variations in morphology of major river deltas as functions of ocean wave and river discharge regimes. *Bull. Am. Assoc. Petrol. Geol.* **57**, 370–398.

Spec. Publs int. Ass. Sediment. (1990) **10**, 29–73

Aspects of sediment movement on steep delta slopes

W. NEMEC

Geological Institute (A), University of Bergen, 5007 Bergen, Norway

ABSTRACT

Depositional characteristics of coarse-grained, sandy and gravelly steep-face deltas are summarized, and the gravity-driven sedimentation processes operative on the subaqueous slopes of such deltas are reviewed. The review further focuses on the mechanics of these processes, by integrating knowledge from modern and ancient delta slopes and relevant laboratory studies. Mechanical aspects of cohesionless sediment mobility and downslope momentum flux are discussed, and distinction is made between frictional debris flows and debris falls. Mechanisms for the transfer of coarsest debris to the delta toe zone are suggested, and various depositional aspects of debris falls and debris flows are explored. The origins and behaviour of turbidity currents on delta slopes are discussed, and possible consequences of density current's 'slumping' stage are indicated (origin of chutes). Supercritical regime and hydraulic jumps in sediment gravity flows are then considered, with emphasis on their depositional effects; possible origins of 'backsets' are suggested. Depositional characteristics of sediment slides are reviewed, with emphasis on the internal shear-strain pattern in slides or flow slides and the origin of another kind of 'backsets'. It is indicated how several important rheological inferences can be made on the basis of the shear-strain signature in a slide or flow-slide deposit. Some other general sedimentological aspects of delta-foreset deposits are discussed; the need for further detailed research and an integrated analytical approach is emphasized, and specific suggestions are given for the future research.

INTRODUCTION

An alluvial delta is a prism of sediment deposited by an alluvial system, whether a solitary river or an alluvial fan, into a body of standing water. One of the more interesting questions related to the process of delta formation is how the alluvial sediment, after crossing the land–water boundary at a delta front, is transported and deposited on the subaqueous delta face.

Somewhat paradoxically, the entire topic of subaqueous sediment movement appears to have been least explored with respect to some most classical delta varieties, such as the coarse-grained, sandy or sand-gravelly, Gilbert-type systems and other steep-face deltas. In studies of ancient deltas, the recognition of large-scale deltaic foresets and the analysis of their varied internal geometries seem to have been sufficient to satisfy the interest of many sedimentologists. The origin of foresets has been attributed almost universally to sediment 'avalanches', with little concern given to the actual mechanism, or range of mechanisms, involved in such downslope sediment movements. This somewhat superficial approach seems to arise from the fact that the presence of (mega-)foresets, especially in association with an alluvial topset, usually makes the deltaic nature of a deposit self-evident. Other varieties of ancient deltas lack this type of direct diagnostic evidence and require detailed facies/process analysis simply as the means of their recognition. As a consequence, the general sedimentological knowledge of the sediment transport mechanisms operative on steep subaqueous slopes of coarse-grained deltas appears to have increased surprisingly little since the early, classical concepts (Gilbert, 1883; Davis, 1890) and laboratory studies (Nevin & Trainer, 1927; Albertson *et al.*, 1950). The steep subaqueous slopes of modern, highly active coarse-grained deltas attracted little attention, as they are virtually inaccessible to direct sedimentological observation, and are hazardous and expensive to study by other means.

Only very recently, the underwater slopes of a number of modern steep-face deltas have been surveyed (see review by Prior & Bornhold, 1988, 1989, this volume) and researchers are beginning to analyze and understand the range of slope facies/processes in similar, coarse-grained ancient deltas (Postma, 1984a,b; Postma & Roep, 1985; Colella *et al.*, 1987; Ori & Roveri, 1987; Postma & Cruickshank, 1988; Postma, Babić *et al.*, 1988).

From quite limited observational material, the author attempts here to review the processes that operate on steep subaqueous slopes of coarse-grained, sandy and sand-gravelly deltas. Many of the inferences and suggestions are necessarily hypothetical and should be regarded with caution, not least because some of them are based on laboratory observations or assumed process analogues from subaerial depositional slopes. Although the review refers to the existing literature on laboratory deltas, scale factors render impossible many meaningful comparisons between laboratory deltas and their far more complicated natural counterparts (see Johansson, 1976, p. 214). Also, for the sake of simplicity, the processes of delta-face reworking by waves or tides are rather ignored. The review focuses on gravity-driven sediment transport, which generally dominates on the steep slopes of coarse-grained deltas.

The author makes numerous references to modern engineering literature, even though experimental evidence on the mechanical behaviour of granular materials is often inconclusive. The behaviour of these materials is complex and varied, and it is difficult to design suitable experiments and to interpret their results. Despite considerable recent progress in kinetic theories on the deformation and flow of granular materials, it remains difficult to form a clear picture of the detailed mechanics of these processes. Although the engineering literature abounds with accounts of tests on natural and artificial frictional sediment (or 'soil') samples under various conditions, most of these tests pertain directly to subaerial settings and their extrapolation to underwater slopes can be no more than qualitative.

Accordingly, the purpose of this chapter is not to undertake a comprehensive review of the sedimentation processes on steep delta slopes, but rather to discuss some selected, important aspects of these processes, with reference to the available field evidence. A picture of the underwater dynamics of steep and coarse-grained, large delta slopes has just started to emerge, and the primary aim of this review is to stimulate interest and further research in the topic.

STEEP-FACE DELTAS

Steep subaqueous depositional slopes considered in this chapter are those associated with the following varieties of coarse-grained deltas: (1) conical, deep-water deltas lacking subaerial distributary plains, and (2) Gilbert-type deltas (Fig. 1). Some relevant characteristics of these systems are outlined briefly below (see also Fig. 2).

Underwater conical deltas

Coarse-grained deltas that lack subaerial distributary plains (Fig. 1A) develop where an alluvial bedload river or fan debouches directly into an excessively deep coastal water; for example, into a deep fjord or a tectonically created, cliff-bounded, deep nearshore environment (see Prior & Bornhold, 1988, 1989, this volume; Ethridge & Wescott, 1984; Massari, 1984; Porębski, 1984; Surlyk, 1984; Leeder & Gawthorpe, 1987, fig. 5; Busby-Spera, 1988). Subaerial delta-plain components are very minor or absent, and virtually all land-derived sediment is deposited beneath the basinal water level. The development and progradation of such deltas, even if very large, have little or no subaerial expression. This is probably why these deltas, until quite recently, have been largely ignored by researchers in modern environments, and have been rarely recognized in ancient basins. However, their occurrences may be quite common, as deltas of this type often precede the development of Gilbert-type systems along cliffed, deep-water coasts. Although these deltas essentially belong to the 'slope-type' (shelf-margin) delta category of Ethridge and Wescott (1984), they may be common features along the cliffed coasts of epicontinental seas or even deep inland lakes. Where developed close to one another, such deltas may coalesce and form underwater slope aprons (e.g. Porębski, 1984; Surlyk, 1984; Busby-Spera, 1988).

The development of such steep, totally subaqueous deltas (Fig. 2A) reflects a negative imbalance between the sediment supply and the effects of basinal water depth (excessive delta-slope height) on downslope sediment transfer. The gravity-driven transport of sediment proceeds directly from the coast edge, and the steep, long underwater slope effectively causes the entire sediment to be distributed subaqueously, by its transfer to the lower slope segment, rather than segregation into Gilbert-type morphological zones (cf. Fig. 2A,

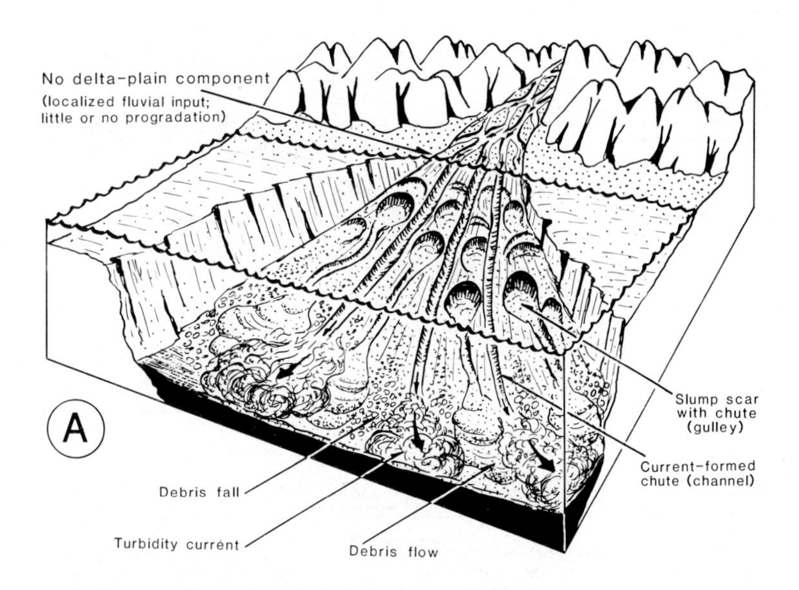

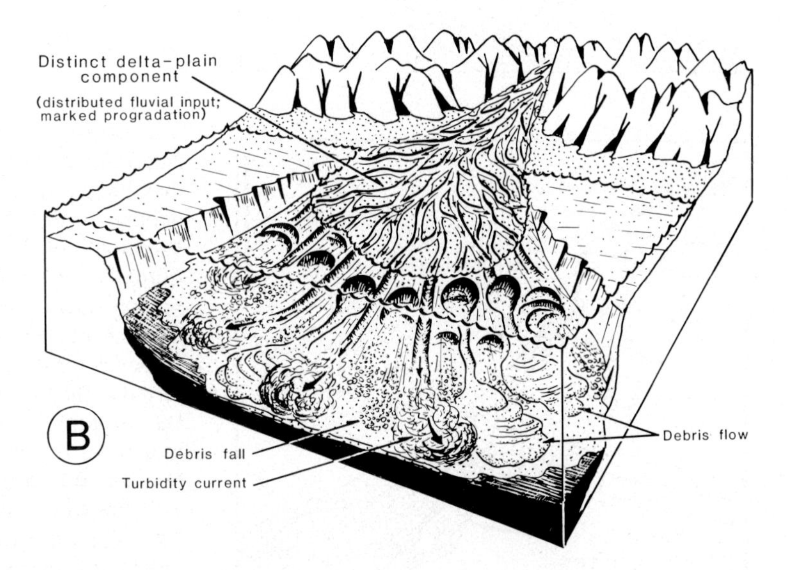

Fig. 1. Idealized diagrams portraying two basic varieties of steep-face, coarse-grained deltas: (A) a conical underwater delta lacking subaerial distributary plain, and (B) a Gilbert-type delta. Such deltas can prograde in underwater topographic confinements or be virtually unconfined. Schematic longitudinal cross-sections are shown in Fig. 2.

B). These deltas are regarded as 'unsteady-state' systems (Prior & Bornhold, 1988). Even with perfectly constant external controlling factors, the growth of the subaqueous delta cone causes progressive shallowing of the basin, thus leading to conditions under which the delta may eventually turn into a Gilbert-type system (see below). Indeed, such an evolutionary transition is recognizable in some well-exposed Gilbert-type deltas that abut against basin-margin submarine palaeoscarps (e.g. Colella, 1988a,c; R. Gawthorpe, pers. comm. 1989).

The slopes of underwater delta cones are steep, exceeding 27° in gravelly sands and reaching 35° in gravels. Even steeper slopes, in excess of 40°, can be expected for very angular debris. As the cone aggrades (Fig. 2A), the mean gradient of its concave slope decreases to some 10–20° and the upper, steeper part of the slope then gradually evolves into

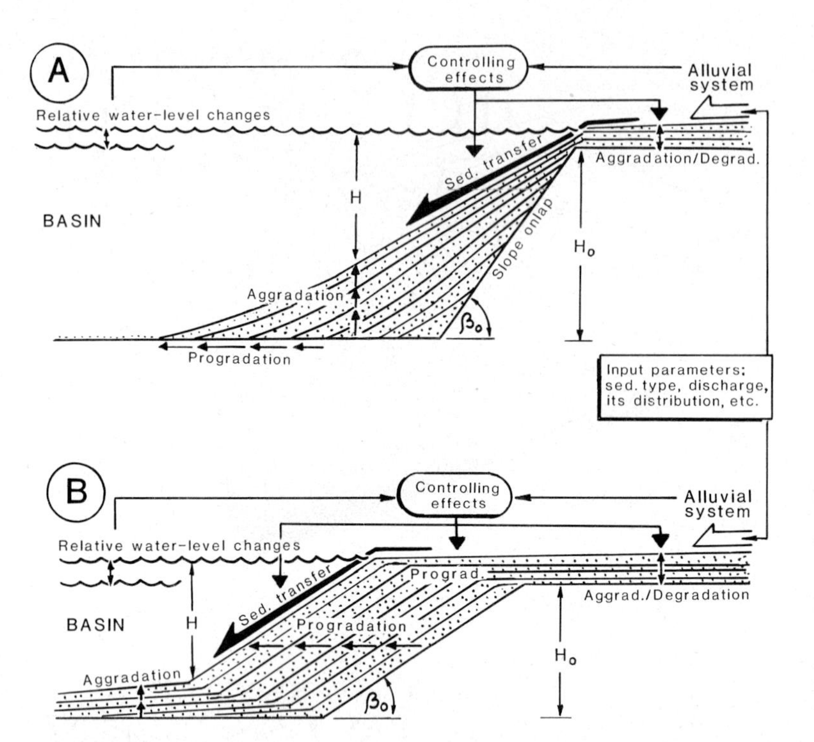

Fig. 2. Schematic summary of the developmental patterns of the two varieties of coarse-grained deltas shown in Fig. 1: (A) a conical underwater delta lacking subaerial distributary plain, and (B) Gilbert-type delta. H_o and β_o are the original height and inclination of basin-margin slope; H is the varied water depth of the delta slope.

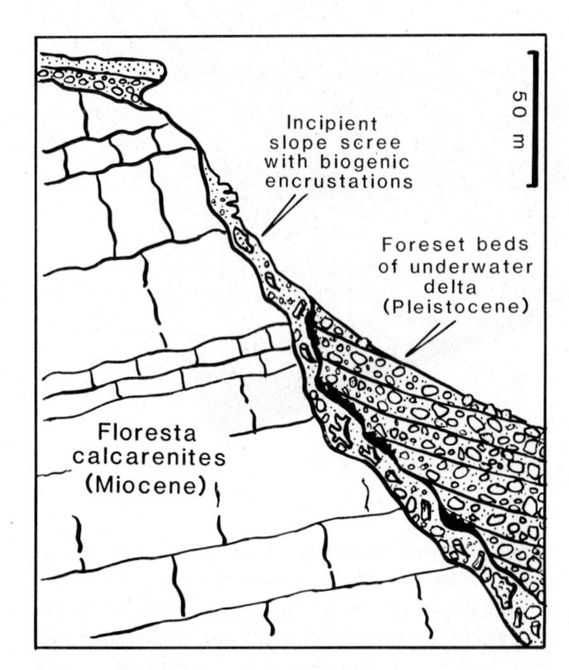

Fig. 3. Outcrop sketch of the most proximal ('back-edge') part of an underwater conical delta abutting against a palaeofault scarp of the Capo dell'Armi horst, Messina Strait, Italy. (Modified from Colella, 1988c, fig. 5; originally from Barrier, 1984.)

a convex-concave, sigmoidal Gilbert-type slope (see Prior & Bornhold, 1988, fig. 11). Cone slopes may be very long, depending on the basinal water depth. In the Bear Bay, for example (Prior & Bornhold, 1988, 1989), such a cone is built into a fjord 375–400 m deep and covers a seafloor area of several square kilometres. The deposits may range from chaotic to well-bedded, and there is a strong tendency for the coarse sand and gravel to be transferred to the lower segment of the slope. In the Bear Bay example, cobble-sized material reaches areas more than 1·4 km from the shoreline; in other cases, boulders 0·5–5 m in size are carried to delta toes 1–2 km from shoreline (Prior & Bornhold, 1988, 1989, this volume). Otherwise, the gross scenario of sediment transport processes resembles that envisaged for the steep faces of large Gilbert-type deltas (see below), although the relative contribution of individual processes appears to be much more varied and somewhat different. These aspects, however, are difficult to assess fully as yet, because of the very limited observational material.

Gilbert-type deltas

These are the classical deltas with tripartite depositional geometry (Figs 1B, 2B). Their gross pattern of development involves sediment transfer across a prograding distributary plain (delta topset segment) on to a relatively steep, prograding delta face (foreset segment), and farther out into a low-gradient delta-toe zone (bottomset segment). The mode of sediment movement changes dramatically as the land-derived, coarse clastic material is transferred from the delta-plain realm, where alluvial processes dominate, into the underwater realm, where gravity-driven processes take over. However, the three delta segments are integral, coexisting and mutually related parts of the system.

Gilbert-type deltas can thus be considered as 'steady-state' systems (Prior & Bornhold, 1988), since the same basic geometric relationship of constituent parts, or delta segments, is maintained with delta growth (Fig. 2B). It does not follow, of course, that the characteristics of the delta segments will remain unchanged. As the distributary delta plain grows in size, the amount and grain size of sediment supplied to the delta front will gradually decrease. Longer-term progradation will thus cause gradual changes in delta-slope characteristics, and will usually be further affected by variations in the 'external' controlling factors (e.g. by relative sea-level variations or tectono-climatic changes in the alluvial system).

The bulk sediment supply effectively determines the size/volume of a delta, whereas the water depth determines the delta thickness, and thus also the height of its subaqueous slope. Water depths must necessarily be much greater than the depth of the alluvial distributary channels. The progradation of a Gilbert-type delta reflects some kind of positive imbalance between the sediment supply and the effects of basinal water depth. The sediment storage on the delta slope must exceed, on a longer term, the capacity for sediment removal and downslope transfer on the slope of given height. This capacity is apparently greater on higher slopes.

Fig. 4. Coarse-grained foresets of Gilbert-type fan deltas, Pleistocene, New Zealand (photos by D.W Lewis). Note the well-defined to diffuse bedding and steep primary dips (in A & B), and the steep imbrication of clast *a*-axes (in B).

The thicknesses of natural Gilbert-type deltas are known to range from a few metres to more than 100 m (see following references). Their subaqueous slopes are steep, commonly up to 20°, reaching 24–27° in sandy deposits and 30–35° in gravels. More continuous avalanches, wave action and other processes can reduce this inclination. The slope (foreset) deposits tend to be well bedded, with bed thicknesses typically around 1 m and often less than 50 cm (Fig. 4). The foreset beds may show tangential transition into bottomset strata, accompanied by downslope fining of grain sizes, or may display downslope coarsening and angular bottom contacts (e.g. Postma & Roep, 1985; Colella *et al.*, 1987; Colella, 1988a,b). Internally, the beds range from stratified to unstratified (massive), and the latter may be ungraded or variously graded. Clast fabric varies from *a*(p) or *a*(p)*a*(i) types in debris-flow beds (Fig. 4B) to an *a*(t)*b*(i) type in tractional deposits, although the latter fabric may be rearranged when exposed to subsequent turbulent flows (see also Johansson, 1976).

Some beds extend over the entire downslope length of the delta face, whereas others are wedge-shaped and pinch out either upslope or downslope. Their strike-parallel extent varies greatly: some beds are broad sheets; others are lobate or constitute thicker, composite lobes; yet others occur as strongly elongate tongues (e.g. Postma, 1984a,b; Postma & Roep, 1985; Colella *et al.*, 1987; Colella, 1988a,b; Postma & Cruickshank, 1988). Slump scars and long, downslope-running scours (chutes) are among common features, and abound in many modern (Prior *et al.*, 1981; Kostaschuk & McCann, 1987; Prior & Bornhold, 1988, 1989, this volume; Syvitski & Farrow, 1989) and ancient deltas (Postma, 1984a,b; Postma & Roep, 1985; Colella *et al.*, 1987; Postma & Cruickshank, 1988). The same characteristics essentially pertain to underwater conical deltas, reviewed above, although the respective observational material is much more limited.

SLOPE REGIME

On steep delta slopes, the potential for downslope transport of sediment intermittently exceeds the rate of sediment supply. The slopes are conducive to small- and large-scale instability, which makes them subject to intermittent local retreat by mass failure and to more continuous processes of mass movement. Settling from suspension and dumping from bedload traction are the principal modes of sediment supply by stream effluents, whereas mass movement is the main conveyor of the sediment on the subaqueous slope itself. Used in a very general sense, sediment 'avalanching' has been invoked as the dominant mass-transport process. The basic concept and nature of the process are reviewed by Allen (1984, vol. II, pp. 148–158), whose considerations are summarized in the next section. Some other relevant, laboratory studies are reported by Johansson (1975, 1976), Hunter (1985) and Hunter and Kocurek (1986).

Avalanching

Transport of sediment over the brink (frontal crest) of a steep-face delta creates a continuous influx of particles to the subaqueous delta slope. Because the rate of deposition resulting from this influx declines downslope, often quite rapidly, the bedload dumping and suspension settling progressively steepen the subaqueous slope. Steepening cannot proceed indefinitely, as there is a limiting slope inclination, the angle of sediment yield, above which the slope surface is unstable in the gravity field. An 'avalanche' of grains will consequently descend the slope from the place where this angle is exceeded, leaving some or all of the slope surface inclined at a lower angle, called the residual angle after failure (see p. 35). Avalanches thus effectively transfer the sediment from the upper part to the lower part and toe of the delta slope.

The process of differential deposition of effluent-derived sediment followed by failure and avalanching will persist for as long as the supply of grains from the delta brink continues. The tendency to steepen the slope by the primary influx of grains is opposed by mass failure and avalanching. Because avalanches have a limited capacity to remove sediment, having finite volumes and travelling at finite velocities, there is a critical sediment-discharge rate at the brink of the delta above which avalanching becomes continuous. For example, when channel mouth bars rapidly prograde on to a steep delta face, they usually contribute to semicontinuous failure and avalanching (Syvitski & Farrow, 1989). The delta slope may thus build forward either discontinuously, as one discrete avalanche after another descends, or in a continuous manner, as a result of sustained and general downslope movement of sediment. The mode can vary in both time and space, even on a local scale along the strike of the slope.

The slope surface, whether viewed as a whole or on a local scale, thus tends to oscillate between a certain higher angle, the *angle of initial yield* (ϕ_i), and a certain lower angle — the *residual angle after failure* (ϕ_r). The latter is commonly referred to as the angle of repose (Van Burkalow, 1945). The difference between the two angles (Allen, 1970a)

$$\Delta\phi = \phi_i - \phi_r \quad (1)$$

is of particular importance as regards the physical character and behaviour of steep natural slopes fed with loose, 'frictional' (cohesionless) sediment, such as the delta slopes considered here.

Allen (1970a, p. 13) asserts that the slope surface can collapse to form avalanches if subjected to sufficiently large downslope acceleration (a). If ϕ is the actual angle of a slope, and $\phi_i > \phi > \phi_r$, then:

$$a = a_{max} \frac{\phi_i - \phi}{\Delta\phi} \quad (2)$$

in which:

$$a_{max} = g(\tan \phi_i - \tan \phi_r),$$

wherein g = acceleration due to gravity. The a value is the critical value of downslope acceleration, which is just sufficient to cause an avalanche on a slope of initial angle ϕ, and a_{max} is the maximum possible downslope acceleration that will initiate a free-running avalanche. As shown by Allen (*op. cit.*), the value of ϕ_r is very nearly constant for a given material regardless of the conditions of avalanching, while ϕ_i is subject to large variations, controlled primarily by the volumetric concentration, or packing, of the sediment grains (Fig. 5). Therefore, $\Delta\phi$ (see above) is not a constant. It approaches 8° for naturally deposited quartz sands, and can be as large as 14° for unsorted sands composed of very irregular-shaped grains.

Allen (*op. cit.*) further shows that equation (2) can be reduced to

$$a \approx 1/\Delta\phi \quad (3)$$

for roughly constant values of the other quantities involved. His experiments confirm that $\Delta\phi$ is a direct function of the volumetric concentration of the sediment grains (C):

$$\tan \Delta\phi = f_1 (C),$$

and hence for a given sediment ($\phi_r \approx$ const.):

$$\tan \phi_i = f_2 (C),$$

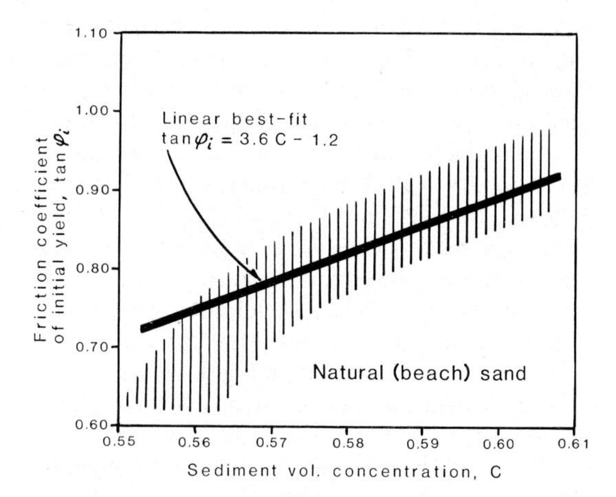

Fig. 5. Relationship between the friction coefficient of initial yield and the fractional volume concentration for naturally occurring, fine-grained beach sand (simplified from Allen, 1970a, fig. 7). Similar sediment behaviour has been predicted theoretically by Allen (*op. cit.*), and demonstrated by his laboratory experiments with glass beads.

as shown by the example in Fig. 5.

Allen's study thus emphasizes the important role of sediment texture and packing in controlling the failure potential of a depositional slope. The volume concentration of grains will depend on the sediment sorting, its mode of deposition (whether from stream bedload traction, suspension settling or mass flow), and degree of subsequent compaction. Accordingly, Allen (*op. cit.*) points out that the magnitude of $\Delta\phi$ is an important index of the stability characteristics of a depositional slope. The $\Delta\phi$ value indicates the probability of an avalanche occurring on a slope. Equations (2) and (3) are thus criteria for the response of the slope to externally applied disturbing forces. Such forces, in the form of shear stress, may arise directly from the downslope component of sediment weight; from loading and shearing by a moving, overpassing sediment mass; from the passage of a local or regional shock wave (produced by sudden slope failure, an avalanche or a seismic tremor); or from cyclic loading by tides or sea waves (for review of these factors, see Prior & Coleman, 1984).

Range of processes

From laboratory studies (Nevin & Trainer, 1927; Albertson *et al.*, 1950; Jopling, 1963, 1965; Johansson, 1975, 1976; Allen, 1984, vol. II, ch. 4;

Hunter, 1985; Hunter & Kocurek, 1986) and observations on steep slopes of modern (Prior *et al.*, 1981; Syvitski & Farrow, 1983, 1989; Kostaschuk & McCann, 1987; Prior & Bornhold, 1988, 1989) and ancient delta (Clemmensen & Houmark-Nielsen, 1981; Colella *et al.*, 1987; Postma, 1984a,b; Postma & Roep, 1985; Postma & Cruick-shank 1988), it is quite clear that 'avalanche', as used in the previous section, is a loose term encompassing a wide range of relatively *rapid* (hence sometimes called 'catastrophic') mass-movement processes (Fig. 6). Usage of this term is similar, if not less consistent, in studies of mass transport on steep subaerial slopes (cf. Mudge, 1965; La Chapelle, 1977; Voight, 1978; Hopfinger, 1983; Eisbacher & Clague, 1984; Nyberg, 1985; Cruden & Hungr, 1986).

In the published sedimentological studies of delta-foreset deposits, some processes are given specific names (see Fig. 6), whereas other, largely unrecognized processes are referred to jointly as avalanches. Unfortunately, the interpretations too often remain on this general level only. Moreover, the term avalanche does not seem to have exactly the same meaning to different authors. Although many researchers treat it as a useful general term (cf. Fig. 6), some others consider it synonymous to debris fall or cohesionless debris flow (sandflow) only. The loose, general meaning of the term is much more convenient; however, it should not be used in delta studies as a mere substitute for the actual recognition of delta-face processes.

The range of processes envisaged for steep, coarse-grained delta slopes (Fig. 6) is defined here rather tentatively, as the available direct observational evidence is still quite limited. In other words, this spectrum of processes is certainly relevant, although the contribution of individual processes to their bulk plexus may vary in time and space, also from one delta slope to another. The suggested range of resedimentation processes (Fig. 6) includes those currently recognized as pertaining to sediment transport and deposition on subaqueous slopes (e.g. see reviews by Rupke, 1978; Stow, 1986), and essentially represents a process continuum. For example (Fig. 6): slope creep may lead to translational sliding when detachment (slip) occurs; a translational slide may turn into a slump, and the latter into a plastic flow as the shear becomes pervasive and remoulding occurs; liquefied mud-free sediment will rapidly turn into a debris flow or a turbidity current; a debris flow may collapse and continue to move in a sliding fashion (as a 'flow slide'), or become fully turbulent; similarly, a voluminous debris fall may collapse and continue to move as a cohesionless debris flow. All such transformations can be expected to occur within the lengths of the subaqueous slopes of natural deltas considered here.

A review of field criteria for the recognition of the depositional products of the various processes listed in Fig. 6 is beyond the immediate scope of this chapter. The relevant existing literature includes Carter (1975), Stanley *et al.* (1978), Nemec *et al.* (1980), Stanley (1980), Lowe (1982, 1988), Massari (1984), Nemec and Steel (1984), Nemec *et al.* (1984), Postma (1984b), Surlyk (1984, 1987), Walker (1984), Postma and Roep (1985), Eyles *et al.* (1987), Nemec *et al.* (1988), Postma, Babić *et al.* (1988), Postma, Nemec and Kleinspehn (1988), Farrel and Eaton (1988) and Martinsen (1989) among others. The following discussion focuses on selected aspects of these processes and their products, with direct reference to steep, sandy or sand-gravelly delta slopes.

SEDIMENT DYNAMICS

Shear rate and flow behaviour

Steep slopes of coarse-grained deltas are dominated by cohesionless sediments, ranging from sand to gravel. Interbeds of mud or mud-rich coarse sediment may occur as well, but these are not the principal type of delta-face deposits and are largely ignored in the present discussion; their mechanical role in delta slopes and modes of movement are relatively well known (e.g. Roberts *et al.*, 1980; Karlsrud & Edgers, 1982; Prior & Coleman, 1982, 1984; Schwarz, 1982; Mandl, 1988).

When an appropriate state of stress is applied to a static, granular (cohesionless) material, the material will yield along stress characteristics as a plastic substance (Savage, 1979, 1983; Vermeer & Luger, 1982). If the strain rate is low enough, the sediment will deform and move downslope by means of very slow, intergranular frictional sliding, with quasi-static grain-to-grain contacts and no major slip surfaces (sediment *creep*, Fig. 6). A slowly deforming slope sediment may otherwise move as a slice of coherent material; there may be one major slip surface and little or no internal deformation (slow, non-avalanching *slide*, Fig. 6), or a multitude of slip surfaces and considerable internal deformation (slow, non-avalanching *slump*, Fig. 6). In

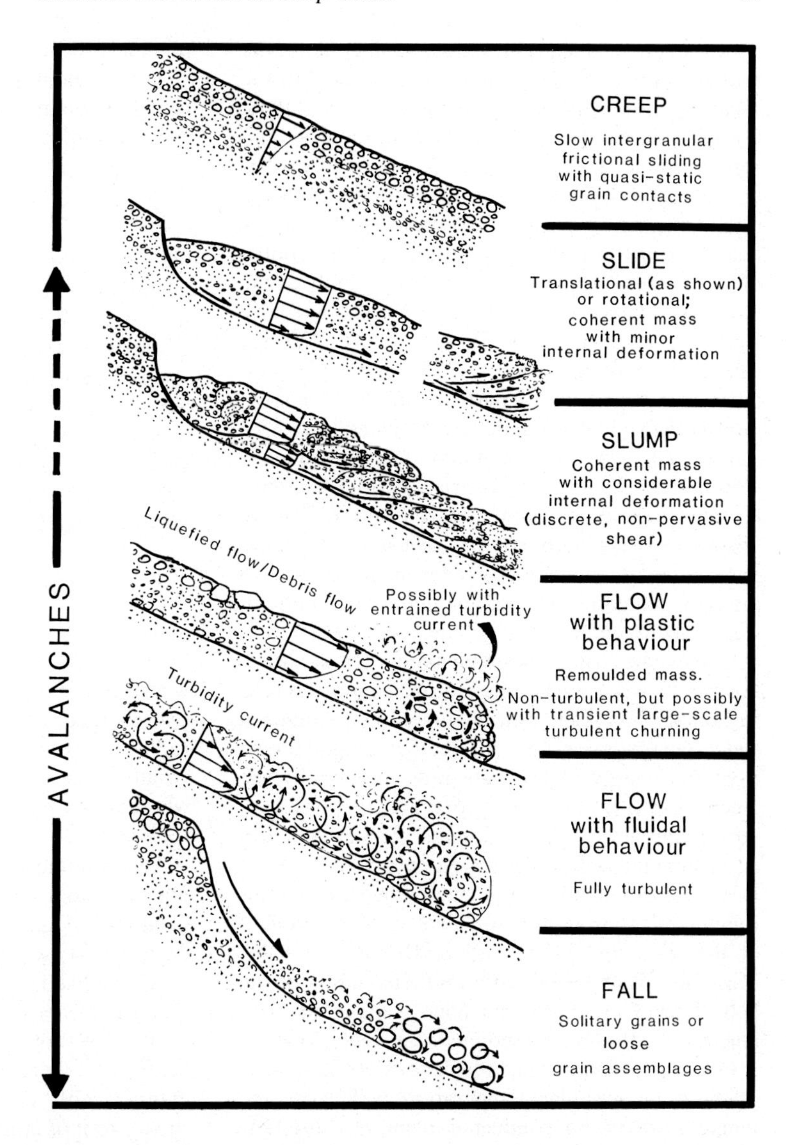

Fig. 6. Range of gravity-driven, sediment-transport processes operative on steep slopes of coarse-grained deltas (inspired by Rupke, 1978, fig. 12.3; Stow, 1986, fig. 12.3). Note the suggested range of 'avalanches' (rapid mass movements on steep slopes), which may include also some slumps and slides. The velocity profiles are schematic, not to scale.

engineering parlance, all of these are cases of the *quasi-static regime* of granular flow (Mróz, 1980; Spencer, 1981). In the sedimentological terminology, however, these modes of sediment movement are not regarded as 'flow' (Fig. 6).

If the shear-strain rate is high and sediment moves rapidly, the impact between particles along the slip surfaces will be sufficient to dislodge individual particles from their parent, coherent 'blocks' of grains; consequently, the intervening slip zones will be enlarged until the entire mass of sediment is moving as dispersed, semi-independent grains, each in relative motion with its neighbours. In engineering terminology, this is the *rapid-flow* (or grain-inertia) *regime* (Savage, 1983; Campbell & Brennen, 1983, 1985b). The state of remoulding and grain dispersion may be pervasive, comprising the entire mass of moving sediment (as in *flow* and *fall* processes, Fig. 6), or be restricted to the basal zone only, as discussed by Campbell (1989b); this latter case would pertain to some rapidly moving, 'avalanching' slides or slumps (cf. Fig. 6).

In a rapidly shearing sediment, any contact between particles is momentary, as the relative motion that drives particles together will immediately force them apart. Individual particles will appear to move in a quasi-random manner about the average motion vector or mean velocity field of the shearing sediment mass (Savage, 1979, 1983; Campbell & Brennen, 1983). This mechanical concept, introduced by Bagnold (1954), strongly evokes an analogy with the thermal motion of molecules in the kinetic theory of gases. Accordingly, the energy associated with the random motions (or vibrations) of the sediment particles has been dubbed the 'granular temperature' (Haff, 1983; Campbell & Brennen, 1983, 1985a,b; Savage, 1983). This so-called 'temperature', as a purely kinetic phenomenon, is a byproduct of particle collisions; simply, interparticle collisions make the individual particles 'rattle' (i.e. experience random, higher-frequency vibrations). The pressure associated with the granular temperature is commonly referred to as the 'dispersive pressure (or stress)', as it acts to force the particles apart. In macroscopic terms, the dispersive pressure maintains the shearing material in a liquidized, viscous state.

Unlike the true, thermodynamic temperature, however, the granular temperature cannot be self-sustaining. To maintain the sediment in a liquidized state, energy must be continually pumped down from the energy of the bulk mass motion by the mechanism of shear-strain work and be converted into grain energy. Thus, granular temperature can be maintained only when driven by gradients in the mean velocity field of the mass flow, and varies roughly as the square of the local velocity gradient of the mass flow (Campbell & Brennen, 1985b). This concept is the key to an understanding of the behaviour of rapid granular flows, such as the avalanches of cohesionless sediment on steep slopes.

One of the important implications is that strong gradients in granular temperature will cause the temperature to be conducted from the intensely shearing, 'high-temperature' zones into 'low-temperature' or non-shearing zones. Hence, the granular temperature may be non-zero inside a 'rigid plug', such as the non-shearing upper or middle part of a mass flow (Campbell & Brennen, 1985a,b); this would mean some additional dilation and thus increased mobility of the sediment mass. Low-magnitude temperatures and the temperature flux to non-shearing zones will tend to be dampened by intergranular viscous liquids, which may range from clean water to silt- or mud-rich slurries. However, a high-magnitude temperature flux to the upper part of a mass flow will cause 'thermal' dispersion (saltation) of grains at the top and thus enhance mixing with the ambient water.

The rheological behaviour of a simple Newtonian liquid is defined by the well-known constitutive relationship:

$$\tau_s = \mu \frac{dU}{dy}, \tag{4}$$

wherein τ_s is the applied shear stress; μ is the coefficient of viscosity (liquid constant); and dU/dy is the rate of shear strain, defined by the local velocity gradient. By analogy, the behaviour of a rapid, *fully shearing* granular flow (state above yield-strength limit) can be considered in terms of a 'self-exciting' Newtonian liquid, with an appropriately modified viscosity coefficient (see Bagnold, 1954; Savage, 1979, 1983; Haff, 1983; Jenkins & Savage, 1983; Campbell & Brennen, 1983, 1985b):

$$\tau_s = \underbrace{\left[\rho \cdot R^2 \cdot f(C) \cdot \frac{dU}{dy}\right]}_{\text{apparent viscosity}} \frac{dU}{dy}, \tag{5}$$

wherein: ρ = specific density of sediment particles ('submerged' density if in water); R = particle radius; $f(C)$ = dimensionless function of sediment volume concentration (solid fraction); and $\tau_s > k$, where k = Coulomb yield strength of the frictional sediment.

This relationship means that, although the yielding behaviour of cohesionless sediments is of the Coulomb–Mohr type (with shear stress proportional to the normal stress), the actual behaviour of a liquidized, flowing sediment is effectively non-Newtonian, as its apparent viscosity is not a constant. The apparent viscosity (or flow's mobility coefficient) here varies in a linearly proportional manner with the shear-strain rate (or with the square root of granular temperature; see Haff, 1983; Campbell & Brennen, 1983, 1985b). In other words, the apparent viscosity of liquidized cohesionless sediment should not be considered as a 'material constant' (cf. Allen, 1985, pp. 162–163), since the plastic flowage of such materials does not represent the ideal, Bingham-type behaviour (see also McTigue, 1982; Savage, 1983). Apparent viscosity (also called the coefficient of viscosity) is the ratio of the applied shear stress to the rate of shear strain, and there is no reason why this ratio should be constant for a particular granular material (see Bagnold, 1954; Rutgers, 1962; Iverson, 1985). Some further implications of this rheological property are discussed below.

Flow mobility

One of the intriguing questions regarding cohesionless sediment movement is how a liquefied sediment flow (*sensu* Lowe, 1982) composed of sand or sandy gravel is able to descend a steep delta slope, often travelling over many hundreds or even thousands of metres, without becoming turbulent and turning into a turbidity current. This particular point was recently addressed by Middleton and Southard (1984, p. 349), who used turbulence criterion for Newtonian liquids (Re > 500) and a Reynolds number for sediment gravity flow defined as:

$$\mathrm{Re} = \frac{\rho_s \cdot U \cdot h}{\mu_s}, \tag{6}$$

wherein: ρ_s = sediment flow density; U = flow mean velocity; h = flow thickness; μ_s = flow apparent viscosity. On this basis, they concluded that liquefied sediment flows, even of quite small scale, are very unlikely to move down any substantial slope without becoming turbulent. This view, and the well-known laboratory evidence that pore water escapes very rapidly from liquefied sands (e.g. Lowe, 1975, 1976; Allen, 1985), have raised much scepticism as to the actual role of liquefied flows as an agent of sediment transport and deposition on steep, relatively coarse-grained delta slopes. Many researchers simply deduce that liquefied sediment on such a slope will either collapse instantly and cease to move, or (if mobile) turn readily into a turbidity current.

Such reasoning, however, is contradicted by observations from steep, modern subaqueous slopes (e.g. Andresen & Bjerrum, 1967; Bjerrum, 1971; Karlsrud & Edgers, 1982; Koning, 1982; Prior & Bornhold, 1986; Schafer & Smith, 1987; Syvitski & Farrow, 1983, 1989) and deductions from ancient slopes (e.g. Nemec *et al.*, 1980; Postma, 1984b; Surlyk, 1984, 1987; Postma & Roep, 1985; Nemec *et al.*, 1988). Liquefied, fully homogenized sands and sand-rich gravels apparently can move as laminar flows over relatively long distances on steep-to-gentle underwater slopes and be deposited without becoming turbulent. For example, Syvitski and Farrow (1989, p. 30) report deposits of cohesionless, non-turbulent, non-erosive sandflows, 3–10 m thick, that moved over 30–40 km in the Cambridge Fjord, Baffin Island. These and similar other deposits (Fig. 7) are massive, lack grading and tractional structures, have sharp boundaries, show water-escape features and some even contain floating megaclasts (see previous references).

Sediment in such settings, once mobilized and slightly dilated by liquefaction (however brief) and low-rate shear strain, may either (1) rapidly accelerate and turn into a turbidity current, as suggested by others, or (2) collapse, *but continue to move* as a cohesionless (frictional) debris flow or a liquefied slide ('flow side'). Similar inferences are presented by Schafer and Smith (1987) and Syvitski and Farrow (1989). In case (2), the movement will be governed by the mechanical laws of plastic flowage (see previous section).

Accordingly, the Reynolds number for a plastic-type sediment gravity flow (Metzner & Reed, 1955; Fredrickson, 1964, p. 230) should be defined as:

$$\mathrm{Re} = \frac{\rho_s \cdot U^{2-n} \cdot h^n}{\mu_s}, \tag{7}$$

wherein: n = flow behaviour index; other symbols as in equation (6).

The index (n) pertains to the viscometric behaviour of the material: $n = 1$ for Newtonian viscometry (i.e. constant, strain rate–independent viscosity), as in an ideal, Bingham plastic; and $n \neq 1$ for non-Newtonian behaviour (i.e. strain rate–dependent viscosity), as in non-Bingham plastics. The value is $n > 1$ for materials whose viscosity *increases* with increasing shear-strain rate ('shear-thickening' materials), and $n < 1$ for those whose viscosity *decreases* with increasing shear-strain rate ('shear-thinning' materials).

As discussed in the previous section, the behaviour of shearing cohesionless sediment flows is expected to be non-Newtonian above their yield-strength limit. A collapsing liquefied sediment will probably commence flow in the 'shear-thinning' fashion, but will soon assume the 'shear-thickening' mode as it changes to a frictional debris flow or 'flow slide' (where dilation is the means of overcoming intergranular friction). The apparent viscosity will be extremely high due to high sediment concentration (cf. Roscoe, 1953; Rutgers, 1962) and such flows may certainly move without becoming turbulent (cf. equation 7 above). The critical Reynolds number for the onset of turbulence in plastic flows will normally be much greater than 500 and can be as high as 50 000 for natural sediments (see Middleton & Southard, 1984, pp. 357–358).

The crucial problem is how a frictional, non-turbulent plastic flow can create its own *mobility* in the absence of a lubricating mud. There seem to be

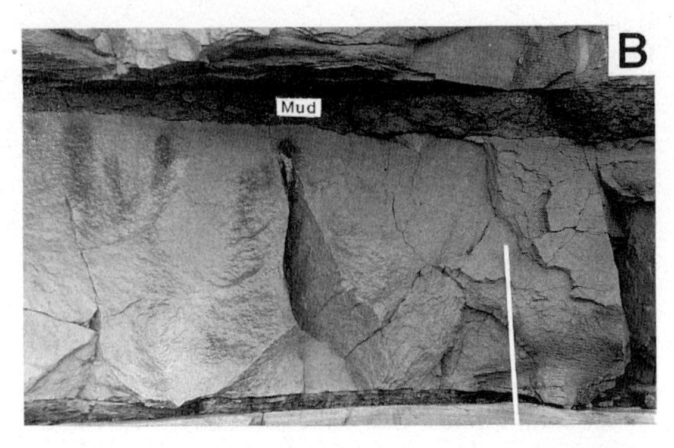

Fig. 7. Thick (*c.* 1 m) massive sandstone beds with sharp bases and tops, deposited in the toe zone of a steep local slope of a braidplain delta (Cretaceous Helvetiafjellet Fm, Spitsbergen). The emplacement mechanism is interpreted as liquefied sandflows with laminar, cohesionless plastic-flow behaviour (for details, see Facies C in Nemec *et al.*, 1988). In (A), note the overturned, sheared large mud-flame (thick arrow) and the associated, syndepositional thrust; movement was to the right, at 50° out of the outcrop. In (B), note the massive structure, sharp base and top, and 'ghosts' of vertical water-escape pipes. In (C), note the upslope-dipping shear-bands (arrowed) and syndepositional thrust surface; movement was to the left, at 30° out of the picture. The associated deposits are turbidites.

several mechanical possibilities, depending on the magnitude and pervasiveness of shear strain.

1 A brief phase of sediment liquefaction on a steep slope may be sufficient to cause instantaneous, high-rate laminar shearing, which disperses the grains. If significant granular temperature results from the collisional stresses (see p. 38), this process may provide mobility and delay collapse of the granular dispersion when it arrives on a gentler slope. Importantly, the bulk frictional coefficient of a collapsing dispersion may considerably decrease.

A standard test in soil mechanics is to shear a static sediment sample and measre the friction coefficient (i.e. the ratio of shear force to normal force) at which the sample yields. The results are well known, and show that the friction coefficient increases with increasing density of the initial sediment grain packing (Fig. 5). Experimental studies of fully developed granular flows (Savage & Sayed, 1984; Campbell & Brennen, 1985b; Campbell & Gong, 1986) show the opposite to be true: the frictional coefficient *decreases* with increasing volumetric grain concentration. This relationship has been attributed to anisotropies in the distribution of grain collision angle, induced by the formation of a dynamic layered microstructure (Campbell & Brennen, 1985b; Campbell, 1986).

2 The shearing material on a slope close to the angle of repose may move as an 'immature sliding flow' (Savage, 1983, p. 266). A stationary, essentially inactive zone can develop in the lower part of the issuing laminar flow (cf. Takahashi, 1978); such an inactive basal layer will be thicker in the upslope than in the downslope direction (cf. Middleton, 1967; Postma, Nemec & Kleinspehn, 1988, fig. 2). An effective slope surface will thus form *within* the flow body, whereby the inclination of the zero-velocity surface will be steeper than the original slope itself. The velocities near the base of flow are quite low in such a case, and the overall velocity profile has a concave shape (Savage, 1983).

3 The shearing material may collapse into a flow slide, riding on a thin 'active' layer of dispersed particles, as in simple translational slides (Campbell, 1989b). The energy that is dissipated in the active layer is constantly resupplied by the kinetic energy of the bulk sliding mass. If the mass of the slide sheet is much greater than the mass of the active layer, the energy dissipation rate may be low and the slide will be supported for a time period long enough for the sediment to travel a long distance (Karlsrud & By, 1981; Campbell, 1989b). The run-out distance, d, depends directly on the initial kinetic energy of the slide (Campbell, 1989b):

$$d = U_o^2 \cdot g^{-1} \cdot f(\lambda, \beta) \quad (8)$$

wherein: U_o = initial velocity; g = acceleration due to gravity; $f(\lambda, \beta)$ is a dimensionless function of $(\bar{\lambda}-\beta)$, with $\bar{\lambda}$ being the particles' mean collision angle (in active layer) and β the slope angle (see definition sketches given as insets in Fig. 8).

This mechanical model predicts that the moving mass will behave in very specific ways: either the run-out distance obeys equation (8) and can be relatively long, or the slide will move and come to rest almost immediately (Fig. 8; note the prominent inflection of the curves). Slope angle appears to be of secondary importance (see Fig. 8). Since the principal way of removing kinetic energy from the slide is by means of the work done by the particles in the active layer, a subaqueous slide on a steep slope may continue to accelerate, inevitably turning into a debris flow or a turbidity current, unless $\bar{\lambda} > \beta$ (if $\bar{\lambda} \approx \beta$, the run-out distance will be 'infinite', as shown by the asymptote in Fig. 8).

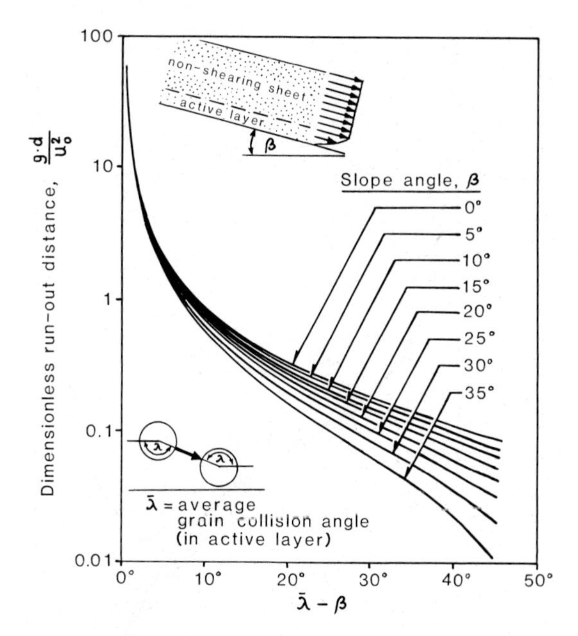

Fig. 8. Predicted dimensionless run-out distances of debris-slides for various slope angles (β), as a function on ($\bar{\lambda}\beta$). Modified from Campbell (1989b, fig. 9).

Downslope momentum flux

In dispersed granular materials moving downslope, the momentum is transferred within the dispersion in two modes (Campbell, 1986, 1989a; Campbell & Gong, 1986). The *streaming* (or 'kinetic') mode refers to the straightforward downslope flux of momentum as the individual particles experience the downslope pull by gravity and move through the dispersed system carrying their momentum with them; the larger the particle, the greater its momentum. The *collisional* mode refers to the transfer of momentum by interparticle collisions. Both mechanisms contribute to the bulk stress tensor in an avalanching granular dispersion, but their relative contributions vary. The streaming mode dominates at low particle concentrations, where collisions of strongly dispersed particles are relatively infrequent. The collisional mode dominates at high particle concentrations, where the particles cannot move far without colliding. The relative importance of the streaming and collisional contributions as a function of particle concentration is shown in Fig. 9. The two stress components have nearly the same magnitude when the concentration is about $C = 0.18$ (18 vol.%) and the streaming component becomes negligible ($\tau_s < 0.1\tau_c$) at about $C = 0.40$ (40 vol.%).

According to the relationships shown in Fig. 9, it is then possible to distinguish two regimes of granular dispersion movement (see also Lun *et al.*, 1984):

1 *cohesionless debris flows* (or 'grainflows'), wherein the high bulk concentrations will render the collisional stress dominant; and

2 *debris falls* (or 'grainfalls', although this term is not used here because of its well-known suspensional connotation introduced by other authors), wherein the high degree of dispersion will normally render the streaming stress dominant.

One regime can easily transform into the other; for example, when a cohesionless debris flow encounters any kind of 'precipice' (e.g., an excessively steep slope or the headscarp of a large slide scar) on its downslope route, or when a debris fall begins to decelerate and collapse on a gentler slope or in the confinement of a gulley. For example, Colella *et al.* (1987, fig. 10) describe lobes of clast-supported, essentially openwork coarse gravel that occur in the lower part of a Gilbert-type delta slope and show a combination of the depositional attributes of debris falls and cohesionless debris flows (see also Colella, 1988b, fig. 13A).

The distinction made between the downslope 'flow' and 'fall' of debris may also be viewed in terms

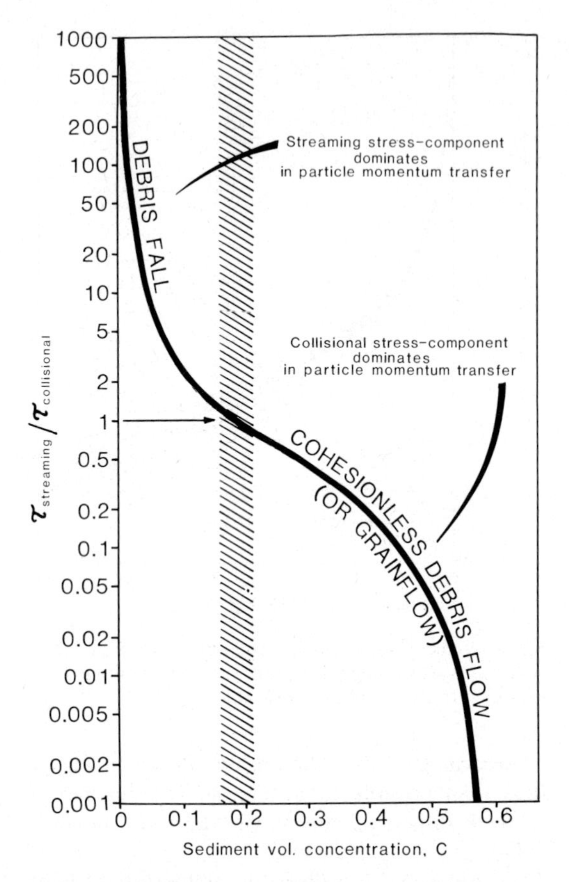

Fig. 9. Schematic plot of the ratio of streaming to collisional components of the bulk stress tensor as a function of the fractional concentration of particles. Modified from Campbell (1989a, fig. 7); process names added.

of the mechanisms of granular temperature generation (see p. 38). The granular temperature, or particle energy, is generated in three ways (Campbell, 1986, 1989a; Campbell & Gong, 1986):

1 as the random component of velocity imparted to a particle by a collision with another particle;

2 by particle movement along the bulk velocity gradients within the shearing granular dispersion (i.e. as a particle moves, by the virtue of its random motions, along the gradient of the flow's mean velocity field, the particle acquires a random velocity component, roughly equal to the difference in mean velocity between the particle's last collision and its present location within the flow); and

3 by the direct pull of gravity that a moving, bouncing and/or rolling particle experiences on a steep slope.

Mechanism 1 is crucial in 'grainflows' (debris flows dominated by dispersive stresses), whereas mechanism 2 becomes more important as the degree of particle dispersion increases. Mechanism 3 will dominate in some extremely dispersed systems, where particles move relatively long distances in isolation. The downslope component of granular temperature will be generated principally by mechanisms 2 and 3, which derive from the streaming mode of downslope momentum flux.

COARSE DEBRIS AHEAD

An interesting aspect of avalanche deposition on the steep slopes of conical underwater deltas and many Gilbert-type deltas is the accumulation of coarsest debris in the lower part and toe zone of the slope (Fig. 10; Colella, 1988a, fig. 13A, B; Colella *et al.*, 1987). This type of segregation of coarse debris on delta slopes, modern and ancient, has been commonly reported, but scarcely explained. Instead, researchers have focused their attention on the tangential foreset beds which show downslope fining, traditionally attributed to particle segregation by settling from stream-derived suspension plumes. In this section mechanisms that produce downslope coarsening in slope deposits are discussed.

Debris falls

Debris-fall processes (term after Holmes, 1965, p. 481) may certainly dominate on slopes of conical underwater deltas (Figs 1A, 2A), especially during the early stage of slope development (Prior & Bornhold, 1988, 1989, this volume). Such processes may also be common on steep slopes of Gilbert-type deltas (Fig. 1B), where rapid textural variations, depositional oversteepening, scouring by waves or currents and mass failure produce local scarps and slope inclinations too steep for the sediment to stay intact or move smoothly, or for a newly supplied sediment to mantle such inclines instantaneously. The steepness of the slope will result in failure and some very rapid avalanches, controlled mainly by debris-fall mechanics (see previous section). The debris may fall as single particles, or as small- to large-volume assemblages (masses) of strongly dispersed particles. Clasts falling from a steep scarp on to a sedimentary slope will bounce, slide and roll over the surface before coming to rest. Larger clasts

Fig. 10. Side-scan sonograph mosaic showing accumulations of coarse debris (including boulder-sized blocks a few metres in diameter) in the lowest slope and toe zone of a modern, conical, underwater delta built into a fjord. Note the elongate tongues of bouldery debris-fall deposits in the middle, and the extensive sheets of turbiditic sand (most recent deposit) to the left, covering the axial sector of the delta cone. Example from the South Bentinck Arm fjord, British Columbia, Canada (see also Prior & Bornhold, this volume, fig. 12). Photo provided by D.B. Prior.

experience a stronger pull of gravity (acquire higher momentum), thus travel faster and outpace the smaller particles. Similarly, large clasts within debris-fall avalanches, if sufficiently dispersed (Fig. 9), will tend to shear (or 'stream') through the system to its front. Although moving as an assemblage, each of the strongly dispersed clasts is driven downslope essentially according to its own individual momentum.

Consequently, the sediment accumulating from debris falls will show a downslope increase in clast size and a corresponding increase in the resultant slope-surface roughness (Fig. 11). Longer slopes and higher degrees of particle dispersion usually increase this clast-size segregation, whereby coarse, openwork gravel may constitute the frontal and medial part of a debris-fall deposit. In a descending debris-fall avalanche, the larger particles will tend to come to rest first, and the finer-grained upslope 'tail' of the avalanche may partly override the coarser debris to create some normal grading within the resulting deposit (Fig. 11). When several debris falls descend the slope at short intervals (so that little or no intervening deposition of 'fines' takes place to smooth the surface), the frictional retardation and grain-trapping effects of the varied surface roughness may lead to an upward coarsening (Fig. 11). Alternatively, the depositional surface may be gradually smoothed because relatively fine-grained, sandy or muddy material from turbidity currents, wave action or suspension settling fills interstitial spaces. Any interstitial accumulation of fine material is very important, because it will influence the slope surface roughness and subsequent depositional processes. Examples of debris-fall deposits are shown in Figures 10–13 (see also Prior & Bornhold, this volume, figs 4, 12; Colella *et al.*, 1987, fig. 10; Postma & Cruickshank, 1988, figs 6B, 8).

Subaqueous debris falls have much in common with 'rockfalls', both subaqueous and subaerial (Rapp, 1960; Gardner, 1970; Carson & Kirkby, 1972; Kirkby & Statham, 1975; Luckman, 1976; Statham, 1976; Carson, 1977; Hoek & Bray, 1977;

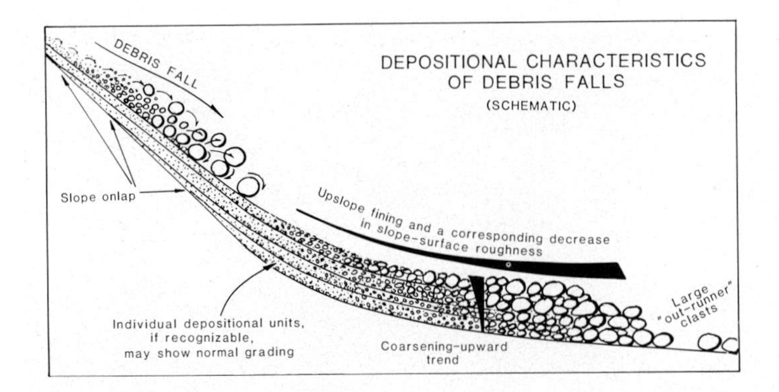

Fig. 11. Depositional characteristics of debris falls (schematic; not to scale).

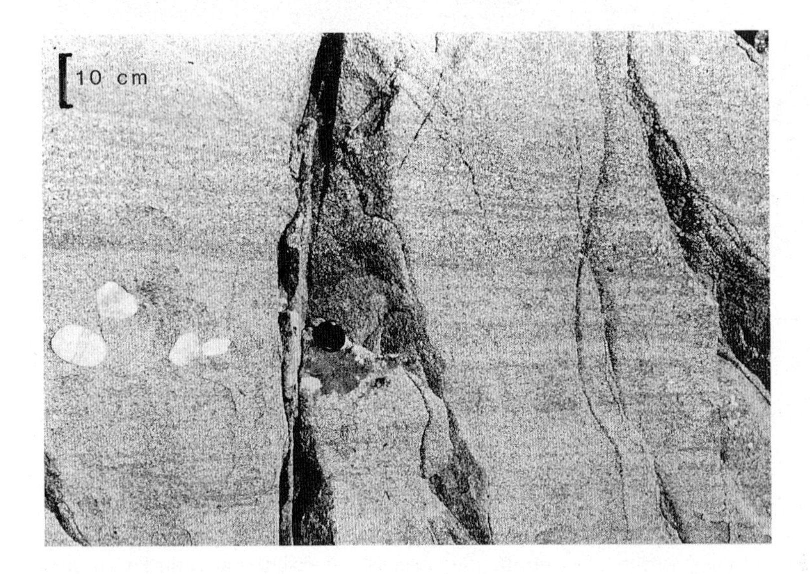

Fig. 12. Scattered pebbles and cobbles within a sequence of thin (2–5 cm) beds comprising sand and granules. The beds show common inverse grading and represent thin, cohesionless debris flows deposited at the toe of a relatively steep subaqueous slope of a lacustrine, gravelly fan delta. The outsized clasts were emplaced as frontal 'out-runners' of debris falls. Palaeotransport direction is to the right, at 45° out of the outcrop face. Example from the Domba fan delta, Devonian Hornelen Basin, Norway (for details, see Facies C in Nemec *et al.*, 1984).

Whalley, 1984; Nyberg, 1985; Statham & Francis, 1986). Unlike rockfalls, however, the debris falls on delta slopes derive from an accumulated *sediment*, variously rounded and presorted, and are part of the general instability phenomena on these subaqueous depositional slopes. Rockfalls, in contrast, are intimately related to free *rocky* headwalls, where the debris is extremely immature and its derivation is controlled by the headwall lithology and weathering pattern. Of course, rockfalls may also contribute to the sediment supply to subaqueous conical deltas (Fig. 1A), many of which abut against underwater rocky cliffs or develop in close proximity to coastal cliffs that shed debris (e.g. Fig. 3; Prior & Bornhold, 1989, fig. 3).

Very few detailed studies of debris-fall and rockfall processes are available, and our knowledge of these processes is much poorer than for other gravity-driven sediment-transport phenomena. To develop an insight into the depositional aspects of subaqueous debris falls, it is instructive to consider some of the relationships established for subaerial scree slopes or cones.

General characteristics of depositional slopes built by debris falls are summarized in Figure 14A. Figure 14 shows some of the quantitative aspects of debris-fall slopes. Mean slope angle, $\bar{\beta}$, appears to increase with decreasing headscarp height (debris free-fall height). The magnitude of debris input, whether as single particles or particle assemblages, is very important: slopes built by voluminous debris falls are much less steep (Fig. 14B). This relationship reflects the greater mobility of larger particle assemblages, wherein a greater part of the momentum is exchanged internally through collisions, which in turn creates higher granular temperature

Fig. 13. (A) Inferred debris-fall deposit within a stratified sandy turbiditic sequence of a steep subaqueous fan-delta slope. Palaeotransport direction is to the left, at 40° out of the picture. Note that the basal clasts tend to occur at progressively higher levels towards the right, suggesting debris-fall emplacement concurrently with the tractional sand deposition from a sustained turbidity current. (B) Inferred debris-fall deposit, comprising large cobbles overlain by pebble gravel. Palaeotransport direction is to the right, at 20° out of the picture. Both examples are from a Devonian lacustrine fan delta, Hovden Island, Hornelen Basin, Norway. Coin (scale) is 3 cm in diameter.

and provides the avalanching dispersion with more 'drive' (Fig. 15). The angle of the initial depositional surface, γ, also influences the slope (Fig. 14A), with steeper mean angles ($\bar{\beta}$) of debris cones developed on inclined ($\gamma > 0$) depositional surfaces.

Observations from laboratory and field experiments have shown that particles move generally with a series of bouncing and rolling impacts on the slope surface, which may retard the particle, accelerate the particle or stop it instantly (Kirkby & Statham, 1975; Luckman, 1976; Statham, 1976; Carson, 1977). In subaqueous conditions, buoyancy and the viscous resistance of ambient water reduce particle momentum. The distance, d, that a particle travels down a subaerial debris slope, after falling a height H, has been approximated by the following relationship (Kirkby & Statham, 1975; Statham & Francis, 1986):

$$d = \frac{H \cdot \sin^2\beta}{\cos\beta \cdot \tan\phi_d - \sin\beta}, \qquad (9)$$

wherein β is the slope angle and ϕ_d is the dynamic angle of sliding friction for the particle. One of the most important controls on ϕ_d for a falling particle is the ratio of its radius R to the radius R_s of static particles on the slope. This relationship is shown in Figure 16, where ϕ_o corresponds to particles that are much larger than those resting on the slope ($R \gg R_s$), and the coefficient b is a dimensionless empirical constant reflecting the properties of the falling particle (density, shape, etc.). The quantity $b \cdot R_s/R$ (see in Fig. 16) effectively defines the tangent of a dilation angle for the particle–slope contact. Experimental b-values vary considerably (Statham & Francis, 1986, fig. 12.4): for angular, inequant, siliceous rock debris (Zingg sphericity of 0.65), falling on a cobble-bouldery scree characterized by a very steep angle of initial yield (ϕ_i=40°), the reported 'best-fit' regression value is b=0.67. For angular limestone debris of more equant shapes (sphericity of 0.80), falling on a gravel scree slope with ϕ_i=35°, the reported best-fit value is b=0.37.

The linear relationship in Fig. 16 shows that as the falling particles become relatively smaller (i.e. as the R_s/R ratio increases), they are more easily retarded by the roughness of the slope surface (ϕ_d increases). This effect is largely responsible for the characteristic downslope segregation of particle sizes on the slopes built by debris falls and rockfalls (see Fig. 11). The particle-size segregation is a self-propagating mechanism, as particles tend to come to rest more rapidly on material of similar or larger size-grade. As a result, debris-fall cones or aprons effectively 'backlap' the original steep slopes on which they develop (Figs 2A, 3, 11), and tend to display coarsening-upward internal trends (Fig. 11).

Debris flows

In debris flows, individual particles are no longer able to 'stream' downslope freely and segregate according to their own momentum. Instead, the

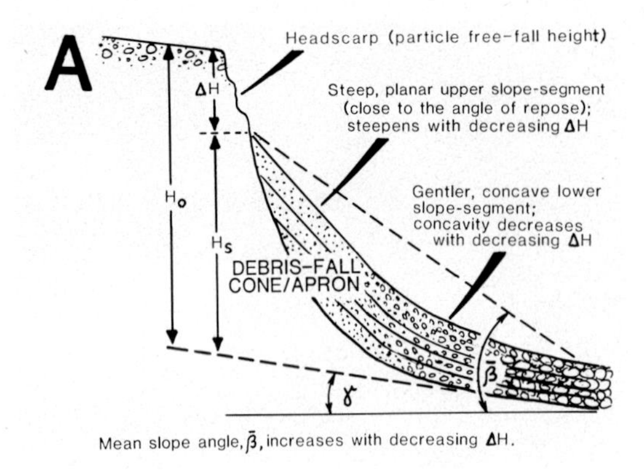

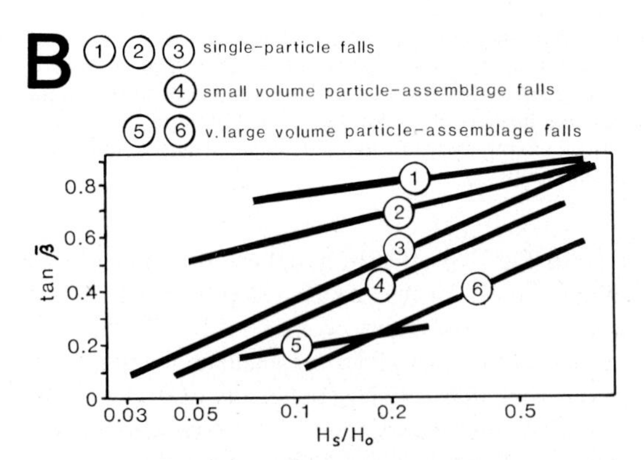

Fig. 14. (A) Geometric relationships and parameters for a depositional slope formed by debris falls. (B) Relationship between mean slope angle and debris free-fall height in simulated, large subaerial scree (rockfall) cones; the initial accumulation slope, γ, was 20° in case 1, 10° in case 2 and 0° in the other cases. (Simplified from Statham & Francis, 1986, fig. 12.5.)

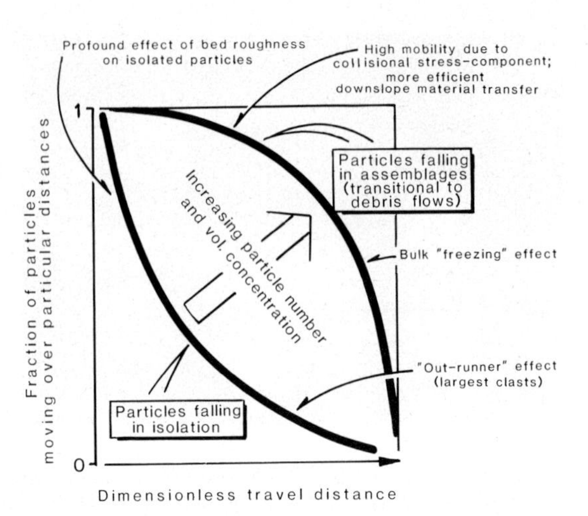

Fig. 15. Cumulative frequency of sediment travel distances on slopes built by debris falls (schematic dimensionless summary based on field and laboratory data for a variety of particle sizes, free-fall heights and scree-slope angles, reviewed by Statham & Francis, 1986).

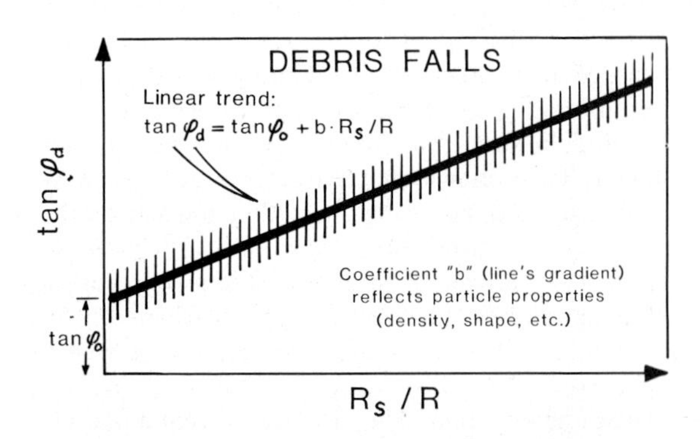

Fig. 16. Generalized relationship between the relative size of a falling particle and the dynamic angle of the particle's sliding friction (ϕ_d) against the slope-surface debris. R is the particle radius and R_s is the radius of static particles on the slope. For comment on tan ϕ_o and b-values, see text. Based on observational data reviewed by Statham & Francis (1986, p. 255).

high volumetric concentration of such dispersion (Fig. 9) causes the particles to share their momentum in generating dispersive pressure and creating the mass-flow mobility. Under such conditions, larger particles will tend to be pushed upwards, towards the free upper surface of the flow (Bagnold, 1954; Walton, 1983). The smaller particles may also percolate downwards, to the base of the flow, by kinematic sieving (Middleton, 1970; Scott & Bridgwater, 1975). In either case, the result will be an upward coarsening (inverse grading) of particle sizes. In addition, coarse debris tends to migrate towards the lateral margins of a debris flow, forming levees and constricting the effective width of the flow (e.g. Pierson, 1980). These and related aspects of particle-size grading in granular-flow avalanches have been reviewed recently by Allen (1984, vol. II, pp. 152–158).

The migration of coarser particles within a debris flow may lead also to a concentration of large clasts in the frontal part of the flow. Frontal concentrations of large clasts have been reported from both subaerial and subaqueous debris flows, with clasts up to large-cobble or boulder-size grade (Pierson, 1980, 1986; Takahashi, 1980; Suwa & Okuda, 1983;

Johnson & Rodine, 1984; see also Allen, 1984, vol. II, pp. 152–158). There are at least two ways in which coarse clasts can be concentrated in the frontal part of a debris flow:

1 Debris flows that have evolved from decelerating debris falls (Fig. 9), or have experienced episodes of strong particle dispersion when accelerating on local steep slopes, may inherently contain coarsest debris in the frontal parts (see previous section). Indeed, debris flows originating on the steep slopes of mountain valleys or ravines and moving as rapid surges have been reported to have coarse, bouldery fronts (Okuda *et al.*, 1980; Takahashi, 1981; Suwa & Okuda, 1983; Johnson & Rodine, 1984; Pierson, 1986; cf. also Colella *et al.*, 1987, fig. 10).

2 Frontal concentration of large clasts may be due to some kind of 'conveyor-belt' mechanism (Allen, 1984, vol. II, pp. 157–158; Pierson, 1986), whereby the upper zone of a debris flow, by the virtue of its relatively high velocity (Fig. 6), moves to the flow's front and feeds it with coarsest debris. The upper zone would carry the coarsest clasts as a result of the kinematic segregation (inverse grading) of particles as the debris acquires granular temperature (see p. 46). Even when riding as a non-shearing 'rigid plug', with zero or non-zero granular temperature, the upper zone would carry the coarsest primary debris, whereas large clasts in the lower, shearing zone would tend to be dropped from the flow, or be pushed upwards, depending on the magnitude of dispersive pressure (Fig. 17; cf. Naylor, 1980). In either case, the coarse clasts will find themselves shearing past a grain layer which, because it comprises smaller particles, appears to them to be comparatively smooth. If the lower layer were to stop, the coarse particles above would tend to travel onwards, because the surface of the lower layer might be insufficiently rough to arrest them (Allen, 1984, vol. II, p. 158).

This mechanism of overpassing may be effective only if the frontal (head) part of a debris flow is not moving faster than the body of the flow (i.e., when the debris flow is not subject to frontal acceleration or rapid spreading). The mechanism does not apply also to more 'watery' flows subject to turbulent churning, wherein large clasts will tend to settle and be outpaced by the finer-grained upper part of the flow, especially when turbulent; however, the flow's head may still be relatively coarse grained in such cases (see Lawson, 1982).

When transferred to the front in the 'conveyor-belt' fashion, the large clasts will tumble down the steep leading edge of the debris flow. Clasts smaller than the frontal thickness of the lower, shearing zone of the flow (Fig. 6) will be overridden, and either left behind as a lag or reincorporated into the flow. Larger clasts will tend to be pushed ahead, bulldozed along by the flow. With this kind of

Fig. 17. Debris-flow beds deposited on the subaqueous slopes of coarse-grained fan deltas. (A) Fine-grained conglomerate bed showing distribution inverse grading in the lower part (inferred zone of intense shear and particle collisions) and large, outsized clasts floating in the upper, ungraded part (inferred 'rigid-plug' zone); example from a Miocene fan-delta slope, Tabernas Basin, southern Spain. (B) Coarse-grained conglomerate bed showing crude, coarse-tail, inverse grading defined by the presence of large clasts in the upper part of the bed (inferred result of mild, laminar shear limited to the lower part of the flow, whereby largest clasts would settle from this sheared part and be dropped from the flow); lens cap (scale) is 5·5 cm; example from a Devonian fan delta, Hornelen Basin, Norway.

frontal entrapment of large clasts, the bouldery front will tend to grow bigger with distance downslope. This tendency, however, will normally be countered by the debris-flow body pushing from behind and shouldering aside the frontal boulders (Sharp, 1942). The latter process, as well as the slow secondary circulation of debris in chute-confined flows (Savage, 1979, fig. 5b), contribute to the formation of lateral levees rich in coarse debris (see also Sharp, 1942; Johnson & Rodine, 1984, figs 8.21, 8.22; Pierson, 1986).

Bouldery fronts in fact appear to act as moving dams, which effectively slow down the debris-flow front and momentarily jam it. This process leads to debris-flow surges, when the debris built up behind the jammed front overtops or pushes through it, and races downslope ahead of the stagnant bouldery front.

The tendency for large clasts to concentrate in the frontal parts of debris flows may thus lead to a preferential accumulation of such coarse debris in the delta toe zone, where the bouldery fronts of debris flows will 'freeze' because of slope flattening. If debris flows prevail on a longer term and their rheological behaviour is roughly the same, the aforementioned process may determine the gross pattern of clast-size segregation in delta-foreset deposits. Otherwise, debris flows deposited on delta slopes may show little or no downslope coarsening, and their inherent longer-term tendency may be, in fact, to carry finer debris over longer distances. This is because many debris flows experience frontal acceleration and spreading. The mechanical behaviour may also vary from one debris flow to another, and more mobile debris flows will spread and tend to be thinner and relatively less competent (cf. Nemec & Steel, 1984; see also Larsen & Steel, 1978; Porębski, 1984; Nemec *et al.*, 1984, figs 26, 27).

Other processes

Turbidity currents on delta slopes tend to deposit their coarsest load near the source and carry fine material over longer distances (e.g. Massari, 1984). However, high-density turbidity currents on steep slopes may trap large clasts in traction in the frontal part (Postma, Nemec & Kleinspehn, 1988), then drop them preferentially in the hydraulic jump conditions of a slope break or delta-toe zone (e.g. Postma *et al.*, *op. cit.*, fig. 4).

Even on delta slopes with strong downslope fining of material, the coarse debris from the upper slope may be transferred to the lower slope and toe zone by means of sliding or slumping. Bowl-shaped slump scars on the upper slope may transfer their coarse sediment downslope as debris flows through chute-and-lobe systems (Prior *et al.*, 1981; Postma & Cruickshank, 1988; Prior & Bornhold, 1988, 1989, this volume). Such processes will cause recognizable textural anomalies within the delta slope (e.g. see Fig. 36, p. 67).

DEBRIS-FLOW THICKNESS AND COMPETENCE

Based on the empirical notion of Bluck (1967), the concept of maximum particle size (MPS) and bed thickness (BTh) correlation has been developed and suggested as a useful semi-quantitative technique for the analysis of debris-flow deposits (Nemec & Steel, 1984). The two parameters have been used as the estimators of flows, thickness and clast-support competence, respectively. It might then be worthwhile to try the method in the analysis of delta foresets, which generally abound in various debris-flow deposits and tend to be well bedded. However, the application of MPS/BTh plots to such depositional settings would require much caution, at least for the following reasons (see also Nemec & Steel, 1984, p. 22):

1 The MPS/BTh concept relies heavily on the assumption of 'unlimited' coarse-particle sizes, such that the actual competence of debris flows can be *satisfied* in terms of the maximum particle sizes carried by the flows. This requirement may not be fulfilled, even in the statistical sense of the concept, in a delta-slope setting. The debris supplied to the delta face is usually presorted, and is likely to have an upper size limit imposed, on a shorter- or longer-term scale, by the transport competence of feeder streams. As a result, the debris flows originating on delta slopes may appear 'too thick' relative to their maximum particle sizes (cases of unsatisfied flow competence). Sandy, gravel-free debris flows (Fig. 7) may serve as an extreme example.

2 Steep delta slopes are inherently unstable, whereby packets of earlier deposited debris-flow units (with particular MPS characteristics) can be remobilized by slumping and resedimented, or 'recycled', as new thicker flows. The actual competence of such flows may then hardly be matched by the available coarsest particles (unsatisfied flow competence).

3 Many subaqueous debris flows originate as 'inertia-flow carpets' driven by turbidity currents, or travel with an entrained turbidity current on top (e.g. Massari, 1984; Postma, Nemec & Kleinspehn, 1988, fig. 1). The turbulent shear stresses transmitted from above may effectively *increase* the debris-flow competence (by increasing the dispersive pressure in such a 'carpet') or *decrease* the competence (by merely destroying the strength of matrix), depending on the magnitude of bulk shear stresses and sediment properties. In the former case, the deposit may appear 'too coarse' (MPS) relative to its thickness (as the flow might retain large clasts that would otherwise be dropped), and in the latter case — 'too thick' relative to its MPS (as the flow might drop the largest clasts that could otherwise be supported in a 'rigid plug'). Indeed, there is some indication that the gravelly deposits of inferred inertia-flow carpets or shearing debris flows driven by turbulent currents show the effect of the flow's 'enhanced' competence (Todd, 1989, fig. 4; see also Nemec & Steel, 1984, fig. 24; Marzo & Anadón, 1988, fig. 14); their MPS/BTh plots may show elevated MPS-values, and if not adequately linked with the flow mechanism, may easily be misinterpreted as an indication of flow cohesiveness (cf. Nemec & Steel, 1984, fig. 19).

4 Reworking by waves or currents may cause bed-top erosion, until armouring by a pavement of large clasts prevents further removal of finer sediment. This process will render debris-flow beds 'too thin' (BTh) relative to their MPS characteristics (see Marzo & Anadón, 1988, p. 333 and fig. 14).

5 On steep slopes, large clasts can be added to freezing debris flows by concurrent debris-fall processes (e.g. Fig. 12; see also Postma, Nemec & Kleinspehn, 1988, p. 57), and such 'outsized' particles will certainly affect the MPS/BTh plots. Outsized clasts may be emplaced in freezing debris flows also by other means (Postma *et al.*, *op. cit.*).

These cautionary remarks are not meant, however, to discourage sedimentologists from applying the method in delta-foreset studies. The intention of the remarks is simply to prevent thoughtless applications. Since the character of the material supplied to a delta front can vary greatly, in both time and space, the rheology of the resulting debris flows cannot be expected to be invariable. The MPS/BTh data must then be collected carefully and grouped according to recognizable bed types (or facies), for which the flow behaviour and clast-support mechanisms ought to be inferred in accordance with the bed texture, grading and clast fabric (see Nemec & Steel, 1984). In other words, one should avoid mixing 'apples' with 'oranges'. When skilfully applied, the method may lead to interesting and important inferences (e.g. Nemec & Steel, 1984, pp. 23–29); even the lack of MPS/BTh correlation for particular debris-flow deposits may be very meaningful (see preceding remarks; for example cf. Surlyk, 1984).

Johnson and Rodine (1984, pp. 335–347) have shown how other important mechanical inferences can be made, in a semi-quantitative way, for active or 'frozen' debris flows (see also Innes, 1983). However, their analytical approach pertains to cohesive debris flows and essentially to subaerial settings; also, it involves a number of physical parameters, whose values are difficult to estimate for a deposit and have to be assumed. Interesting theoretical considerations of debris-flow mechanics are also presented, in terms of relatively simple mathematical equations, by Takahashi (1980, 1981).

TURBIDITY CURRENTS

Turbidity currents are among the important mechanisms of sediment transport and deposition on steep slopes of coarse-grained deltas (e.g. Massari, 1984; Prior & Bornhold, 1988, 1989, this volume). These density currents on delta slopes are known to be generated in at least three different ways:

1 They may derive from the dispersed head-parts of subaqueous debris flows (Fig. 6), or simply evolve from such flows whenever these accelerate sufficiently to become fully turbulent (cf. Hampton, 1972; Ferentinos *et al.*, 1988; Weirich, 1989).

2 They may be generated directly by sediment-laden stream effluent in the form of 'hyperpycnal' density underflow (Bates, 1953), carrying the stream load that bypasses the brink zone of a delta and is pulled downslope by gravity (cf. Wright *et al.*, 1986, 1988; Prior, Bornhold & Johns, 1986; Kostaschuk & McCann, 1987; Prior *et al.*, 1987; Bornhold & Prior, this volume).

3 They may derive from flood-stage stream effluent less directly, by intense sediment fallout from the suspension plume produced by stream outflow jet to blanket large areas of the subaqueous slope with dense, mobile suspension which may evolve into a sheet-like underflow (cf. Hay *et al.*, 1982; Wright *et al.*, 1986; Syvitski & Farrow, 1989; Bornhold & Prior, this volume).

The turbidity currents in case (1) will tend to occur as short-lived surges, whereas in case (2) and (3) they may range from brief surges to more continuous (flowing for hours or days) density 'streams' (Wright *et al.*, 1986, 1988; Hay, 1987a, b; Prior, Bornhold *et al.*, 1987). The behaviour of turbidity currents on delta slopes seems to be rather specific, as these currents descend from the very shallow-water depths of the upper part of a delta slope into progressively deeper water. This aspect and some of its implications are discussed in the next section.

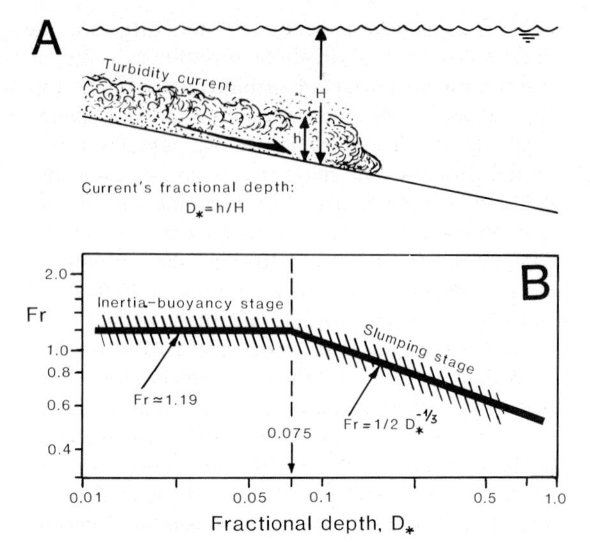

Fig. 18. (A) Definition sketch of fractional depth, and (B) functional relationship between the Froude number of a density current's head region and the fractional depth. Modified from Huppert & Simpson, 1980, fig. 1.

Slumping of turbidity currents

As a turbidity current descends the subaqueous delta slope, the current propagates under large, though rapidly decreasing, fractional depths (see definition in Fig. 18A). The influence of fractional depth on such an intruding density current of 'infinite' length has been considered theoretically by Benjamin (1968), and the influence on the Froude-number condition at the current's head region has been studied experimentally by Simpson and Britter (1979), Huppert and Simpson (1980) and Britter and Simpson (1981). The key relationship is that, until the fractional depth (Fig. 18A) becomes smaller than a critical value (see below), the horizontal (upslope-acting) buoyancy forces established by the density current remain unbalanced by the inertia forces of the current. In other words, the relative motion of the fluid surrounding the intruding density current causes the current to propagate less rapidly than predicted by the well-known classical theories pertaining to deep-water conditions (see also Middleton, 1966).

This 'shallow' stage of density current movement, in which fractional-depth effects are important, has been termed the *slumping stage* by Huppert and Simpson (1980). During this stage, the density current 'slumps' through a series of equal-area rectangles (two-dimensional view parallel to flow) or axisymmetric discs of equal volume (three-dimensional view), and the thickness of the current does not vary significantly along its length. The movement is controlled at the head of the current and can be approximated by a steady-state Froude number (Fr) relationship as a function of the fractional depth (Huppert & Simpson, 1980).

This latter relationship is shown schematically in Figure 18B. The linear-fit model of Huppert and Simpson (1980) is meant to emphasize the fact that the variation of Fr with fractional depth or D_* less than 0.075 is considerably smaller than for D_* greater than 0.075; no real discontinuity in the derivative of Fr against D_* at $D_* = 0.075$ is claimed by Huppert and Simpson. The experimental model predicts that the rate of advance of the current for D_* greater than 0.075 is approximately constant, and is not controlled by $t^{-1/3}$ (where t = propagation time) as in the classical deep-water conditions where fractional depths are assumed to be very small. The 'slumping' stage may be preceded by a brief initial stage during which the head of the density current accelerates to achieve the propagation rate dictated by its density and thickness.

As the fractional depth decreases (Fig. 18A), the horizontal buoyancy forces set up by the intruding density current will eventually be balanced by the inertia forces within the current; the density current will then enter the *inertia-buoyancy stage* (Fig. 18B) of Huppert and Simpson (1980). This relationship holds until the current finally becomes so thin that viscous, rather than inertia forces begin to counterbalance the buoyancy forces; this is the *viscous-buoyancy stage* of density current progradation.

Implications

An important implication is that the intruding turbidity current on a delta slope will tend to behave in the 'slumping' fashion until the depth of the surrounding water is approximately 13 times greater than the thickness of the current in its head region (Fig. 18). This means that even some relatively small (few metres thick) natural density currents may propagate in the slumping stage over considerable downslope distances. Depending upon the actual depth of delta toe and the current thickness, some turbidity currents may behave in the 'slumping' fashion over the entire downslope length of the delta face, or even for much longer distances across the prodelta zone (provided the delta slope flattens out very gradually, such that the current is not drastically altered by a slope break).

The Froude number (Fr) of a turbidity current defines the ratio of its inertia force (momentum transfer to the substratum) to its gravity force (dynamic weight transfer to the substratum). Low values of Froude number ($Fr < 1$) mean that the normal stress exerted by the weight force of the moving current per unit area of the substratum is high relative to the normal momentum lost over this unit area. In other words, the work done by the weight force of such a current will be relatively high.

Low values of Fr further mean that buoyancy forces dominate (see p. 50), whereby the turbidity current issuing from a 'point' source (such as a channel mouth, localized slope failure or debris-flow snout) will travel downslope as a narrow, very slowly expanding underflow. Apart from its straightforward downslope propagation, the 'slumping' current may experience very little bulk expansion by dilution (see p. 50), and Huppert & Simpson, 1980), in which case the 'ploughing' work of the current's weight force will be maximized. The elongate current will tend to scour the slope in two ways: by eroding (incorporating) sediment, and by destabilizing slope sediment through loading and shearing, in such a way that the liquidized material will escape downslope leaving a gulley.

As noted by Huppert & Simpson (*op. cit.*), the 'slumping' stage occurs over a sufficiently short time scale that the effects of mixing between the density current and the ambient water are likely to be of secondary importance. The mixing will be important, however, during the inertia–buoyancy stage of current propagation (Fig. 18B).

Slumping of turbidity currents may then serve as an explanation for the origin of *chutes* on delta slopes (Fig. 19). These large, trough-like scours on modern delta slopes are known to extend downslope from bowl-shaped (though often coalescing) slump scars or bottle-neck failures, and from stream channel mouths (Hoskin & Burrel, 1972; Prior *et al.*, 1981; Syvitski & Farrow, 1983, 1989; Prior, Yang *et al.*, 1986; Wright *et al.*, 1986; Kostaschuk & McCann, 1987; Prior & Bornhold, 1988, 1989, this volume; Weirich, 1989, Bornhold & Prior, this volume). The chutes range in width from 10–30 m to 200 m, and their depths vary from 2–5 m to 20 m. Distinct levees are generally lacking. Some chutes are relatively short (200–500 m), straight and include simple gullies left by liquidized sediment flows (deposited as distinct lobes at the gulley mouths). Many such chutes have lobate turbiditic 'splays' off their mouths. Other chutes are long (more than 2 km) and have distinct bends, apparently due to modification by a large number of turbidity currents which have formed and used such chutes as channels. Large distal turbiditic splays emanate from their mouths. Small chutes occasionally occur within larger chutes. These characteristics correspond well with the theoretical predictions of the model presented above. Similar chutes and associated downslope deposits, in the form of sediment-flow lobes and turbiditic splays, have been recognized or inferred in many ancient deltas and delta-related systems (Fig. 20; see also Collinson, 1970, 1986; Massari, 1984; Postma, 1984a, b; Surlyk, 1984, 1987; Heller & Dickinson, 1985; Colella *et al.*, 1987; Colella, 1988b; Nemec *et al.*, 1988; Postma & Cruickshank, 1988). The chutes on delta slopes become infilled by deposits of smaller, less-robust turbidity currents; by debris flows spawned by slumping of the chute walls and upslope areas; or by suspension settling related to the stream channel mouths, from which the chutes often emanate.

The model in Figure 18B may then be used to make some valuable quantitative estimates or predictions. For example: On the subaqueous slope of the Noeick fan delta (Bornhold & Prior, this volume), chutes formed by turbidity currents appear to die out at water depths between 120 m and 190 m. At these depths, the turbidity currents expand rapidly and attain $Fr \geq 1$ (as indicated by abundant, long-crested antidunes farther downslope), hence probably passing the critical fractional depth of *c.* 0.075 (Fig. 18B). This would imply (Fig. 18A) turbidity current thicknesses between 9 m and 15 m (which match well the range inferred on other grounds by Bornhold & Prior, *op. cit.*).

Two other points pertaining to turbidite deposition on steep delta slopes need a comment. First,

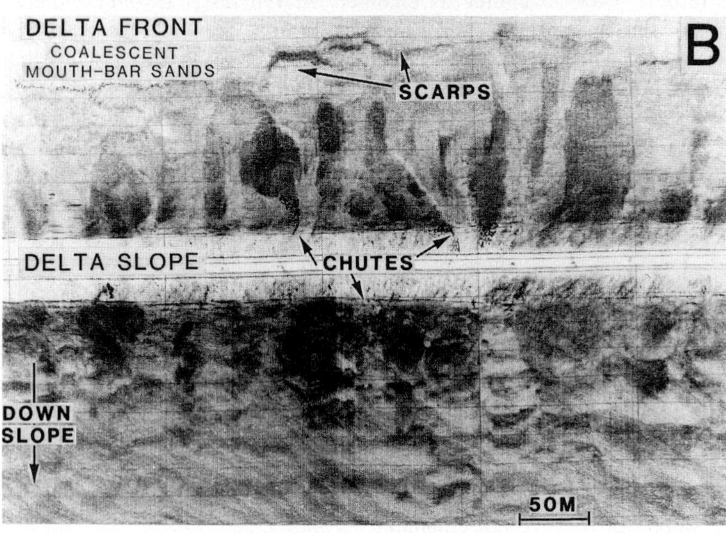

Fig. 19. Side-scan sonograph mosaics showing chutes developed on a steep (12–16°) subaqueous slope of a coarse sandy, valley-sandur delta, Adventfjord, Spitsbergen. Chute varieties include channels formed by density underflow currents (darkest-toned chutes in A), and gullies formed by slope failure (light-toned chutes in B and in the left-hand part of A). The dark-toned sonograph record is inferred to represent freshly deposited coarse sediment, whereas the light-toned record probably represents fine-grained suspension deposits (for further details, see Prior *et al.*, 1981). Photos provided by D.B. Prior.

how a turbidity current, whether elongate or more sheet-like, is able to deposit some or most of its load on the steep slope, while remaining competent enough to carry much load across the flat prodelta zone. Second, if the Froude number of such turbidity currents tends to be low ($Fr < 1$, meaning subcritical flow regime; Fig. 18B), one might expect the turbidites on delta slopes to be devoid of upper (supercritical) flow-regime features. The latter supposition is apparently contradicted by what one can see in delta-slope turbidites (e.g. Massari, 1984), which abound in upper flow-regime stratification types (Figs 21, 22; see also sands in Fig. 13A).

Turbidity currents descending a delta face travel on a progressively gentler slope and their velocities tend to be checked in the head region (Middleton, 1966; Huppert & Simpson, 1980). Moreover, the energy expenditure due to friction on rough slope surfaces and due to the development of internal waves on the upper surface of density current (Benjamin, 1968; Wright *et al.*, 1988) will obviously vary, and will be greater for thin and sheet-like currents. This variation will affect also the flow velocity. It has been argued in the literature (e.g. Dżutyński & Sanders, 1962; Middleton, 1970; Lowe, 1982, 1988) that the collapse of a high-density suspension can be triggered by relatively minor changes

Fig. 20. Examples of chutes in the Pliocene Gilbert-type Abrioja delta, southeastern Spain (for details, see Postma, 1984b). (A) Coarse gravel-filled chutes (arrows), exposed by weathering as elongate ridges 6–8 m wide, on an exhumed lower delta slope; the delta face prograded to the left, and the chutes had been filled with debris-flow deposits. (B) Section through the marginal part of a gravel-filled chute (view roughly parallel to the delta-face strike); the composite debris-flow sequence and irregular relief of the chute wall indicate multiple infill, with episodes of rescouring by debris flows and bypassing turbidity currents. Photos by G. Postma.

in flow velocity. Once some grains begin to settle and the particle concentration decreases, the 'snowballing' effect of declining concentration results in the rapid collapse and deposition of an entire coarse-particle population (cf. Fig. 21B). This deposition of coarse-grained bedload on delta slope will be accompanied by a downslope bypassing of most of the finer-grained suspended load, which will be subject to progressive collapsing and more gradual deposition. As a result, the relatively coarse-grained, incomplete ('top cut-out') turbidites tend to dominate on delta slopes (Figs 21, 22), giving way to the finer-grained, more 'classical' turbidites in the delta toe and prodelta zone. The transition can be quite rapid and rather complex if further affected by the hydraulic jump conditions of a slope break.

The collapsing of suspension and incorporation of slope sediment will both cause bedload concentration in a head-checked turbidity current to be very high. Consequently, the apparent viscosity of the bedload layer will be relatively high (cf. Roscoe, 1953; Rutgers, 1962), and that will increase the flow strength and may suppress the near-bed turbulence (see Middleton, 1967, 1970; Postma *et al.* 1988). Therefore, deposition will be from traction carpets, direct suspension dumping, and the upper-stage, plane-bed tractional regime, even though the actual flow intensity may be subcritical (Allen & Leeder, 1980; Lowe, 1988).

Fig. 21. Details of foreset deposits of a Pleistocene Gilbert-type delta, Skien, southern Norway. (A) Two debris-flow gravel beds separated by a composite unit comprising tractional turbiditic sand and debris-fall gravel (emplaced concurrently); note the normal grading (darkening-upward tone) and ubiquitous plane-parallel stratification in the sand. (B) Coarse debris-flow gravel overlain by a normally-graded turbidite unit comprising finer gravel and plane-stratified sand; thin debris-fall deposit at the top. Bar (scale) is 10 cm.

Fig. 22. Amalgamated T_{ab}, T_b and minor T_{bc} sandy turbidites (5–25 cm thick) deposited by surging and more sustained currents on a steep slope of a braidplain delta (Cretaceous Helvetiafjellet Fm., Spitsbergen); scale is 1 m. The delta lobe (mouth-bar system *c*. 1 km wide) prograded on to steep scarp of a large (*c*. 2 km wide and 50 m deep) slide scar, thus developing a steep depositional slope dominated by sediment gravity flows. Associated non-turbiditic deposits are shown in Figure 7; for details, see Nemec *et al.* (1988).

SUPERCRITICAL FLOWS

Froude number and flow density

The theory of density-current behaviour discussed in the previous section was developed and tested experimentally for Newtonian liquids, and thus should apply reasonably well to low-density turbidity currents (Benjamin, 1968; Simpson & Britter, 1979; Huppert & Simpson, 1980). However, this theory of Froude-number control by fractional depth (Fig. 18) cannot be applied to high-density turbidity currents and non-turbulent granular flows without considerable modification. 'High-density' turbidity currents (Lowe, 1982) are those in which high particle concentration causes non-Newtonian flow behaviour (for $C > 20-25$ vol.% according to Rutgers, 1962) and turbulence begins to be dampened in the lower part of a flow. (This meaning of the term is not strictly the same as the looser usage by some other authors; e.g., Wright *et al.*, 1986, 1988.) Such high-density flows on steep slopes may probably be supercritical essentially irrespective of their fractional depth. High particle concentration increases the flow strength and decreases its boundary resistance, such that the flow's behaviour can effectively be supercritical even though its conventional, 'Newtonian' Froude number:

$$\mathrm{Fr} = \frac{U}{(g \cdot h)^{1/2}} \tag{10}$$

may appear to be somewhat lower than unity (see discussion by Lowe, 1988). The parameters in equation (10) are: U = mean (thickness-averaged) flow velocity; g = acceleration due to gravity; h = flow thickness.

In contrast to the open-channel flow of a Newtonian liquid (whose density would be known and constant), the average sediment concentration or flow density at a particular location within a high-density, flowing sediment–water mixture cannot be assumed to be known, or to be constant. Moreover, the actual density of such a mixture, especially when flowing on a natural slope, will be subject to considerable fluctuations; the density will increase or decrease as sediment is eroded by, or deposited from, the flow. As a result, the velocity profiles of high-density flows are extremely varied (cf. Postma, Nemec & Kleinspehn, 1988), and representative mean velocities are difficult to estimate. All of these reasons render equation (10) a somewhat inadequate estimate as far as the high-density sediment flows are concerned.

For the purpose of laboratory estimations, Brennen *et al.* (1983, p. 32) converted the mass flow rate (discharge), m, and flow thickness, h, to dimensionless Froude numbers defined as:

$$\mathrm{Fr} = \frac{m}{\rho' \cdot C \cdot w \cdot h\,(g \cdot h)^{1/2}} \qquad (11)$$

and

$$\mathrm{Fr}^* = \frac{\mathrm{Fr}}{(\cos\beta)^{1/2}}, \qquad (12)$$

where the other parameters are: ρ' = particles specific gravity (ratio of the specific weight of a given volume of the solid fraction to the weight of an equal volume of the interparticle liquid); C = particle concentration (solid fraction); w = flow width; g = acceleration due to gravity; β = slope angle. Following the semantics for open-channel flows of Newtonian liquids, these authors (*op. cit.*) consider sediment flows with $\mathrm{Fr}^* < 1$ as 'subcritical' and those with $\mathrm{Fr}^* > 1$ as 'supercritical'. The mass-flow rate is readily measurable in laboratory channels, by means of timed collections of material discharging from the channel. However, particle concentrations cannot be measured (suitable techniques are lacking) and have to be assumed.

In summary, the high-density sediment flows, as well as the lower-density turbidity currents if not particularly thick (i.e. those propagating at relatively low fractional depths; Fig. 18), can all be supercritical on delta slopes. This conclusion brings us to the problem of hydraulic jumps.

Hydraulic jumps

On their way downslope, sediment gravity flows on steep delta slopes will commonly encounter obstructions in the form of various 'slope breaks' (sites of abrupt flattening or even upflow inclination of the slope surface). In addition to the general slope break at the toe of a delta, there will be local breaks produced by slides, slumps and 'frozen' debris-flow bodies. When a supercritical flow encounters such an obstruction, a *hydraulic jump* may occur within the flow upcurrent from the obstruction. The phenomenon is illustrated by Fig. 23, although that particular sketch pertains specifically to cohesionless debris-flow avalanches (discussed on p. 56). In the case of a turbulent flow, the effect will be roughly analogous to the formation of an antidune (see Gilbert, 1914; Middleton, 1965; Middleton & Southard, 1984, p. 259). Sediment will be deposited at the obstruction, forming an upflow-dipping slipface; this depositional face, coupled with a similar (though much less steep) 'step' at the flow's upper surface, will tend to accrete and migrate in the upflow direction.

The occurrence of hydraulic jumps may thus explain some of the 'backsets' (sets of upslope-dipping cross-strata) observed in ancient steep-face deltas. Examples are shown in Figs 24 and 25. Davis (1890, p. 197) was probably the first to recognize the occurrence of backsets in steep-face deltas (glaciofluvial in his case). Since then, a variety of backsets have been reported from Gilbert-type deltas (e.g. Postma, 1979, 1984b; Postma & Roep, 1985; Colella *et al.*, 1987; Colella, 1988b; Postma & Cruickshank, 1988). Some of these features may have another origin, as discussed later in this section and in the following section, or may even represent

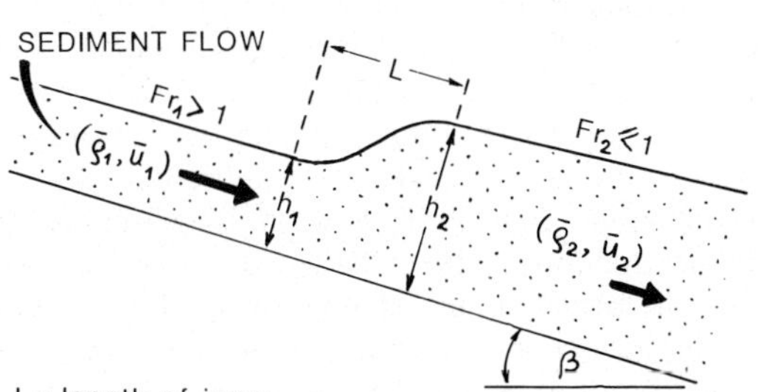

Fig. 23. Definition sketch of hydraulic jump (modified from Savage, 1979). The meaning of parameters is explained in the figure; ρ_i and $\bar{U}_i$ are thickness-averaged densities and velocities, and Fr_i is the flow Froude number, where subscripts i = 1 and i = 2 refer respectively to regions upflow and downflow of the jump.

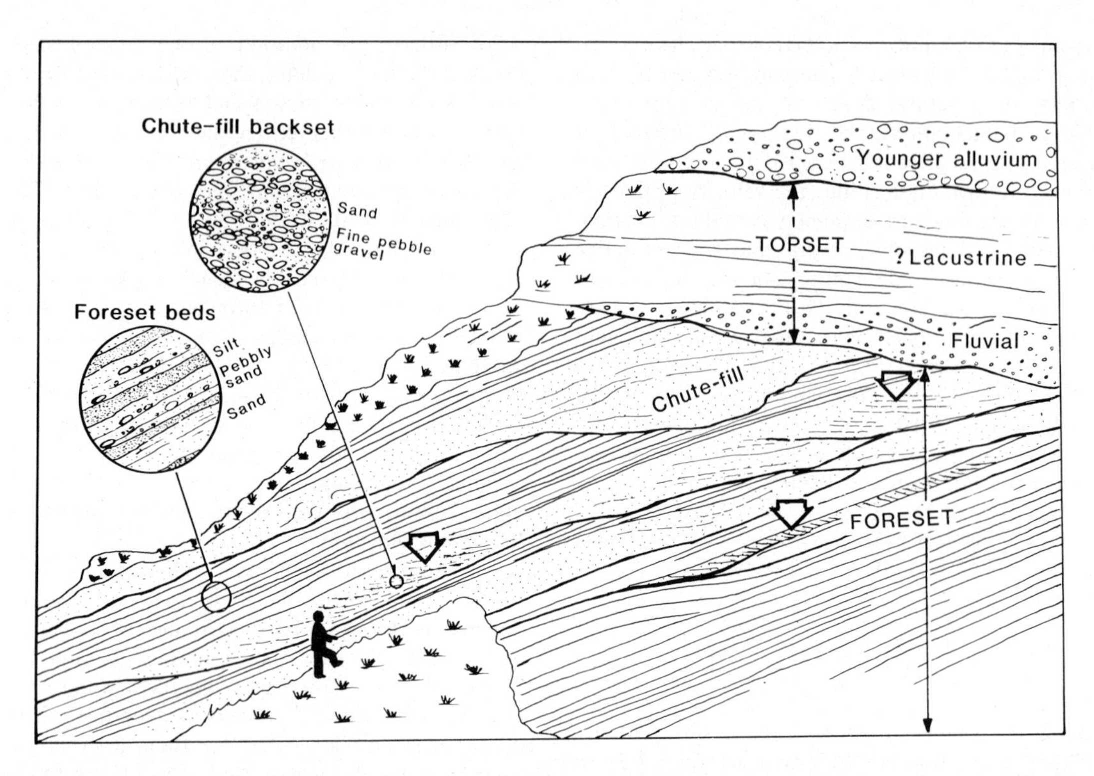

Fig. 24. Examples of 'backsets' (indicated by the large arrows) in the Pleistocene Gilbert-type Ferrocinto delta, Calabria, southern Italy (see also Colella, 1988b, p. 67). The thicker backsets occur as relatively coarse-grained infills of chutes. The delta prograded somewhat obliquely to the left. (Sketch by G. Postma.)

simple antidunes (Fig. 26; cf. Fig. 25B and see also Massari, 1984, pp. 271–272; Bornhold & Prior, this volume).

In turbulent flow conditions, backset cross-strata were shown by Simons & Richardson (1963) and Middleton (1965) to develop on the stoss-sides of antidune bedforms. Also, Jopling & Richardson (1966) produced backset stratification by a 'shooting' (supercritical) flow in laboratory conditions. Johansson (1975) discussed some aspects of backset formation in relation to stream effluent conditions, emphasizing the role of water and sediment discharge, sediment segregation and bed stability. His discussion pertains mainly to the stream-jet separation, bulk eddying and creation of an upslope-directed return flow (see also recent review of 'flow-separation' phenomena by Allen, 1984), hence is relevant only to small laboratory-type deltas, where backflow ripples can develop on the delta face. In a later paper, Johansson (1976) presented the results of a detailed laboratory study of the development and clast-fabric characteristics of coarse-grained backsets at the mouth of a small laboratory tunnel (meant to simulate subglacial outflow conditions).

Granular jumps

Hydraulic jumps can occur also in granular material flows, such as rapid cohesionless debris flows, including their dry subaerial varieties. The phenomenon has been studied theoretically and experimentally by several authors (Morrison & Richmond, 1976; Savage, 1979; Brennen *et al.*, 1983), and termed *granular jump* (Fig. 23).

For a subcritical ($Fr^* < 1$) granular flow, even a minor obstruction or slope break causes the flow to come to a halt. However, an obstruction placed in a supercritical flow ($Fr^* > 1$) can produce one of the following two effects. A small obstruction (relative to the flow thickness) can simply cause a local acceleration of the flow in the vicinity of the obstruction and thus generate a local depression in the free surface of the flow. A larger obstruction to flow discharge will cause the formation of a hydraulic

Fig. 25. Examples of 'backsets'. (A) Thick unit of pebbly sand (arrowed) with upslope-dipping cross-strata in the lower, gentler (11° dip) segment of a foreset of subaqueous outwash delta (the Pleistocene Drente Fm., Netherlands). The delta prograded to the left, at *c*. 40° away from the viewer. The backset unit grades from pebbly sand (MPS = 2·5 cm) near the base to medium/fine sand at the top; an analogous, but thinner backset occurs in the underlying unit. (B) Pebbly sand unit with upslope-dipping cross-strata in the gentle lower segment of a foreset of subaqueous outwash delta; the delta prograded to the left, at *c*. 30° away from the viewer (same formation as above; for details, see Postma *et al.*, 1983). Photos by G. Postma.

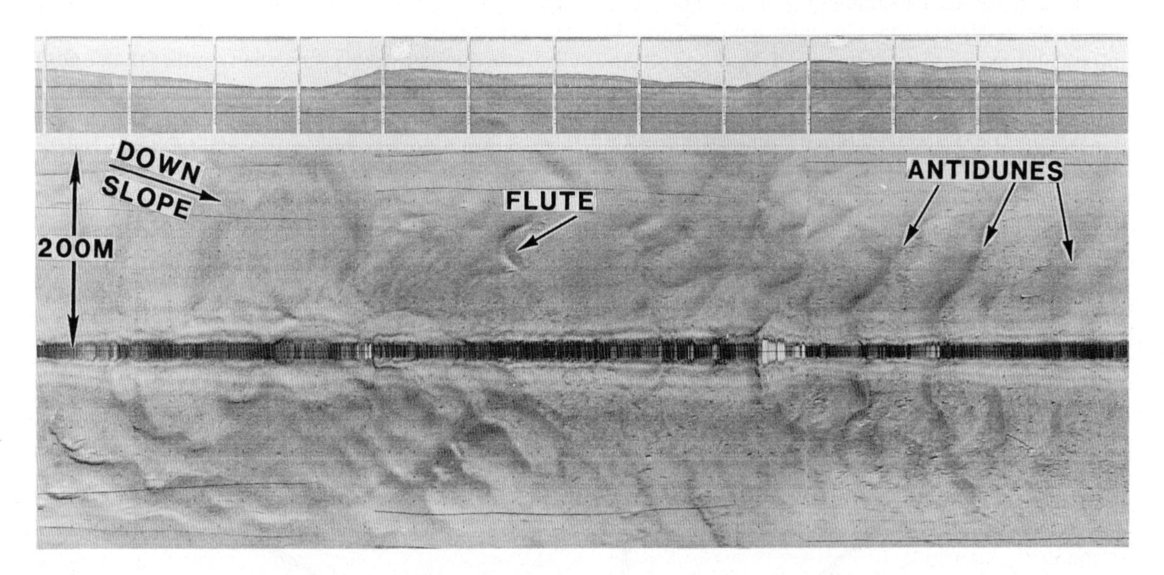

Fig. 26. Side-scan sonograph image of a portion of the lower slope of modern fan delta, showing antidunes and large crescentic flutes formed by stream-generated turbidity currents. Example from the Noeick fan delta, British Columbia, Canada; for details, see Bornhold & Prior (this volume). Photo provided by D.B. Prior.

jump (Fig. 23) which tends to propagate in the upslope direction. Depending on the position and size of the obstruction, the jump will either find an equilibrium position some short distance upslope from the obstruction, or propagate a long distance upslope, provided the flow is of sufficient duration and uniformity (Brennen *et al.*, 1983).

Unlike the highly unsteady mixing in a hydraulic jump in a Newtonian liquid at high Reynolds numbers (see antidunes p. 55), the structure of a granular jump is steady, experimentally repeatable, and readily recognizable. The substantially greater capacity for energy dissipation in shearing granular flows causes the granular jumps to be more like hydraulic jumps of Newtonian liquids at low Reynolds numbers (Brennen *et al.*, 1983). In granular jumps (Fig. 23), the length of a jump is small, making the 'step' in flow thickness well pronounced, with h_2 possibly several times greater than h_1. The steep, upslope-dipping face at the free surface of the flow (Fig. 23) may be coupled with, or accompanied by, a similar steep face formed by the sediment that is deposited because of the jump.

Three types of granular-jump behaviour can be distinguished (Brennen *et al.*, 1983), which depend largely on the slope inclination (Fig. 27). On slopes less than about the angle of repose of the flowing material, the subcritical flow downslope of the jump tends to shear over its entire thickness (type A in Fig. 27); deposition occurs near the obstacle only, where some sediment will normally be trapped and this localized accretion will eventually cancel the obstacle relief. On slopes slightly steeper than the angle of repose, the flow downslope of the jump shows a more complex behaviour (type B in Fig. 27). A wedge of sediment will be deposited, with its steep depositional face migrating upslope to an equilibrium location determined by the obstruction (Fig. 27; for the detailed geometric relationship, see Brennen *et al.*, 1983). The upslope distance over which the depositional face migrates will be fixed by the character of the obstruction, although the jump itself may migrate farther upslope to reach its own equilibrium position. Between the two locations, the flow is fully shearing as in Type A (Fig. 27). The upper surface of the depositional wedge will be flat and inclined at the angle of repose, with the Froude number of the overpassing flow close to unity. On slopes a few degrees steeper than the angle of repose, the granular jump begins to behave in yet another way (type C in Fig. 27). The jump migrates upslope and is followed closely by the steep accreting face of the depositional wedge. As long as the flow continues and the slope is uniform, the distance of upslope migration of these coupled features seems to be unlimited. Savage (1979) noted that the steepness of a granular jump (h_2/h_1 relative to L, in Fig. 23) increases with the upslope Froude number (Fr_1) and with the slope gradient. His theoretical and experimental data (Savage, 1979, fig. 32) further show that a granular jump can occur even if the flow's conventional Froude number (Fr) is somewhat smaller than unity; the same is indicated by equation (12) (p. 55).

Granular jumps can thus be expected to occur on steep delta slopes, and can probably create 'back-

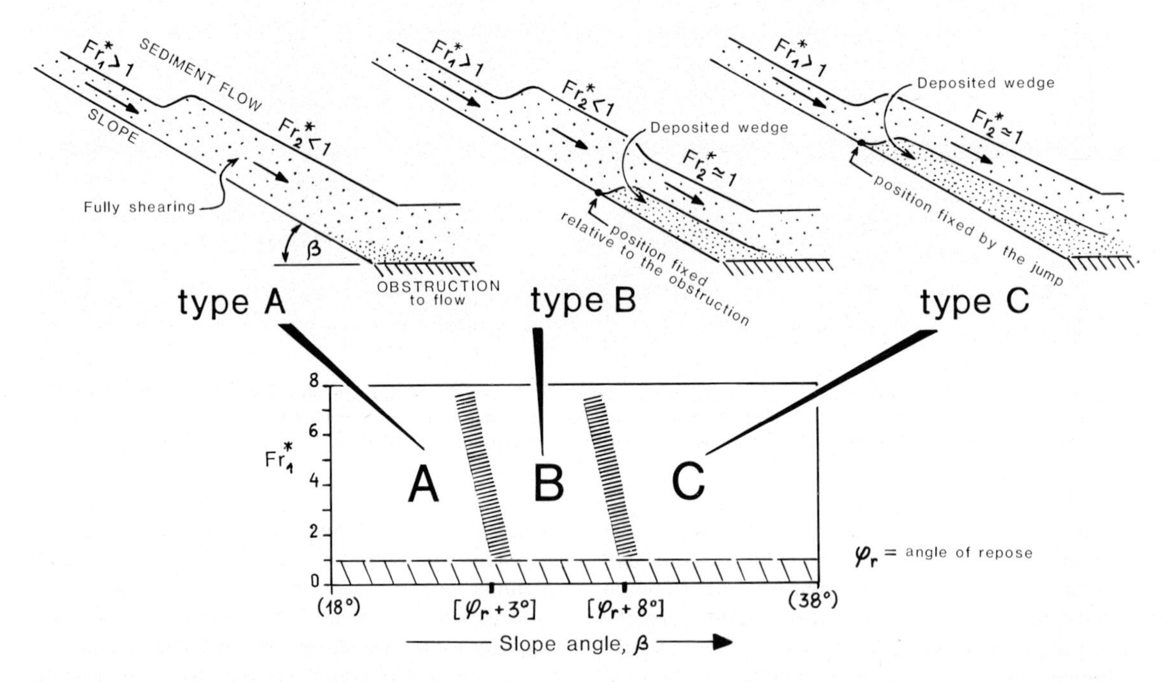

Fig. 27. Main types of granular-jump behaviour as a function of slope angle; the Froude number (Fr*) here is as defined by equation (12) in the text. (Schematic summary of the results of laboratory experiments with dry granular flows performed by Brennen *et al.*, 1983.)

sets' as the depositional result (see above). Such backsets would range from localized accretionary 'smooth-out' features (cf. Colella *et al.*, 1987, fig. 12) to relatively long tabular sets of upslope-dipping cross-strata (cf. Fig. 24), as in the case of simple hydraulic jumps discussed on p. 55. The slope on which mass flow of a particular sediment occurs will usually be higher than the sediment's angle of repose. For example, a stable steep slope made of gravelly material, even if sand-bearing or smoothed by entrapped sand, may be much steeper than the angle of repose of a sandflow that descends this slope. Such textural variations will certainly be important factors in promoting jumps (see also Fig. 5 and earlier discussion on p. 42).

The behaviour of non-Newtonian granular flows appears to bear some further resemblance to the antidune regime in Newtonian liquids. In his experiments with non-turbulent rapid chute flows, Savage (1979) noticed that, even in the absence of any obstruction, occasionally 'surge waves' were spontaneously generated at the downstream end of the laboratory chute; similar spontaneous waves were generated by granular jumps themselves. These waves moved upslope for considerable distances (15 times greater than the chute width and 60 times greater than the flow thickness), and were swept back downslope again. The exact mechanism responsible for the generation of these spontaneous waves and their depositional effects remain unknown (for theoretical discussion of surface waves in turbulent density currents, see Benjamin, 1968).

STRAIN SIGNATURES OF SLIDING

Sediment slides

Sediment sliding (Fig. 6) is one of the gravity-driven processes commonly encountered on steep delta slopes (Prior & Bornhold, 1988, figs 18, 19; Syvitski *et al.*, 1988). The downslope movement and deformation in sediment slides are, in general, neither uniform (in space) nor steady (in time). The unsteady, non-uniform behaviour of sediment slides is crucial for an understanding of some fundamental aspects of slide dynamics that are not apparent in simple static or steady-state considerations based on laboratory shear-box tests.

Some important general characteristics of translational slide sheets derived from steep slopes can be summarized as follows (see Iverson, 1986; Jones & Preston, 1987; Mandl, 1988; Martinsen, 1989; Martinsen & Bakken, 1990).

1 The topographic surface of a sediment slide tends to have a broadly uniform slope. Slopes of local undulations or ridges (Figs 28, 29) may depart by as much as 20° from the average, but only in very localized zones.

2 The thickness of a slide sheet tends to be roughly uniform, and depends on the slope gradient and sediment properties. There is a tendency for a slide sheet to be thinner near its downslope toe, although on steep underwater slopes this primary trend is commonly modified or even reversed by secondary, contemporaneous deformation of the slide sheet (see discussion further below).

3 Most shear deformation occurs in the basal zone of a slide sheet. The behaviour of the basal shear zone may range from quasi-static to rapid shear-regime (see earlier text, p. 37); in this latter case, the shear zone can expand vertically to comprise as much as 20–30% of the slide-sheet thickness. Discrete shear planes are scarcely detectible in the basal shear zone. Lateral shear zones, along the margins of the slide body (Fig. 28), are usually thinner, but may be as thick (wide) as the basal zone.

4 Surface transverse ridges (Figs 28, 29) form as an expression of longitudinal compressional deformation, whereas surface lows or scarps (e.g. Martinsen & Bakken, 1990) are the normal expression of extensional deformation. Both features are short-lived, due to surficial modification by subsequent sediment transport processes (cf. Fig. 29). The expression of compressional and extensional zones may vary from one slide sheet to another, also because some sedimentary materials fail more readily in compression, others in extension.

The surface ridges formed in the frontal part of a slide indicate that the leading edge of the slide or flow slide initially ground to a halt and the rest of the material piled-up behind it. Deformation experienced by the slide sheet will then be somewhat analogous to that observed in the wedges of deforming sediment in front of moving bulldozers (cf. Hubbert & Rubey, 1959; Shreve, 1968). The surface ridges or undulations that form are merely a topographic expression of the internal deformation of the slide sheet. Depending on the material properties and its stress state, this internal deformation may range from the movement of irregular rigid zones partitioned by narrow shear zones, to a quite pervasive shear strain (bulk plastic yield) if the state of applied stress has caused the entire thickness of

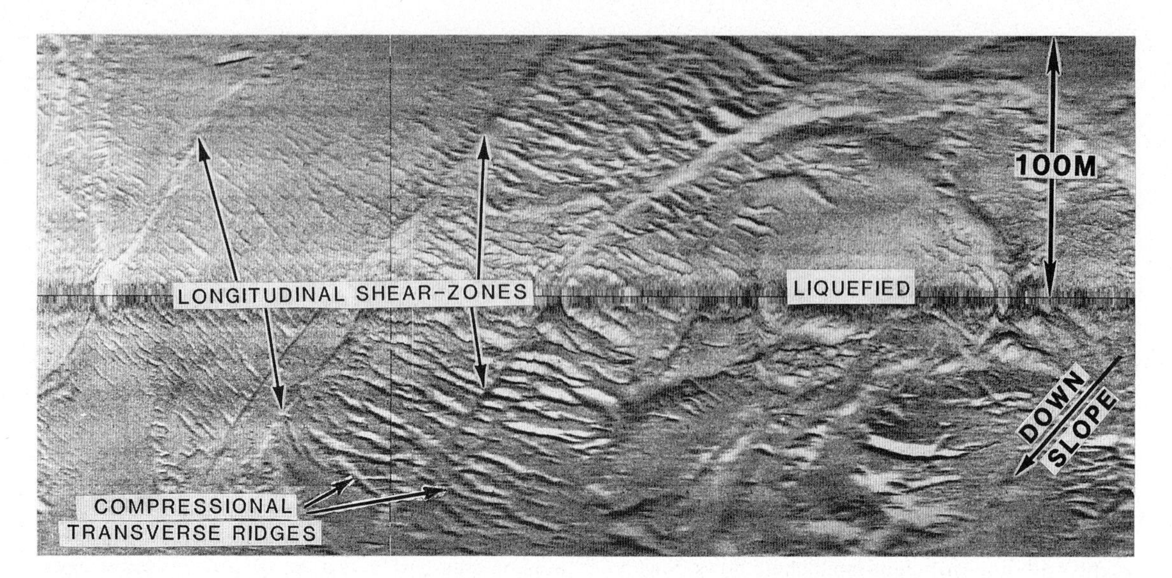

Fig. 28. Side-scan sonograph mosaic of a modern submarine 'flow-slide' deposit, showing compressional transverse ridges and longitudinal shear zones. The deposit represents a delta slope-derived, sandy debris flow whose frontal part, after initial freezing, continued to move and deform in a 'rigid', sliding fashion; non-uniform, differential movement split the sheet into narrower longitudinal belts separated by lateral shear zones. Example from Kitimat Arm fjord, British Columbia, Canada (for details, see Prior *et al.*, 1982). Photo provided by D.B. Prior.

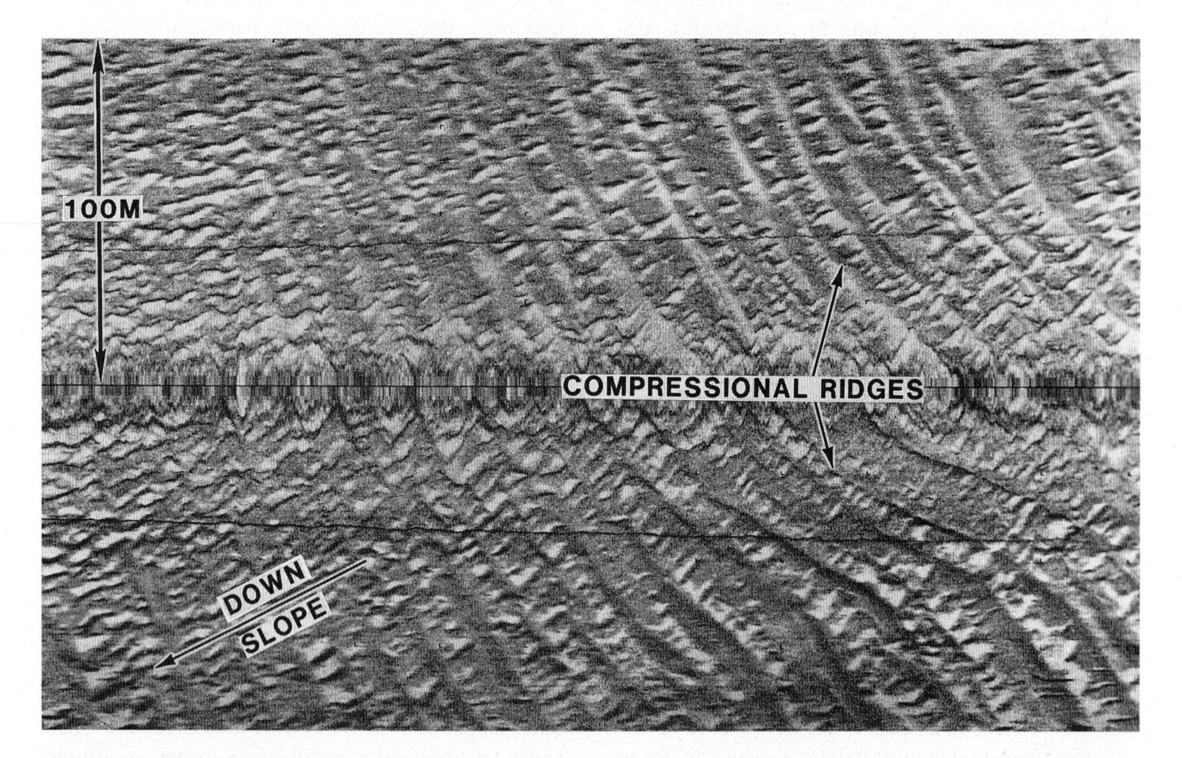

Fig. 29. Side-scan sonograph mosaic showing compressional transverse ridges on the surface of a modern submarine 'flow slide' (debris flow which completed its downslope movement in a sliding fashion). Note the curvilinear, broadly arcuate geometry of the ridges, and the superimposed megaripples formed by longshore currents in the inter-ridge swales; note also how the longshore sand transport tends to obliterate the ridge-and-swale topography. Example from Kitimat Arm fjord, British Columbia, Canada (for details, see Prior *et al.*, 1982). Photo provided by D.B. Prior.

the material to be on the verge of Navier–Coulomb shear failure. Similar deformation can occur in debris flows, whose frontal parts often turn into flow slides when 'freezing' and losing their Navier–Coulomb state (Figs 28, 29). Arcuate transverse ridges and highs, with upslope-concave plan-view shape, have been reported from both subaerial (Voight, 1978; Keefer & Johnson, 1983) and subaqueous debris flows (Prior *et al.*, 1984; Kastens & Shor, 1985) that have been frontally retarded or halted by morphological obstacles (Fig. 30). The following discussion is focused on this particular aspect of translational slide or flow-slide deformation, and on its structural signature in the resulting deposits.

Shear-strain pattern

A slice of sediment detached from a steep delta slope slides down under its own weight. The sliding would turn into a 'catastrophic', rapidly accelerating mass movement, with complete remoulding and turbulence, if the slide motion were not checked (cf. Fig. 8). Arriving at a lower or rougher slope, the frontal part of the sliding sheet often abruptly decelerates, thus acting as an effective brake. This frontal retardation causes stress build-up, with compressional shortening and thickening of the frontal part as it is pushed downslope. In other words, when the slide or freezing debris-flow body encounters resistance to slip on its sole, the downslope component of stress will induce strain within the body itself (Raleigh & Griggs, 1963; Shreve, 1968; Price, 1973, 1977).

In a granular material, this process will initially cause some frictional dilation, either limited to certain zones or more pervasive (depending on shear-strain distribution), so that the pore spaces increase and accommodate more interparticle liquid. As the particles are rearranged because of the shear strain, the pore spaces tend to close and assume an even tighter packing. The effect will be an increase in the pore-liquid pressure. If the strain occurs more rapidly than the excess pore pressure can dissipate by upward escape of the pore liquid from the slide sheet, the pore pressure within a large part of the deforming sheet may increase. The process will begin in the basal shear zone, which collapses when the slide's 'braking' begins, and then will rapidly propagate to higher levels as the internal strain develops. The short-term rise of pore pressure may be dissipated in various ways (see inset diagrams

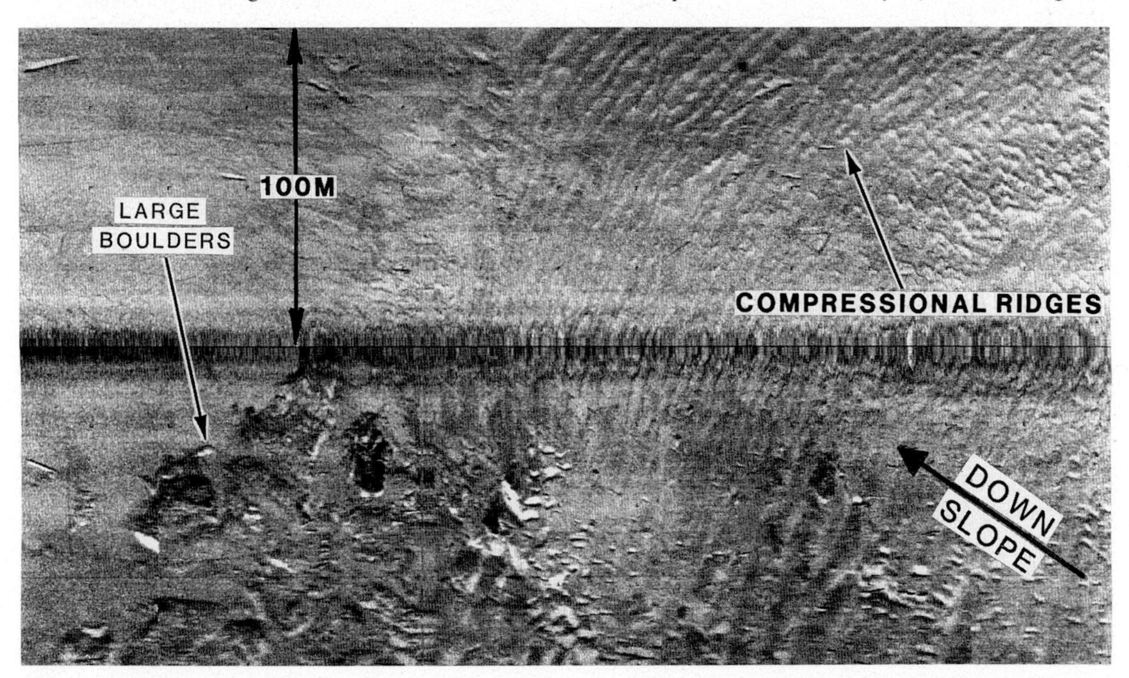

Fig. 30. Side-scan sonograph mosaic of the frontal part of a modern, submarine debris-flow deposit, showing arcuate compressional ridges. The delta-slope-derived, cohesive debris flow was frontally retarded by the low-gradient morphology of delta toe. Note the large, boulder-sized blocks emplaced by debris-fall avalanches to the delta toe zone. Example from Kitimat Arm fjord, British Columbia, Canada (for details, see Prior *et al.*, 1984). Photo provided by D.B. Prior.

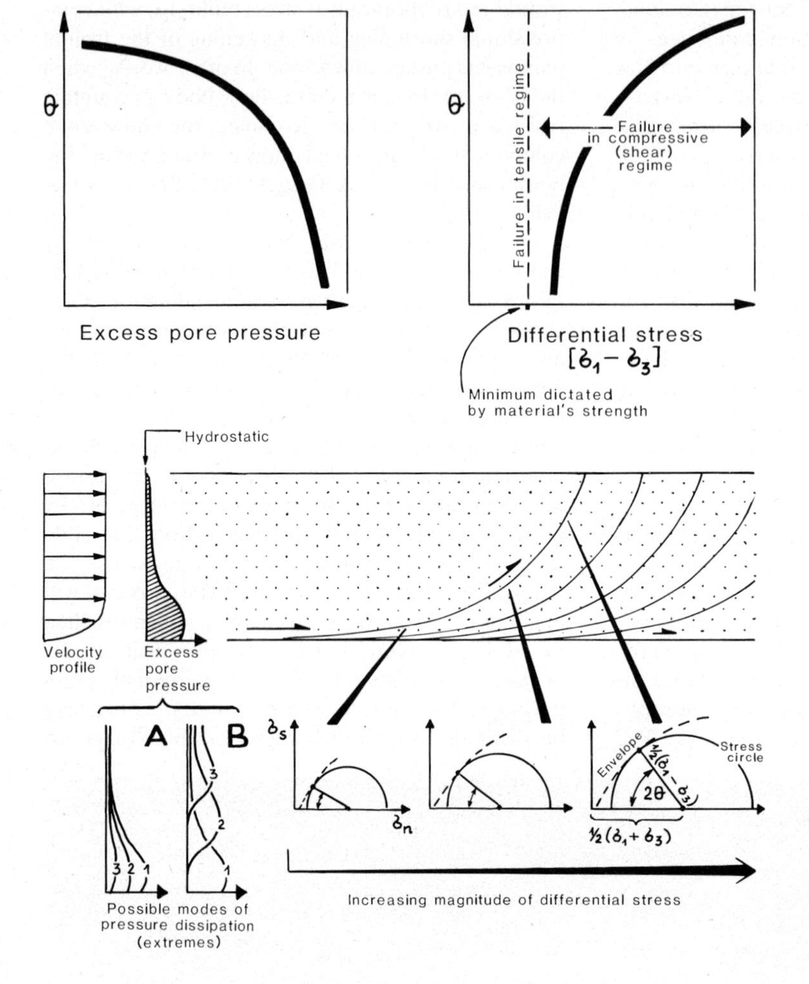

Fig. 31. Relationships between the angle θ and pore-liquid pressure, and between θ and differential stress (upper diagrams); θ is the angle that the internal shear surface makes with the axis of maximum principal stress. Lower diagram illustrates sediment failure at upward increasing differential stress and/or decreasing pore-liquid pressure, with approximate failure conditions and angles of shear; σ_1 is the maximum compressional stress and σ_3 is the normal stress. Excess pore pressure may or may not accompany this deformation. Graphical summary based on Price (1977, figs 11, 12).

A & B in Fig. 31, bottom left), again depending on the strain distribution and material properties (porosity, effective permeability, their vertical distribution, etc.).

The diagram in Fig. 31 (centre) shows a sediment slide whose forward motion has been retarded, so that the consequent downslope component of stress initially gives rise to an increase in the differential stress in the slide's interior. The differential stress along the base concurrently will decrease, because the downslope push of the slide is now taken-up in the slide's interior. The elastic deformation of the sediment (pore-space compression) that results from this build-up of downslope stress, causes a concomitant increase in pore-liquid pressure. As a consequence the pore-pressure ratio (Hubbert & Rubey, 1959), r, within the slide increases, according to the formula (modified by Davis *et al.*, 1983):

$$r = (p - \gamma_w \cdot D)/(\sigma_n - \gamma_w \cdot D) \quad (13)$$

The parameters in this equation are: p = pore-liquid pressure; σ_n = total vertical stress (including the overlying column of water) at the point of interest; γ_w = specific weight of water; D = water depth.

As shown in Fig. 31 (see first the upper two diagrams), a relatively low differential stress and elevated pore pressure in the lower part of the slide will give rise to a shear surface (practically a shear zone) that makes a small angle, θ, with the axis of maximum principal stress, and hence roughly with the slide's sole (see p. 64). The angle between the shear surface and the sole will be later referred to as the 'basal step-up angle' (α_F) of the forward-verging shear surface. Roscoe (1970) estimated the thickness of such a localized shear zone, or 'surface', in granular material to be in the order of 10 grain diameters; this figure has been confirmed both theoretically (Bridgwater, 1980) and experimentally (Scrapelli & Wood, 1982). For pure convenience, however, the geometric term 'surface' is retained in the following text.

As may be seen from Fig. 31, the development of the initial, forward-verging shear surface is the mechanical response of a slide to the frontal retardation, whereby the slide, to continue its movement, is forced to overcome its frictional strength by reducing its own effective thickness. The developing shear surface thus becomes a temporal 'new sole' of the slide. The pushing mass, and hence the downslope component of force, remains unchanged. When the effective thickness of the slide is reduced, the differential stress acting on the 'thrust-wedge' part of the slide must increase (Fig. 31, bottom right). Moreover, the pore-liquid pressure in this (still unsheared) part is likely to be lower than in the freshly sheared portion of the slide. The higher differential stress and lower pore-liquid pressure will then result in a forward propagation of the shear surface, but at a higher angle α_F (Fig. 31, upper diagrams).

By extending this argument (after Price, 1977), a higher differential stress and a lower pore pressure can be expected in the upper levels of the slide sheet, so that the shear surface would propagate forwards at a progressively higher angle (Fig. 31). The result would then be the development of a forward-verging, concave-upward listric ('sledge-runner') shear surface.

As mentioned already, the forward-verging shear surface acts as a new, local sole within the slide sheet. The conditions outlined earlier may thus be repeated and propagated upslope in a retrogressive fashion, with the development of successive shear surfaces in the upslope direction (Fig. 31). The successive surfaces (shear zones) will tend to be progressively steeper.

Alternatively, one may argue that once the excess pore pressure in the listric shear-zone has been dissipated, the pore pressure ratio (r) in this zone is likely to become less than unity (cf. equation 13. Frictional shear stress will then build up rapidly within the listric shear zone and subsequent ones may develop, one after the other, in an analogous

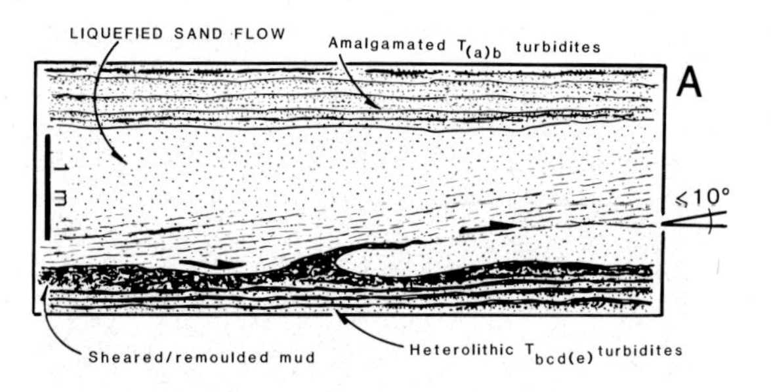

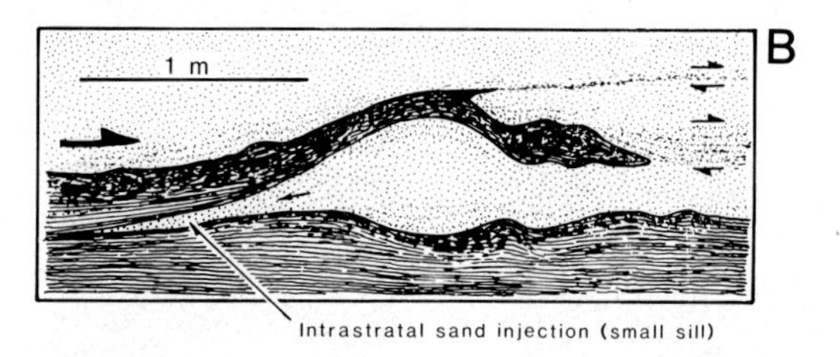

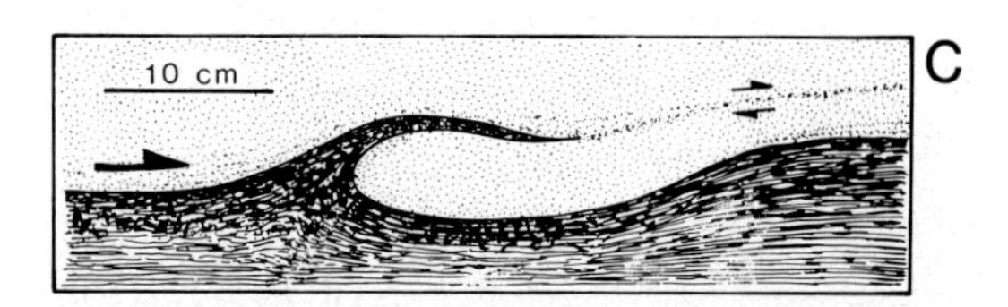

Fig. 32. Field example of upslope-dipping shear bands in the lower part of a cohesionless liquefied sandflow which completed its movement in a sliding fashion (A). The base of the sandstone unit shows overturned muddy load-flames, as illustrated also by sketches (B, C) from adjacent stations. The deposition took place in the lower part of a steep, sandy delta slope (for details and other examples, see Fig. 7C and Nemec *et al.*, 1988).

'retrogressive' fashion. The initial listric shear zone will develop in the front of the slide sheet, where the sediment mass first meets resistance to movement, and the subsequent listric zones will form progressively backwards.

In either case, the entire process may be very rapid in a steep delta-slope setting, also because the sliding, deforming sediment will be freshly deposited and undercompacted. The result will be yet another type of 'backset', produced by compressive shear strain in the present case. The upslope-dipping shear zones (or 'bands') may be isolated or closely spaced, and their step-up angle may vary greatly, as discussed previously. This diversity is illustrated by examples in Figs 7C, 32, 33D; see also Postma (1984b, fig. 20). In other cases, the shear bands may merge and be indistinguishable, or the shear deformation may appear to be virtually pervasive if the

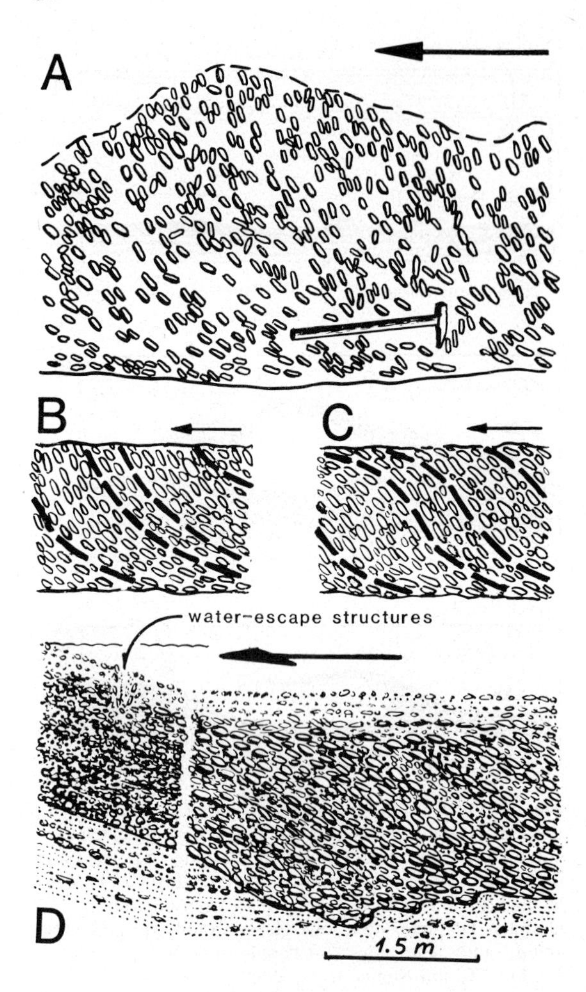

Fig. 33. Field examples of inferred syndepositional band-shear features: (A) steepening-upward clast fabric in a submarine mass-flow conglomerate deposited on a steep fan-delta slope (redrawn from Massari, 1984); (B) and (C) are schematic sketches of clast fabrics in mass-flow conglomerates described by Massari (*op. cit.*); (D) 'backset bedding' in a fan-delta slope sequence (redrawn from Postma, 1979; for other examples see Postma, 1984b). Arrows indicate downslope direction.

retarded, 'bulldozed' sediment mass has been in the state of bulk plastic flowage or subject to concurrent shearing by an overpassing flow. In a deforming gravel, the shear-bands will be quite thick (see estimates p. 62) and hence may tend to coalesce (Fig. 33D). Figure 33 (A–C) shows examples of steepclast fabric, dipping in an upslope direction, defined by preferred dimensional orientation of inequant particles. These beds (Massari, 1984) are ungraded and occur in the 'proximal' parts of a fan-delta palaeoslope, implying deposition from debris flows of relatively low mobility. The *a*(p) fabric type is a result of laminar shear. The shear apparently continued during and immediately after the flow's stop-page, as there is an upward steepening of the fabric above the base, clearly due to rotation of the longest axes of particles towards the vertical. Fabric flattening near the top (Fig. 33C) may be due to shear strain exerted by a concurrent mass flow moving on top of the freezing and deforming gravel layer.

Price (1973, 1977) has also pointed out that if the first upward-verging shear zone occurs as the result of a relatively minor obstacle to the forward motion of the slide sheet, the increase in vertical loading of this new sole (due to thrusting) and the possible further increase in pore pressure within this sole zone and beneath it, may lead to *overriding* (as illustrated by the syndepositional thrusts in Figs 7A, C; see also Nemec *et al.*, 1988; Postma, 1984b; Postma & Roep, 1985). Similar aspects of plastic deformation and overthrusting have been discussed extensively by other authors (e.g. Hubbert & Rubey, 1959; Raleigh & Griggs, 1963; Hsü, 1969; Mandl, 1988).

Rheological inferences

It will be shown how the state of stress and the mechanical properties of the deforming sediment slide or flow slide can be inferred from the observed internal strain pattern. The considerations refer to the deformation associated with the 'bulldozering' process discussed in the previous section.

The Navier–Coulomb failure criterion in such case (Fig. 34B) is most easily satisfied along two surfaces which contain the intermediate principal stress axis and are inclined at an angle $\theta = 45° - \phi/2$ (see e.g. Mandl, 1988, p. 17) to the axis of the maximum compressional stress, σ_1, where ϕ is the angle of effective internal friction; the coefficient of effective friction is $f = \tan\phi$. The σ_1 axis must dip forwards (Fig. 34B) if there is a resisting shear traction along the basal slip zone (Fig. 34A). In the present case, however, it can be argued that the inclination, ω, of the σ_1 axis (Fig. 34B) will be negligibly small (probably in the order of 1–3°). The depositional slope (β) is assumed to be relatively steep, the slide-sheet thickness is likely to be small relative to its upslope extent, and the pore-pressure ratio (see equation 13, p. 62) may approach $r = 1$. Therefore, the maximum compressional stress (σ_1)

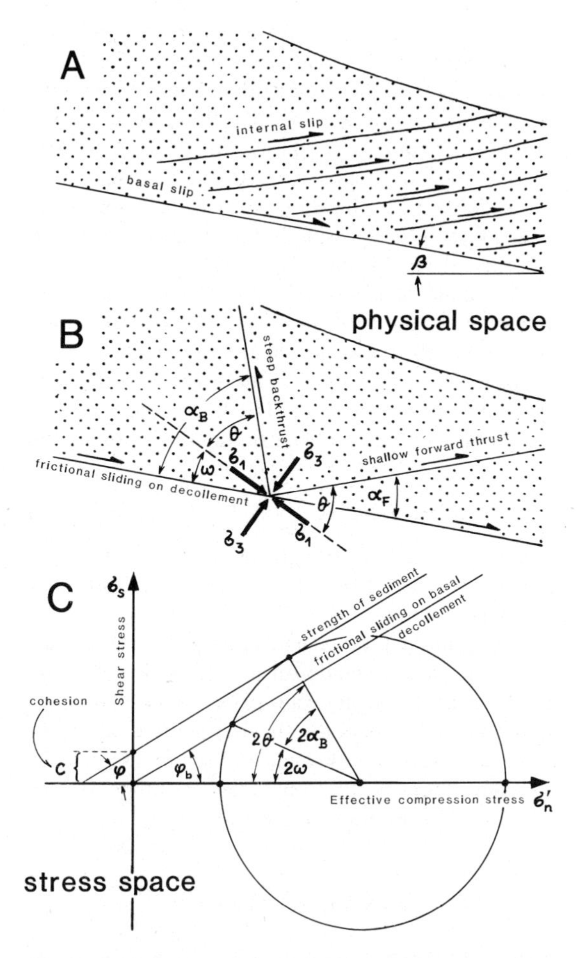

Fig. 34. Stress state and development of forward-verging slip zones in a sediment slide, shown in terms of two-dimensional physical space (A and B) and Mohr stress circle (C). Cohesion may or may not be involved. The angle ω (in B, C) is grossly exaggerated for graphical purposes, as its value in the present case is expected to be negligibly small (see text).

can be expected to act nearly parallel to the base of the slide sheet (see Raleigh & Griggs, 1963; Price, 1977).

The forward-verging shear surfaces and the backward-verging ones (Fig. 34B) should have dips, relative to the basal slip zone, of

$$\alpha_F = 45° - \phi/2 - \omega \qquad (14)$$

and

$$\alpha_B = 45° - \phi/2 + \omega, \qquad (15)$$

respectively. The dip angle ω of the σ_1 stress axis is then:

$$\omega = (\alpha_B - \alpha_F)/2 \qquad (16)$$

These simple relationships (see e.g. Dahlen, 1984; Dahlen & Suppe, 1984; Davis & von Huene, 1987; Mandl, 1988) can be used in interpreting the observed step-up angles, α_F and α_B, of coupled, conjugate internal slip surfaces developed in a deforming sediment unit (Fig. 34B). Although such coupled surfaces are known to accompany some slump folds (Fig. 35) and may form in sedimentary slides (e.g. Scrapelli & Wood, 1982; Martinsen & Bakken, 1990, figs 8, 10a), the deformation of a relatively thin body of soft frictional sediment on a steep slope creates strong mechanical preference for the forward-verging slip surfaces (Fig. 34A, B). Knowing the angle α_F (measured in outcrop), the estimation of effective frictional properties can be based on equation (14), which gives after rearrangement:

$$\phi = 90° - 2\alpha_F - 2\omega, \qquad (17)$$

wherein ω is thought to be negligibly small (see argument on p. 64).

Relationships among the state of stress, the coefficient of internal friction and the orientation of internal slip (or thrust) surfaces have been discussed

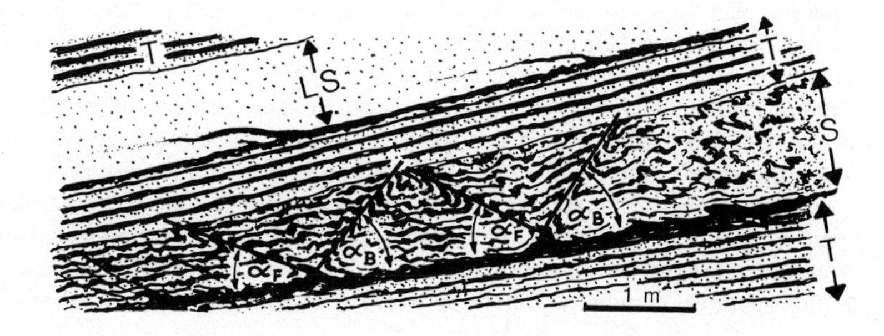

Fig. 35. Example of slump folds with coupled, conjugate shear zones (cf. Fig. 34B). The deposits are: T = thinly bedded turbidites; S = slump in a heterolithic package of thin muddy turbidites; LS = liquefied sandflow (cohesionless debris-flow) unit. From the Cretaceous Helvetiafjellet delta front, eastern Spitsbergen (for details, see explanation to Fig. 22 and Nemec *et al.*, 1988).

by numerous authors (e.g. Hubbert & Rubey, 1959; Hsü, 1969; Dahlen, 1984; Dahlen & Suppe, 1984; Mandl, 1988, pp. 157–187). By using a Mohr circle construction (Fig. 34C), it can be shown (Dahlen, 1984) that the σ_1-axis dip angle (ω) relative to the basal slip zone is related to the sediment cohesion and to the friction coefficients of the interior and basal zone (f and f_b, respectively) of the sediment slide, by the following equation

$$[f(1-r) + c/\gamma \cdot h] \sin 2\omega = f_b\,(1-r_b)\,[(1+f^2)^{1/2} - f \cdot \cos 2\omega], \quad (18)$$

wherein: f, f_b = coefficients of effective friction ($\tan\phi$) of the slide's interior and basal zone, respectively; c = sediment cohesion; h = local thickness of the sediment slide; γ = mean specific weight of the sediment; r, r_b = pore-pressure ratios (see equation 13) within the deforming slide and along its basal zone, respectively.

For negligible cohesion, such that $c \ll \gamma \cdot h \cdot f\,(1-r)$, equation (18) reduces to:

$$\omega = 1/2\,[\arcsin(\sin\phi'_b/\sin\phi)] - \phi'_b/2, \quad (19)$$

where:

$$\phi'_b = \arctan f'_b = \arctan[f_b\,(1-r_b)/(1-r)] \quad (20)$$

The angle ϕ'_b is the effective friction angle ϕ_b for shearing along the basal slip zone, modified to account for any possible difference between the pore-pressure ratios within the slide and along its base (Davis et al., 1983).

The forward-vergent slip surfaces (shear zones) may range from nearly planar to steeply listric ('sledge-runner') varieties, one of the principal controlling factors being the coefficient of effective friction. For example, if the sediment slide's interior is not appreciably stronger than its basal zone ($f \approx f_b$), the slip surfaces will tend to be listric and steep, because the mechanically identical basal and stepping-up slip zones will be simultaneously in a state of Navier–Coulomb failure. The backward propagation of successive shear zones, if closely spaced, will lead to additional steepening (see p. 63). Sediment cohesion also promotes listric and relatively steep slip zones (cf. Dahlen & Suppe, 1984). Planar and low-angle slip zones will form where the basal slip zone is considerably weaker than the slide's interior ($f > f_b$) and cohesion is negligible; this is illustrated by the example in Fig. 32. For the two cases, the former ($f \approx f_b$) would tend to promote more pervasive internal strain, due to merging of multiple shear zones, whereas the latter ($f > f_b$) would rather promote discrete or virtually isolated shear zones. Sigmoidal shear zones (like those inferred in Figs 33C, D) may be due to considerable vertical differences in the effective frictional resistance to shear (see also Fig. 31, top), or to an accompanying deformation of the slide's top by the shear stresses imposed by an overriding or overpassing sediment gravity flow (cf. Massari, 1984; Postma, 1984b, figs 20, 29).

The angle of dip of the axis of maximum compressional stress (angle ω in Fig. 34B) depends upon the ratio of the Coulomb shear strength of the basal slip zone $[f_b\,(1-r_b)\,\gamma \cdot h]$ to the strength of the slide interior $[f(1-r)\,\gamma \cdot h + c]$. To obtain an estimate of this ratio, equation (18) can be rewritten in the form:

$$S_b/S \approx \sin 2\omega/[(1 + f^2)^{1/2} - f \cdot \cos 2\omega], \quad (21)$$

where: S_b = shear strength in the basal slip zone; S = shear strength in the slide's interior; and other symbols as above.

Important qualitative and semi-quantitative inferences can thus be made for a slide or flow slide on the basis of the shear-strain pattern recorded in the deposit. This renders the shear-generated 'backsets' both interesting and useful, and sedimentologically as significant as the backsets of other origins (discussed in the previous section).

FINAL REMARKS

The sedimentological research on delta foresets during the last few years has been a reaction against some of the simplistic views that evolved from the early, classical studies on laboratory deltas (Nevin & Trainer, 1927; Albertson *et al.*, 1950; Jopling, 1963, 1965). Descriptions of the development of laboratory 'microdeltas', which many would regard as an analogue for Gilbert-type deltas, led to the promotion of the view that the steep slopes of natural deltas are much like the slipfaces of megaripples, where one sheet of avalanching grains descends after another, covering the subaqueous slope more-or-less uniformly and causing its bulk, smooth accretion (progradation). This simplistic notion persisted in the imagination of many sedimentologists until quite recently. As the large underwater slopes of modern steep-face deltas began to be surveyed in increasing detail, so it became clear that the sedimentation on such slopes involves a remarkably wide range of gravity-driven processes and the progradation of the delta face is neither 'smooth' nor uniform. Upon closer examination, ancient steep-face deltas appear to reveal the same picture.

The preliminary review of the processes operative on steep delta slopes presented here is by no means exhaustive, and is probably incomplete. Its aim is merely to signal the large diversity of delta-face processes, and to highlight some of their interesting aspects that seemed to have been largely ignored or overlooked by the sedimentologists. Many of the suggestions are necessarily hypothetical or purely speculative, and will require careful verification in the field or laboratory. Manageable, large-size laboratory analogues can be designed to simulate some of the processes and to examine or improve our knowledge of their mechanics. Computer simulations begin to play a significant role in this latter area of research, and the results of some relevant studies have been included in the present review.

The considerations have been focused on frictional (cohesionless) sediment movement, but it should be clear to the reader that cohesive sediments may be involved as well in the processes on steep delta faces. Although the mechanics of muddy sediment movement is much better understood (some key references are given in text), the role of cohesive mud on steep, underwater slopes spasmodically loaded with coarse granular material remains rather poorly explored. The preservation potential for mud may be low in such settings, but the mechanical role of muddy material may be profound. For example, the intervening deposition of mud on steep, coarse-grained delta slopes will enhance rapid failures (e.g. Postma, Babić *et al.*, 1988) and promote textural inversions (e.g. Larsen & Steel, 1978; Nemec *et al.*, 1984). When mixed with mud, coarse granular materials will also change their rheological properties.

Another poorly explored topic worth laboratory research is the actual potential for pore-pressure build-up in various textural mixtures of clay-free granular materials under shear in subaqueous conditions. Steep slopes of gravelly deltas are susceptible to 'gravitational winnowing' (Postma, 1984b; see also Nemec & Steel, 1984; Nemec *et al.*, 1984), and this process also deserves more attention and further laboratory investigation.

The sedimentological picture of steep delta slopes has just started to develop, clearly offering a large scope for future research. Suffice it to say that foreset-bed facies are among the least studied. The present review emphasizes the need for detailed,

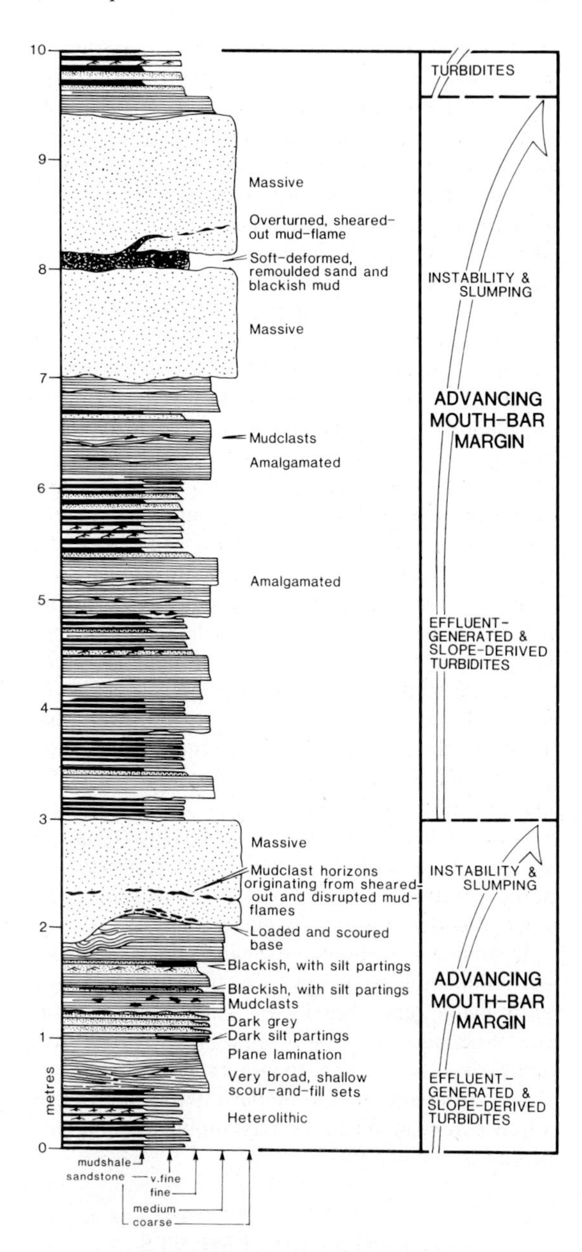

Fig. 36. Repetitive depositional sequences related to mouth-bar progradation and failure on a steep local slope of sandy braidplain delta (Cretaceous, eastern coast of Spitsbergen). For facies details and additional comments, see Figs 7, 22, 32. Note the textural anomalies created by coarse-sand transfer, as debris-flow lobes, from the upper slope to delta toe zone. (Slightly modified from Nemec *et al.*, 1988, fig. 14.)

'bed-by-bed' analysis of delta foreset deposits, as this is the way the individual depositional processes can be recognized and their relative role assessed for a particular ancient delta. Apparently, few authors have attempted such a detailed analysis; instead, delta foresets are commonly described *en bloc*, as if they were homogeneous packets of identical beds. The distinction of various foreset-bed facies (as genetic varieties) and analysis of their spatial organization shall be the means of recognizing the actual patterns of underwater sediment dispersal. Again, the varied styles of dispersal observed on modern delta slopes (e.g. Prior & Bornhold, 1988, figs 17–19) may serve as excellent analogues. It should be possible to recognize sequences of foreset-bed facies attributable to single events of local slope failure (see example in Fig. 36). Some characteristic or diagnostic sequences can then be distinguished, also to facilitate comparisons of different slopes or slope instability regimes. To follow this methodological path will be an exciting, yet challenging task, as one must take also into consideration the effects of basinal processes (waves or tides) and external factors (e.g. sea-level variations) on the foreset facies and geometrical aspects. Comparative studies of foresets in carefully selected marine and lacustrine Gilbert-type deltas might be very instructive. The same pertains to the computer simulations of cross-sectional, delta-progradation patterns, especially when verified against some well-studied modern systems (Syvitski *et al.*, 1988)

In summary, it should be clear to the reader that our sedimentological knowledge in the present area of research relies heavily on studies of modern, coarse-grained, steep-face deltas. This review thus emphasizes the necessity to integrate detailed observations from ancient and modern deltas in order to advance or examine our understanding of delta-face processes and deposits.

ACKNOWLEDGEMENTS

The manuscript was typed by A. Pedersen, and was reviewed by J.S. Collinson (Shrewsbury), D.W. Lewis (Christchurch), O.J. Martinsen (Bergen), G. Postma (Utrecht) and D.B. Prior (Dartmouth), who gave many useful comments and kindly provided some of the illustrations. Their help is greatly appreciated by the author.

REFERENCES

Albertson, M.L., Dai, Y.B., Jensen, R.A. & Rouse, H. (1950) Diffusion of submerged jets. *Trans. Am. Soc. civ. Engrs.* **115**, 639–697.

Allen, J.R.L. (1970a) The angle of initial yield of haphazard assemblages of equal spheres in bulk. *Geol. Mijnb.* **49**, 13–22.

Allen, J.R.L. (1970b) Avalanching of granular solids on dune and similar slopes. *J. Geol.* **78**, 326–351.

Allen, J.R.L. (1984) *Sedimentary Structures – Their Character and Physical Basis*. Elsevier, Amsterdam, 663 pp.

Allen, J.R.L. (1985) *Principles of Physical Sedimentology*. Allen & Unwin, London, 288 pp.

Allen, J.R.L. & Leeder, M.R. (1980) Criteria for the instability of upper-stage plane beds. *Sedimentology* **27**, 209–217.

Andresen, A. & Bjerrum, L. (1967) Slides on subaqueous slopes in loose sands and silts. In: *Marine Geotechnique* (Ed. by A.F. Richards), pp. 221–239. University of Illinois Press, Urbana.

Bagnold, R.A. (1954) Experiments on a gravity-free dispersion of large solid spheres in a Newtonian fluid under shear. *Proc. R. Soc. London, Ser.* **A255**, 49–63.

Barrier, P. (1984) Evolution téctono-sédimentaire Pliocène et Pléistocène du Détroit de Messine (Italie). Unpublished, Doctoral dissertation, Univer. de Marseille-Luminy, France, 270 pp.

Bates, C.C. (1953) Rational theory of delta formation. *Bull. Am. Assoc. Petrol. Geol.* **37**, 2119–2162.

Benjamin, T.B. (1968) Gravity currents and related phenomena. *J. Fluid Mech.* **31**, 209–248.

Bjerrum, L. (1971) Subaqueous slope failures in Norwegian fjords. *Publ. Norw. geotech. Inst.* **88**, 1–8.

Bluck, B.J. (1967) Deposition of some Upper Old Red Sandstone conglomerates in the Clyde area: a study in the significance of bedding. *Scott. J. Geol.* **3**, 139–167.

Brennen, C.E., Sieck, K. & Paslaski, J. (1983) Hydraulic jumps in granular material flow. *Powder Technol.* **35**, 31–37.

Bridgwater, J. (1980) On the width of failure zones. *Geotechnique* **30**, 533–536.

Britter, R.E. & Simpson, J.E. (1981) A note on the structure of the head of an intrusive gravity current. *J. Fluid Mech.* **112**, 459–466.

Busby-Spera, C.J. (1988) Development of fan-deltoid slope aprons in a convergent-margin tectonic setting: Mesozoic, Baja California, Mexico. In: *Fan Deltas: Sedimentology and Tectonic Settings* (Ed. by W. Nemec and R.J. Steel), pp. 419–429. Blackie and Son, London.

Campbell, C.S. (1986) The effect of microstructure development on the collisional stress tensor in a granular flow. *Acta Mech.* **63**, 61–72.

Campbell, C.S. (1989a) The stress tensor for simple shear flows of a granular material. *J. Fluid Mech.* **203**, 449–473.

Campbell, C.S. (1989b) Self-lubrication of long runout landslides. *J. Geol.* **97**, 653–665.

Campbell, C.S. & Brennen, C.E. (1983) Computer simulation of shear flows of granular material. In: *Mechanics of Granular Materials: New Models and Constitutive Relations* (Ed. by J.T. Jenkins and M. Satake),

pp. 313–326. Elsevier, Amsterdam.
Campbell, C.S. & Brennen, C.E. (1985a) Chute flows of granular material: some computer simulations. *J. appl. Mech.* **52**, 172–178.
Campbell, C.S. & Brennen, C.E. (1985b) Computer simulation of granular shear flows. *J. Fluid Mech.* **151**, 167–188.
Campbell, C.S. & Gong, A. (1986) The stress tensor in a two-dimensional granular shear flow. *J. Fluid Mech.* **164**, 107–125.
Carson, M.A. (1977) Angles of repose, angles of shearing resistance and angles of talus slopes. *Earth Surf. Proc.* **2**, 363–380.
Carson, M.A. & Kirkby, M.J. (1972) *Hillslope Form and Processes*. Cambridge University Press, Cambridge, 475 pp.
Carter, R.M. (1975) A discussion and classification of subaqueous mass-transport with particular application to grain-flow, slurry-flow, and fluxoturbidites. *Earth-Sci. Rev.* **11**, 145–177.
Clemmensen, L.B. & Houmark-Nielsen, M. (1981) Sedimentary features of a Weichselian glaciolacustrine delta. *Boreas* **10**, 229–245.
Colella, A. (1988a) Pliocene–Holocene fan deltas and braid deltas in the Crati Basin, southern Italy: a consequence of varying tectonic conditions. In: *Fan Deltas: Sedimentology and Tectonic Settings* (Ed. by W. Nemec and R.J. Steel), pp. 50–74. Blackie and Son, London.
Colella, A. (1988b) Gilbert-type fan deltas in the Crati Basin (Pliocene–Holocene, southern Italy). In: *Int. Works. Fan Deltas (1988) – Excursion Guidebook* (Ed. by A. Colella), pp. 19–77. Universita della Calabria, Cosenza.
Colella, A. (1988c) Quaternary Gilbert-type fan deltas of the Messina Strait. In: *International Workshop on Fan Deltas (1988) – Excursion Guidebook* (Ed. by A. Colella), pp. 139–152. Universita della Calabria, Cosenza.
Colella, A., De Boer, P.L. & Nio, S.D. (1987) Sedimentology of a marine intermontane Pleistocene Gilbert-type fan-delta complex in the Crati Basin, southern Italy. *Sedimentology* **34**, 721–736.
Collinson, J.D. (1970) Deep channels, massive beds and turbidity-current genesis in the Central Pennine Basin. *Proc. Yorkshire geol. Soc.* **37**, 495–519.
Collinson, J.D. (1986) Submarine ramp facies model for delta-fed, sand-rich turbidite systems: Discussion. *Bull. Am. Assoc. Petrol. Geol.* **70**, 1742–1743.
Cruden, D.M. & Hungr, O. (1986) The debris of the Frank Slide and theories of rockslide-avalanche mobility. *Can. J. Earth Sci.* **23**, 425–432.
Dahlen, F.A. (1984) Noncohesive critical Coulomb wedges: an exact solution. *J. geophys. Res.* **89**, 10125–10133.
Dahlen, F.A. & Suppe, J. (1984) Mechanics of fold-and-thrust belts and accretionary wedges: cohesive Coulomb theory. *J. geophys. Res.* **89**, 10087–10101.
Davis, D.M., Suppe, J. & Dahlen, F.A. (1983) Mechanics of fold-and-thrust belts and accretionary wedges. *J. geophys. Res.* **88**, 1153–1172.
Davis, D.M. & von Huene, R. (1987) Inferences on sediment strength and fault friction from structures at the Aleutian Trench. *Geology* **15**, 517–522.
Davis, W.M. (1890) Structure and origin of glacial sand plains. *Bull. geol. Soc. Am.* **1**, 195–202.
Dżułyński, S. & Sanders, J.E. (1962) Current marks on firm mud bottom. *Trans. Conn. Acad. Arts Sci.* **42**, 57–96.
Eisbacher, G.H. & Clague, J.J. (1984) Destructive mass movements in high mountains: hazard and management. Paper Geol. Surv. Can. 230 pp.
Ethridge, F.G. & Wescott, W.A. (1984) Tectonic setting, recognition and hydrocarbon reservoir potential of fan-delta deposits. In: *Sedimentology of Gravels and Conglomerates* (Ed. by E.H. Koster and R.J. Steel). Mem. Can. Soc. Petrol. Geol. **10**, pp. 217–235.
Eyles, N., Clark, B.M. & Clague, J.J. (1987) Coarse-grained sediment gravity flow facies in a large supraglacial lake. *Sedimentology* **34**, 193–216.
Farrel, S.G. & Eaton, S. (1988) Foliations developed during slump deformation of Miocene marine sediments, Cyprus. *J. struct. Geol.* **10**, 567–576.
Ferentinos, G., Papatheodorou, G. & Collins, M.B. (1988) Sediment transport processes on an active submarine fault escarpment, Gulf of Corinth, Greece. *Mar. Geol.* **83**, 43–61.
Fredrickson, A.G. (1964) *Principles and Applications of Rheology*. Prentice-Hall, Englewood Cliffs (N.J.), 326 pp.
Gardner, J.S. (1970) Rockfall: a geomorphic process in high mountain terrain. *Albertan Geogr.* **6**, 15–20.
Gilbert, G.K. (1883) The topographic features of lake shores. *Ann. Rep. U.S. geol. Surv.* **5**, 69–123.
Gilbert, G.K. (1914) The transportation of debris by running water. *Prof. Pap. U.S. geol. Surv.* **86**, 263 pp.
Haff, P.K. (1983) Grain flow as a fluid-mechanical phenomenon. *J. Fluid Mech.* **134**, 401–430.
Hampton, M.A. (1972) The role of subaqueous debris flow in generating turbidity currents. *J. sedim. Petrol.* **42**, 775–793.
Hay, A.E. (1987a) Turbidity currents and submarine channel formation in Rupert Inlet, British Columbia: 1. Surge observations. *J. geophys. Res.* **92**, 2875–2881.
Hay, A.E. (1987b) Turbidity currents and submarine channel formation in Rupert Inlet, British Columbia: 2. The roles of continuous and surge-type flow. *J. geophys. Res.* **92**, 2883–2900.
Hay, A.E., Burling, R.W. & Murray, J.W. (1982) Remote acoustic detection of a turbidity current surge. *Science* **217**, 833–835.
Heller, P.L. & Dickinson, W.R. (1985) Submarine ramp facies model for delta-fed, sand-rich turbidite systems. *Bull. Am. Assoc. Petrol. Geol.* **69**, 960–976.
Hoek, E. & Bray, J. (1977) *Rock Slope Engineering*. Revised 2nd edn. Institute of Mining and Metallurgy, London, 402 pp.
Holmes, A. (1965) *Principles of Physical Geology*. 2nd edn. Thomas Nelson, London, 1288 pp.
Hopfinger, E.J. (1983) Snow avalanche motion and related phenomena. *Ann. Rev. Fluid Mech.* **15**, 47–76.
Hoskin, C.M. & Burrel, D.C. (1972) Sediment transport and accumulation in a fjord basin, Glacier Bay, Alaska. *J. Geol.* **80**, 539–551.
Hsü, K.J. (1969) Role of cohesive strength in the mechanics of overthrust faulting. *Bull. geol. Soc. Am.* **80**, 927–952.

HUBBERT, M.K. & RUBEY, W.W. (1959) Role of fluid pressure in mechanics of overthrust faulting. *Bull. geol. Soc. Am.* **70**, 115–166.

HUNTER, R.E. (1985) Subaqueous sandflow cross-strata. *J. sedim. Petrol.* **55**, 886–894.

HUNTER, R.E. & KOCUREK, G. (1986) An experimental study of subaqueous slipface deposition. *J. sedim. Petrol.* **56**, 387–394.

HUPPERT, H.E. & SIMPSON, J.E. (1980) The slumping of gravity currents. *J. Fluid Mech.* **99**, 785–799.

INNES, J.L. (1983) Debris flows. *Prog. phys. Geogr.* **7**, 469–501.

IVERSON, R.M. (1985) A constitutive equation for mass-movement behavior. *J. Geol.* **93**, 143–160.

IVERSON, R.M. (1986) Unsteady, non-uniform landslide motion: 1. Theoretical dynamics and the study datum state. *J. Geol.* **94**, 1–15.

JENKINS, J.T. & SAVAGE, S.B. (1983) A theory for the rapid flow of identical, smooth, nearly elastic, spherical particles. *J. Fluid Mech.* **130**, 187–202.

JOHANSSON, C.E. (1975) Some aspects of delta structures. *Svensk geogr. Årsbok* **51**, 87–99.

JOHANSSON, C.E. (1976) Structural studies of frictional sediments. *Geogr. Ann.* **58**, 201–301.

JOHNSON, A.M. & RODINE, J.R. (1984) Debris flow. In: *Slope Instability* (Ed. by D. Brunsden and D.B. Prior), pp. 257–361. Wiley, Chichester.

JONES, M.E. & PRESTON, R.M.F. (Eds.) (1987) *Deformation of Sediments and Sedimentary Rocks*. Spec. Publ. geol. Soc. London **29**, 350 pp.

JOPLING, A.V. (1963) Hydraulic studies on the origin of bedding. *Sedimentology* **2**, 115–121.

JOPLING, A.V. (1965) Hydraulic factors controlling the shape of laminae in laboratory deltas. *J. sedim. Petrol.* **35**, 777–791.

JOPLING, A.V. & RICHARDSON, E.V. (1966) Backset bedding developed in shooting flow in laboratory experiments. *J. sedim. Petrol.* **36**, 821–825.

KARLSRUD, K. & BY, T. (1981) Stability evaluations for submarine slopes: data on run-out distance and velocity of soil flows generated by subaqueous slides and quick-clay slides. *Rept. Norw. geotech. Inst.* **52207**–7, 67 pp.

KARLSRUD, K. & EDGERS, L. (1982) Some aspects of submarine slope stability. In: *Marine Slides and Other Mass Movements* (Ed. by S. Saxov and J.K. Nieuwenhuis), pp. 61–81. Plenum Press, New York.

KASTENS, K.A. & SHOR, A.N. (1985) Depositional processes of a meandering channel on Mississippi fan. *Bull. Am. Assoc. Petrol. Geol.* **69**, 190–202.

KEEFER, D.K. & JOHNSON, A.M. (1983) Earth flows: morphology, mobilization, and movement. *Prof. Pap. U.S. geol. Surv.* **1264**, 56 pp.

KIRKBY, M.J. & STATHAM, I. (1975) Surface stone movement and scree formation. *J. Geol.* **83**, 349–362.

KONING, H.L. (1982) On an explanation of marine flow slides in sand. In: *Marine Slides and Other Mass Movements* (Ed. by S. Saxov and J.K. Nieuwenhuis), pp. 83–94. Plenum Press, New York.

KOSTASCHUK, R.A. & MCCANN, S.B. (1987) Subaqueous morphology and slope processes in a fjord delta, Bella Coola, British Columbia. *Can. J. Earth Sci.* **24**, 52–59.

LA CHAPELLE, E.R. (1977) Snow avalanches: a review of current research and applications. *J. Glaciol.* **19**, 313–324.

LARSEN, V. & STEEL, R.J. (1978) The sedimentary history of a debris flow-dominated alluvial fan: a study of textural inversion. *Sedimentology* **25**, 37–59.

LAWSON, D.E. (1982) Mobilization, movement and deposition of active subaerial sediment flows, Matanuska Glacier, Alaska. *J. Geol.* **90**, 279–300.

LEEDER, M.R. & GAWTHORPE, R.L. (1987) Sedimentary models for extensional tilt-block/half-graben basins. In: *Continental Extensional Tectonics* (Ed. by M.P. Coward, J.F. Dewey and P.L. Hancock). Spec. Publ. geol. Soc. London 28, pp. 139–152.

LOWE, D.R. (1975) Water escape structures in coarse-grained sediments. *Sedimentology* **22**, 157–204.

LOWE, D.R. (1976) Subaqueous liquefied and fluidized sediment flows and their deposits. *Sedimentology* **23**, 285–308.

LOWE, D.R. (1982) Sediment gravity flows: II. Depositional models with special reference to the deposits of high-density turbidity currents. *J. sedim. Petrol.* **52**, 279–297.

LOWE, D.R. (1988) Suspended-load fallout rate as an independent variable in the analysis of current structures. *Sedimentology* **35**, 765–776.

LUCKMAN, B.H. (1976) Rockfalls and rockfall inventory data: some observations from Surprise Valley, Jasper National Park, Canada. *Earth Surf. Proc.* **1**, 287–298.

LUN, C.K.K., SAVAGE, S.B., JEFFREY, D.J. & CHEPURNIY, N. (1984) Kinetic theories for granular flow: inelastic particles in Couette flow and slightly inelastic particles in a general flowfield. *J. Fluid Mech.* **140**, 223–256.

MANDL, G. (1988) *Mechanics of Tectonic Faulting*. Elsevier, Amsterdam, 407 pp.

MARTINSEN, O.J. (1989) Styles of soft-sediment deformation on a Namurian (Carboniferous) delta slope, Western Irish Namurian Basin, Ireland. In: *Deltas: Sites and Traps for Fossil Fuels* (Ed. by M.G.K. Whateley and K.T. Pickering). Spec. Publ. geol. Soc. London 41, pp. 167–177.

MARTINSEN, O.J. & BAKKEN, B. 1990 Extensional and compressional zones in slumps and slides in the Namurian of County Clare, Ireland. *J. geol. Soc. London* **147**, 153–164.

MARZO, M. & ANADÓN, P. (1988) Anatomy of a conglomeratic fan-delta complex: the Eocene Montserrat Conglomerat, Ebro Basin, northeastern Spain. In: *Fan Deltas: Sedimentology and Tectonic Settings* (Ed. by W. Nemec and R.J. Steel), pp. 318–340. Blackie and Son, London.

MASSARI, F. (1984) Resedimented conglomerates of a Miocene fan-delta complex, Southern Alps, Italy. In: *Sedimentology of Gravels and Conglomerates* (Ed. by E.H. Koster and R.J. Steel). Mem. Can. Soc. Petrol. Geol. 10, pp. 259–278.

MCTIGUE, D.F. (1982) A nonlinear constitutive model for granular materials: applications to gravity flow. *J. appl. Mech.* **49**, 291–296.

METZNER, A.B. & REED, J.C. (1955) Flow of non-Newtonian fluids: correlation of the laminar, transition, and turbulent flow regimes. *J. Am. Inst. chem. Engrs.* **1**, 434–439.

MIDDLETON, G.V. (1965) Antidune cross-bedding in a large flume. *J. sedim. Petrol.* **35**, 922–927.

MIDDLETON, G.V. (1966) Experiments on density and turbidity currents. I: Motion of the head. *Can. J. Earth*

Sci. **3**, 523–546.

MIDDLETON, G.V. (1967) Experiments on density and turbidity currents. III: Deposition of sediment. *Can. J. Earth Sci.* **4**, 475–505.

MIDDLETON, G.V. (1970) Experimental studies related to flysch sedimentation. In: *Flysch Sedimentology in North America* (Ed. by J. Lajoie). Spec. Pap. geol. Assoc. Can. **7**, pp. 253–272.

MIDDLETON, G.V. & SOUTHARD, J.B. (1984) *Mechanics of Sediment Movement.* SEPM Short Course no. 3, 2nd edn. Soc. econ. Paleont. Mineral., Tulsa, 401 pp.

MORRISON, H.L. & RICHMOND, O. (1976) Application of Spencer's ideal soil model to granular material flow. *J. appl. Mech.* **43**, 49–53.

MRÓZ, Z. (1980) On hypoelasticity and plasticity approaches to constitutive modelling of inelastic behaviour of soils. *Int. J. numer. anal. Meth. Geomech.* **4**, 45–55.

MUDGE, N.R. (1965) Rockfall-avalanche and rockslide-avalanche deposits at Sawtooth Ridge, Montana. *Bull. geol. Soc. Am.* **76**, 1003–1014.

NAYLOR, M.A. (1980) The origin of inverse grading in muddy debris flow deposits — a review. *J. sedim. Petrol.* **50**, 1111–1116.

NEMEC, W., PORĘBSKI, S.J. & STEEL, R.J. (1980) Texture and structure of resedimented conglomerates: examples from Książ Formation (Famennian-Tournaisian), southwestern Poland. *Sedimentology* **27**, 519–538.

NEMEC, W. & STEEL, R.J. (1984) Alluvial and coastal conglomerates: their significant features and some comments on gravelly mass-flow deposits. In: *Sedimentology of Gravels and Conglomerates* (Ed. by E.H. Koster and R.J. Steel). Mem. Can. Soc. Petrol. Geol. 10, pp. 1–31.

NEMEC, W., STEEL, R.J., GJELBERG, J., COLLINSON, J.D., PRESTHOLM, E. & ØXNEVAD, I.E. (1988) Anatomy of collapsed and re-established delta front in Lower Cretaceous of eastern Spitsbergen: gravitational sliding and sedimentation processes. *Bull. Amer. Assoc. Petrol. Geol.* **72**, 454–476.

NEMEC, W., STEEL, R.J., PORĘBSKI, S.J. & SPINNANGR, Å. (1984) Domba Conglomerate, Devonian, Norway: process and lateral variability in a mass flow-dominated, lacustrine fan-delta. In: *Sedimentology of Gravels and Conglomerates* (Ed. by E.H. Koster and R.J. Steel). Mem. Can. Soc. Petrol. Geol. 10, pp. 295–320.

NEVIN, C.M. & TRAINER, D.W., JR. (1927) Laboratory study in delta-building. *Bull. geol. Soc. Am.* **38**, 451–458.

NYBERG, R. (1985) Debris flows and slush avalanches in northern Swedish Lappland: Distribution and geomorphological significance. *Avhandl. Medd. Lunds Univ. geogr. Inst.* **97**, 222 pp.

OKUDA, S., SUWA, H., OKUNISHI, K., YOKOYAMA, K. & NAKANO, M. (1980) Observations on the motion of a debris flow and its geomorphological effects. *Z. Geomorph. N.F.* **35**, 142–163.

ORI, G.G. & ROVERI, M. (1987) Geometries of Gilbert-type deltas and large channels in the Meteora Conglomerate, meso-Hellenic basin (Oligo-Miocene), central Greece. *Sedimentology* **34**, 845–859.

PIERSON, T.C. (1980) Erosion and deposition by debris flows at Mt. Thomas, North Canterbury, New Zealand. *Earth Surf. Proc.* **5**, 227–247.

PIERSON, T.C. (1986) Flow behavior of channelized debris flows, Mount St. Helens, Washington. In: *Hillslope Processes* (Ed. by A.D. Abrahams), pp. 269–296. Allen and Unwin, Boston.

PORĘBSKI, S.J. (1984) Clast size and bed thickness trends in resedimented conglomerates: example from a Devonian fan-delta succession, southwest Poland. In: *Sedimentology of Gravels and Conglomerates* (Ed. by E.M. Koster and R.J. Steel). Mem. Can. Soc. Petrol. Geol. 10, pp. 399–411.

POSTMA, G. (1979) Preliminary note on a significant sequence in conglomeratic flows of a mass transport-dominated fan-delta (lower Pliocene, Almeria Basin, SE Spain). *Konin. Neder. Akad. Wetensch.* **B 82**, 465–471.

POSTMA, G. (1984a) Slumps and their deposits on fan delta front and slope. *Geology* **12**, 27–30.

POSTMA, G. (1984b) Mass-flow conglomerates in a submarine canyon: Abrioja fan-delta, Pliocene, southeast Spain. In: *Sedimentology of Gravels and Conglomerates* (Ed. by E.H. Koster and R.J. Steel). Mem. Can. Soc. Petrol. Geol. 10, pp. 237–258.

POSTMA, G., BABIĆ, L., ZUPANIČ, J. & RØE, S.-L. (1988) Delta-front failure and associated bottomset deformation in a marine, gravelly Gilbert-type fan delta. In: *Fan Deltas: Sedimentology and Tectonic Settings* (Ed. by W. Nemec and R.J. Steel), pp. 91–102. Blackie and Son, London.

POSTMA, G. & CRUICKSHANK, C. (1988) Sedimentology of a late Weichselian to Holocene terraced fan delta, Varangerfjord, northern Norway. In: *Fan Deltas: Sedimentology and Tectonic Settings* (Ed. by W. Nemec and R.J. Steel), pp. 144–157. Blackie and Son, London.

POSTMA, G., NEMEC, W. & KLEINSPEHN, K.L. (1988) Large floating clasts in turbidites: a mechanism for their emplacement. *Sedim. Geol.* **58**, 47–61.

POSTMA, G. & ROEP, T.B. (1985) Resedimented conglomerates in the bottomsets of Gilbert-type gravel deltas. *J. sedim. Petrol.* **55**, 874–885.

POSTMA, G., ROEP, T.B. & RUEGG, G.J.H. (1983) Sandy-gravelly mass-flow deposits in an ice-marginal lake (Saalian, Leuvenumsche Beek Valley, Veluwe, The Netherlands), with emphasis on plug-flow deposits. *Sedim. Geol.* **34**, 59–82.

PRICE, N.J. (1973) A note on geological and engineering strain-rates, hydraulic fracturing and slope stability. *Geol. appl. Idregeol.* **8**, 57–63.

PRICE, N.J. (1977) Aspects of gravity tectonics and the development of listric faults. *J. geol. Soc. London* **133**, 311–327.

PRIOR, D.B. & BORNHOLD, B.D. (1986) Sediment transport on subaqueous fan delta slopes, Britannia Beach, British Columbia. *Geo-Mar. Lett.* **5**, 217–224.

PRIOR, D.B. & BORNHOLD, B.D. (1988) Submarine morphology and processes of fjord fan deltas and related high-gradient systems: modern examples from British Columbia. In: *Fan Deltas: Sedimentology and Tectonic Settings* (Ed. by W. Nemec and R.J. Steel), pp. 125–143. Blackie and Son, London.

PRIOR, D.B. & BORNHOLD, B.D. (1989) Submarine sedimentation on a developing Holocene fan delta. *Sedimentology* **36**.

PRIOR, D.B., BORNHOLD, B.D., COLEMAN, J.M. & BRYANT, W.R. (1982) The morphology of a submarine slide, Kitimat Arm, British Columbia. *Geology* **10**, 588–592.

PRIOR, D.B., BORNHOLD, B.D. & JOHNS, M.W. (1984)

Depositional characteristics of a submarine debris flow. *J. Geol.* **92**, 707–727.

PRIOR, D.B., BORNHOLD, B.D. & JOHNS, M.W. (1986) Active sand transport along a fjord-bottom channel, Bute Inlet, British Columbia. *Geology* **14**, 581–584.

PRIOR, D.B., BORNHOLD, B.D., WISEMAN, W.J., JR. & LOWE, D.R. (1987) Turbidity current activity in a British Columbia fjord. *Science* **237**, 1330–1333.

PRIOR., D.B. & COLEMAN, J.M. (1982) Active slides and flows in underconsolidated marine sediments on the slopes of the Mississippi Delta. In: *Marine Slides and Other Mass Movements* (Ed. by S. Saxov and J.K. Nieuwenhuis), pp. 21–49. Plenum Press, New York.

PRIOR, D.B. & COLEMAN, J.M. (1984) Submarine slope instability. In: *Slope Instability* (Ed. by D. Brunsden and D.B. Prior), pp. 419–455. Wiley, Chichester.

PRIOR, D.B., WISEMAN, Wm.J., JR. & BRYANT, W.R. (1981) Submarine chutes on the slopes of fjord deltas. *Nature* **290**, 326–328.

PRIOR, D.B., YANG, Z.-S., BORNHOLD, B.D., KELLER, G.H., LU, N.Z., WISEMAN, W.J., JR., WRIGHT, L.D. & ZHANG, J. (1986) Active slope failure, sediment collapse, and silt flows on the modern subaqueous Huanghe (Yellow River) delta. In: *Huanghe (Yellow River) Delta* (Ed. by G.H. Keller and D.B. Prior). Geo-Mar. Lett. 6(2), pp. 85–95.

RALEIGH, B. & GRIGGS, D. (1963) Effects of the toe of the thrust plane. *Bull. geol. Soc. Am.* **74**, 819–838.

RAPP, A. (1960) Recent developments of mountain slopes in Karkevagge and surroundings, northern Scandinavia. *Geogr. Ann.* **42**, 65–200.

ROBERTS, H.H., SUHAYDA, J.N. & COLEMAN, J.M. (1980) Sediment deformation and transport on low angle slopes: Mississippi River Delta. In: *Thresholds in Geomorphology* (Ed. by D.R. Coates and J. D. Vitek), pp. 131–167. Allen and Unwin, Boston.

ROSCOE, R. (1953) Suspensions. In: *Flow Properties of Disperse Systems* (Ed. by J.J. Hermans), pp. 1–38. Interscience, New York.

ROSCOE, R. (1970) The influence of strains in soil mechanics. *Geotechnique* **20**, 129–170.

RUPKE, N.A. (1978) Deep clastic seas. In: *Sedimentary Environments and Facies* (Ed. by H.G. Reading), pp. 372–415. 1st edn. Blackwell Scientific Publications, Oxford.

RUTGERS, Ir.R. (1962) Relative viscosity of suspensions of rigid spheres in Newtonian liquids. *Rheol. Acta.* **2**, 202–210.

SAVAGE, S.B. (1979) Gravity flow of cohesionless granular materials in chutes and channels. *J. Fluid Mech.* **92**, 53–96.

SAVAGE, S.B. (1983) Granular flows down rough inclines — review and extension. In: *Mechanics of Granular Materials: New Models and Constitutive Relations* (Ed. by J.T. Jenkins and M. Satake), pp. 261–282. Elsevier, Amsterdam.

SAVAGE, S.B. & SAYED, M. (1984) Stress developed by dry cohesionless granular materials in an annular shear cell. *J. Fluid Mech.* **142**, 391–430.

SCHAFER, C.T. & SMITH, J.N. (1987) Hypothesis for a submarine landslide and cohesionless sediment flows resulting from a 17th-century earthquake-triggered landslide in Quebec, Canada. *Geo-Mar. Lett.* **7**, 31–37.

SCHWARZ, H.-U. (1982) *Subaqueous Slope Failure – Experiments and Modern Occurrences*. Contrib. Sedim. **11**, 432 pp. Schweizerbartsche Verlagbuchhand, Stuttgart.

SCOTT, A.M. & BRIDGWATER, J. (1975) Interparticle percolation: a fundamental solids mixing mechanism. *Ind. eng. Chem., Fundam.* **14**, 22–26.

SCRAPELLI, G. & WOOD, D.M. (1982) Experimental observations of shear band patterns in direct shear tests. In: *Deformation and Failure of Granular Materials* (Ed. by P.A. Vermeer and H.J. Luger), pp. 473–484. Balkema, Rotterdam.

SHARP, R.P. (1942) Mudflow levees. *J. Geomorph.* **5**, 222–227.

SHREVE, R.L. (1968) *The Blackhawk Landslide*. Spec. Pap. geol. Soc. Am. 108, 47 pp.

SIMONS, D.B. & RICHARDSON, E.V. (1963) Forms of bed roughness in alluvial channels. *Trans. Am. Soc. civ. Engrs.* **128**, 284–323.

SIMPSON, J.E. & BRITTER, R.E. (1979) The dynamics of the head of a gravity current advancing over a horizontal surface. *J. Fluid Mech.* **94**, 477–495.

SPENCER, A.J.M. (1981) Deformation of ideal granular materials. In: *Mechanics of Solids* (Ed. by H.G. Hopkins and M.J. Swell), pp. 607–651. Pergamon Press, Oxford.

STANLEY, D.J. (1980) The Saint-Antonin Conglomerate in the Maritime Alps: a model for coarse sedimentation on a submarine slope. *Smithn. Contrib. mar. Sci.* **5**, 1–25.

STANLEY, D.J., PALMER, H.D. & DILL, R.F. (1978) Coarse sediment transport by mass flow and turbidity current processes and downslope transformations in Annot Sandstone canyon-fan valley systems. In: *Sedimentation in Submarine Canyons, Fans and Trenches* (Ed. by D.J. Stanley and G. Kelling), pp. 85–115. Dowden, Hutchinson and Ross, Stroudsburg (Pa.).

STATHAM, I. (1976) A scree slope rockfall model. *Earth Surf. Proc.* **1**, 43–62.

STATHAM, I. & FRANCIS, S.C. (1986) Influence of scree accumulation and weathering on the development of steep mountain slopes. In: *Hillslope Processes* (Ed. by A.D. Abrahams), pp. 245–267. Allen and Unwin, Boston.

STOW, D.A.V. (1986) Deep clastic seas. In: *Sedimentary Environments and Facies* (Ed. by H.G. Reading), pp. 399–444. 2nd edn. Blackwell Scientific Publications, Oxford.

SURLYK, F. (1984) Fan-delta to submarine fan conglomerates of the Volgian–Valanginian Wollaston Forland Group, East Greenland. In: *Sedimentology of Gravels and Conglomerates* (Ed. by E.H. Koster and R.J. Steel). Mem. Can. Soc. Petrol. Geol. 10, pp. 359–382.

SURLYK, F. (1987) Slope and deep shelf gully sandstones, Upper Jurassic, East Greenland. *Bull. Am. Assoc. Petrol. Geol.* **71**, 464–475.

SUWA, H. & OKUDA, S. (1983) Deposition of debris flows on a fan surface, Mt. Yakedake, Japan. *Z. Geomorph. N.F.* **46**, 79–101.

SYVITSKI, J.P.M. & FARROW, G.E. (1983) Structures and processes in bay head deltas: Knight and Bute inlets, British Columbia. *Sedim. Geol.* **30**, 217–244.

SYVITSKI, J.P.M. & FARROW, G.E. (1989) Fjord sedimentation as an analogue for small hydrocarbon-bearing fan deltas. In: *Deltas: Sites and Traps for Fossil Fuels* (Ed.

by M.K.G. Whateley and K.T. Pickering). Spec. Publ. geol. Soc. London 41, pp. 21–43.

SYVITSKI, J.P.M., SMITH, J.N., CALABRESE, E.A. & BOUDREAU, B.P. (1988) Basin sedimentation and the growth of prograding deltas. *J. geophys. Res.* **93**, 6895–6908.

TAKAHASHI, T. (1978) Mechanical characteristics of debris flow. *Proc. Am. Soc. civ. Engrs., J. hydr. Div.* **104** (HY8), 1153–1169.

TAKAHASHI, T. (1980) Debris flow in prismatic open channel. *Proc. Am. Soc. civ. Engrs., J. hydr. Div.* **106** (HY3), 381–396.

TAKAHASHI, T. (1981) Debris flow. *Ann. Rev. Fluid Mech.* **13**, 57–77.

TODD, S.P. (1989) Stream-driven, high-density gravelly traction carpets: possible deposits in the Trabeg Conglomerate Formation, SW Ireland and some theoretical considerations of their origin. *Sedimentology* **36**, 513–530.

VAN BURKALOW, A. (1945) Angle of repose and angle of sliding friction: an experimental study. *Bull. geol. Soc. Am.* **56**, 669–708.

VERMEER, P.A. & LUGER, H.J. (Eds.) (1982) *Deformation and Failure of Granular Materials*. Balkema, Rotterdam, 673 pp.

VOIGHT, B. (Ed.) (1978) *Rockslides and Avalanches, 1. Natural Phenomena*. Elsevier, Amsterdam, 833 pp.

WALKER, R.G. (1984) Turbidites and associated coarse clastic deposits. In: *Facies Models* (Ed. by R.G. Walker). Geosci. Can. Reprint Series 1, 2nd edn, 171–188.

WALTON, O.R. (1983) Particle-dynamics calculations of shear flow. In: *Mechanics of Granular Materials: New Models and Constitutive Relations* (Ed. by J.T. Jenkins and M. Satake), pp. 327–338. Elsevier, Amsterdam.

WEIRICH, F.H. (1989) The generation of turbidity currents by subaerial debris flows, California. *Bull. geol. Soc. Am.* **101**, 278–291.

WHALLEY, W.B. (1984) Rockfalls. In: *Slope Instability* (Ed. by D. Brunsden and D.B. Prior), pp. 217–256. Wiley, Chichester.

WRIGHT, L.D., WISEMAN, W.J., JR., BORNHOLD, B.D., PRIOR, D.B., SUHAYDA, J.N., KELLER, G.M., YANG, Z.-S. & FAN, Y.B. (1988) Marine dispersal and deposition of Yellow River silts by gravity-driven underflows. *Nature* **332**(6164), 679–682.

WRIGHT, L.D., YANG, Z.-S., BORNHOLD, B.D., KELLER, G.H., PRIOR, D.B. & WISEMAN, W.J., JR. (1986) Hyperpycnal plumes and plume fronts over the Huanghe (Yellow River) delta front. In: *Huanghe (Yellow River) Delta* (Ed. by G.H. Keller and D.B. Prior). Geo-Mar. Lett. 6 (2), pp. 97–105.

Spec. Publs int. Ass. Sediment. (1990) **10**, 75–90

The underwater development of Holocene fan deltas

D. B. PRIOR* *and* B. D. BORNHOLD†

* *Geological Survey of Canada, Atlantic Geoscience Centre, Dartmouth, Nova Scotia, Canada, B2Y 4A2*
† *Geological Survey of Canada, Pacific Geoscience Centre, Sidney, British Columbia, Canada, V8L 4B2*

ABSTRACT

The subaqueous morphology and geometry of developing Holocene fan deltas in fjords in British Columbia are used to interpret underwater sediment-dispersal processes. The fans are constructed by combinations of processes occurring with various frequencies and magnitudes, including subaqueous debris avalanching, inertia flows, turbidity flows, slope failure and settling of suspensions from buoyant plumes. Long-term fan development responds to changes in sediment supply to the coastline and reduction in underwater relief and slope gradients. A four-stage synthetic evolutionary sequence reflects differences in process combinations as the fan deltas develop.

INTRODUCTION

Side-scan sonar imagery, sub-bottom profiles, and sediment samples from Holocene fan deltas and related high-energy depositional systems suggest that underwater components are constructed by several different sediment-dispersal processes. Settling of suspended sediment from plumes combines with bottom or near-bottom transport processes, including boulder avalanching, coarse-grained inertia flows, high- and low-density turbidity currents, and translational and rotational sliding (e.g. Prior *et al.*, 1981a, 1981b, 1984, 1986, 1987; Prior & Bornhold, 1986, 1988, 1989; Kostaschuk & McCann, 1987; Corner *et al.*, 1988, and this volume; Ferentinos *et al.*, 1988; Syvitski *et al.*, 1988; Syvitski & Farrow, 1989; Bornhold & Prior, this volume).

There is great variety in underwater fan-delta architecture, depending on the process combinations and the relative dominance of each. Differences in types and rates of sediment supply influence the underwater dispersal processes. Offshore-relief and bottom-slope gradients are important controls. As entire subaerial/subaqueous systems develop, the roles of the various underwater processes change. Subaerial alluvial-fan growth and nearshore deltaic progradation alter the volumes and textures of sediment arriving on the underwater fan and delta slopes. Underwater fan aggradation, progradation and basin filling reduce offshore relief, bottom slopes and gravitational stresses.

The present paper uses side-scan imagery from a number of different fan deltas in British Columbia fjords including Bute Inlet, Knight Inlet, Howe Sound, Phillips Arm and Bentinck Arm, to characterize underwater fan-delta morphology. Form/process interpretations are made from the remotely acquired imagery, supported by observations from submersible and the results of bottom sampling and coring. A synthetic model is constructed to illustrate some of the underwater form/process changes that occur as the depositional systems evolve.

British Columbia fan deltas are especially instructive because they are forming today, having begun to build following regional deglaciation in late Pleistocene times. The fjord fan deltas are at various stages of development, but when compared to ancient, tectonically controlled fan deltas they are within one uplift/sedimentation cycle.

HIGH-ENERGY DEPOSITIONAL SYSTEMS IN FJORDS

Fjords contain excellent examples of active, high-energy depositional systems, with rapid, episodic, coarse-grained sediment input to high-relief, steep offshore slopes. Fan deltas, generally built of coarse gravel and sand, develop along fjord sidewalls, where supply catchments are small and steep, but offshore relief is greatest (Fig. 1). Fjord-head deltas generally introduce finer-grained gravel, sand and silt, usually in large volumes because of extensive subaerial catchments. Because of the confining effects of valley walls, fjord-head deltas are rarely fan shaped, usually having relatively straight, across-fjord progradational fronts (Fig. 1).

Acoustic-survey data have been acquired for more than 20 sidewall fan deltas and fjord-head deltas in British Columbia (e.g., Prior *et al.*, 1981b; Syvitski & Farrow, 1983; Prior *et al.*, 1984; Prior & Bornhold, 1986, 1988, 1989; Kostaschuk & McCann, 1987; Syvitski & Farrow, 1989; Syvitski *et al.*, 1988). Interpretation of fan-delta building processes combines the information from published examples from Bute Inlet, Howe Sound and Knight Inlet with newly acquired data from Phillips and Bentinck Arms. The latter are used primarily for illustrations for the present paper.

In British Columbia, delta growth began as the fjords and neighbouring mountain drainage basins became generally ice free, dated regionally to about 10 000–12 000 years ago. Fjord-head deltas began their latest progradational phase as sea level reached its present relative position about 8000–9000 years ago (Clague, 1981). Delta progradation has resulted in enlargement of the subaerial deltaic plains and downfjord delta-front deposition into progressively deeper water. Subaqueous delta-front relief in the fjord-head systems has generally increased and offshore slopes have become steeper, with down-fjord advance.

Accompanying regional deglaciation, subaqueous, sidewall fan deltas began prograding over and interdigitating with late Pleistocene and early Holocene glaciomarine sediments accumulating as fjord-bottom infill. Large volumes of sediment have been deposited along the fjord walls. Sidewall fan aggradation and gradual fjord-wide sedimentation have led to reduction of underwater relief and progressively lower sidewall fan gradients. Where sufficient sediment is retained near the shoreline for subaerial development, the underwater fans have not reached the point of aggradation (Fig. 2). Other underwater fans have aggraded sufficiently to form subaerial alluvial fans or small fan-shaped deltas (e.g. Prior & Bornhold, 1988, 1989) (Figs 1, 2). Continued alluvial-fan and fan-delta growth involves enlargement of the subaerial component as the entire system progrades across-fjord. Because the deeper fjord cross-profiles are generally flat, further sidewall fan-delta growth involves no increase in total relief. Rather, as the fjord bottom fills, the water depths and subaqueous fan relief decline.

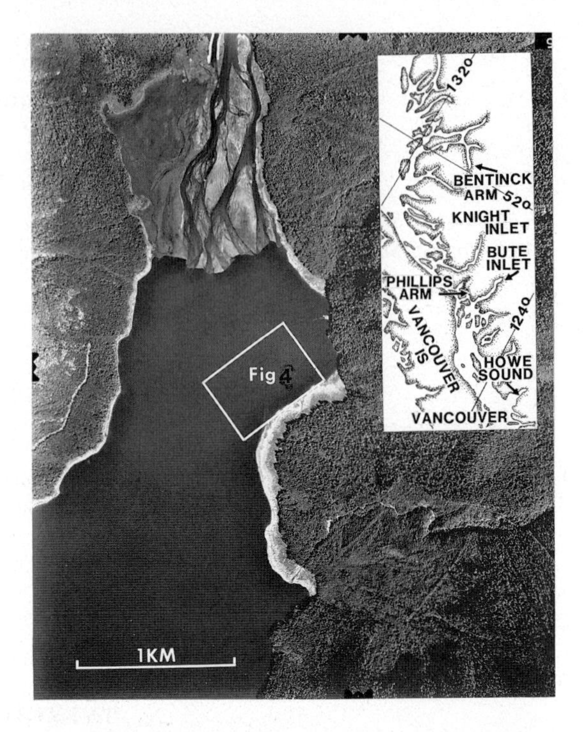

Fig. 1. Fjord-head delta and sidewall fan delta, Phillips Arm, British Columbia (inset — location map). Note location of data presented in Fig. 4.

Thus, although the absolute time available for sidewall fan-delta and fjord-head delta construction is roughly the same regionally, actual submarine geometries vary widely. Offshore relief and sediment supply are important variables, but a major control on development is the efficiency of underwater sediment-dispersal processes (Prior & Bornhold, 1988, 1989).

UNDERWATER SEDIMENT DISPERSAL AND MORPHOLOGY

The rates of sediment supply into British Columbia fjords vary greatly. Low-frequency, high-magnitude events such as ice-dammed lake outbursts occur on

Fig. 2. Fjord-side features in South Bentinck Arm and locations of data presented in Figs 3, 6, and 8.

time scales of tens to hundreds of years (Bornhold & Prior, this volume). Similarly, very infrequent, high-magnitude snowmelt floods may occur, such as the 1986 Bute Inlet flood during which discharge rose from < 200 m^3/sec to more than 1600 m^3/sec in 24 hours, exceeding the average summer maximum by about 800 m^3/sec. There is also a seasonal discharge pattern. Winter river flows are generally low, and little sediment is contributed to the offshore fans. In summer, fluvial discharges are high because of seasonal rainfall. Additionally, within the rainy and snowmelt seasons, short-term floods involve waxing, peak and waning sediment supply to the fjord sides, on time scales of hours to days. Many of the smaller fjord-side streams and rivers are not monitored, but there is evidence of extraordinarily high sediment loads in events lasting only a few hours (e.g. Lister *et al.*, 1984).

Sediment transport away from the river mouths and across the subaqueous slopes is directly related to fluvial sediment-supply regimes. Because of the variability of sediment input, particularly variations in sediment concentrations, textures, supply rates and durations, there are several different ways in which sediment is distributed offshore. The fan-delta morphologies and sediments contain the signatures of a spectrum of processes, ranging from large boulder transport over the seafloor to slow settling of particles from suspension in the water column. Different types of sediment transport mechanisms may coexist and interrelate, causing superimposition of process signatures on the fan surfaces. Also, the roles of the various processes may change within an individual flood, both temporally and spatially. As sediment-supply rates increase and decline, the relative importance of different transport mechanisms may change at any point on the fan surface. The distance of transport, for example, out across the fan, changes with waxing and waning energy. Nevertheless, the sonar imagery of underwater fan morphology provides considerable insight into the dominant modes of fan construction. Distinctive bottom features result from the principal sediment transport mechanisms, as follows:

Debris avalanching

Some underwater fans exhibit accumulations of very coarse debris composed of large boulders (0·5–5 m across), small subangular rock fragments, cobbles and gravel. The bouldery debris may be arranged in linear block 'trains' or boulder 'streams' radiating outward and downslope from the fan apex. In some very steep systems (> 20°) the bouldery debris dominates fan morphology, with numerous closely spaced systems carrying boulders to downslope fan margins 1·0–1·5 km from the shoreline (Prior & Bornhold, 1988).

In cone-shaped underwater fans lacking coastal deltas the boulder and gravel accumulations are localized. The coarsest bouldery debris covers near-shore, underwater, cone-apex regions and also extends downslope along the outer cone flanks (Fig. 3). By comparison, the central parts of the cones are relatively smooth and are composed of finer sediments. The fan-flank 'streams' radiate outward, and near the outer margins of the fans there are scattered, isolated boulders and local boulder piles.

Fan-deltas that have well-developed subaerial alluvial/deltaic components may also have underwater boulder streams. They are usually found directly seaward of the mouths of the principal river distributaries. Very localized, elongate gravel and boulder lobes extend across the fan surfaces, which otherwise are smooth and featureless. The river-mouth debris is extremely coarse-textured; individual lobes can be very large, but downslope transport distances are relatively small (Fig. 4).

The processes responsible for underwater transport of boulder-rich sediment appear to be high-energy, cohesionless debris avalanching, involving rolling, cascading and collision of large rock fragments over steep (> 20°) underwater slopes. The subaqueous debris avalanches are generated by extremely energetic subaerial debris flows and debris torrents. Large quantities of boulders, gravel, sand, logs and organic mulch are delivered to the coastline during some floods. Where these high-energy flows traverse coastal alluvial fans or deltaic plains, they can cause extensive damage to roads and buildings, and even loss of life (Eisbacher & Clague, 1981;

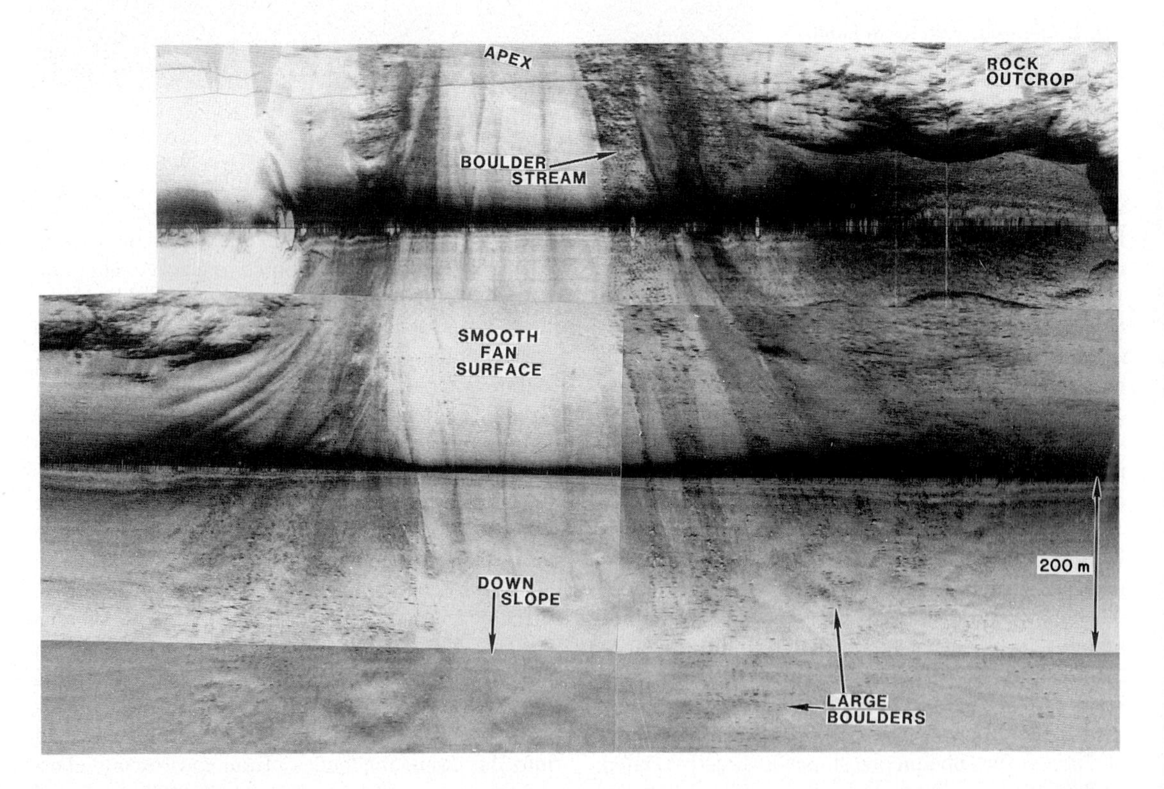

Fig. 3. Side-scan sonar mosaic of boulder streams and scattered boulders on a fan in South Bentinck Arm (location in Fig. 2).

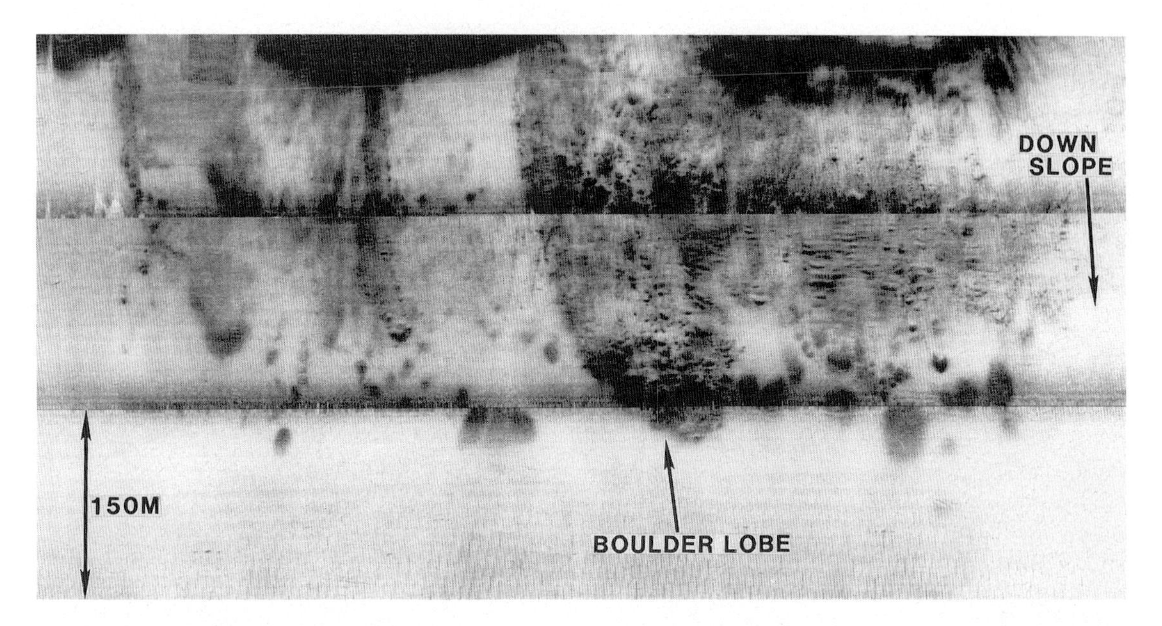

Fig. 4. Side-scan sonar mosaic of boulder lobes near a river-mouth distributary on a sidewall fan delta in Phillips Arm (location in Fig. 1).

Lister *et al.*, 1984). Alternatively, the debris-charged floods discharge directly from high-relief catchments into the fjord sides.

Apparently, the coarse-textured debris reaches the shoreline with sufficient momentum to plunge below the sea surface and start down the underwater slopes. The fast-moving, dense, boulder and gravel flows underflow the seawater and begin to spread and diffuse across the fan apex. Downslope movement of the larger particles by rolling and avalanching follows the steeper underwater fan gradients, particularly along their flanks. Some boulders outrun the main boulder trains, finally coming to rest around the outer fan margins.

Evidence for underwater debris avalanching of coarse-textured sediment has been recognized elsewhere. Postma and Roep (1985) report 'foreset avalanching' from ancient sediments. Progradational deposition of late Pleistocene gravel- and boulder-dominated deltas probably requires this process (e.g. Elfström, 1989).

Inertia flows

In all of the fan deltas surveyed, gravel and coarse sand are transported for long distances from the shoreline. Coarse gravel has been observed from a submersible a distance of 1800 m from a river-mouth source, over an average bottom gradient of 12° (Prior & Bornhold, 1989).

Side-scan sonar data show that the gravel and sand are commonly arranged in divergent and anastomosing patterns away from the river mouths over the steep, cone-shaped fan apexes (Fig. 5). Shallow, linear depressions or 'swales' are separated by low, rounded elongate ridges. The divergent and anastomosing swale and ridge morphology on fan apexes can range in relief from 10 to 20 m (Bornhold & Prior, this volume) to less than 2 m, with swale widths ranging from 10 to 200 m. The swales are smooth-floored, rising gradually onto the neighbouring constructional ridges. Dark tones on the sonographs indicate the presence of gravel and sand in the swales. Submersible observations confirm that swales are usually filled with undulating sheets of sand and gravel. Some cobbles are arranged in slope-parallel lines, with shallow, moat-like scours, around individual clasts (Prior & Bornhold, 1989).

At the downslope ends of the swales, two contrasting morphologies can occur. In some fan systems the swales become shallower and less distinct as they widen downslope (Fig. 5). Individual swales terminate as thin, indistinct splays of sand and fine gravel. Alternatively, the divergent swales may converge, with neighbouring swales merging in a downslope direction into incised, steep-walled chutes

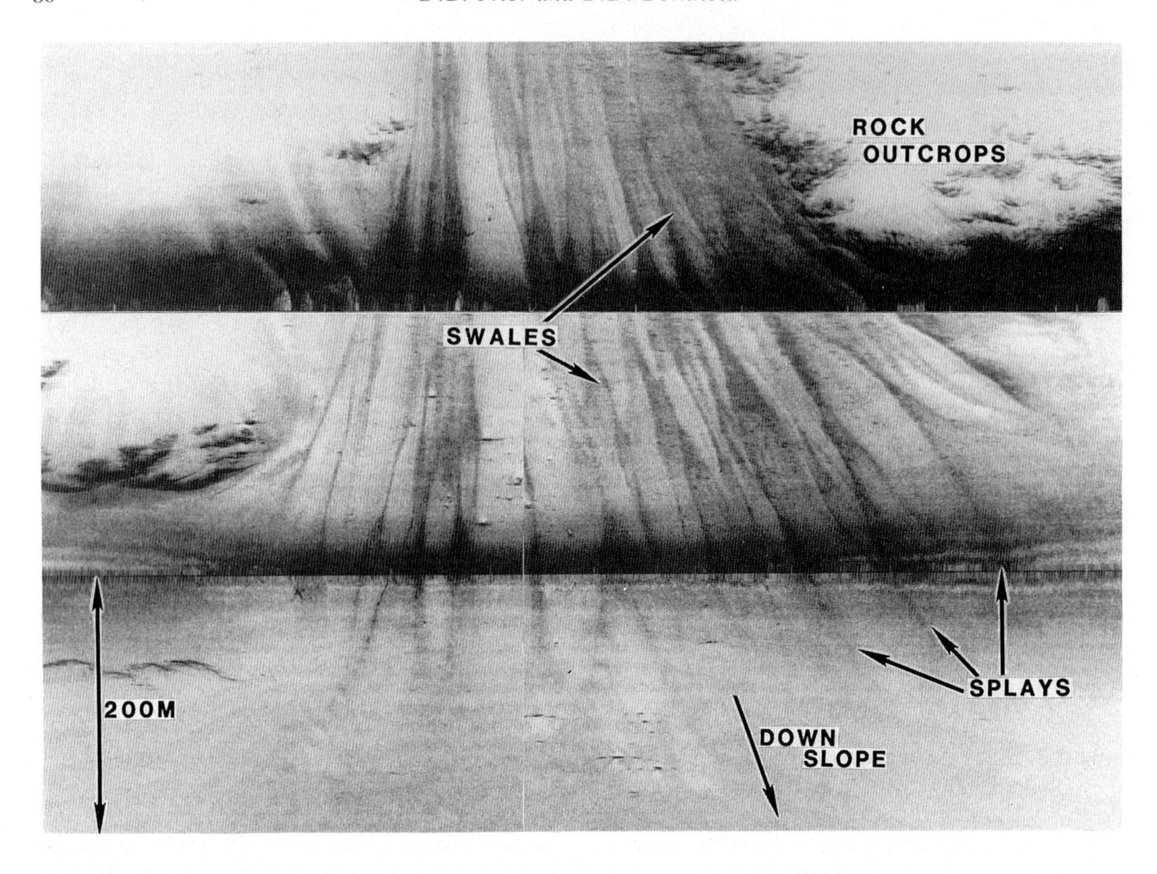

Fig. 5. Side-scan sonar swath of gravel and sand swales on a small fan (South Bentinck Arm).

or erosional conduits (Prior & Bornhold, 1989; Bornhold & Prior, this volume). In this context the sand and gravel swales appear as dendritic tributaries, feeding into selected pathways for continued transport farther downfan.

Large gravel- and sand-filled erosional chutes are commonly found seaward of the main river distributaries on well-developed, subaerial alluvial fans or fan-shaped deltas (Figs 3, 6) (e.g. Prior & Bornhold, 1989). The chutes extend directly from the river mouths as incised linear features, up to 200 m wide for distances of up to 1 km down the otherwise smooth fan surface (Fig. 6). The gravel and sand are arranged as linear 'streams' within the chute floor, as subtle, low-relief undulations, elongate downslope. Such large chutes usually reach the fjord floor at the distal fan margin, where the sand and gravel spread out as low-relief splays.

The gravel and coarse sand in the swales and chutes are introduced to the subaqueous fans by flooding rivers. During high-energy floods the gravel and sand riverborne load appears to have sufficient density and velocity to overcome buoyancy and frictional effects at the river mouth. The river channel may continue across the apex of the subaqueous fan because of flood scouring of the river-mouth bar. The confined, channelized flow thus reduces river-mouth, dispersive energy loss. The near-bed or 'moving-bed' concentrations of sediment are largely unaffected by seawater density and thus begin moving across the subaqueous slope as hyperpycnal flows. Where the bottom gradient of the nearshore fan is greater than the slope of the river thalweg, the offshore transport of gravel and sand is enhanced. River-driven transport is replaced by gravity-driven sediment motion over the underwater slopes.

Gravel/coarse sand transport downslope over the fan appears to be achieved by high-density, pseudo-laminar 'inertia flows' described experimentally by Postma *et al.* (1988). These authors illustrate a mechanism by which coarse particles move over the bottom supported by a combination of dispersive

Fig. 6. Side-scan sonar mosaic of a gravel-filled chute off a fan-delta distributary, South Bentinck Arm (location in Fig. 2).

pressure, hindered settling and enhanced buoyant lift. Additional evidence for such gravel and sand transport processes has recently been provided by Ferentinos *et al.* (1988), who reported subaqueous fan chutes carrying coarse sediment fed by a combination of earthquake-triggered sediment instability and underwater sliding.

Turbidity flows

Medium to fine sand is widely distributed over the underwater fans, and transport distances can be large, even over low-gradient, distal fan areas. For example, sand dispersal has been detected on the outer edge of one fan a distance of 2700 m from the coastline over slopes as low as 1.5° (Prior & Bornhold, 1989). Two quite different sand distribution patterns are recognized: linear sand bodies arranged normal to fan contours, and widespread, thin sand sheets and layers interbedded by silts and muds.

Medium to fine sand occurs within the floors of incised chutes, which radiate outward across the fan surfaces (Figs 7, 8). These chutes range in width from 10 m to 100 m and are generally straight, with subparallel walls cut 3–5 m into the fan surfaces. Sidewalls are erosional, exposing gently dipping layers of sand, silt or mud, which comprise the interchute parts of the fans. Chute walls may be stepped, but there are no chute-flanking levees. Typically the chutes are supplied with sediment from upslope from converging linear swales (Fig. 7). The

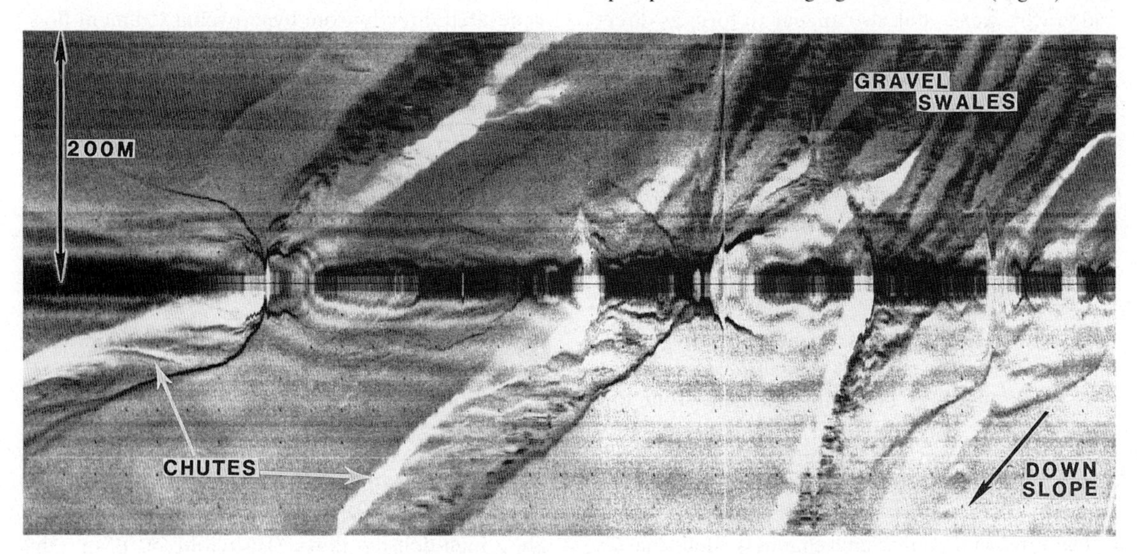

Fig. 7. Side-scan sonar swath of gravel swales leading into incised chutes (Bear Bay, Bute Inlet).

Fig. 8. Side-scan sonar swath of radiating sandy chutes across an upper fan surface, South Bentinck Arm (location in Fig. 2).

chutes terminate near the distal fan margins with locally divergent sand splays. The longitudinal profiles of the chutes are highly irregular. Arcuate steps occur along the chute floors; they are commonly regularly spaced, with concave-downslope plan forms. The greatest relief is towards the chute axes, declining towards the sidewalls. These features, with their characteristic 'heelmark' geometry, are interpreted as erosional flutes (Prior & Bornhold, 1989; Bornhold & Prior, this volume).

Flutes and sand are also found within very large, sinuous channels, which occur on some fan deltas (e.g. Bornhold & Prior, this volume), and seaward of many of the fjord-head systems (Syvitski & Farrow, 1983; Prior *et al.*, 1986, 1987; Prior & Bornhold, 1988). The very large channels are sometimes occupied near their heads by converging sand and gravel swales, but also appear to form as direct seaward extensions of major river distributary channels. The subaqueous channels lead downslope from the river mouths and are incised deeply into the underwater slopes. An individual channel may continue as a highly sinuous, sand-transporting system eroded into low-gradient fjord floors for distances of up to 30 km (Prior *et al.*, 1986).

Sand is also generally spread as thin sheets over the interchute/interchannel areas. The thin sand layers, set within sequences of muds and silts, may be massive and structureless, but usually are graded, fining upward (Fig. 9). The fan-surface sands can also be arranged as bedforms, ranging from low ripples (Prior & Bornhold, 1989) to antidunes (Bornhold & Prior, this volume), which are commonly associated with scattered, isolated flutes (Fig. 10).

The erosional chutes and channels, flute marks, ripples/dunes and sand distributions are all the result of downfan turbidity currents. Active turbidity flow has been monitored within one fjord-head channel system (Prior *et al.*, 1987), and there is growing evidence that turbidity flows occur frequently in British Columbia fjords (e.g. Hay *et al.*, 1982; Syvitski & Farrow, 1983; Syvitski *et al.*, 1988). For example, between October 1988 and March 1989 a turbidity-event detector moored just above the seafloor in Bute Inlet recorded 18 discrete events.

Turbidity currents are generally believed to be due to local, slope-instability processes on some of the fjord-head, delta-front slopes (e.g. Prior *et al.*, 1987; Syvitski *et al.*, 1988). However, on many sidewall fan deltas there is little evidence for present-day instability processes (e.g. Prior & Bornhold, 1989; Bornhold & Prior, this volume). Rather, the bottom features strongly suggest that turbidity currents are generated directly from hyperpycnal sediment flows introduced by the rivers during floods. Turbidity currents could accompany inertia flows, as documented by Postma *et al.* (1988). The direct supply of inertia-flow sediments into the chutes suggests that inertia flows and turbidity currents may coexist. Alternatively, an individual flood may carry sufficient sediment for suspended-sediment turbulent flows continuing downslope from the river mouths but without the necessity for inertia flows. Conceivably turbidity currents may precede inertia flows during waxing floods and outlast them during waning conditions.

Turbidity-current erosion of chutes and transport of sands along them is probably due to high-density flows. Flutes are also consistent with high-energy turbulent conditions. Similarly, scattered flutes and ripples over fan surfaces suggest widespread, sheet-like, high-density flows (Bornhold & Prior, this volume). However, fine sand layers within bedded

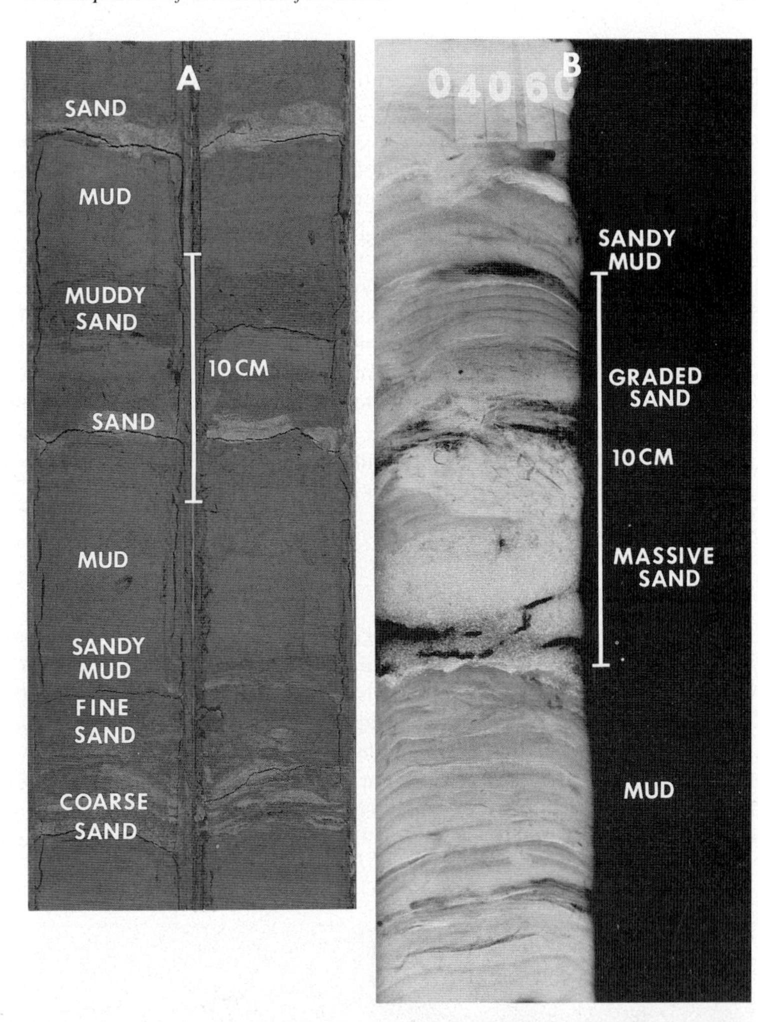

Fig. 9. Examples of piston cores from the lower slopes of Bear Bay fan in Bute Inlet; A — core photographic; B — core X-ray radiograph.

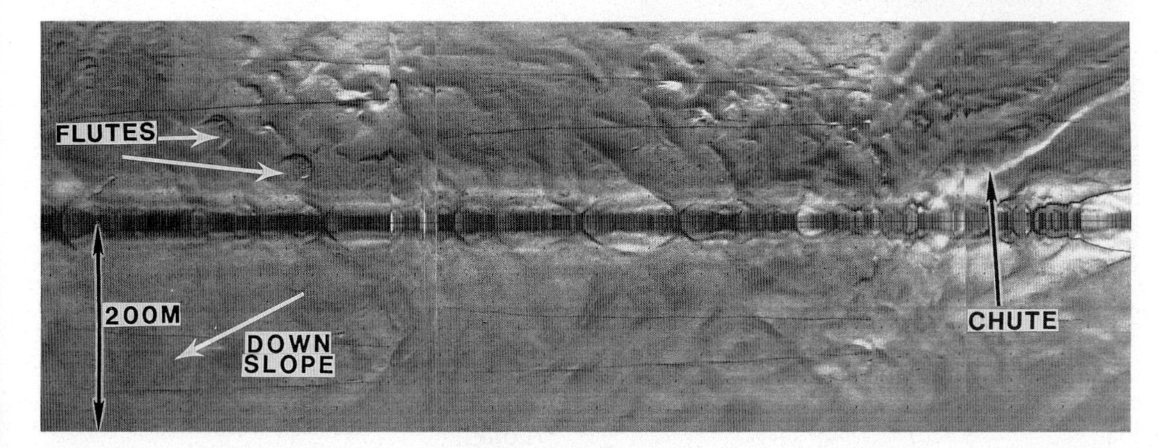

Fig. 10. Side-scan sonar swath showing the irregular surface of the Noeick River fan delta (South Bentinck Arm) with chutes and scattered isolated flutes (see also, Bornhold & Prior, this volume).

mud sequences in interchute areas suggest that low-energy turbidity flows may cover entire fans in a manner similar to the blanketing effects of turbidity flows over fjord floors (e.g. Hay *et al.*, 1982).

Fan-delta sediment transport by turbidity currents has recently been reported by Ferentinos *et al.* (1988). Turbidity currents are responsible for turbidite sequences at the mouths of fan-delta chutes and damage to offshore cables. Ferentinos *et al.* (1988) conclude that 'during the winter density currents can be initiated at the mouths of rivers'. Turbidity-current channels seaward of modern, high-energy deltas have also been documented by Hoskin and Burrell (1972) and Weirich (1989), while similar processes are believed to be responsible for channelized, graded, coarse-gravel sequences in ancient fan-delta deposits (e. g. Surlyk, 1984).

Slope instability

Large-scale landslide features are not common on the subaqueous slopes of the gravel- and sand-dominated, sidewall, fan deltas in British Columbia. By comparison, rotational and translational sliding is known from several fjord-head delta — from slopes composed of clays and silts. For example, at Kitimat there is a history of repeated delta-front failure, and the latest debris flow, in 1975, extended downfjord a distance of 5 km over bottom slopes of 0.4–0.5° (Prior *et al.*, 1984; Johns *et al.*, 1986). Similarly, other fine-grained, fjord-head deltas, such as in Bute Inlet, are heavily dissected and large segments have been displaced or removed by underwater sliding (Syvitski & Farrow, 1983; Prior *et al.*, 1986).

On sidewall fan deltas composed of coarse-grained sediments, slope failure appears to be a rather local occurrence. A failure event is known from Woodfibre, Howe Sound, where a wood-treatment factory and its wharfs were built on the shoreline of a small gravelly fan delta. In 1955, just after a very low tide, parts of the factory and docks were destroyed by an underwater slide 10 m deep, which extended downslope to water depths of 200–210 m over bottom slopes of 27–28° (Terzaghi, 1956; Prior *et al.*, 1981b).

Elsewhere the evidence from bottom morphology is difficult to interpret, and some features could also be explained by some of the other high-energy, fan-surface processes. For example, on the flanks of some steep-gradient fans (> 15°) there are short troughs, which begin abruptly on the fan surfaces as bowl-shaped source regions, sometimes arranged as *en echelon* depressions. The troughs generally widen downslope, becoming shallower, terminating in low-relief lobes of debris (Fig. 11). This source bowl, trough and lobes morphology has been recognized

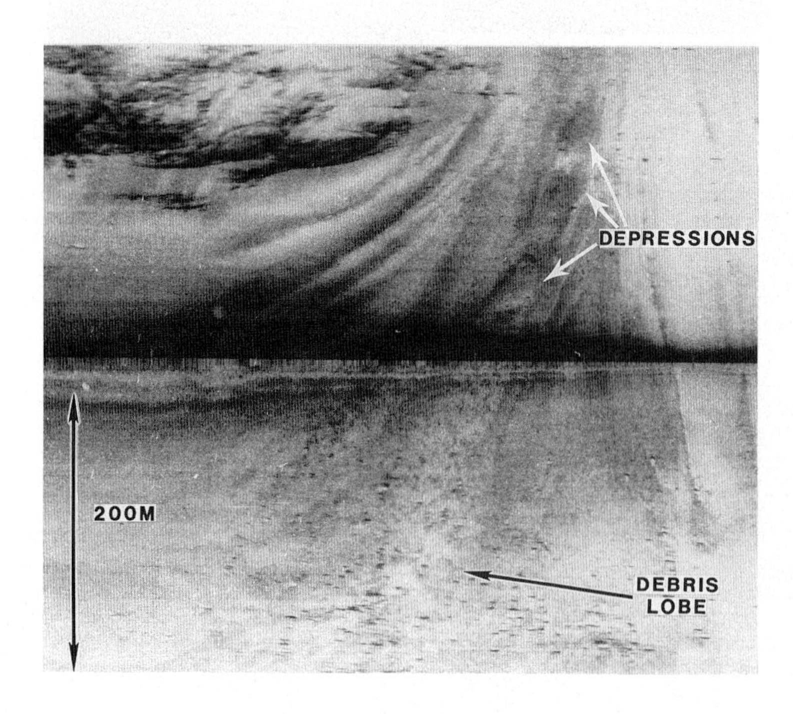

Fig. 11. Side-scan sonar swath showing a shallow trough, source bowls and debris lobe resulting from fan-slope instability (South Bentinck Arm).

as typical of instability processes in both modern and ancient depositional settings (e.g. Prior *et al.*, 1981a; Postma, 1984).

However, there are some similarities between landslide troughs and the chutes eroded by turbidity currents. The latter are usually long and narrow, lie downslope from convergent gravelly swales, contain flutes and end downslope with divergent thin, sandy splays (e.g. Prior & Bornhold, 1989). There are also some similarities between arcuate depressions caused by successive retrogressional sliding and flutes (e.g. Normark *et al.*, 1979). But instability processes usually result in an association of arcuate depressions and downslope depositional lobes on moderate slopes. Flutes are most commonly open, empty depressions with almost total removal of eroded debris on a wide range of bottom gradients, including flat seafloor.

Instability-induced arcuate scarps are widespread on the lower slopes of the Britannia Beach fan delta (Howe Sound) (Prior & Bornhold, 1986). The lower fan, composed primarily of silts and muds with an average gradient of 7–8°, has numerous sinuous and arcuate seafloor scarps bounding detached slabs of sediment. Further downslope there is a roughly lobate area with scattered blocks of debris and discontinuous pressure ridges. Retrogressive failure of silty mud sediments of the lower fan appears to be related to loading by sand and gravel from upslope during fan progradation.

Settling from plumes

All of the fan deltas contain fine-grained sand, silt and clay. These sediments are generally dispersed widely over all parts of the fans and over the neighbouring fjord floors, where clay and silt concentrations usually average about 30–40% and 60–70%, respectively. Clay and silt are generally absent from those parts of the fans where there are active boulder avalanching and inertia flows. However, after each particular event silt and clay will begin to accumulate as a draping blanket over the coarser-grained sediments. The smooth, featureless, fan seafloor between the boulder streams and gravel/sand-filled swales and chutes is composed of sequences of clay and silt, with sand and muddy sand (Fig. 9). These clay and silt sequences are interrupted by sand introduced by turbidity flows. Those fans that have large, well-developed, subaerial, alluvial or deltaic plains are characterized underwater by extensive featureless areas composed of downfan-dipping silt and mud beds. The thickest bedded sequences are on the lower fan slopes but thin distally to the fan margins.

The fine-sediment fractions in fan deltas are mainly deposited by buoyant, hypopycnal plumes. They are an important mechanism in very widespread sediment dispersal off river mouths (e.g. Wright *et al.*, 1988) and are particularly important in fjord sedimentation (e.g. Syvitski *et al.*, 1988).

LONG-TERM FAN-DELTA DEVELOPMENT

The different types of distinctive fan-delta morphology associated with various underwater sediment-dispersal processes in fjord sidewall systems are summarized schematically in Figs 12, 13, 14 and 15. Each individual fan delta displays its own blend of processes and associated features. Total fan architecture reflects the contributions made by each process in different parts of the fan over different time scales, ranging from long-term fan evolution to the impacts of individual dispersal events.

There is a very large range of possible form/process combinations, depending particularly upon the type of sediment input, ranging from, at one extreme, fans dominated by boulders and gravel to those receiving only fine sand, silt and clay. Despite the obvious complexity, it is possible to compare the evidence from many different fans. General patterns of long-term fan-delta development and evolution can be examined in relation to two principal controls: changes in sediment supply and changes in subaqueous relief and fan gradient.

Long-term sediment supply

Long-term sediment supply to underwater fans over the past 10 000–12 000 years began with very high-energy meltwater floods accompanying deglaciation of fjord-side drainage basins. Medium-penetration seismic profiles show chaotic mounded sediments at the inner bases of the fans, indicating that the fans began as fjord-wall bouldery debris deposited by avalanching (Prior & Bornhold, 1989).

Since late glacial times there have undoubtedly been changes in the drainage basins that affect sediment supply, particularly weathering rates, relief changes, natural afforestation and industrial deforestation. A major control on the arrival of sediment

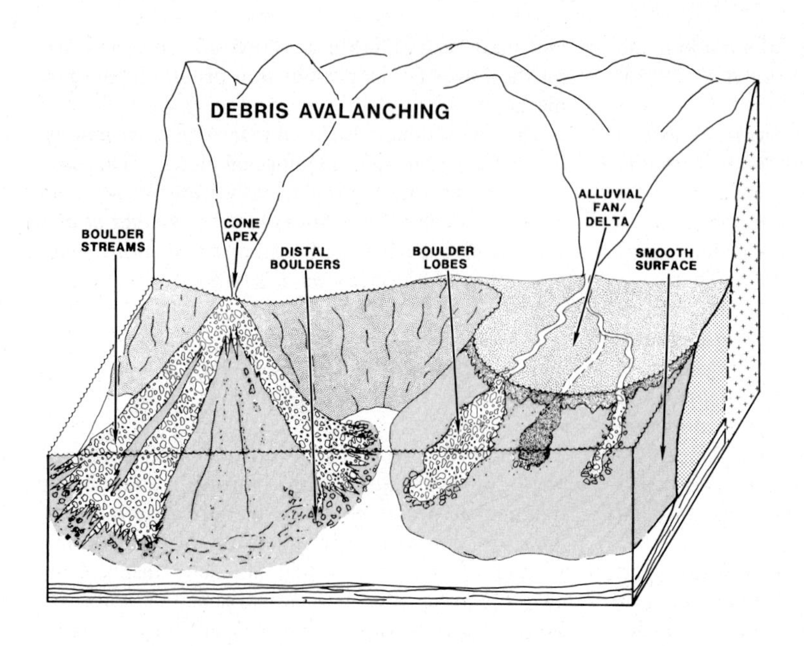

Fig. 12. Schematic three-dimensional perspective of underwater-debris avalanche morphology on a cone-shaped fan and seaward of an alluvial fan/ deltaic plain.

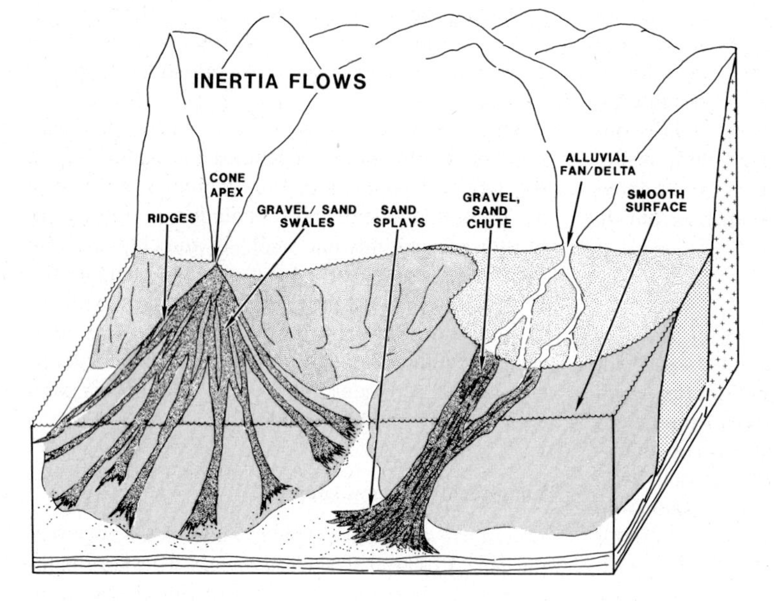

Fig. 13. Schematic, three-dimensional perspective of underwater inertia-flow morphology on a cone-shaped fan and seaward of an alluvial fan/deltaic plain.

at the coastline appears to be the growth of subaerial alluvial fans and deltaic plains. In order for these features to develop, sediment is retained at the shoreline and in the lower reaches of the rivers. Alluvial-fan growth means a decline in the supply of coarser sediment to the offshore fan. Further, alluvial and deltaic features often develop numerous distributary channels, subdividing the total river load and reducing the amounts reaching the shoreline at any single point. Channels also change location, altering the position of each multiple-sediment entry point.

Subaqueous relief and fan gradients

The subaqueous fjord-side relief (between shoreline and fjord floor) was initially large as deglaciation took place, and fluctuated temporarily because of

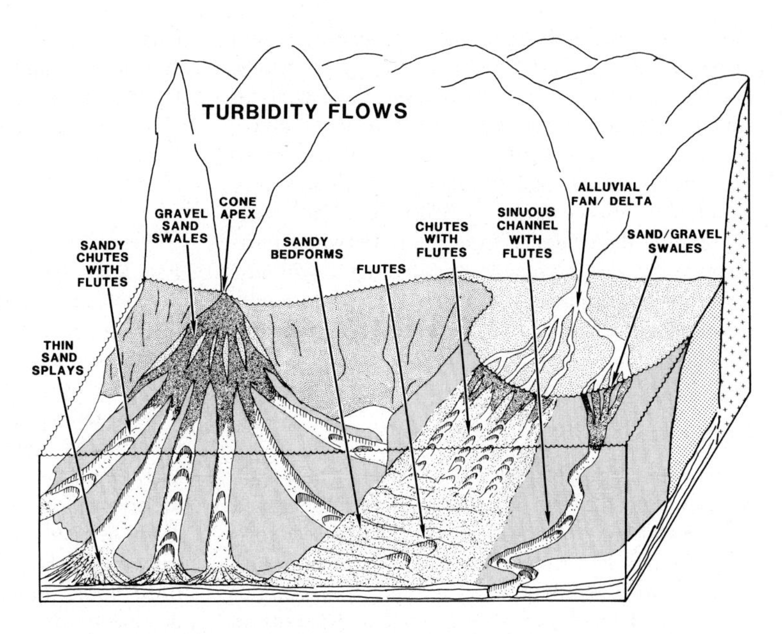

Fig. 14. Schematic, three-dimensional perspective of underwater turbidity-flow morphology of a cone-shaped fan and seaward of an alluvial fan/deltaic plain.

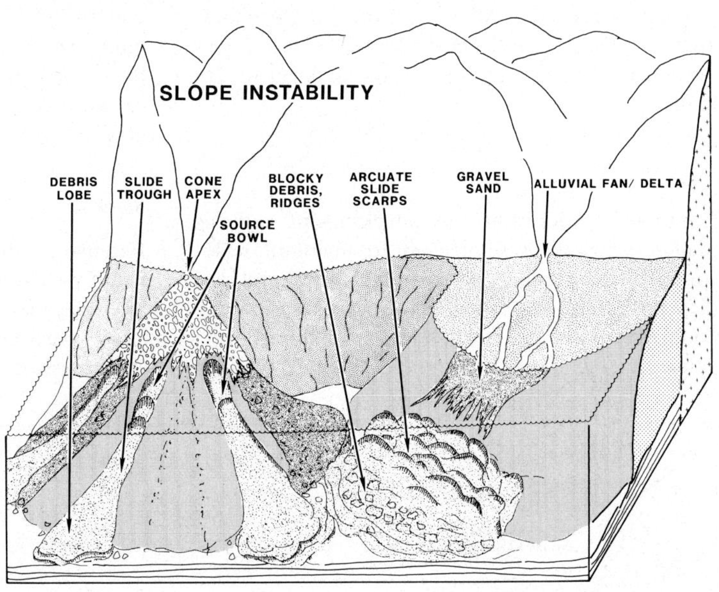

Fig. 15. Schematic, three-dimensional perspective of underwater slope-instability morphology on a cone-shaped fan and seaward of an alluvial fan/deltaic plain.

isostatic and eustatic effects. Since about 8000–9000 years ago, prisms of sediment, emplaced subaqueously, have gradually grown by fan aggradation and distal progradation, accompanied by fjord-wide bottom aggradation. The distal fan- and fjord-bottom sediments interdigitate. Most of the fjord sidewall depositional systems have built large cones of sediment with their apexes close to the shoreline, while others have aggraded sufficiently to form subaerial alluvial fans or deltaic plains of various sizes. Fjord-bottom sedimentation reduces total available offshore relief, while fan aggradation and progradation result in overall decline of underwater fan gradients. Reduced relief and reduced average slopes mean smaller gravitational stresses and a major control on avalanching, inertia and turbidity flows.

Together, long-term reduction in average sediment supply and reduction in gravitational stress mean an overall tendency for higher-energy sediment dispersal to be replaced by lower-energy processes. Sedimentation up and out from the fjord walls as subaqueous cone-shaped fans, aggradation above sea level to construct true 'fan deltas' and continued growth subaerially and subaqueously involve progressively lower-energy subaqueous processes. Combining the evidence from the underwater parts of many fans, it is possible to construct a synthetic evolutionary sequence in which several fan-delta development stages can be conceptualized (Fig. 16), as follows:

Stage 1

Under conditions of high offshore relief and sediment supply of very coarse boulders and gravel, underwater fan growth will begin primarily by debris avalanching. Fan geometry from the river-mouth point sources is dominated initially by boulder streams and blocky underwater talus with bottom gradients exceeding 20° (e.g. Prior & Bornhold, 1988). As the fan grows and gradients decline, debris avalanching occurs primarily on the steeper fan flanks.

Stage 2

Continued underwater fan development reduces gradients below those needed to maintain widespread avalanching. Coarse gravel and sand are dispersed from the fan apex outwards across the underwater cone by inertia flows, over slopes averaging 12° (e.g. Prior & Bornhold, 1989). Turbidity currents accompany inertia flows, with vertical flow stratification and downslope segregation as observed by Postma *et al.* (1988). Turbidity currents erode chutes, which transport sand downslope to the distal fan and fjord floor (Prior & Bornhold, 1989).

Stage 3

Fan aggradation results in subaerial alluvial fan/deltaic development, accompanied by progressive filtering out of the coarser sediment fractions reaching the shoreline. Underwater slope gradients average 4–5° (e.g. Bornhold & Prior, this volume). Coarse sediment, transported mainly by gravel and sand inertia flows and infrequently by avalanching, is confined to the upper subaqueous slopes immediately seaward of the major active distributaries. Turbidity currents from river-mouth underflows locally erode chutes and channels carrying sand to the fjord floor. Subaqueous fan gradients continue to decline by distal progradation. Large areas of the underwater fans between chutes are blanketed with muds distributed by surface plumes. Upper delta-front sand progradation and loading of the muds cause slope instability and sliding, even though fan gradients are moderate.

Stage 4

Extensive subaerial, alluvial fan or deltaic plains result in many shifting distributaries functioning as multiple, temporary point sources of sediment. The size of particles, sediment volumes and thus suspension densities reaching the distributary mouths are

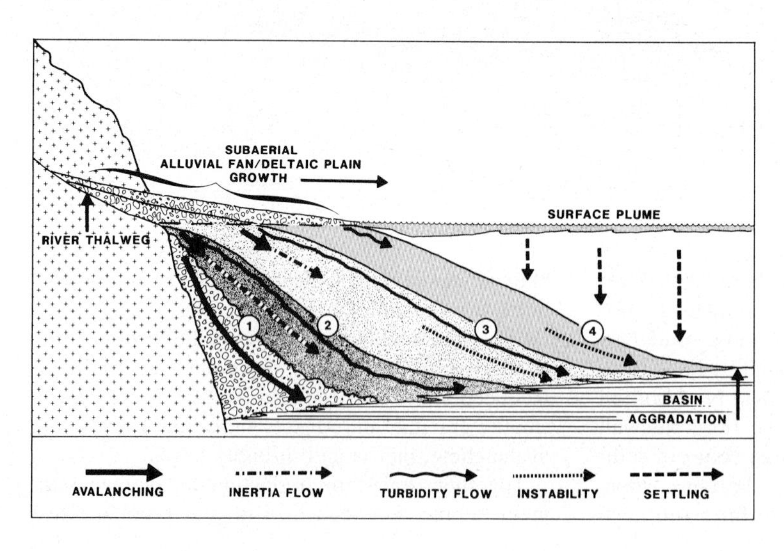

Fig. 16. Conceptual fan-delta development sequence reflecting reductions in sediment supply to the underwater slopes as the subaerial fan component grows and offshore relief declines.

considerably reduced. Deltaic progradation is principally by distributary mouth-bar growth or limited-distance turbid flows of fine sands over the upper subaqueous slopes. The prodelta slopes are mantled by thick sequences of bedded muds deposited from surface plumes. Progradational loading can cause sliding.

SUMMARY AND CONCLUSIONS

1 Comparisons of fan-delta morphology and sediment patterns on developing Holocene fan deltas show that there are five principal modes of submarine sediment dispersal. Each fan preserves its own combination of process signatures, resulting in a great variety of fan-delta architecture.

2 Boulder- and gravel-fed fans develop by debris avalanching; gravel and sand are dispersed downfan by inertia flows and turbidity currents. Low-energy dispersal of silt and clay over entire fans and neighbouring fjord floors occurs from settling from buoyant plumes. Fan-delta slope instability is apparently not common on coarse-textured fans, but may be difficult to recognize morphologically. Fan slopes in fine sands and muds preserve slide scars and debris lobes resulting from local instability.

3 Sediment dispersal underwater is directly related to supply by rivers. Fluvial regimes vary widely, involving large contrasts in magnitudes and frequencies of sediment input to the fjords. There are low-frequency, very high-magnitude floods, seasonal fluctuations and short-term fluctuations in sediment supply within individual floods.

4 River floods bring sediment to the fan in sufficient concentrations and densities to underflow seawater, and continue downslope, particularly as inertia and turbidity flows.

5 Long-term fan development reflects a wide range of possible process combinations, but changes in sediment supply and underwater fan gradient appear to be major controls. The growth of the subaerial component of the systems reduces the supply of coarser sediment to the shoreline. Multiple distributaries subdivide the total load, reducing the amounts arriving at delivery points to the subaqueous slopes. Offshore fan aggradation combined with fjord-bottom sedimentation progressively reduces relief and slope inclinations.

6 The evidence from many developing fan deltas provides the basis for a conceptual four-stage development sequence in which the relative roles of fan processes evolve in response to changes in sediment supply and relief.

ACKNOWLEDGEMENTS

This research has been supported by the Geological Survey of Canada and by the National Science Foundation, Grant DPP-850118. The master and crew of the *Vector* are recognized for their important role in data acquisition. Ivan Frydecky, Bill Hill, Kim Conway, Bertrand Blaise, Gail Jewsbury, Greg Liebzeit and Floyd DeMers provided invaluable field technical support. We are very grateful to Ralph Currie, George Postma, Wojtek Nemec, Mike Leeder and Albina Colella for enthusiastic discussions about modern fan deltas. The manuscript also benefited greatly from the constructive reviews of Geoff Corner and Brian McCann. This report is Contribution Number 39389 of the Geological Survey of Canada.

REFERENCES

Clague, J.J. (1981) Late Quaternary geology and geochronology of British Columbia, Part II, Summary and discussion of radiocarbon dated Quaternary history. *Paper Geol. Surv. Can. 80–35,* 41 pp.

Corner, G.D., Nordhal, E., Munch-Ellingsen, K. & Robertsen, K.R. (1988) Sedimentology of the Holocene Alta delta, northern Norway. *Abstr. Int. Works. Fan Deltas*, Calabria.

Eisbacher, G.H. & Clague, J.J. (1981) Urban landslides in the vicinity of Vancouver, British Columbia, with special reference to the December 1979 rainstorm. *Can. geotech. J.* **18**, 205–216.

Elfström, A. (1989) The upper Pite River valley. UNGI Rapport Nr. 70, Uppsala University Naturgeografiska Institutionen.

Ferentinos, G., Papatheodorou, G. & Collins, M.B. (1988) Sediment transport processes on an active submarine fault escarpment, Gulf of Corinth, Greece. *Mar. Geol.* **83**, 43–61.

Hay, A.E., Burling, R.W. & Murray, J.W. (1982) Remote acoustic detection of a turbidity current surge. *Science* **217**, 833–835.

Hoskin, C.M. & Burrell, D.C. (1972) Sediment transport and accumulation in a fjord basin, Glacier Bay, Alaska. *J. Geol.* **80**, 539–551.

Johns, M.W., Prior, D.B., Bornhold, B.D., Coleman, J.M. & Bryant, W.R. (1986) Geotechnical aspects of a submarine landslide. *Mar. Geotechn.* **6** (3), 243–279.

Kostaschuk, R.A. & McCann, S.B. (1987) Subaqueous morphology and slope processes in a fjord delta, Bella Coola, British Columbia. *Can. J. Earth Sci.* **24**, 52–59.

Lister, D.R., Kerr, J.W.G., Morgan, G.C. & Vandine, D.F. (1984) Debris torrents along Howe Sound, British

Columbia. *IV Int. Symp. on Landslides, Toronto* **1**, 649–654.

Normark, W.R., Piper, D.J.W. & Hess, G.R. (1979) Distributary channels, sand lobes and mesotopography of Navy submarine fan, California Borderland, with application to ancient fan sediments. *Sedimentology* **26**, 749–774.

Postma, G. (1984) Slumps and their deposits in fan-delta front and slope. *Geology* **12**, 27–30.

Postma, G., Nemec, W. & Kleinspehn, K.L. (1988) Large floating clasts in turbidites: a mechanism for their emplacement. *Sedim. Geol.* **58**, 47–61.

Postma, G. & Roep, R.B. (1985) Resedimented conglomerates in the bottomsets of Gilbert-type gravel deltas. *J. sedim. Petrol.* **55**, 874–885.

Prior, D.B. & Bornhold, B.D. (1986) Sediment transport on subaqueous fan-delta slopes, Britannia Beach, British Columbia. *Geo-Mar. Lett.* **58**, 57–61.

Prior, D.B. & Bornhold, B.D. (1988) Submarine morphology and processes of fjord fan deltas and related high-gradient systems: modern examples from British Columbia. In: *Fan Deltas: Sedimentology and Tectonic Settings* (Ed. by W. Nemec and R.J. Steel), pp. 125–143. Blackie and Son, London.

Prior, D.B. & Bornhold, B.D. (1989) Submarine sedimentation on a developing Holocene fan delta. *Sedimentology* **36**, 1053–1076.

Prior, D.B., Bornhold, B.D. & Johns, M.W. (1984) Depositional characteristics of a submarine debris flow. *J. Geol.* **92**, 707–727.

Prior, D.B., Bornhold, B.D. & Johns, M.W. (1986) Active sand transport along a fjord-bottom channel, Bute Inlet, British Columbia. *Geology* **14**, 581–584.

Prior, D.B., Bornhold, B.D., Wiseman, W.J., Jr. & Lowe, D.R. (1987) Turbidity current activity in a British Columbia fjord. *Science* **237**, 1330–1333.

Prior, D.B., Wiseman, W.J., Jr. & Bryant, W.R. (1981a) Submarine chutes on the slopes of fjord deltas. *Nature* **290**, 326–328.

Prior, D.B., Wiseman, W.J. & Gilbert, R. (1981b) Submarine slope processes on a fan delta, Howe Sound, British Columbia. *Geo-Mar. Lett.* **1**, 85–90.

Surlyk, F. (1984) Fan-delta to submarine-fan conglomerates of the Volgian–Velanginian Wollaston Forland Group, East Greenland. In: *Sedimentology of Gravels and Conglomerates* (Ed. by E.H. Koster and R.J. Steel) Mem. Can. Soc. Petrol. Geol. 10, pp. 359–382.

Syvitski, J.P. & Farrow, G.E. (1983) Structures and processes in Bay Head deltas: Knight and Bute Inlets, British Columbia. *Sedim. Geol.* **30**, 217–244.

Syvitski, J.P.M. & Farrow, G.E. (1989) Fjord sedimentation as an analogue for small hydrocarbon-bearing fan deltas. In: *Deltas: Sites and Traps for Fossil Fuels* (Ed. by M.K.G. Whateley and K.T. Pickering), pp. 21–44. Blackwell Scientific Publications, Oxford.

Syvitski, J.P.M., Smith, J.N., Calabrese, E.A. & Boudreau, B.P. (1988) Basin sedimentation and the growth of prograding deltas. *J. geophys. Res.* **3**(C-6), 6895–7908.

Terzaghi, K. (1956) Varieties of submarine slope failures. *Proc. Texas Soil Mechan. and Found. Engr. Conf., 8th, Austin*, 1–41.

Weirich, F.H. (1989) The generation of turbidity currents by subaerial debris flows, California. *Bull. geol. Soc. Amer.* **101**, 278–291.

Wright, L.D., Wiseman, W.J., Bornhold, B.D., Prior, D.B., Suhayda, J.N., Keller, G.H., Yang, Z.-S. & Fan, Y.B. (1988) Marine dispersal and deposition of Yellow River silts by gravity-driven underflows. *Nature* **332**(6164), 679–682.

Spec. Publs int. Ass. Sediment. (1990) **10**, 91–111

Fan-delta facies associations in late Neogene and Quaternary basins of southeastern Spain

C. J. DABRIO

Departamento de Estratigrafía, Facultad de Geológicas, Universidad Complutense, 28040-Madrid, Spain

ABSTRACT

Fan deltas occur in a wide range of tectonic and sedimentary settings that are likely to generate varied associations of facies. This is the case in some late Neogene basins of the Betic Cordillera (southeastern Spain) which are described in terms of increasing subsidence. A better understanding of the geodynamics of basins can be obtained from a careful study of the facies associations found in these examples.

In the Cope Basin, most of the coarse sediment of the delta front was captured in beaches of reflective type. Pleistocene changes of sea level coupled with low rates of subsidence produced offlapping coastal units on a scale of a few metres. In the Almanzora River Basin, corals and algae repeatedly colonized the shallow delta fronts forming patch reefs despite the large amounts of conglomerate and micaceous sandstone supplied by fan-delta processes. Megasequences on a decameter scale were generated due to lateral shifting of the active delta lobes, under an assumed tectonic control. In the Carrascoy range (Murcia Basin), lateral displacement of depocentres of coarse sediments generated both coarsening and fining-upwards megasequences tens to hundreds of metres thick under prominent synsedimentary tectonic activity. Progradation of channel-dominated, fan-delta lobes generated lenticular units of kilometric radius that related laterally to active slopes and submarine fans. In the Sorbas (Tabernas) Basin, tectonics was the most prominent control on sedimentation, masking the effects of sea-level changes and the lateral migration of fan deltas along the hundreds to thousand metres thick filling of the basin. Large-scale sliding off the steep basin margin produced huge megabreccia beds that incorporated a mixture of debris from various sources.

INTRODUCTION

Holmes (1965) defined a fan delta as an alluvial fan prograding directly into a standing body of water from an adjacent highland. A relatively large number of papers dealing with fan deltas have been published after the discovery of hydrocarbons in sediments of fan-delta-related origin, creating some confusion about the precise meaning of the term. McPherson *et al.* (1987) used the term fan delta in the original Holmes' (1965) way, introducing a new name (braid deltas) for gravel-rich deltas that form where a braided fluvial system progrades into a standing body of water. Braid deltas have no necessary relationship with alluvial fans. Nemec and Steel proposed a revised terminology where a fan delta is 'a coastal prism of sediments delivered by an alluvial-fan system and deposited, mainly or entirely subaqueously, at the interface between the active fan and a standing body of water. Fan deltas represent interaction between heavily sediment-laden alluvial-fan systems and marine or lacustrine processes' (Nemec & Steel, 1988). This is the use adopted in this paper.

Fan-delta facies associations are likely to record very precisely the geological events that happened along the basin margin and especially the reciprocal effects of tectonics, climate, sea-level changes, variations of alluvial sediment input and marine or lacustrine processes. As fan deltas occur in a wide range of tectonic and sedimentary contexts, it is assumed that they can produce a varied mosaic of sedimentary facies which are still poorly known. More information on these relationships is needed for future palaeogeographical reconstructions. The aim of this synthesis is to extend the knowledge about fan-delta sequences and facies patterns from the compilation of many data on fan deltas in several of the late Neogene basins of the Betic Cordillera (Spain). The examples are described in order of inferred increasing subsidence.

GEOLOGICAL SETTING

The late Neogene and Quaternary sedimentation in the eastern Betic Cordillera (Fig. 1) took place in a complex tectonic and palaeogeographic framework related to the Alpine Orogeny of large islands surrounded by interconnected depressions (Montenat, 1977). Most of these islands correspond to present-day mountainous ranges (sierras), but some others disappeared following tectonic subsidence.

Earlier tectonic interpretations assumed that the geodynamic history of these basins included an extensional stage (Tortonian–early Pleistocene), directed E–W and NE–SW with large vertical movements, and a compressional stage (early Pleistocene–present) along a N–S direction during which some of the former normal faults underwent strike-slip deformation (Bousquet & Philip, 1976; Bousquet *et al.*, 1976; Baena *et al.*, 1982; Pineda *et al.*, 1983a,b; Bordet, 1985). In recent times the role of horizontal movements directed N–S, SE–NW, NW–SE and E–W has been widely recognized, assuming major dextral strike-slip faults (Bernini *et al.*, 1983; Sanz de Galdeano *et al.*, 1984; Ott d'Estevou, 1980; Sanz de Galdeano *et al.*, 1986). Montenat *et al.* (1987) interpreted these basins as located inside a wide left-lateral shear zone trending NE–SW. The system, mainly inherited from previous structural stages, was affected by N–S compression inducing a slight perpendicular extension. During the late Neogene, the stress field rotated with a direction of regional shortening shifting from NW–SE (Tortonian) to N–S (late Tortonian–Pliocene) and again NW–SE (late Pliocene–Holocene). These rotations had a notable effect upon the kinematics of faulting and consequently on the geometry and sedimentary filling of the basins. Two types of basins evolved simultaneously: wrench furrows and grabens. The marine and terrestrial sedimentation of the Neogene to Quaternary basins in southeast Spain took place in this complex pattern of partly interconnected basins.

The substratum of the basins and the adjacent mountain ranges (sierras) that acted as source rocks is made up of metamorphic and metasedimentary rocks (micaschists, phyllites, dolostones and quartzites) of the Internal Zone of the Betic Cordillera (Fig. 1). Erosion of these rocks supplied the sediments to the fan deltas which partly infilled the basins. Textural and compositional properties are inherited from the schistose rocks forming the adjacent mountains: schistosity strongly conditioned the flat, tabular shape of pebbles whereas the high content of mica in source areas is likely to have influenced the rheology of the masses of removed detritus.

SEDIMENTARY FACIES

A large spectrum of sedimentary facies have been distinguished in late Neogene fan deltas of southern Spain. In this paper, Dabrio & Polo's (1988) informal terminology will be followed, with the same process interpretations. A synoptic list is given:

C-1. Conglomeratic channel-fills, marine fauna. Submarine channels.

C-2. Conglomeratic layers with marine fauna: (a) wave-produced lag deposits after fan-derived gravels; (b) reworking of shallow-marine (shoreface) sediments.

C-3. Parallel-stratified clast-supported conglomerates, imbricated clasts, marine fauna. Coastal to sublittoral deposits.

C-4. Unstratified matrix-supported conglomerates. Subaqueous debris-flow deposits.

S-1. Parallel-stratified coarse sandstones to fine pebble conglomerates, marine fauna. Reworking of fan-derived sediments by waves and/or currents in shallow-marine and shoreface zones of fan deltas.

S-2. Parallel-laminated fine-grained micaceous sandstones to siltstones. Suspensional settling of fines on shallow-marine fan-delta slopes, near or below fair-weather wave-base.

S-3. Parallel-laminated cross-bedded and wave-ripple cross-laminated sandstones to micaceous siltstones. Lower foreshore to transition zones of shallow-marine fan deltas.

M. Massive red sandy mudstone, scattered clasts. Settling of fines after flooding.

COPE BASIN (MURCIA PROVINCE)

General

Cope Basin (Fig. 2) was formed during the late Miocene along fractures directed N 60° E and N 120° E (Larouziére *et al.*, 1987). Fan deltas and shallow-marine sediments filled the basin during early and

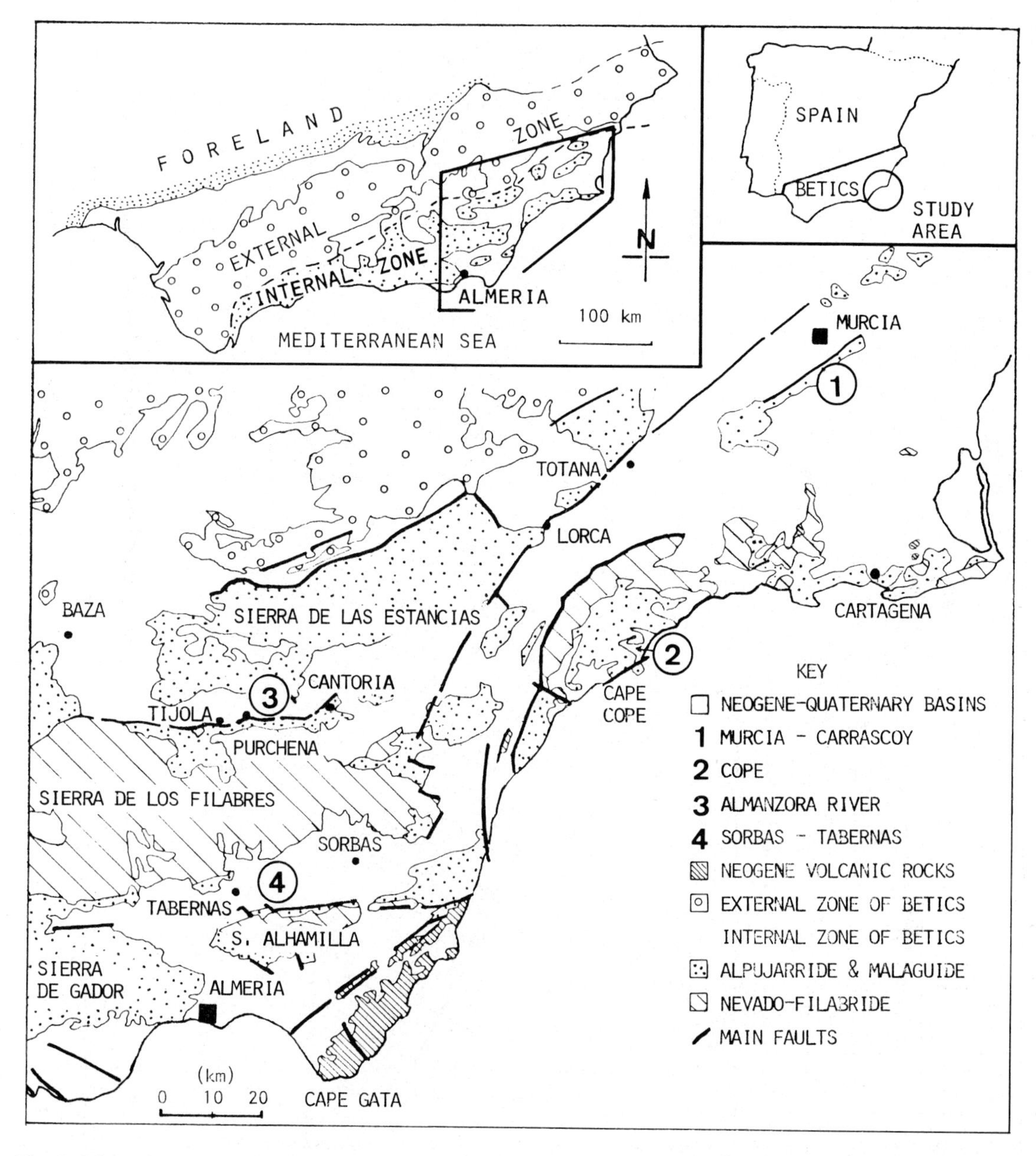

Fig. 1. Schematic geologic map of the eastern Betic Cordillera and location of the basins described in this paper.

middle Pleistocene. Sedimentation proceeded until middle Pleistocene when tectonic uplifting triggered a regressive tendency, interrupted by several marine transgressions, a trend that continues at the present time (Bardají *et al.*, 1987).

Sedimentation of fan deltas occurred in a complex realm due to the coincidence of: (1) movements of active faults, that generated a gentle subsidence; (2) successive eustatic sea-level changes; and (3) various sedimentary processes acting both in the fan deltas and the neighbouring environments. Mutual interference resulted in the deposition of complex offlapping sequences which are commonly referred to as 'sequences of marine and terrestrial levels' or

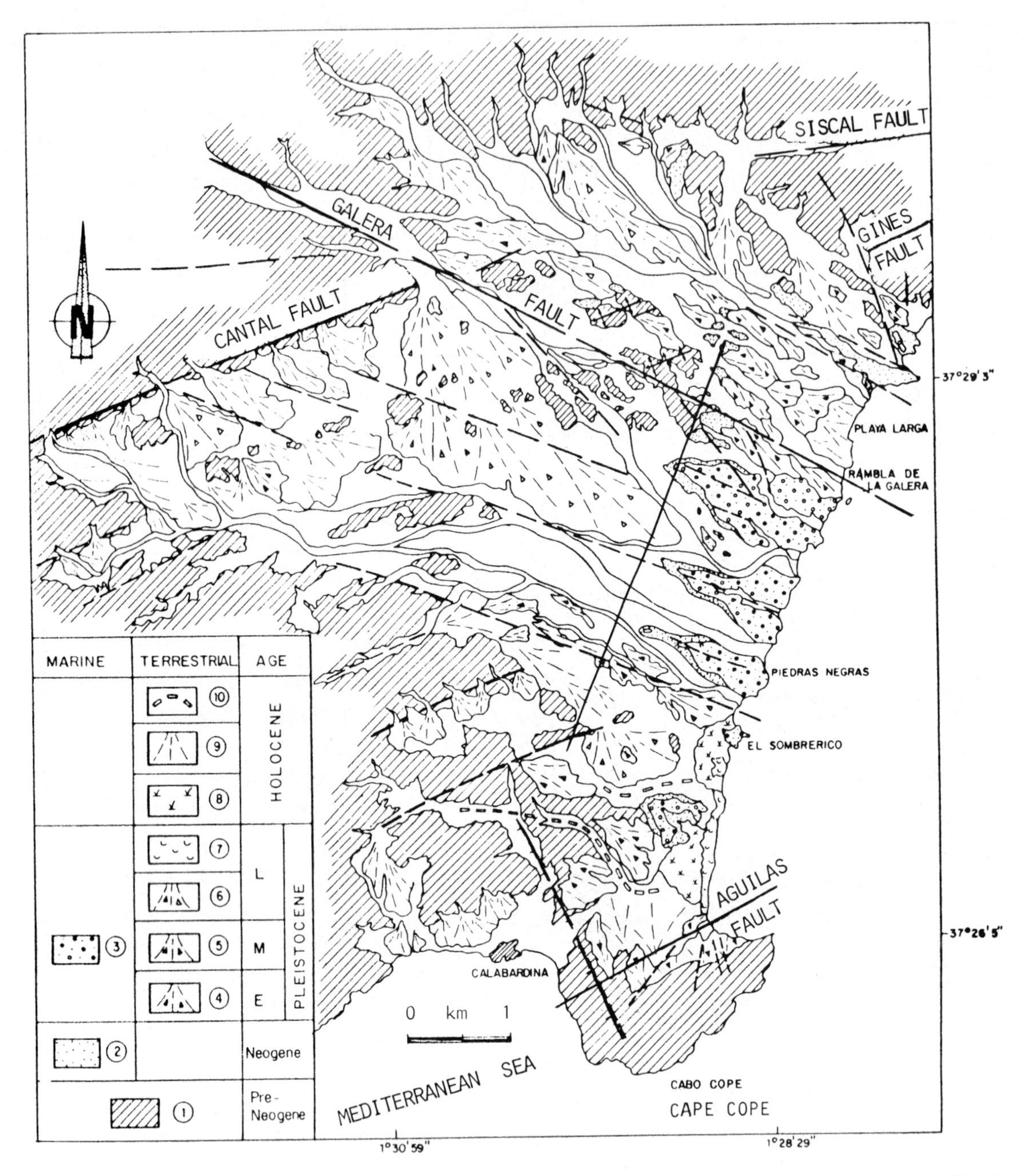

Fig. 2. Map of the Cope Basin to show the control exerted by fault systems on the morphology of fan-delta deposits. Key: 1. metamorphic rocks of the Internal Zones of the Betics; 2. fossiliferous sandstones and marlstones; 3. conglomerates and sandstones (prograding units made up of coastal fan-delta deposits; (4, 5, 6, and 9) alluvial fans (subaerial fan-delta deposits); 7. aeolian dunes; 8. fine-grained lagoonal deposits; 10. abandoned channels; (L) lower; (M) middle; (E) early. Modified after Bardají *et al.* (1986).

'marine terraces' (Goy *et al.*, 1986a) (Fig. 3). A much more detailed report can be found in Bardají *et al.* (this volume).

Controls on sedimentation

Fault movements

Gentle subsidence along N 60°-directed fractures occurred in the Cope Basin during the Neogene. The fractures were reactivated as dextral strike-slip faults during the Quaternary due to compression oriented N 150° E, triggering two secondary systems of fractures: N 120° E and N 30–40° E in the basin interior (Fig. 2). These two systems form a square-like pattern where differential subsidence took place during sedimentation of fan deltas. The orientation and geometry of the fan-delta bodies were consequently controlled by the directions of the active fractures.

Subsidence along the system N 120° E controlled the direction of the feeding channels and, in this way, the elongation of the sedimentary bodies. The system N 30–40° E (at nearly right angles to the former) conditioned the orientation of Pleistocene shorelines. As the system N 30–40° E acted also as a tilting line in response to the compressional stress it strongly controlled the sedimentation of Pleistocene sequences, particularly the rate of uplift of the margin of the basin.

Subsidence was not very prominent during the Pleistocene, although the positive relief of the mountains was pronounced. This conclusion is supported by the geometry of the resulting sedimentary units: they are thin wedges which slope gently towards the sea, showing offlap, with a tendency to toplap. In my opinion, strong subsidence generates thick accumulations of sediments. A continuous, gradual uplift would result in a long-term relative fall of sea level able to generate a single sequence of offlapping deposits. However, the occurrence of erosional surfaces separating successive marine and terrestrial deposits witnesses the superimposed shorter-term eustatic changes of sea level (Fig. 3). These features are very much like those active today in the basin and allowed to differentiate between the effects of tectonic subsidence and eustatic sea-level changes (Bardají *et al.*, this volume).

Dynamics of fan deltas

In this paper, it is assumed that the dynamic and climatic conditions recorded in late Pleistocene fan deltas of southeastern Spain were essentially the same as those presently found in this area. This is supported by the remarkable parallelism observed

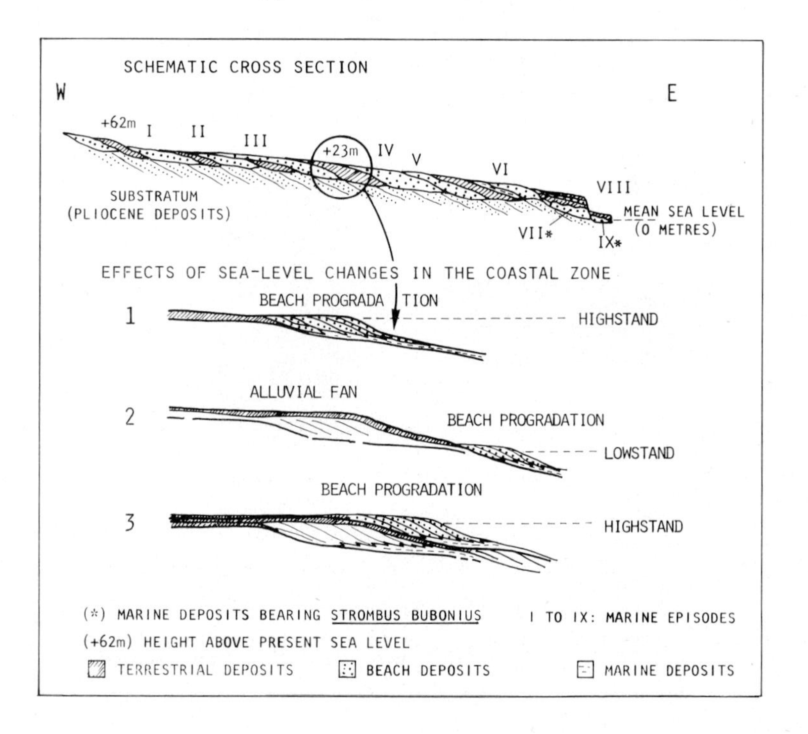

Fig. 3. Development of offlapping marine (dots) and terrestrial (inclined lines) units in Cope Basin in response to relative sea-level changes. The area pictured in steps 1, 2 and 3 is the same. Approximate scale of the schematic section *c.* 5 km. Vertical scale indicated by heights above mean sea-level.

in the sedimentary successions deposited in coastal environments of both ages (Dabrio *et al.*, 1984; Bardají *et al.*, 1986; Goy *et al.*, 1986b).

The dynamic regime of present-day fan deltas in southeastern Spain is controlled by episodic discharge and flooding of fans (very often of catastrophic nature) during the heavy rains which bring very large amounts of sediment in a few hours or days time. They are followed by long (several years) periods of inactivity when coastal and shallow-marine reworking accumulates sediment in beaches and spit bars. Thus, the coarse sediments of fan deltas feed the coastal zone and the resulting deposit exhibits a large-scale progradational geometry and a vertical sequence of facies (Fig. 4) in response to the coastal dynamics (Dabrio *et al.*, 1985).

The most characteristic exposed facies of the Pleistocene fan deltas of the Cope Basin are the coastal deposits which are virtually identical to those fed by rivers or by longshore drift but not directly related to fan deltas. They are comparable to sheltered and accretional beaches (as described by Bryant, 1983 and Short & Wright, 1983).

Eustatic changes of sea level

Pleistocene sea-level changes generated a succession of phases of highstand and lowstand in Cope Basin, which involved shifting of the coastline.

During highstands most of the coarse sediments remained in fan deltas developed next to the fractures and the areal extent of the subaerial fans was very restricted. Peripheral beaches prograded actively under high input of sediment. However, for any individual unit a stable, or somewhat falling sea level is deduced from the toplap of clinoforms observed in the sedimentary units (Fig. 3).

Lowstands caused subaerial exposure, weathering and erosion of a large part of the margin of the basin including the fan deltas deposited during former

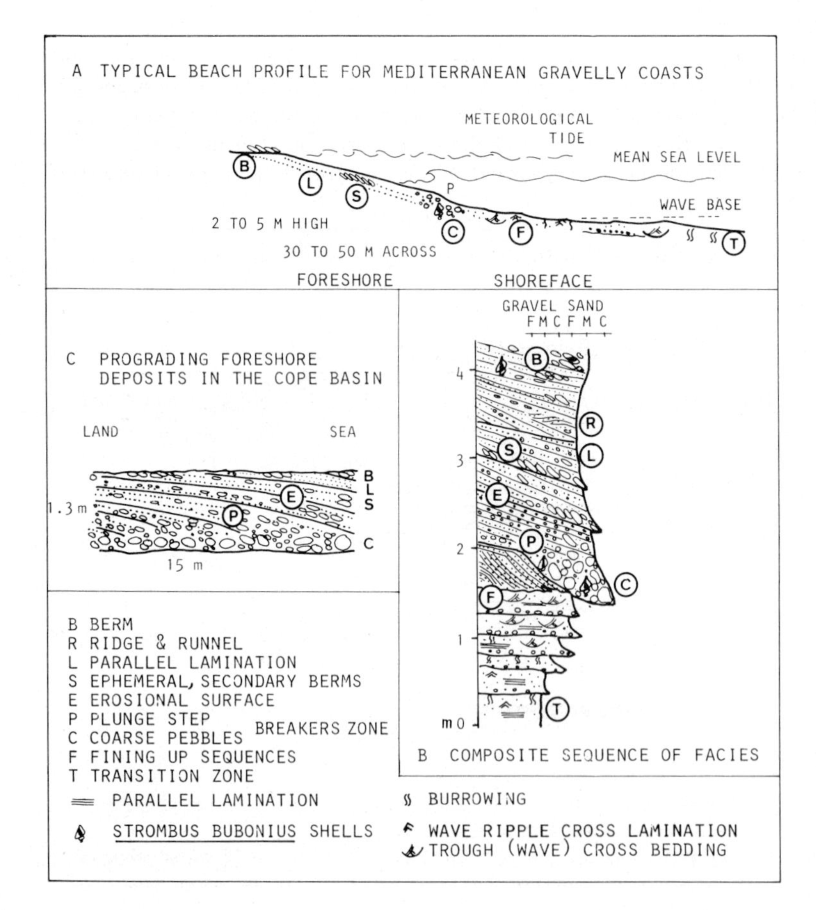

Fig. 4. A. Beach profile of gravelly coasts in the tideless southeastern coast of Spain, used as a model for Pleistocene gravelly beaches. B. Ideal sequence generated by progradation of late Pleistocene and present gravelly beaches of southeast Spain (modified from Dabrio *et al.*, 1985). C. Field drawing of the foreset facies in prograding gravelly beaches of the Cope Basin.

highstands. Most of the new coarse-grained input concentrated into telescopic fans (deposited at the toe of the exposed fans formed during highstands) fed by deeply incised channels. Almost no coarse sediment accumulated outside these active lobes; the most common deposit in exposed areas is massive red mudstone (M).

These large-scale changes of sea level can be easily deduced from careful analysis of morphology of the depositional sequence and the preserved clinoforms. In any case, topographic heights must not be used as equivalent to stratigraphic position because the magnitude of the relative oscillations may be very variable and, furthermore, younger tectonics may have changed the original relative heights.

But, in addition to the geometry of units and lateral relationships of clinoforms, the study of the lower-foreshore facies marked by coarse grain sizes and/or cross-bedding (Dabrio *et al.*, 1985) gives valuable data about small-scale changes of sea level during deposition of the larger-scale units (Bardají *et al.*, 1987 and this volume). Even minor fluctuations of sea level will produce lateral and vertical shifting of the inflexion point.

Conclusion

In the Cope Basin thin, sedimentary units fill a more-or-less rhomboidal pattern of uplifted and downed fault blocks. Those blocks experiencing positive subsidence were infilled by fan-delta sediments, whereas raised blocks did not collect much fan-delta sediments (except for the red mudstone facies). Continuous, gradual uplifting during the Pleistocene resulted in a gentle relative fall of sea level which generated an offlapping sequence of successive units of fan-delta deposits interrupted by eustatic sea-level changes. Subsidence was not very prominent as demonstrated by the geometry of the resulting sedimentary units.

THE ALMANZORA BASIN

General

The basin of the Almanzora River (Almería Province) is one of the Neogene narrow troughs of tectonic origin (Fig. 1), related to the western termination of the sinistral Alhama de Murcia fault system (Montenat *et al.*, 1985). The basin trends approximately E–W and has a width of about 2–5 km. Maximum subsidence took place along the southern margin of the basin, generating an asymmetric basin (Fig. 5). This morphology strongly influenced the repartition of sedimentary environments and also the geometry of the resulting sedimentary units (a detailed review can be found in Dabrio & Polo, 1988). During late Neogene time it served as a temporary marine connection between the Mediterranean Sea in the east and the largely confined Baza Basin situated a few kilometres to the west (Fig. 1).

Fan-delta deposits

Three informal lithostratigraphic units are distinguished (Fig. 5) in the mostly siliciclastic basin-fill (Dabrio & Polo, 1988). The lower unit I is made up of alluvial-fan deposits consisting of reddish channel-fill and debris-flow facies. A part of the middle unit II corresponds to fan-delta deposits (Figs 6, 7), with well-preserved depositional morphologies and varied sedimentary facies, including carbonate reef deposits. The upper unit III includes basinal deposits consisting mostly of white marly deposits and sandy and gravelly turbidites.

Despite the considerable siliciclastic input to the nearshore environments, corals (massive and branching *Porites* spp., *Tarbellastrea* spp. and hemispheroid *Diploria* spp.), oysters and red algae were able to colonize some areas of the shallow-marine slopes of the prograding fan deltas of Unit II, forming patch reefs that became successively buried by siliciclastic sediments. They formed also fringing reefs on erosional topographic heights of rocky coastal zones (Fig. 7). In both cases a low-diversity coral fauna is a distinct feature.

Characteristic successions (Fig. 8) are made up of grey micaceous sandstones and lutites with interbedded conglomerates and layers of skeletal carbonate debris (Maldonado, 1970; Martín García, 1972; Voezman *et al.*, 1978). Megasequences of decametric scale are interpreted as being generated by lateral shifting of the active delta lobes, under an assumed tectonic control. The role played by sea-level changes has not yet been evaluated.

The organic versus siliciclastic content of the coral–algal build-ups appears to have been dependent of their actual position with respect to the shoreline and open sea according to the 'corridor-like' nature of the basin. In the western sector, the sedimentary succession includes several episodes of

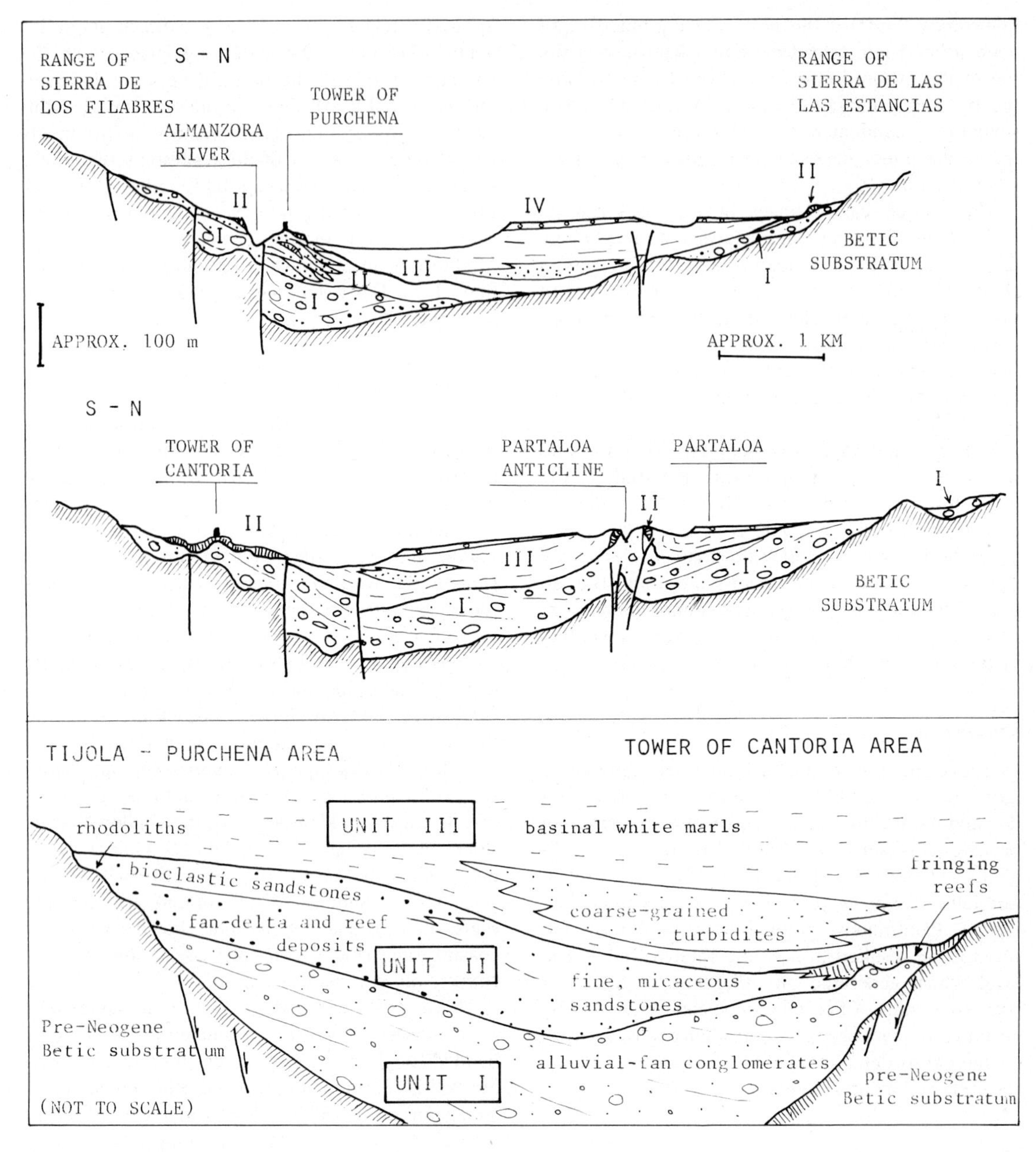

Fig. 5. Schematic cross-sections (above) and ideal stratigraphic framework (below) of the late Neogene sedimentary infill of the Almanzora Basin. IV: younger (Pleistocene?) alluvial-fan deposits.

colonization by patch reefs, with large amounts of fine sediments. This peculiar association is thought to be a result of the sheltered nature of this area with respect to the open sea, whereby the winnowing action of waves becomes considerably reduced. Higher energy levels were characteristic for the eastern segment of the basin, close to the seaward entrance of this marine corridor favouring more diverse reef-forming associations (Dabrio & Polo, 1988).

Close analogues for this specific association of coarse terrigenous clastics and organic carbonates

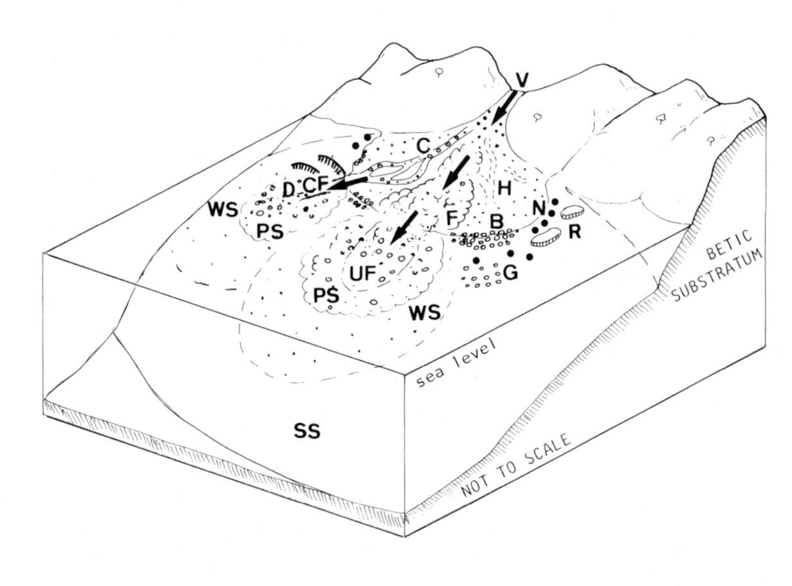

Fig. 6. Conceptual model of sedimentary processes and resulting facies in the fan deltas of the Almanzora Basin (it is not implied that all processes took place at the same time). Key: (V) alluvial valley; (C) fan channel; (CF) submarine channel-fill; (D) subaqueous debris flows and turbidity currents; (F) flash floods and debris flows entering the sea, eroding parts of gravelly beaches (B) and incorporating wave-produced 'lags' of gravels (G); (UF) floods continued as underwater flows; (PS) proximal slope (laminated sandstones and conglomeratic 'lag'); (WS) wave-rippled and (or) parallel-laminated micaceous sandstones; (SS) parallel-laminated, fine micaceous sandstones. Off inactive areas (H), patch reefs (R) and (or) rhodoliths (N) developed. Scales are only indicative. After Dabrio & Polo (1988).

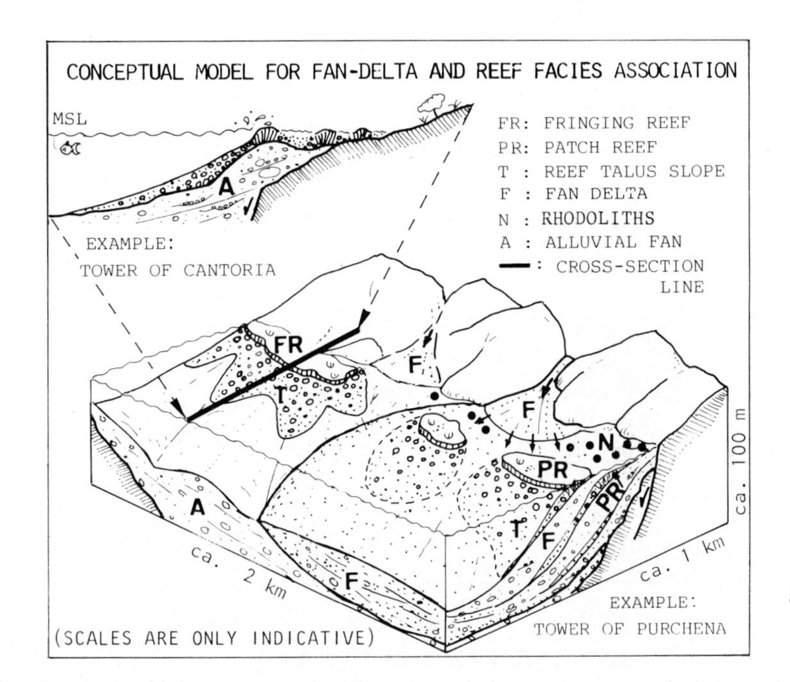

Fig. 7. Conceptual models of fan-delta and reef facies associations in the Almanzora Basin. Patch reefs formed on prograding fan deltas, whereas fringing reefs developed on abandoned or drowned erosional surfaces. The relationship between rhodolith development and patch reefs is largely speculative and cannot be supported by field evidence in the present case. MSL: mean sea level. Not to scale but the area picture is about 2 km across. After Dabrio & Polo (1988).

have been described in the Gulf of Aqaba (Hayward, 1985) and Gulf of Suez (Sellwood & Netherwood, 1984; Roberts, 1987), where lateral transitions from siliciclastic to carbonate facies occur within distances of a few tens of metres. The scale and internal facies organization of some of the fan deltas described from the Gulf of Aqaba type III of Hayward (1985), defined as medium to small alluvial fans prograding directly into the sea, provide a good analogue for the examples summarized here. Some differences have been noted as well, namely the growth of red-algal rhodoliths in the pools of reef flats, the higher coral diversity cited and the lower content of fine-grained siliciclastics in Middle East examples.

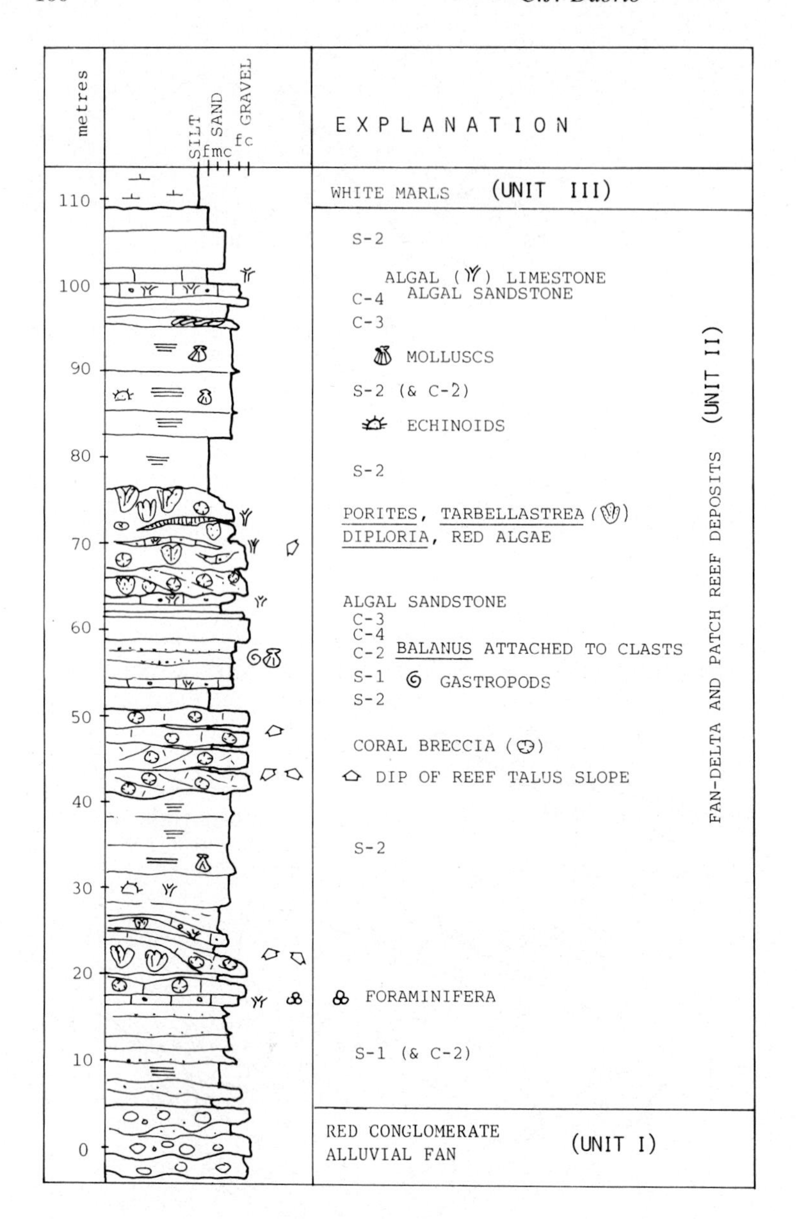

Fig. 8. Stratigraphic section of the Torre de Purchena fan-delta deposits. Same letters as in the text. After Dabrio & Polo (1988).

LATE MIOCENE FAN DELTAS IN CARRASCOY – MURCIA BASIN

General

The complex of fan deltas of Carrascoy (Fig. 1) is a part of the basin of Murcia placed on top of two sinistral strike-slip faults oriented N 40–50° E and N 80° E, with a complex, syntectonic infill underlined by many unconformities and notable wedging-outs (Montenat *et al.*, 1985; Montenat *et al.*, 1987). One of the most prominent features is the complete wedging-out of the approximately 1200 m thick 'Tortonian I' succession over less than 10 km measured parallel to the northern edge of the basin (Montenat, 1973). The regional stratigraphic framework proposed in this paper for the late Miocene of Carrascoy (Fig. 9) reflects the complex nature of the assumed synsedimentary tectonic activity. The active strike-slip faults induced sinistral horizontal displacement, but also triggered a prominent, rapid subsidence of the southern block.

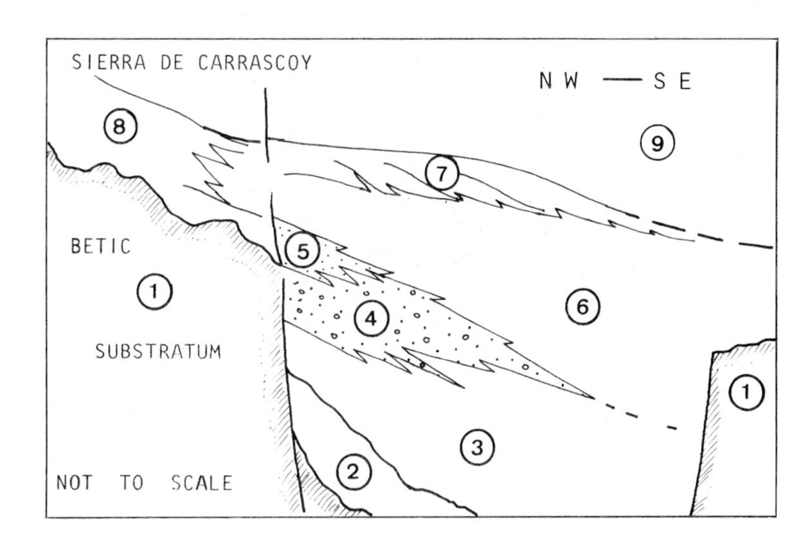

Fig. 9. Lithostratigraphic framework of the Murcia (Carrascoy) basin. Key: 1. Betic substratum; 2. conglomerate (sedimentary breccia) and turbidite (Tortonian-I); 3.–7. Tortonian-II to Messinian-I units; 3. marlstones, turbidites and megabreccia ('La Naveta marlstones'); 4. fan-delta deposits ('Puerto de la Cadena conglomerate'); 5. shelf conglomerates and calcarenites, lateral equivalent to fan-delta deposits; 6. basinal marlstones and turbidites ('Torremendo marlstones'); 7. shallow-marine calcarenites ('La Virgen limestones'); 8. upper-most Miocene–Pliocene terrestrial conglomerates; 9. Mio–Pliocene marlstones and conglomerates. Horizontal scale: about 10 km; vertical scale: some hundreds of metres.

Facies associations

Alluvial-fan (subaerial fan delta) subaqueous fan delta (both proximal and distal) and distal talus-slope and basinal facies have been recognized. *Subaerial fan-delta facies* are very similar to those of previously described alluvial fans. *Proximal subaqueous fan-delta facies* include channelized conglomerates (C-1), disorganized conglomerate (C-2, C-3, C-4) and laminated sandstones (S-1). *Distal subaqueous fan-delta facies* are marked by yellowish sandstone and siltstones (S-1 to S-2) with interbedded conglomerate (C-1 and C-4) layers. *Distal talus-slope and basinal facies* include hemipelagic light-grey silty marlstones and decametric sequences of turbidites.

Interpretation and discussion

Sedimentary model of fan deltas

These facies associations record deposition on a tectonically active basin margin. Differential subsidence (accomodated along faults) is responsible for the asymmetry of sedimentary bodies on both sides of the Sierra de Carrascoy (Montenat, 1973): thin onlapping, proximal, mostly terrestrial, deposits in the north and thick marine successions towards the southeast which generated a coastal onlap following a relative rise of sea level (Fig. 9).

The proposed sedimentary model (Fig. 10) consists of a basin margin with narrow, sloping shelf that was connected distally to an abrupt slope. Off the mouth of the mountain valleys draining the palaeo-Sierra de Carrascoy fan deltas developed prograding upon the narrow shelf and slope system. Sedimentary processes in these fan deltas consisted of shifting, channelized flows and flash floods, including a component of mass transport. Detailed mapping and correlation suggest that the radius of the coarse-grained facies generated by these fan deltas ranged from 4–5 km, but the sediment input clearly surpassed this distance, mostly as sediment gravity flows fed by partial destruction of shallow-water, shelf and fan-delta slope deposits. Well-rounded clasts suggest reworking of previous Neogene deposits, evidencing the cannibalistic nature of the basin.

Interpretation of sedimentary megasequences: the tectonic control

Up to 12–14 km (measured stratigraphically, not vertically) of fan-delta sediments were deposited along the active margin of the basin. In detail, individual units of alluvial-fan and fan-delta deposits can be distinguished inside this huge pile of sediments. These units show sequences of various scales which are thought to record diverse controls of sedimentation. The existence of active wrench faults and the overall organization of deposits very much resembles

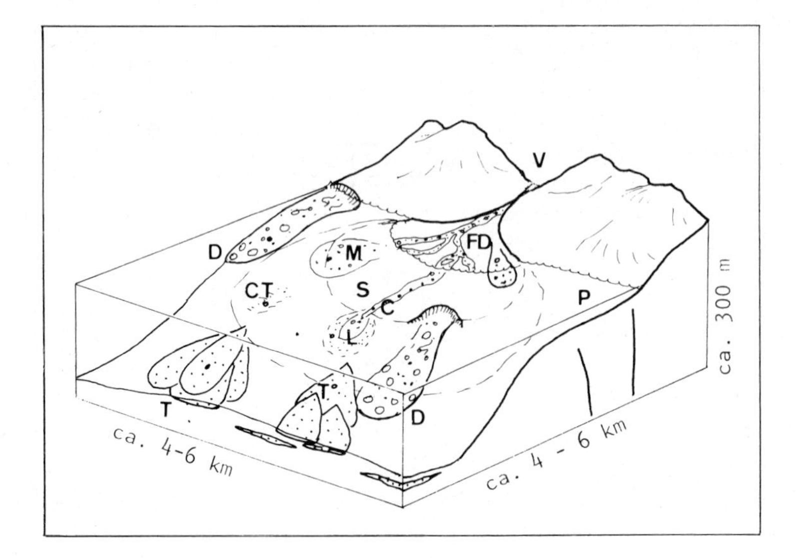

Fig. 10. Model of a fan-delta lobe in El Puerto de la Cadena. Explanation in the text. Key: (V) fluvial valley; (FD) fan delta (letters on the subaerial zone); (P) narrow shelf; (S) prodelta and slope; (C) channel; (L) coarse-grained lobe at the end of a submarine channel; (M) wave-winnowed gravels; (D) debris flows; (T) turbidites; (CT) turbidite channels.

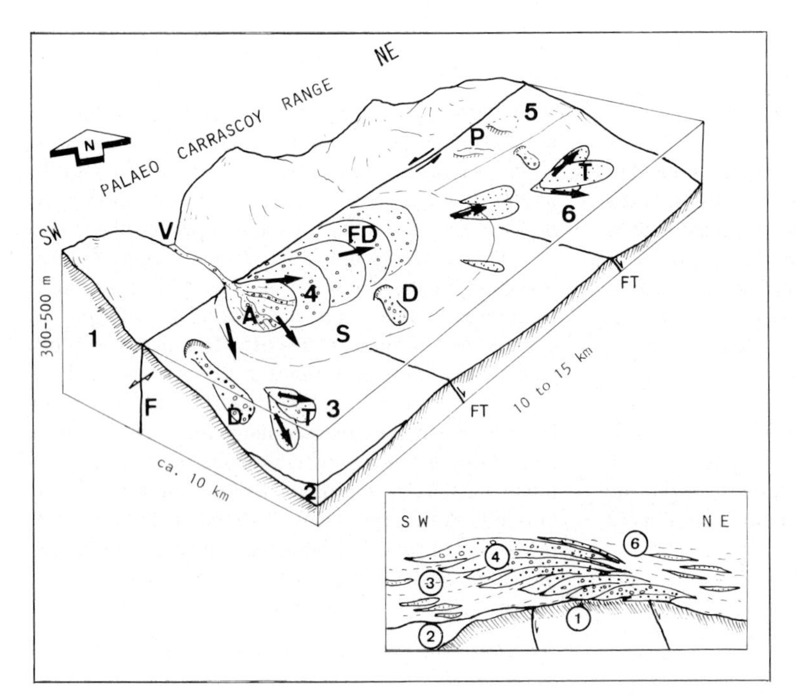

Fig. 11. Model of margin of basin related to a synsedimentary strike-slip fault (F) in Carrascoy range. Below, dynamic interpretation of lithostratigraphic units distinguished in Fig. 10 (numbers have the same meaning): (FD), successive fan-delta lobes having a channelized subaerial part (A); (D) debris flows; (S) prodelta and slope; (T) turbidite deposits (including channels); (FT) transverse faults generating additional subsidence in various zones of the basin margin.

the model of the basin margin described in the Devonian Hornelen Basin by Steel *et al.* (1984). A similar mechanism is invoked in this paper: the relative movement of the blocks defining the active border of the basin forced the points of sediment input (i.e. the apical zones of fans next to the mountain valleys but separated from them by the wrench fault) to move laterally. The lateral offset of units (mostly deposits of fan deltas having only radii of about 4–5 km) produced the huge stratigraphic thickness (Fig. 11).

It is logical to think that lateral movements along the offset fault will be recorded in the sediments. I assume that the lateral migration of depocentres (related to the displacement of the shelf towards the northeast) generated various types of megasequences. Coarsening- and thickening-upwards developed in areas placed to the southwest of the points of sediment supply as they progressively approached them (Fig. 12). However, fining- and thinning-upward megasequences developed in those areas progressively moved away from the emitting valleys (Fig. 13). Increased rates of deposition off these points favoured the progradation of fan-delta complexes (including well-developed subaerial facies) that were subsequently drowned when subsidence prevailed over deposition once they moved away from the valleys. Sequences of smaller order (briefly described in the next section) are found inside the former evidencing the diversity of the sedimentary processes involved and the repeated lateral migration of subenvironments.

A later relative rise of sea level resulted in a hemipelagic drape of several of the shallower-water units (Fig. 9) but no precise estimate of the relative influence of eustatic and tectonic causes can be made yet.

Morphologies inherited from former stages and differential subsidence favoured by transverse faults which were very active since the late Miocene (Nuñez *et al.*, 1974) strongly influenced sedimentation during this displacement of sedimentary environments. This interpretation is supported by the progressive change of palaeocurrent directions. In subaerial

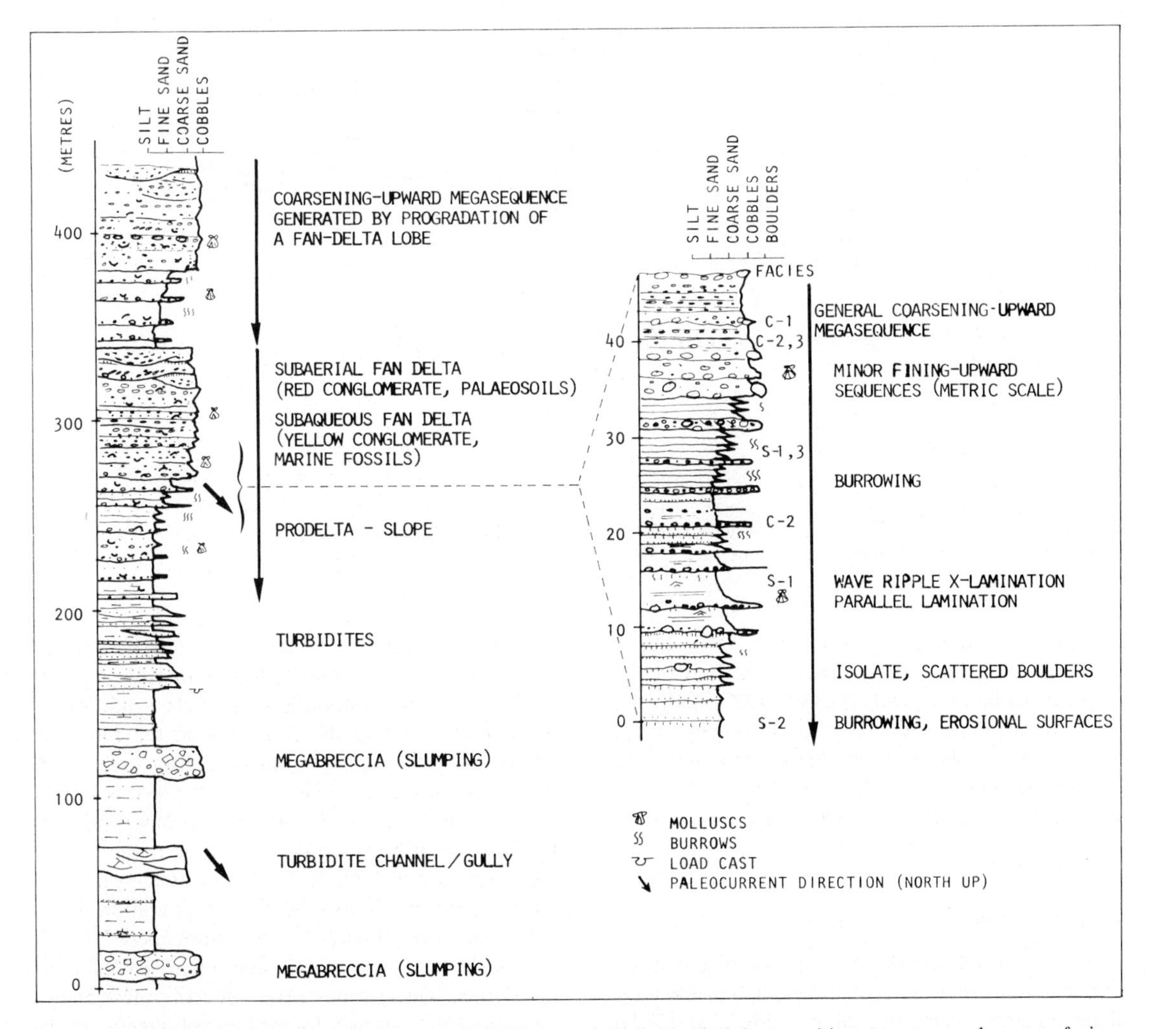

Fig. 12. Stratigraphic succession of La Naveta (SW Carrascoy) with detail of the transition zone to conglomerate facies. Fan-delta deposits generate coarsening-upwards megasequences (indicated by downward-pointing arrows) in the upper half of the succession.

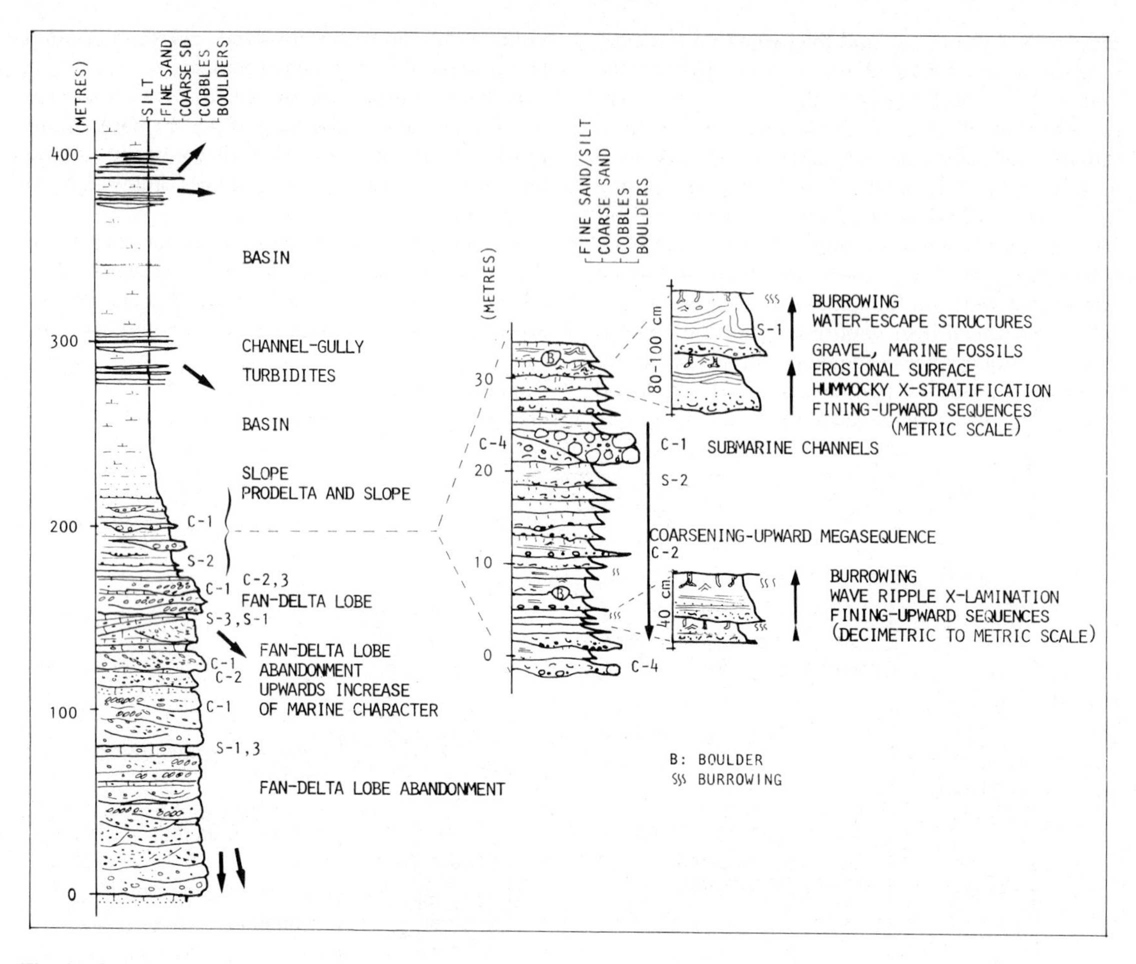

Fig. 13. Stratigraphic succession of Puerto de la Cadena on the road from Murcia to Cartagena (central Carrascoy mountains) with detail of prodelta and slope facies. (B) scattered boulders in sandy deposits.

fans they are consistently towards N 150–180° E, but they turn progressively towards the east in subaqueous fan-delta deposits (N 120–130° E) and still more in turbidites (N 110–120° E in channel-fill deposits, N 30–100° E in overbank turbidites). This is probably related to deflection of currents by the positive reliefs induced by underlying sedimentary bodies (Fig. 11).

Second-order megasequences (decametric scale)

Sequences of decametric scale are measured in the lenticular units that can be distinguished inside those of hectometric scale cited above (Figs 12, 13). In the area of La Naveta these second-order megasequences show coarsening- and thickening-upwards (Fig. 12). They consist of a lower (up to 50 m thick) unit of sandstones (interpreted as marine deposits on the fan slope), a middle conglomeratic and sandy unit (interpreted as the transition to the fan-delta front) and an upper conglomeratic unit, with marine fossils and evidences of wave-reworking. Variable thicknesses of reddish conglomeratic nonfossiliferous alluvial-fan deposits overlay the former sequence in many places. Such an arrangement of facies is interpreted as reflecting progradation of the delta front and alluvial-fan environments of individual fan-delta lobes inside the main complex of fan deltas. These lobes prograded or retrograded, adapting to previously formed morphologies, probably in response to relative sea-level changes.

Fan abandonment took place when subsidence

did not compensate deposition, which is the case when lobes are moved away from the emitting valleys. This is well represented in the area of Puerto de la Cadena (Fig. 13). The resulting fining and thinning megasequences consist of three units (not always easy to distinguish) separated by sharp vertical changes. The lower unit is channel-fill-dominated conglomerate (C-1) with scattered remains of marine fossils. In the middle unit, alternations of fossiliferous calcarenites (roughly described as S-1 and S-3) and unconfined conglomerate (C-2) are thought to record the active fan-delta front with incised channels and poorly developed sublittoral deposits which may witness to rapid deposition. The upper unit consists of tabular to gently wedge-shaped layers of calcarenites and conglomerate with various vertical motifs (C-2, C-3, C-4). They correspond to littoral and sublittoral deposits on the subsiding, transgressed fan-delta lobe.

Decimetric to metric *third-order sequences* are found inside the previously described ones. They are interpreted in terms of diverse sedimentary processes: filling of channels, rapid deposition on the shelf after floods, settling of sediments after storms, etc. (Figs 12, 13).

Comparison

The association of fan deltas and submarine fans is common in relation with tectonically active edges of basins. This is the case of the Eocene–Oligocene deposits of the Santa Ynez Mountains (Van de Kamp *et al.*, 1974) where the so-called episodes B (unstable slope with delta progradation) and D (unstable slope and fracturation of the margin of the basin) are similar to those invoked for Carrascoy. In some aspects the model of Carrascoy strongly resembles the fractured slope aprons (Stow, 1985, 1986) with fan deltas building an active slope which feeds sediment gravity flows.

Surlyk (1978, 1984) interpreted the late Jurassic deposits of the Wollaston Forland (Greenland) as formed on slopes fed by fan deltas, related to large normal faults. The 4 km-thick succession includes four fining-upward megasequences (several hundred metres thick) thought to be related to tectonic activity. Smaller metric to decametric, second-order fining-upwards are interpreted as filling and abandonment of channels or surging flows, later covered by hemipelagites. Sequences and controls of sedimentation are comparable to Carrascoy, although the tectonic movements involved and the sedimentary processes interpreted are somewhat different.

The model of displacement of depocentres by strike-slip faults was proposed by Gloppen and Steel (1981) and Steel *et al.* (1984) to explain the huge stratigraphic thicknesses (more than 25 km) of fan-delta, fluvial and lacustrine sediments in the Devonian Hornelen Basin (Norway). Steel (1988) documented the difficulty of distinguishing vertical motifs caused by progradation from those caused by offset faults. The described association of fan delta and steep slope can be compared also to the Eocene Wagwater Group of Jamaica described by Wescott and Ethridge (1980).

LATE NEOGENE FAN DELTAS OF TABERNAS-SORBAS BASIN

General

The basin of Tabernas-Sorbas (Fig. 1) is a narrow trough elongate in a N 110° E direction. It is an asymmetric basin with the thickest accumulations located along the southern margin (Jacquin, 1970; García Monzón *et al.*, 1973; García Monzón *et al.*, 1974; Ott d'Estevou, 1980).

Montenat *et al.* (1987) interpreted the basin as a wrench-fault furrow with active subsidence caused because the movement of the fault blocks was not completely horizontal. This type of furrow allows the accumulation of large amounts of siliciclastic sediments as sediment gravity flows due to the rapid denudation of the margins of the basin and their collapse caused by the progressive uplift. The basin shows a characteristic cannibalistic behaviour.

The late Tortonian deposits filling the Tabernas-Sorbas Basin illustrate the processes of emplacement and later remobilization of the huge volumes eroded from the margins of the basin. Two lithostratigraphic units are distinguished (Fig. 14).

Unit I is a succession, more than 50 m thick, of red, heterometric conglomerates, with boulders up to 3 m in length, resting unconformably on top of the Betic substratum. No palaeontological dating could be made but Montenat (1973) assumed a Tortonian age based on regional criteria. Most of the sediment was supplied by the metamorphic rocks derived from the Nevado-filabride Complex of the Betics (as defined by Egeler & Simon, 1969) cropping out in Sierra de los Filabres, the mountainous ranges located to the north (García Monzón *et al.*, 1973).

Unit II lies unconformably on top of Unit I and is

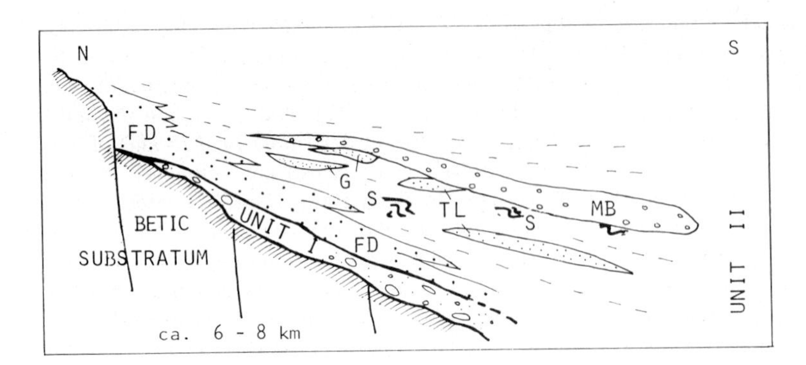

Fig. 14. Lithostratigraphic units distinguished in the basin of Tabernas-Sorbas. (Unit I) alluvial-fan red conglomerates, Tortonian; (Unit II) Tortonian-II and Messinian-I deposits. Key: (FD) fan-delta conglomerates; (G) sandstones and conglomerates in turbidite channels; (TL) sandstones (turbidite lobes); (S) slump; (MB) megabreccia.

made up of grey-greenish conglomerates and sandstones that change laterally and upwards into a thick packet of micaceous marls and sandstones with layers of megabreccia, up to 40 m thick. The breccia is formed by fragments of nevado-filabride metamorphic rocks embedded in fine, micaceous sandstones. The lowermost 50–60 m of conglomerates of Unit II are relatively parallel bedded. They include remains of shallow-marine fossils such as oysters, pectinids and large barnacles (Balanidae) attached to rock fragments. They are interpreted as submarine fan-delta deposits (FD in Fig. 14). The rest of Unit II is interpreted as slope and deep-water basinal turbidite deposits with a major contribution of other types of gravitational processes (slides and slumps) as documented by Roep and Kleverlaan (1983).

It is interesting to note that the most prominent present-day relief (the Sierra de Alhamilla, which limits the basin towards the south) did not supply appreciable amounts of sediments to the basin. This indicates a rapid, deep erosion of the northern hinterlands and a palaeogeographic model different to that of the present.

Facies associations

The *alluvial-fan facies association* is found only in Unit I. Predominant facies are reddish disorganized, heterometric conglomerates (mostly of the debris-flow type) and crudely bedded conglomerates (which could be referred to as Miall's (1978) Gms facies) with a more or less erosional lower boundary.

The *subaqueous fan-delta facies association* has all the siliciclastic facies described from Almanzora basin except for C-1. Large channel-shaped units of trough cross-bedded sandstones show palaeocurrent flows towards the south-southeast. Layers of non-confined conglomerates include marine fauna, in particular Balanidae which, in some places, form individual layers.

The *slope to basin facies association* forms the most important part of the upper Miocene filling of the basin ('Tortonian-2', Montenat, 1973). According to Roep and Kleverlaan (1983) it consists of a thick succession of turbidites deposited in a deep-sea fan and mudstone basin plain facies. Kleverlaan (1987) calculated the slope of the basin as 6° towards the south and southwest. The steep slope induced sliding with prominent slump scars filled up with chaotic masses and vergent folds in the frontal part of the slumped masses.

Several layers of laterally continuous megabreccia and large embedded blocks of collapsed talus-basin deposits crop out in the basin. The occurrence of well-preserved, shallow-marine to sublittoral fauna in these layers suggests a mechanism of emplacement which did not cause much erosion of the fossil remains (probably a mass transport). The largest layer of megabreccia is the 'Gordo Megabed', which seems to be the result of a single depositional event interpreted as a 'seismite' (*sensu* Mutti *et al.*, 1984) (Kleverlaan, 1987).

Interpretation and discussion

The proposed model for the Tortonian deposits of Sorbas-Tabernas Basin integrates the effects of rapid subsidence with repeated collapse of coarse-grained shallow-marine deposits which were incorporated in to the basinal and turbidite successions being deposited at the base of the slope (Fig. 15). The most prominent features of the megabreccias are the northern source area, the chaotic nature, the coeval turbidite sedimentation and the various scales of slumps.

A problem that remains obscure is the precise genesis of the materials forming the breccias and its

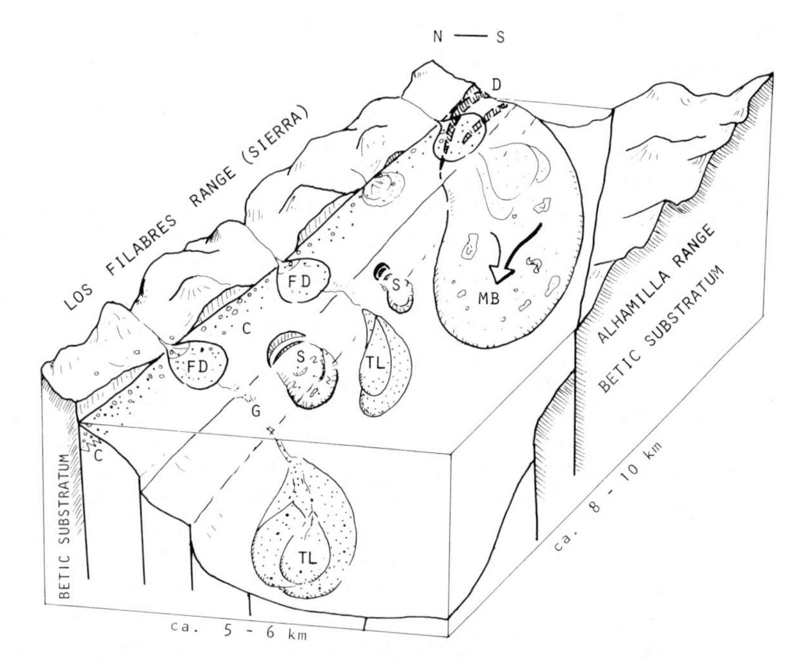

Fig. 15. Idealized model of the basin of Tabernas during deposition of the Tortonian sediments (vertical scale exaggerated to emphasize the important role of differential subsidence). Key: (FD) fan-delta deposits along the fractured margin of the basin; (C) coarse-grained deposits near cliff and escarpments; (D) large-scale sliding responsible for deposition of Gordo Megabreccia (MB); (S) slumps; (GL) turbidite channels (gullies); (TL) turbidite deposits.

relationship with the fan deltas. From a genetic point of view, the only original features preserved are the lithology and the very coarse grain size of the (usually) poorly rounded and sorted boulders. Other interesting features are the occurrence of: (1) layers almost exclusively made up of barnacles (Balanidae); (2) blocks with attached isolated or grouped barnacles (all these fossils hardly show any trace of erosion); and (3) sandstone beds with marine fauna. According to these data, Roep & Kleverlaan (1983) suggested that the edge of the basin should be formed by small cliffs related to active faults. Montenat *et al.* (1987) indicated the existence of fractures running along the northern border of the basin, relating some of their movements to shearing along N 70° E and N 40° E directions.

In this paper it is assumed that the mountainous northern margin was related to active faulting (Fig. 15). The steep slopes may have favoured the accumulations of large boulders near cliffs and also the development of debris-flow-dominated fan deltas. Sliding of these coarse-grained masses of sediment (that were temporarily stored in the narrow shelf) is supposed to have generated the megabreccia. Probably the shallow-marine debris moved as mass flows mixing with large removed pieces of (deeper) turbiditic deposits (Fig. 15).

Similar examples are found in the northern margin of the Granada Basin (Dabrio *et al.*, 1978) where slumps and large boulders also indicate a rather abrupt slope. Present analogues with identical source areas are found along the Mediterranean coast of the Granada Province.

In short, in the Tabernas-Sorbas Basin, rapid subsidence related to strike-slip faults caused deposition of coarse-grained, debris-flow-dominated fan deltas along a narrow shelf. The shelf changed laterally to an active slope incised by gullies that fed submarine fans and slumps of various scales. Sliding along the active, steep margin of the basin prompted deposition of huge layers of megabreccia that incorporated a mixture of coastal, fan-delta, slope and fan deposits. Successions up to 1 km thick were deposited. The most prominent control of sedimentation was tectonism, masking the effects of sea-level changes and lateral migration of fan deltas.

CONCLUSIONS

The described examples illustrate various facies associations in fan deltas of the Almería region.

In Cope Basin, low rates of subsidence coupled with significant topographic differences induced coalescing fan deltas to produce offlapping coastal units on a scale of metres, very sensitive to Pleistocene changes of sea level. Most of the coarse sediment of the delta front was captured in beaches of reflective type.

In the Almanzora River Basin, channels and mass flows of fault-related fan deltas supplied large amounts of conglomerate and micaceous sandstone but corals and algae repeatedly colonized the shallow delta fronts forming patch reefs. Lateral shifting of the active delta lobes, under an assumed tectonic control, generated megasequences of decametric scale. The contribution of sea-level changes has not been evaluated.

In Carrascoy — Murcia Basin — active synsedimentary strike-slip faults displaced laterally the depocentres, so that coarse sediments generated both coarsening and fining-upwards megasequences on a scale of tens to hundreds of metres under a prominent tectonic control. Progradation of channel-dominated fan-delta lobes generated lenticular units of kilometric radius that related laterally to active slopes and submarine fans.

In Sorbas (Tabernas) Basin, high subsidence related to strike-slip faults favoured the deposition of thick sedimentary successions on a scale of hundreds of metres to 1 km. Coarse-grained debris-flow-dominated fan deltas deposited along a narrow shelf connected laterally to an active slope with gullies that fed submarine fans and slumps of various scales. Sliding along the active steep margin of the basin prompted deposition of huge megabreccias that incorporated a mixture of coastal, fan-delta, slope and fan deposits. The most prominent control of sedimentation was tectonism, masking the effects of sea-level changes and lateral migration of fan deltas.

It is not easy to compare these examples due to the diversity of processes, sedimentary facies, scales and the geodynamic regime of basins involved, but the main aim has been to focus on some of the defining features and to widen the current points of view for more precise future observations. From this perspective, the more relevant conclusions are given in Table 1 and in the models of basin margins (Fig. 16).

It is proposed here that careful study of sedimentary features recorded in stratigraphic successions and the lateral and vertical facies relationships provide useful data regarding the geodynamic behaviour

Table 1. Defining features of various fan deltas

	Cope	Almanzora	Murcia	Sorbas
Subsidence	Small	Moderate	High	Very High
Fan morphology	Bajada, coalescent in proximal zones, control by fractures	Isolated, locally coalescent	Isolated, coalescence in talus slope	Probably coalescent
Fan radius	2–3 km	3–5 km	5–6 km	?
Sedimentary controls	Sum of fan + coastal processes + sea-level changes	Channel + debris flow + marine + organic growth	Channelized fans + steep slope + strike-slip faulting	Gravity on steep slope
Dispersion of coarse sediment	Accumulated along the coast	To subaqueous fan delta	To subaqueous fan delta and slope	To the basin (megabreccia)
Response to sea-level changes	Prograding coastal wedges and erosion	Prograding fan delta lobes with patch reefs	Progradation and retrogradation of delta lobes	Masked by other factors; progradation of turbiditic lobes?
Subaerial fan delta	Braided rivers with channels overbank	Alluvial: mass flow and channel	Alluvial: mass flow and channel	Probable mass-flow-dominated alluvial
Fan delta prodelta	Conglomeratic beaches and fines, gentle slope	Channel, mass flow, reefs, sand waves, gentle/moderate talus slope	Channel, mass flow, mixed shelf, slumps on steep slope	Mixed shelf with slumps: very steep talus slope
Talus basin	?	Fine micaceous sandstone + coarse sandstone (grain flows)	Shallow turbiditic channels with sandstone lobes	Deep sea fans, slumps and megabreccia

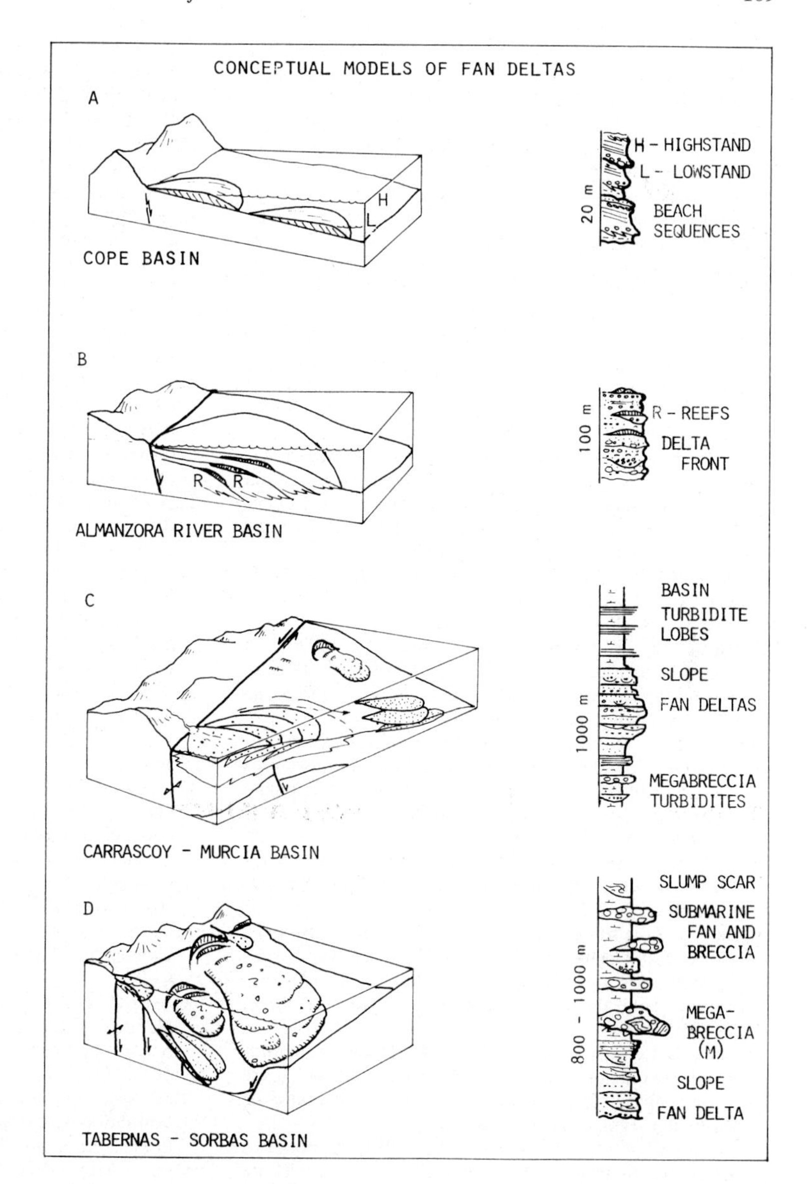

Fig. 16. Simplified conceptual models of various types of fan deltas in eastern Betic Cordillera, presented (from top to bottom) in increasing subsidence, and resulting successions.

of the basins and the major controls of deposition, namely the magnitude and rate of subsidence and the type of source area because it is of vital importance for the development of many sedimentary processes and the resulting facies.

ACKNOWLEDGEMENTS

This paper is a part of a wider research programme with financial support of the 'Programación Científica del CSIC.' 630/070 and the spanish DGICYT Project PB 88-0125. The author is grateful to R. Steel and F. Surlyk who carefully reviewed the manuscript and made many useful scientific and linguistic improvements.

REFERENCES

BAENA, J., GARCIA-RODRIGUEZ, J., MALDONADO, A., UCHUPI, E., UDIAS, A. & WANDOSSELL, J. (1982)

Mapa geológico de la plataforma continental española y zonas adyacentes, Escala 1:200000. Almería-Garrucha, Chella-Los Genoveses. IGME, Madrid.

BARDAJI, T., CIVIS, J., DABRIO, C.J., GOY, J.L., SOMOZA, L. & ZAZO, C. (1986) Geomorfología y estratigrafía de las secuencias marinas y continentales cuaternarias de la Cuenca de Cope (Murcia). In: *Estudios sobre geomorfología del sur de España* (Ed. by F. López Bermúdez and J.B. Thornes). Com. Meas. Theory Applic. Geomorphol., pp. 11–16. Int. Geogr. Union, Universidades de Murcia y Bristol.

BARDAJI, T., DABRIO, C.J., GOY, J.L., SOMOZA, L. & ZAZO, C. (1987) Sedimentologic features related to Pleistocene sea-level changes in the SE Spain. In: *Late Quaternary Sea-level Changes: Measurement, Correlation and Future Applications* (Ed. by C. Zazo). Trab. Neog. Cuat. Museo Nal. Cienc. Nat. 10, pp. 79–93.

BERNINI, M., BOCCALETTI, M., GELATI, R., MORATTI, G. & PAPANI, G. (1983) Fenomeni di trascorrenza nella evoluzione neogenica-quaternaria della Catena Betica. *Alti Riunione 'Meccanismi di avanzamento delle ricerche e problematiche emerse'*. Univ. Firenze, 65–75.

BORDET, P. (1985) *Le volcanisme miocène des Sierras de Gata et de Carboneras (Espagne du Sud-Est)*. Doc. et Trav. IGAL, Paris. 8, 70 pp.

BOUSQUET, J., MONTENAT, C. & PHILIP, H. (1976) La evolución tectónica reciente de las Cordilleras Béticas orientales. *Reunión Geod. Cord. Bética y Mar de Alborán*. Univ. Granada, 59–78.

BOUSQUET, J. & PHILIP, H. (1976) Observations microtectoniques sur la compression Nord-Sud quaternaires des Cordillères Bétiques orientales (Espagne méridionale, Arc de Gibraltar). *Bull. Soc. Géol. Fr.* **7** (XVIII), 711–724.

BRYANT, E. (1983) Sediment characteristics of some Nova Scotian beaches. *Maritime Sediments and Atlantic Geology* **19**, 127–142.

DABRIO, C.J., FERNANDEZ, J., PENA, J.A., RUIZ BUSTOS, A. & SANZ DE GALDEANO, C.M. (1978) Rasgos sedimentarios de los conglomerados miocénicos del borde oriental de la Depresión de Granada. *Est. Geol.* **34**, 89–97.

DABRIO, C.J., GOY, J.L. & ZAZO, C. (1984) Dinámica litoral y evolución costera en el Golfo de Almería desde el 'Tirreniense' a la actualidad. *I. Cong. Español de Geología, Segovia*. **I**, 507–522.

DABRIO, C.J., GOY, J.L. & ZAZO, C. (1985) A model of conglomeratic beaches in tectonically-active areas (Late Pleistocene–Actual, Almería, Spain) *Abstr. Int. Ass. Sedim. 6th Europ. Reg. Mtg.* Lleida, Univ. de Barcelona, 104–105.

DABRIO, C.J. & POLO, M.D. (1988) Late Neogene fan deltas and associated coral reefs in the Almanzora Basin, Almería Province, southeastern Spain. In: *Fan Deltas: Sedimentology and Tectonic Settings* (Ed. by W. Nemec and R.J. Steel), pp. 354–367. Blackie and Son, London.

EGELER, & SIMON, O. (1969) Sur la tectonique de la Zone Bétique (Cordillères Bétiques, Espagne). Etude basée sur les recherches dans le secteur compris entre Almería et Vélez Rubio. *Versl. Kon. Ned. Akad. v. Wetensch., Afd. Natuurk.* **3**, 90 pp.

GARCIA MONZON, G., KAMPSCHUUR, W. & VERBURT, J. (1974) *Mapa y Memoria explicativa de la Hoja 1031 (Sorbas) del MAGNA*. IGME, Madrid, 46 pp.

GARCIA MONZON, G., KAMPSCHUUR, W., VISSERS, R., VERBURT, J. & WOLFF, R. (1973) *Mapa y Memoria explicativa de la Hoja 1030 (Tabernas) del MAGNA*. IGME, Madrid, 31 pp.

GLOPPEN, T.G. & STEEL, R. (1981) The deposits, internal geometry and structure in six alluvial fan-fan delta bodies (Devonian–Norway) — A study on the significance of bedding sequence in conglomerates. *Spec. Pub. Soc. econ. Paleont. Mineral Tulsa*, **31**, 49–69.

GOY, J.L., ZAZO, C., BARDAJI, T. & SOMOZA, L. (1986a) Las terrazas marinas del Cuaternario reciente en los litorales de Murcia y Almeria (España): el control de la neotectónica en la disposición y número de las mismas. *Est. Geol.* **42**, 439–443.

GOY, J.L., ZAZO, C., DABRIO, C.J. & HILLAIRE-MARCEL, CL. (1986b) Evolution des systémes de lagoons-îles barriére du Thyrrenian a l'actualité a Campo de Dalías (Almería, Espagne). *Edit. de l'Orsrom, Coll. Travaux et Documents* **197**, 169–171.

HAYWARD, A.B. (1985) Coastal alluvial fans (fan deltas) of the Gulf of Aqaba (Gulf of Eilat), Red Sea. *Sedim. Geol.* **43**, 241–260.

HOLMES, A. (1965) *Principles of Physical Geology*. Thomas Nelson, London, 1288 pp.

JACQUIN, J.P. (1970) *Contribution a l'ètude geologique et minière de la Sierra de Gador (Almería, Espagne)*. Thèse Univ. Nantes. 501 pp.

KLEVERLAAN, K. (1987) Gordo Megabed: a possible seismite in a Tortonian submarine fan, Tabernas Basin, Province Almería, Southeast Spain. *Sedim. Geol.* **51**, 165–180.

LAROUZIERE, F.D., MONTENAT, C., OTT D'ESTEVOU, P. & GRIVEAUD, P. (1987) Evolution simultanée de bassins néogènes en compression et en extension dans un couloir de décrochement: Hinojar et Mazarrón (sud-est de l'Espagne). *Bull. Centres Rech. Explor.-Prod. Elf-Aquitaine*, **11**, 23–38.

MALDONADO, A. (1970) Estudio geológico de la región Caniles-Serón (Cordillerra Bética). *Bol. Geol. Min.* **56**, 6–22.

MARTIN GARCIA, L. (1972) Estudio litoestratigráfico del Neógeno-Cuaternario del Valle del Almanzora (sector Serón-Purchena). *Cuad. Geol. Univ. Granada*, **3**, 121–132.

MCPHERSON, J.G., SHANMUGAM, G. & MOIOLA, R.J. (1987) Fan deltas and braid deltas: Varieties of coarse-grained deltas. *Bull. geol. Soc. Am.* **99**, 331–340.

MIALL, A.D. (1978) Lithofacies types and vertical profile models in braided rivers: a summary. In: *Fluvial Sedimentology* (Ed. by A.D. Miall). Mem. Can. Soc. Petrol. Geol. 5, pp. 597–604.

MONTENAT, C. (1973) *Les formations neogénes et quaternaires du Levant Espagnol*. Thése d'Etat. Université d'Orsay. 3 vols, 1170 pp.

MONTENAT, C. (1977) *Les bassins neogènes du Levant d'Alicante et de Murcia (Cordilléres Bétiques orientales, Espagne). Stratigraphie, paléogèographie et evolution dynamique*. Doc. Lab. Géol. Fac. Sci. Lyon 69, 345 pp.

MONTENAT, C., OTT D'ESTEVOU, P. & MASSE, P. (1985) Les bassins néogenes des Cordilléres bétiques orientales: génese et évolution dans un couloir de décrochement crustal. *Rés. Comm. Séance Spec. Soc. geol. Fr, Paris*, 2 pp.

MONTENAT, C., OTT D'ESTEVOU, P. & MASSE, P. (1987)

Tectonic–sedimentary characters of the Betic Neogene basins evolving in a crustal transcurrent shear zone (SE Spain). *Bull. Centres Rech. Explor.-Prod. Elf-Aquitaine*, **11**, 1–22.

MUTTI, E., RICCI LUCCHI, F., SEGURET, M. & ZANZUCCHI, G. (1984) Seismoturbidites: a new group of resedimented deposits. *Mar. Geol.* **55**, 103–116.

NEMEC, W. & STEEL, R.J. (1988) What is a fan delta and how do we recognize it?. In: *Fan Deltas: Sedimentology and Tectonic Settings* (Ed. by W. Nemec and R.J. Steel), pp. 2–13. Blackie and Son, London.

NUÑEZ, A., MARTINEZ, W. & COLODRON, I. (1974) *Mapa y Memoria explicativa de la Hoja 934 (Murcia) del MAGNA*. IGME, Madrid, 34 pp.

OTT D'ESTEVOU, P. (1980) *Evolution dynamique du bassin néogène de Sorbas (Cordillères bétiques orientales, Espagne)*. Doc. et Trav. IGAL, Paris. **1**, 264 pp.

PINEDA, A., GINER, J., ZAZO, C. & GOY, J.L. (1983a) *Mapa y Memoria explicativa de la Hoja 1·046 (Carboneras) del MAGNA*. IGME, Madrid, 79 pp.

PINEDA, A., GINER, J., ZAZO, C. & GOY, J.L. (1983b) *Mapa y Memoria explicativa de las Hojas 1·059 y 1078-bis (El Cabo de Gata e Isla de Alborán) del MAGNA*. IGME, Madrid, 79 pp.

ROBERTS, H.H. (1987) Modern carbonate-siliciclastic transitions: humid and arid tropical examples. *Sedim. Geol.* **50**, 25–65.

ROEP, T. & KLEVERLAAN, K. (1983) *Excursion to a submarine-fan complex of Tortonian age and geology of the Tabernas Basin, Almería, SE Spain*. Univ. Amsterdan. 13 pp.

SANZ DE GALDEANO, C., ESTEVEZ, A., LOPEZ GARRIDO, A.C. & RODRIGUEZ FERNANDEZ, J. (1984) La fracturación tardía al SW de Sierra Nevada (terminación occidental del corredor de las Alpujarras, Zona Bética). *Estud. geol.* **40**, 183–191.

SANZ DE GALDEANO, C., RODRIGUEZ FERNANDEZ, J. & LOPEZ GARRIDO, A.C. (1986) Tectonosedimentary evolution of the Alpujarran corridor (Betic Cordilleras, Spain). *Giornale di Geologia*. **48**, 85–90.

SELLWOOD, B.W. & NETHERWOOD, R.E. (1984) Facies evolution in the Gulf of Suez area: sedimentation history as an indicator of rift initiation and development. *Modern Geol.* **9**, 43–69.

SHORT, A.D. & WRIGHT, L.D. (1983) Physical variability of sandy beaches. In: *Sandy Beaches as Ecosystems*. pp. 133–144. Dr W. Junk Publishers, The Hague.

STEEL, R. (1988) Coarsening-upward and skewed fan bodies: symptoms of strike-slip and transfer fault movement in sedimentary basins. In: *Fan Deltas: Sedimentology and Tectonic Settings* (Ed. by W. Nemec, and R.J. Steel), pp. 75–83. Blackie and Son, London.

STEEL, R., SIEDLECKA, A. & ROBERTS, D. (1984) The Old Red Sandstone basins of Norway and their deformation: a review. In: *The Caledonian Orogen: Scandinavia and Related Areas* (Ed. by D.G. Gee and B.A. Stuart), pp. 1–23. J. Wiley and Sons Ltd.

STOW, D.A.W. (1985) Deep-sea clastics: where are we and where are we going? In: *Sedimentology: Recent Developments and Applied Aspects* (Ed. by P.J. Brenchley and B.P.J. Williams). Spec. Publ. geol. Soc. London, 18, pp. 67–93.

STOW, D.A.W. (1986) Chapter 12. Deep clastic seas. In: *Sedimentary Environments and Facies* (Ed. by H.G. Reading), pp. 399–444. Blackwell Scientific Publications, Oxford.

SURLYK, F. (1978) Submarine fan sedimentation along fault scarps on tilted fault blocks (Jurassic–Cretaceous boundary. East Greenland). *Bull. Gronland geol. Unders.* **128**, 108 pp.

SURLYK, F. (1984) Fan-delta to submarine fan conglomerates of the Volgian–Valanginian Wollaston Foreland Group, East Greenland. In: *Sedimentology of Gravels and Conglomerates* (Ed. by E.H. Koster and R.J. Steel). Mem. Can. Soc. Petrol. Geol. 10, pp. 359–382.

VAN DE KAMP, P.C., HARPER, J.D., CONNIF, J.J. & MORRIS, D.A. (1974) Facies relations in the Eocene–Oligocene in the Santa Ynez Mountains, California. *J. geol. Soc.* **130**, 545–565.

VOEZMAN, F.M., MARTIN, L. & GOMEZ PRIETO, J.A. (1978) *Mapa y Memoria explicativa de la Hoja 995 (Cantoria) del MAGNA*. IGME, Madrid, 51 pp.

WESCOTT, W.A. & ETHRIDGE, F.G. (1980) Fan-delta sedimentology and tectonic setting — Yallahs fan delta, southeast Jamaica. *Bull. Am. Assoc. Petrol. Geol.* **64**, 374–399.

Spec. Publs int. Ass. Sediment. (1990) **10**, 113–127

Tectonic controls on coarse-grained delta depositional systems in rift basins

R. L. GAWTHORPE* *and* A. COLELLA†

* *Department of Geology, The University, Manchester M13 9PL, UK*

† *Dipartimento di Scienze della Terra, Università della Calabria, 87030 Arcavacata (Cosenza), Italy*

ABSTRACT

Rift basins (extensional and transtensional) commonly have an asymmetric, half-graben form that results from the throw on one border fault zone being much greater than on the other basin margin. This structural geometry, the detailed linkages between fault segments and the episodic nature of fault activity have a pronounced effect on coarse-grained delta depositional systems. A clear tectonic control can be observed on the location, form and basin-fill architecture and on the internal geometry of individual coarse-grained delta sequences in both modern and ancient rift basins. Delta location is related to structurally controlled topography and bathymetry; lateral (footwall- and hanging-wall-derived) and axial coarse-grained deltas may develop. However, the lower gradients of the hanging wall, and particularly the axial slopes, tend to promote the development of finer-grained deltaic systems. Transfer zones linking major fault segments act as loci for drainage entering the rift basin and hence are commonly the sites of coarse-grained delta deposits.

Due to the steep basin margin and large bathymetric differential across the border fault zone, footwall-derived coarse-grained deltas are generally of Gilbert-type and form isolated fan- to wedge-shaped bodies, occupying an area of $< 10\ km^2$. In contrast, the lower gradient of the hanging wall promotes the development of coarse-grained deltas with a wedge- to sheet-like form which have a low-gradient delta-front. These deltas commonly coalesce and occupy a much larger area than footwall-derived systems. Axial fan deltas are expected to show vertical stacking patterns adjacent to the border fault zone. When deformation involves a significant strike-slip component, the architecture of the basin-fill becomes complex due to lateral offset of feeder channels and depocentres. This effect is most pronounced adjacent to oblique-slip border faults and in strike-slip transfer zones, where asymmetric, off-lapping coarse-grained deltas may develop.

Near-surface coseismic deformation may trigger major failure of the delta front, leading to a complex internal structure of stacked foreset units separated by slide planes. However, where near-surface deformation is less intense, a simpler delta foreset unit may develop, consisting of alternating sigmoidal and oblique clinoforms which reflect episodes of slip and no slip respectively. Coarse-grained deltas deposited during tectonic activity should also display evidence for tilting associated with fault displacements; in the simplest case, the degree of tilt should increase in progressively older deltas. Tectonic effects on the internal structure of fan deltas and evidence for syndepositional tilting are most apparent in systems deposited adjacent to the border fault zone.

INTRODUCTION

Coarse-grained deltas are common depositional systems in many lacustrine and marine rift (extensional and transtensional) sedimentary basins, for example, East Greenland (Surlyk, 1977, 1978), Dead Sea (Sneh, 1979; Manspeizer, 1985), Gulf of Suez (Garfunkel & Bartov, 1977; Sellwood & Netherwood, 1984; Gawthorpe *et al.*, 1990), Crati Basin and Messina Strait (Colella, 1988a, b), Lake Hazar (Dunne & Hempton, 1984), Gulf of Corinth (Leeder *et al.*, 1988; Ori, 1989), Ridge Basin (Crowell, 1975; Link & Osbourne, 1978). The current upsurge of interest in coarse-grained delta depositional systems partly results from their potential use in basin analysis, but also because they act as hydrocarbon reservoirs, form stratigraphic traps and are sites of mineral deposits.

The location of coarse-grained deltas, often adjacent to basin margins, means they have the potential to act as recorders of the tectonics that operated during basin evolution, in particular tilt events and fault-displacement history (Leeder & Gawthorpe, 1987; Colella, 1988a). The area most sensitive to tectonically induced base-level changes is the alluvial to marine/lacustrine transition zone, where fluctuations of a few metres can give rise to a substantial change in the facies of coarse-grained delta deposits (Postma & Cruickshank, 1988; Colella, 1988b). However, coarse-grained deltas are also strongly affected by more regional phenomena, such as climatic/eustatic changes, for example in the Corinth Basin (Collier, 1990) and the Crati Basin and Messina Strait (Colella, 1988b). Hence, unravelling the response of such deltas to tectonic and eustatic control is important in the analysis of many ancient sedimentary basins. This paper concentrates on isolating the tectonic controls on coarse-grained delta depositional systems, paying particular attention to the location and architecture of these systems within the rift environment, their form and internal geometry.

BASIN GEOMETRY AND STRUCTURAL STYLE

Sedimentary basins developed under transtensional or orthogonal extension have a similar geometry, that of a half-graben, where the throw on one border fault is much greater than on the other (Fig. 1). This structural asymmetry is seen in many modern and ancient rift basins throughout the world in different plate-tectonic settings; for example, in extensional basins, the Gulf of Suez (Colletta *et al.*, 1988), UKCS (Harding, 1984; Cheadle *et al.*, 1987), East Greenland (Surlyk, 1977, 1978), East Africa (Rosendahl *et al.*, 1986) and, in transtensional basins, Lake Hazar (Hempton & Dewey, 1981), Ridge Basin (Crowell, 1975; Link & Osbourne, 1978) and the Californian Borderlands (Howell *et al.*, 1980). Thus it appears that rift kinematics are independent of regional dynamics and hence general comments, applicable to a number of tectonic settings can therefore be made regarding rift structure.

Individual half-graben are typically 50–150 km in length (parallel to the border fault) and 15–50 km wide (Fig. 1). In its simplest form, the surface morphology of these basins can be resolved into a footwall scarp slope developed along the border fault margin and a hanging-wall dip slope. The footwall scarp slope dips steeply into the hanging wall, often at angles in excess of 30°, whereas the hanging-wall dip slope has a gentle dip, often of about 10°. The exact form of the hanging-wall dip slope is controlled by the three-dimensional geometry of the border faults and their linkages. In extensional settings, border faults are generally planar down to a depth of about 10 km and dip at 30–60° (e.g. Jackson, 1987; Stein & Barrientos, 1985). However, in rifts with a high strike-slip component, the border faults often have a complex geometry (Fig. 1), being composed of a number of fault segments which have a braided form (strike-slip duplexes) and define flower structures (Tchalenko & Ambraseys, 1970; Aydin & Page, 1984; Crowell, 1975; Woodcock & Fischer, 1986). Subsidiary faulting and associated folding on the hanging-wall dip slope affect its exact form and may create a series of intrabasin horst and graben structures. In extensional basins such structures are generally subparallel to the border faults, whereas in strike-slip basins subsidiary structures are more oblique, initiating as Riedel shears (e.g. Wilcox *et al.*, 1973).

The finite length of both the border and intrabasin fault segments necessitates accommodation of their displacement. This occurs across transfer zones (Fig. 1), which range in width and structural style from a single fault zone to complex flexures of fault arrays tens of kilometres across. Individual fault transfer zones are generally associated with intrabasinal structures, creating a sinuous fault trace of the kind often seen on detailed structure maps. The larger, complex transfer zones are often associated with linkages between individual half-graben of different polarity. This structural style is common in elongate rift basins and is clearly displayed in the East African Rift (Rosendahl *et al.*, 1986), the Gulf of Suez Rift (Colletta *et al.*, 1988, Gawthorpe *et al.*, 1990) and the Italian Apennines (Westaway *et al.*, 1989).

Since many sedimentary processes of coarse-graind deltas are gravity driven, the tectonic slopes created by asymmetric subsidence have a profound effect on erosion, sediment transfer and deposition. In addition, the various fault linkages (faults and flexures) within the basins act as potential conduits for depositional systems to enter rift basins, or to be con-

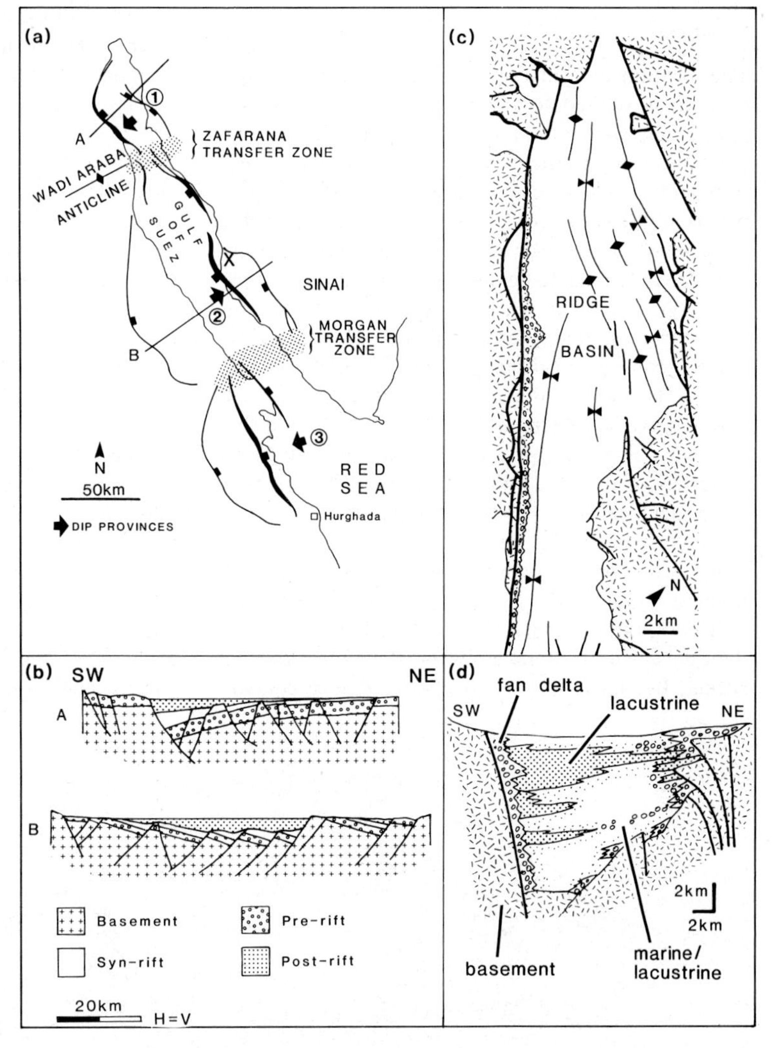

Fig. 1. Structural style of rift basins. (a) Simplified structure map of the Gulf of Suez extensional system, showing the three main half-graben segments linked by major polarity-flipping transfer zones. The letter X indicates the location of Figs 4, 11. (b) Cross-sections across the northern and central half-graben illustrating the asymmetric basin form, border fault zone and synthetic and antithetic faults (after Colletta *et al.*, 1988) (c) Simplified structure map of the Ridge Basin, California, with the main dextral strike-slip fault bounding the west side of the basin. Note the *en echelon* folds in the basin-fill (unornamented) (after Crowell, 1975, Link & Osbourne, 1978). (d) Simplified cross-section through the Ridge Basin showing the asymmetric form, with steep footwall scarp associated with the border fault and the broad lower gradient hanging-wall dip slope (after Crowell, 1975; Link & Osbourne; 1978).

strained within the rift. In the following section the effect of these phenomena on coarse-grained delta sedimentation is examined in detail.

TECTONIC CONTROLS ON COARSE-GRAINED DELTAS

Coarse-grained delta depositional systems are affected by synsedimentary tectonism on a variety of scales (Leeder & Gawthorpe, 1987). First, on a basin scale, footwall uplift and hanging-wall subsidence create sediment sources and depocentres, whilst the detailed fault geometry acts to influence the location of drainage systems. This large-scale tectonic control thus has a fundamental influence on the location of coarse-grained deltas in the rift environment. Second, the asymmetric subsidence associated with half-graben development affects the external form of delta sequences and the stratigraphic relationships between successive fan sequences, i.e. basin-fill architecture. Third, the scale of vertical fault displacement during individual slip events and the episodic nature of faulting, characterized by alternating phases of tectonic activity and relative quiescene, have a marked effect, not only on the basin-fill architecture but also on the internal geometry of individual deltas. The following section addresses this hierarchical tectonic control on coarse-grained deltas under the headings: location, form and basin-fill architecture, and internal geometry.

Location

The location of coarse-grained deltas in rift basins (Fig. 2) is related to structurally controlled topography and bathymetry which, in turn, control the location of depocentres, sediment entry points and sediment production potential. Nowhere in rift basins are these controls more apparent than in the zone along the footwall scarp where footwall-derived drainage enters the hanging-wall basin. In this setting coarse-grained delta location is related to streams issuing from drainage basins which developed during rifting: 'consequent drainage basins' (Leeder *et al.*, 1988). The marked decrease in gradient associated with emergence of streams off the footwall onto the hanging wall results in a lowering of the transport efficiency of the flow and deposition of fan bodies. The exact location of the drainage basins and, hence, the delta bodies may be influenced by anisotropy, either structural or compositional, of the footwall uplands.

In addition to the coarse-grained deltas previously described, whose drainage basins developed during rifting, Leeder *et al.* (1988) also recognize coarse-grained deltas which are sourced from drainage basins that formed prior to rifting: 'antecedent drainage'. The drainage system sourcing the present-day coarse-grained delta at Aigion along the south side of the Gulf of Corinth, Greece is an excellent example of antecedent drainage. In this case, antecedence is indicated by the presence of older, Pliocene–Pleistocene fan deltas in the uplifted footwall to the currently active normal fault bounding the southern margin of the gulf. In many ancient basins, however, antecedent coarse-grained deltas may be difficult, if not impossible, to identify, due to the low preservation potential of the drainage basin and associated deposits in the footwall of border faults.

Depositional systems derived from the hanging-wall are generally more diverse than those developed adjacent to the footwall scarp, due mainly to the lower gradient of the hanging-wall dip slope. For example, coarse-grained delta bodies may pass laterally into broad coastal plains and/or shallow-marine sand bodies. A corollary of the lower gradient of the hanging-wall dip slope, and particularly axial slopes, compared to the gradient of the

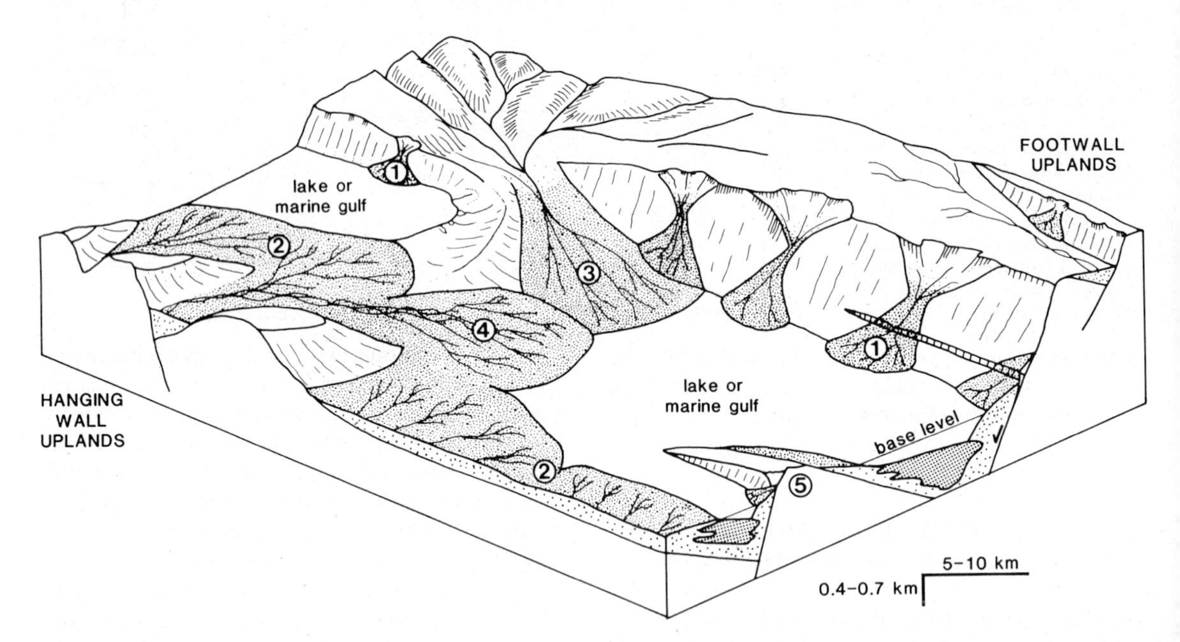

Fig. 2. Schematic block diagram illustrating the characteristic locations of coarse-grained deltas in both modern and ancient rift basins. (1) Footwall-derived coarse-grained deltas sourced from consequent drainage basins developed along the border fault zone. (2) Hanging-wall-derived coarse-grained deltas forming a relatively continuous fringe along the hanging-wall dip slope. (3) Delta sourced from a flexural transfer zone separating two *en echelon*, border-fault segments. (4) Axial delta. (5) Coarse-grained deltas derived from an intrabasin fault block. Note, that in many rift basins, the low gradient of the hanging wall, and especially axial slopes, promotes the development of finer-grained fluviodeltaic depositional systems rather than coarse-grained deltas.

footwall scarp is that the transport capacity of the feeder alluvial system may be much lower. Thus finer-grained, sand/mud deltaic systems may develop in these basin settings, rather than coarse-grained deltas. This difference between footwall-derived fan deltas and axial depositional systems is clearly seen in the Devonian Hornelen Basin, Norway (e.g. Steel & Gloppen, 1980) and the Little Sulphur Creek basins (Nilsen & McLaughlin, 1985).

Excellent examples of hanging-wall-derived fan deltas are present along the northern margin of the Lamia Basin, Greece, along the east side of the Walker Lake Basin, Nevada (Link *et al.*, 1985), in the Gulf of Suez (Fig. 5) and along the eastern side of the Crati Basin (Colella, 1988a). The Crati Basin is an 'L-shaped' Pliocene–Holocene basin which, in its N–S-trending sub-basin, contains two distinctive coarse-grained delta types derived from opposing basin margins (Fig. 3). Deltas that prograded from the Coastal Range to the west, a rapidly uplifted footwall, are mainly conglomeratic Gilbert-type systems. In contrast, the deltas derived from the hanging wall are dominantly finer-grained, shelf-type systems. The different types of delta were chiefly controlled by the different rates of vertical fault motion and hence basin-floor gradient on the opposing basin margins (Colella, 1988a).

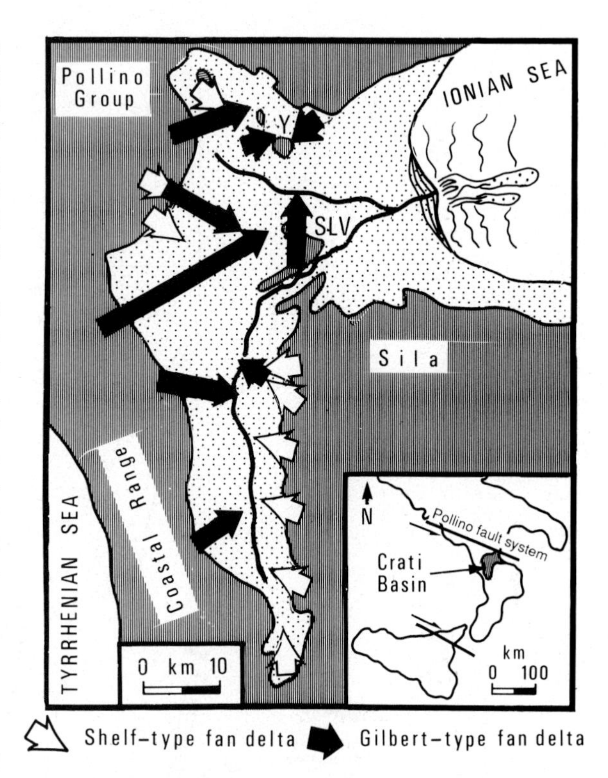

Fig. 3. Map of the Crati Basin, Italy, showing the location and progradation direction of the main types of Pliocene–Holocene coarse-grained deltas within the basin-fill. Gilbert-type bodies were mainly derived from the footwall highlands (Coastal Ranges), whereas lower-gradient, shelf-type deltas were sourced from the hanging-wall uplands (Sila Massif). SLV = San Lorenzo dell Vallo High, Y indicates the position of Fig. 7.

Where the hanging wall is dissected by major syn- and/or antithetic faults, intrabasin highs with a horst or tilt-block morphology may act as local sediment sources for coarse-grained delta systems. The San Lorenzo del Vallo High in the Crati Basin, Italy (Colella, 1988a,b) acted as a local, intrabasin source for Pleistocene, Gilbert-type deltas (Fig. 3). This high, located at the boundary between the N–S- and E–W-trending sectors of the basin, began to uplift at the end of the early Pleistocene, subdividing the Crati Basin into two main gulfs. This uplift caused cannibalization of older (early Pleistocene) deltaic sediments, providing an intrabasin sediment source for delta progradation. Similar structural geometries may occur in the footwall of border faults associated with footwall back-basins, but the preservation potential of these deposits is generally low due to footwall uplift and erosion. Coarse-grained delta deposits in this structural location are clearly seen in the Miocene sequences from the footwall of the main border fault of the central and southern dip-domains of the Gulf of Suez (e.g. Gawthorpe *et al.*, 1990; Burchette, 1988).

Fault linkages, or transfer zones, either as cross-faults or flexures, form loci for drainage development and act as conduits for drainage systems. These structures are of major importance on an inter- and intrabasin scale in locating coarse-grained depositional systems. On an interbasin scale, major transfer zones separating half-graben segments may act as topographic lows and enable drainage to enter the rift basin. Subsurface data from the Morgan Transfer Zone in the Gulf of Suez illustrated in Rine *et al.* (1988) and Smale *et al.* (1988) suggest progradation of a 'larger-than-average' fan-delta along this zone from both rift margins. Similar transfer-zone control on drainage is apparent in the East African Rift (Crossley, 1984), where drainage systems entering at major polarity-flipping transfer zones often become major axial depositional systems. On an intrabasin scale, transfer zones still have a major effect on the course of drainage systems. These zones may link segments of the border fault system

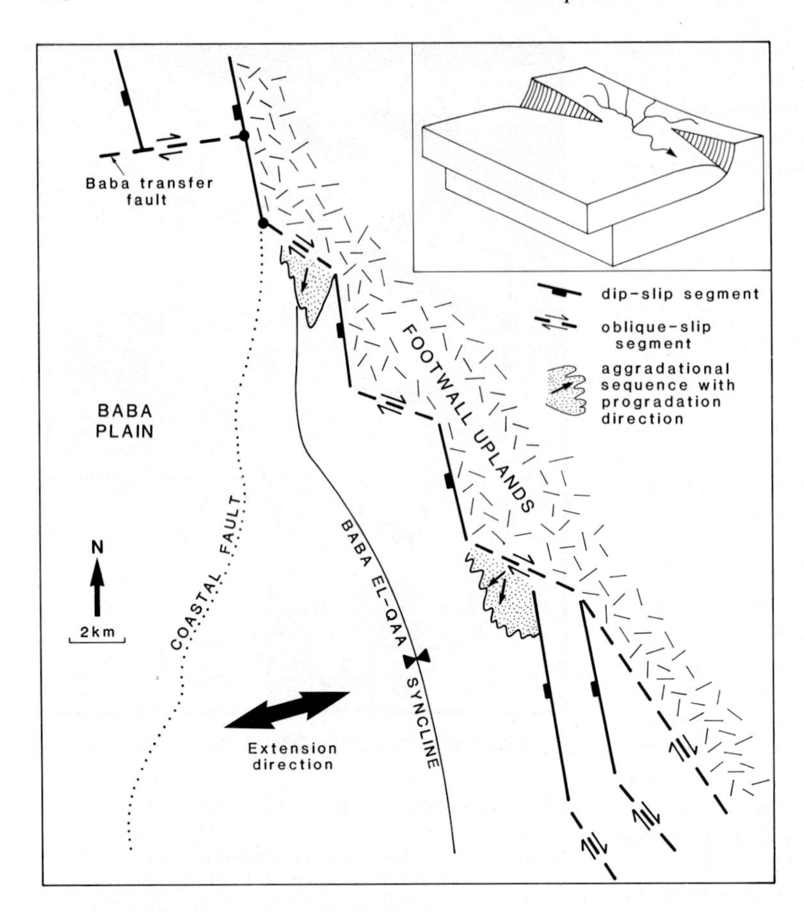

Fig. 4. Simplified structure map showing the location of synextension fan deltas of Miocene age exposed in the footwall of the border fault of the central half-graben, Gulf of Suez (for location see Fig. 1). Note how the deltas are located at strike-slip jogs linking extensional segments of the fault zone. These zones were probably small flexural transfer zones during the early stages of extension (see inset) and localized drainage entering the hanging wall.

within a half-graben and/or intrabasin structures. In the late Jurassic of the Wollaston Forland, East Greenland, flexural transfer zones are associated with the influx of conglomerates and sandstones several kilometres in thickness (Surlyk, 1977, 1978). Small footwall-derived fan deltas associated with the East Margin Fault Zone, Gulf of Suez, are similarly located at small transfer segments (Gawthorpe *et al.*, 1990) (Fig. 4).

Form and basin-fill architecture

The terms 'form' and 'basin-fill architecture' are used here to describe the external size and shape of an individual coarse-grained delta body and the geometrical relationships between delta bodies within the basin-fill. Differences in form and basin-fill architecture result from variations in the rates of three main variables: sedimentation, relative subsidence and tilting.

Form

Two main external forms of coarse-grained delta can be recognized, depending on their structural location. Coarse-grained deltas derived from the footwall generally have a fan-shaped external form (*sensu* Mitchum *et al.*, 1977). Consequent such deltas are generally restricted to within 5–10 km of the footwall scarp and cover an area of < 20 km^2, often < 10 km^2 (Figs 4, 5, 6). The bodies thicken towards the border fault and have a steep basinward margin. Two main factors are responsible for this form. First, the size of the fans is related to the size of the drainage basin (Denney, 1965; Hooke, 1972), so that small consequent drainage basins developed along the narrow footwall scarp produce relatively small coarse-grained deltas. Second, the marked topographic and bathymetric differential across the border fault and the tilt of the hanging wall towards the fault severely limit the amount of basinward

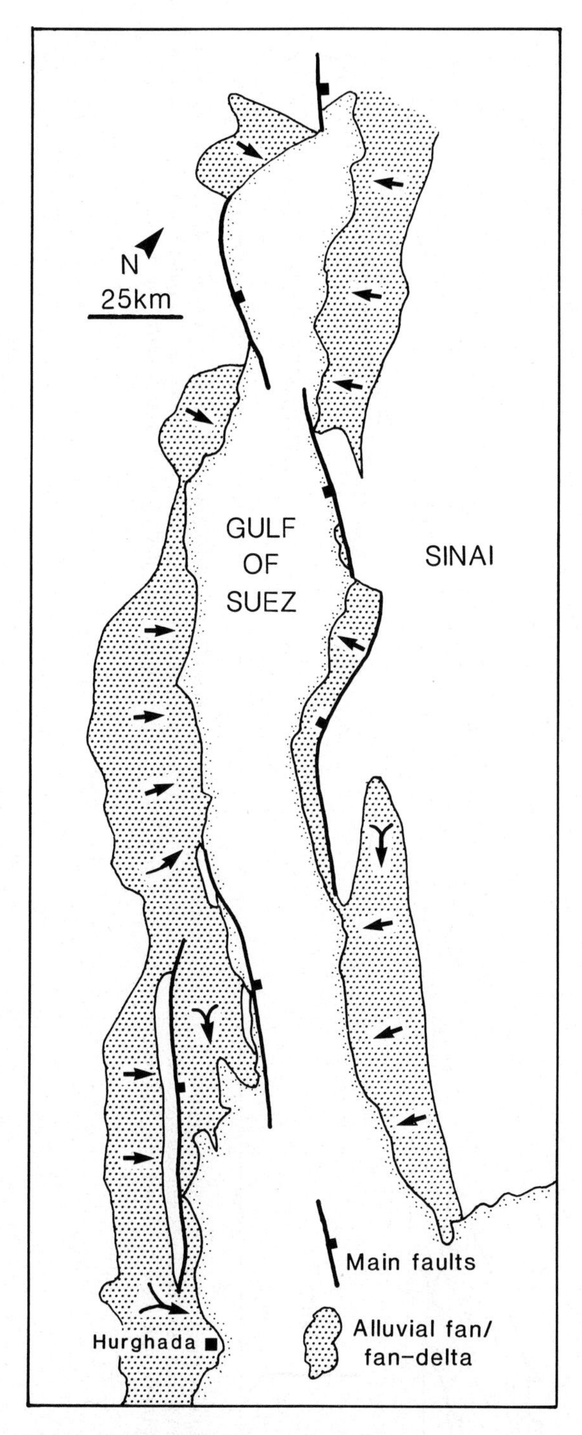

Fig. 5. Simplified map of the Gulf of Suez showing the main border fault segments and the location of Quaternary alluvial and fan-delta deposits. Note the common occurrence of broad, continuous fringes along the hanging wall of the half-graben and the narrow, occasionally coalesced fan deltas along the border faults. Arrows indicate dominant sediment-transport direction.

progradation. In addition, subsidence of the hanging wall allows thick sequences to develop; individual footwall-derived coarse-grained deltas may be several hundred metres in thickness. The Plio–Pleistocene to Recent fan deltas developed along the fault-bounded, southern margin of the Gulf of Corinth, Greece, display many of the morphological features just described (Ferentinos *et al.*, 1988). Near the town of Skinos, Recent fan deltas form isolated bodies 1–2 km^2 in area and are currently being reworked by marine processes (Leeder *et al.*, 1988). Further west, between the towns of Corinth and Aigion, exposed Plio–Pleistocene fan deltas form bodies generally between 150 m and 250 m thick (Fig. 6).

Since the drainage basin area associated with coarse-grained deltas sourced via transfer zones or antecedent channels is more variable and often much larger than that associated with consequent systems (Leeder & Gawthorpe, 1987), the size of these deltas is often much larger than the values just quoted. The complex geometry of transfer zones results in marked variations in the form of coarse-grained deltas from fan-like to more wedge-shaped bodies. Typically, coarse-grained deltas associated with fault-dominated transfers have a fan-like form, whereas fan- to wedge-shaped forms dominate in flexural zones.

Coarse-grained deltas derived from the hanging wall or axially tend to be laterally more extensive than their foot-wall-derived counterparts and have a wedge- to sheet-like form. Examples of hanging-wall-derived coarse-grained deltas from the Gulf of Suez (Fig. 5) suggest that individual bodies cover an area of about 150 km^2 and prograde into the basin by approximately 15 km. In the transtensional Walker Lake Basin, Nevada (Link *et al.*, 1985), hanging-wall-derived fan deltas prograde between 4 km and 8 km down the hanging-wall dip slope, whereas footwall-derived fan deltas prograde only 1–2 km. The larger drainage basin area developed along the hanging-wall dip slope (due to its lower gradient and greater areal extent when compared to the footwall scarp) is the main factor responsible for this difference in form. The exact form of these bodies is strongly influenced by intrabasin structure, although they tend to show a down-slope thickening trend and have low delta-front slopes. These types of depositional system have been termed 'shelf-type' or 'shoal-type' coarse-grained deltas (Etheridge & Wescott, 1984; Leeder *et al.*, 1988).

The episodic nature of rifting also has a marked

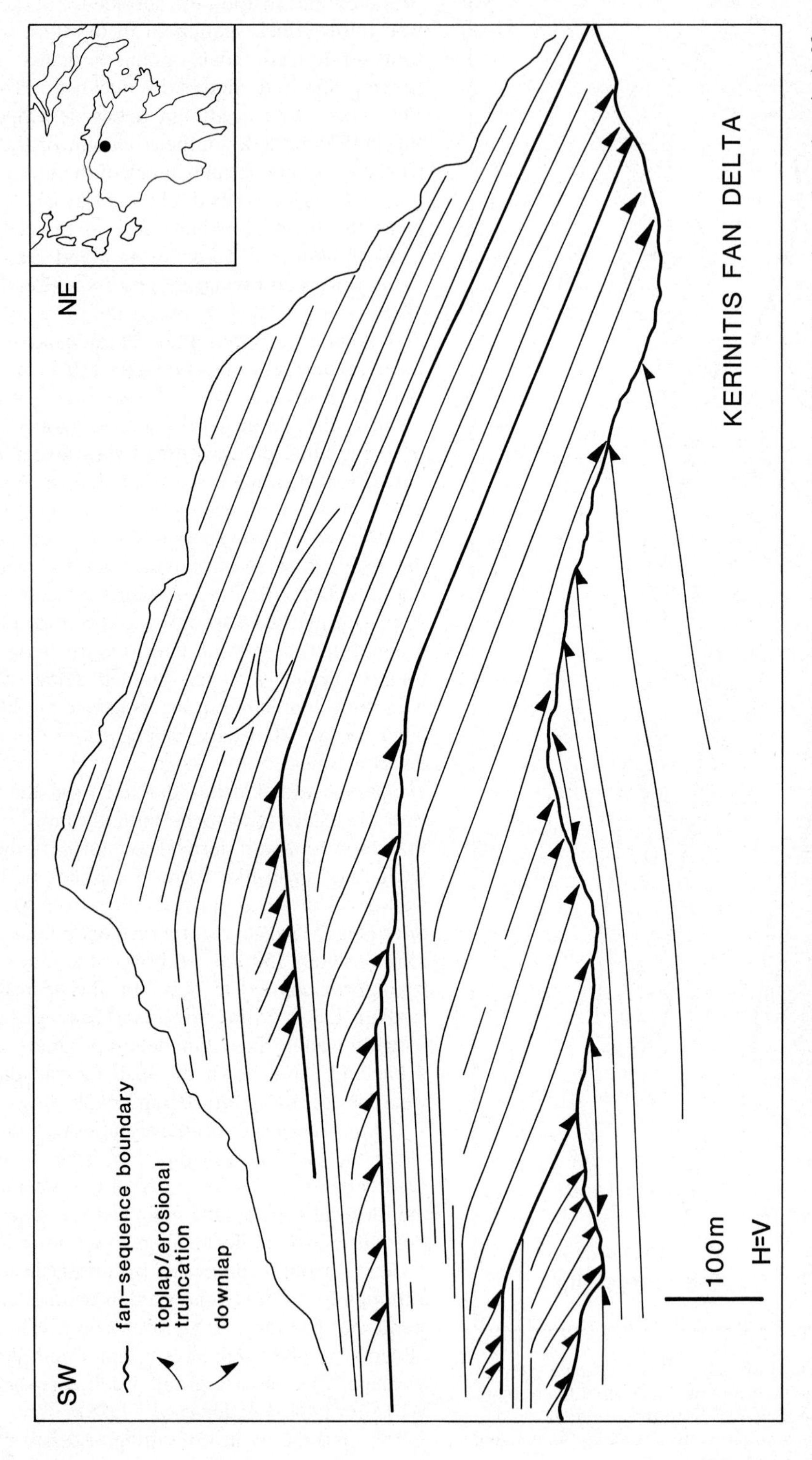

Fig. 6. Simplified line drawing of a large, uplifted, Pliocene–Pleistocene footwall-derived fan-delta near the town of Aigion, Greece (inset shows location). Note the > 600 m thickness of vertically stacked fan-delta sequences; sequence boundaries are marked by thicker black lines. Note the sigmoidal and oblique clinoform geometries within individual fan sequences. The border fault active during Pliocene–Pleistocene times lies a few kilometres to the southwest.

effect on the form of coarse-grained deltas. For example, the fan deltas of the Abu Alaqa Group, Gulf of Suez, show a marked difference in form depending on whether they were deposited during phases of tectonic activity or relative quiescence (Gawthorpe *et al.*, 1990). Tectonically active intervals are characterized by a fan-shaped form, extending only a few kilometres from the footwall scarp, with a progradation distance to thickness ratio of 0.25 to 0.4. In contrast, coarse-grained deltas deposited during quiescent intervals have a progradation distance to thickness ratio of between 0.01 and 0.03, are wedge- to sheet-like in form and cover a much larger area, up to 50 km^2.

Basin-fill architecture

Since coarse-grained deltas are mainly deposited by processes that are gravity driven, these depositional systems will attempt to follow topographic/bathymetric lows and thus basin asymmetry and intrabasin structure will have a marked influence on the basin-fill architecture associated with such deltas. In rift basins undergoing orthogonal extension, differences in architecture are dominated by the position of the delta system with respect to the border fault. Footwall-derived coarse-grained deltas create a series of vertically stacked deltas sequences which, if the feeder conduit is fixed — for example, antecedent drainage or transfer-zone control — may form units ranging from several hundred metres to over a kilometre in thickness. This vertical stacking of deltas is clearly displayed by the Plio–Pleistocene Kerinitis fan-delta system exposed in the footwall of the active fault bounding the south side of the Gulf of Corinth (Fig. 6). This fan-delta system consists of at least four vertically stacked delta sequences, 150–200 m thick, that show limited progradation onto the hanging wall. Although many workers describe laterally extensive bajada along footwall scarps, our observations in basins undergoing orthogonal extension suggest many coarse-grained deltas form isolated bodies and significant amounts of basinal fines may extend right up to the footwall scarp.

In contrast to the architecture developed in areas undergoing orthogonal extension, basins where there is a significant component of strike-slip deformation may have laterally continuous coarse-grained delta fringes. These fringes develop due to the lateral offset of feeder channels with respect to the depocentre, with time creating systematic off-lap and younging trends in the basin-margin fill. Spectacular examples of this type of delta architecture are seen in the Hornelen Basin (e.g. Steel, 1988), the Ridge Basin (Crowell, 1975, 1982) and the Crati Basin (Colella, 1988b). Recognition of these architectural styles is an important constraint in the palaeotectonic analysis of sedimentary basins, but they may be difficult to identify unless matching of 'exotic' clasts from local source areas is possible, or individual delta sequences can be linked to specific drainage systems. An example of the latter is present in the hanging wall to the WNW–ESE-trending Pollino fault bounding the northern margin of the Crati Basin (Colella, 1988b) (Fig. 7). The Pollino fault is an oblique-slip fault, down-throwing to the southwest, with a sinistral strike-slip component. It was active during early Pleistocene times, separating the feeder drainage system (Raganello River) from a series of Gilbert-type fan deltas to the south. The periodic and rapid subsidence of the hanging-wall basin floor, together with substantial lateral movement, gave rise to a series of off-lapping fan-delta sequences as the feeder channel migrated northwestward.

In hanging-wall and axial coarse-grained deltas, the direction of progradation with respect to the tectonic tilt vector has a major control on architecture. Although this effect has received little attention with respect to coarse-grained deltas, parallel studies on alluvial architecture (e.g. Leeder & Gawthorpe, 1987; Bridge & Leeder, 1979; Alexander & Leeder, 1987) suggest how they may react to deformation. Axial systems tend to reoccupy the area of maximum subsidence and thus vertical stacking of delta bodies should be expected adjacent to the border fault. Simulation models for alluvial stratigraphy illustrate how, with increased tilting (fault throw), the connectivity between sand bodies increases and the zone of channel belts (delta lobes) becomes more localized adjacent to the border fault (Fig. 8). However, where footwall-derived coarse-grained deltas occupy the area of maximum subsidence, or where intrabasin graben trap axial coarse-grained deltas, the architecture can be significantly more complex.

Hanging-wall-derived coarse-grained deltas often coalesce to form laterally extensive sheets, as illustrated in the Gulf of Suez (Fig. 5). Consideration of the effects of normal faulting on the hanging wall suggests that hanging-wall slopes will experience a gradient increase (Leeder & Gawthorpe, 1987) during fault activity. The effect of these movements

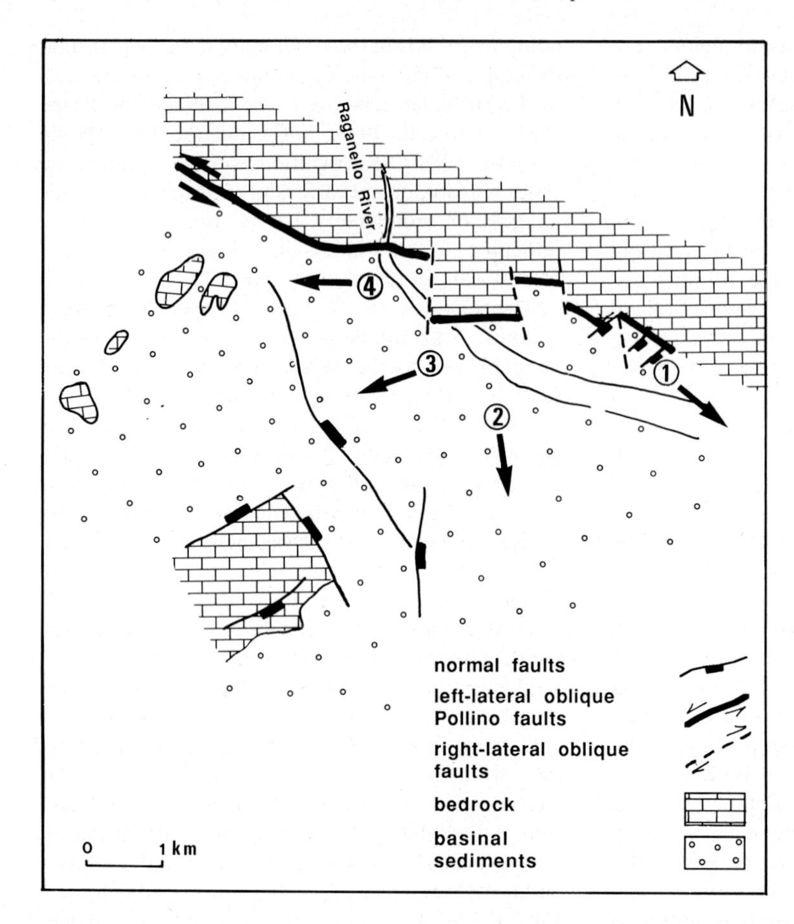

Fig. 7. Sketch map of the Pollino fault zone, Crati Basin, Italy, illustrating the position of four fan-delta sequences (labelled 1–4; no. 1 is the oldest). Note how progressive sinistral strike-slip has moved the Raganello River (feeder system) to the northwest relative to the hanging wall, creating a series of northwest-younging fan-delta sequences. Arrows indicate the progradation direction (see Fig. 3 for location).

on sedimentation has been clearly demonstrated for alluvial fan/fan-delta deposits in Death Valley, California (Hooke, 1972). Hanging-wall-derived fans will undergo fan-head channel incision and a new fan segment may form basinward of the old fan, creating an off-lapping, down-slope thickening and fining wedge. In the Megara Basin, Greece, off-lapping coarse-grained deltas on the hanging-wall dip slope form a body up to 30 m thick, with individual off-lapping, wedge-shaped fan sequences up to 15 m in thickness (Fig. 9). These thicknesses and architecture are very different from the large, footwall-derived, Kerinitis fan delta described on p. 121 (Fig. 6).

Both axial and hanging-wall zones in rifts, because of their low gradient, are particularly prone to base-level changes of either climatic or eustatic origin, which may operate on a Milankovich time scale. Highstands may cause inundation of the coarse-grained delta sytems, and deposition of basinal fines (marine or lacustrine). These form laterally extensive, sand-poor horizons within the basin-fill and may form important regional seals for hydrocarbon accumulation. Alternatively, such changes may cause considerable erosion of coarse-grained deltas during lowstands as seen in the fan deltas on the hanging-wall of the Megara Basin, Greece.

Internal geometry

The internal geometry of coarse-grained deltas can be controlled by faulting and associated surface deformation in rift basins. Since these geometries can be appreciated on both an outcrop and seismic scale, they are of great importance in the analysis of ancient sedimentary basins.

The large bathymetric differential across the footwall border fault and the occurrence of large-scale faulting events encourages the formation of Gilbert-

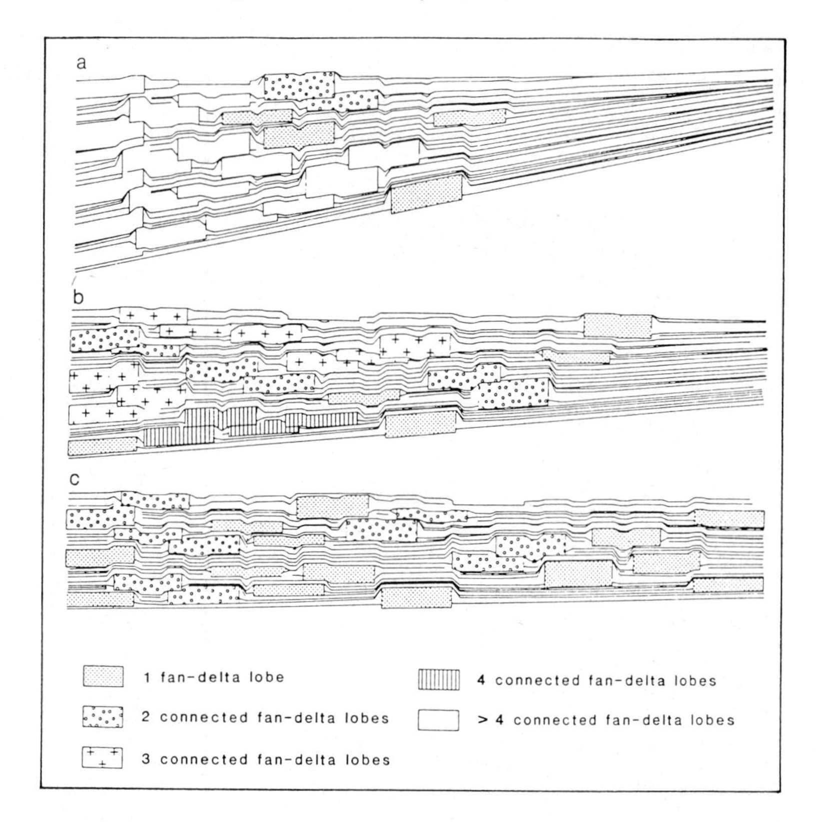

Fig. 8. Computer simulation showing how tilting of the hanging wall influences the position and basin-fill architecture of axial fluvio-delta systems (after Alexander & Leeder, 1987). The cross-sections are perpendicular to a fault which downthrows to the right, and is located at the left of the sections. Note how, with increasing tilt, the connectivity of the fan-delta bodies increases and their distribution becomes concentrated closer to the fault zone.

type deltas rather than other depositional systems characterized by lower delta-front slopes. Fault motion during individual slip events, if located near the shoreline, triggers subaqueous avalanche of the fluvial bedload carried out at the river mouth and causes the development of foreset beds. The back edge of the delta foreset unit (BEFU) is then represented by a synsedimentary fault. The behaviour of this fault has a marked influence on the complexity of the foreset unit. Fault motion characterized by recurrent large-scale slip events causes periodic rejuvenation of the fault scarp and the stacking of several groups of foreset beds (Fig. 10). In contrast, fault displacement histories composed of an initial large-scale slip event but followed by smaller increments of slip and by periods of no slip cause the development of a simple delta foreset unit, showing a single group of foresets (clinoforms) with alternating sigmoidal and oblique clinoform geometries (Colella, 1988a; Gawthorpe *et al.*, 1990). Phases of active faulting and basin subsidence are reflected by sigmoidal clinoforms with delta growth being mainly aggradational. In contrast, periods of relative quiescene and stillstand promote progradational growth, marked in the sedimentary record by the development of oblique clinoforms. In settings with a marked strike-slip component, the internal geometry of footwall-derived coarse-grained deltas is further complicated and a distinct asymmetry and laterally offset stacking pattern of successive clinoform units is developed (e.g. Steel, 1988).

The stacking of foresets and the sigmoidal and oblique geometries of delta clinoforms may be a response to eustatic/climatic fluctuations, rather than tectonically induced changes of base-level. The problem therefore arises of how to recognize the tectonic control on base-level changes and consequently on sedimentation style. In the Crati Basin, where marine, Gilbert-type deltas occur, the following indicators of tectonic control have been found: (1) identification of a synsedimentary fault marking the BEFU; (2) synsedimentary deformation of the foreset beds close to the BEFU; (3) the number of base-level changes recorded in the sediments of a certain age is much greater than the number of eustatic fluctuations known for that age; and (4) evidence for syndepositional tilting.

One of the most reliable indicators of tectonic

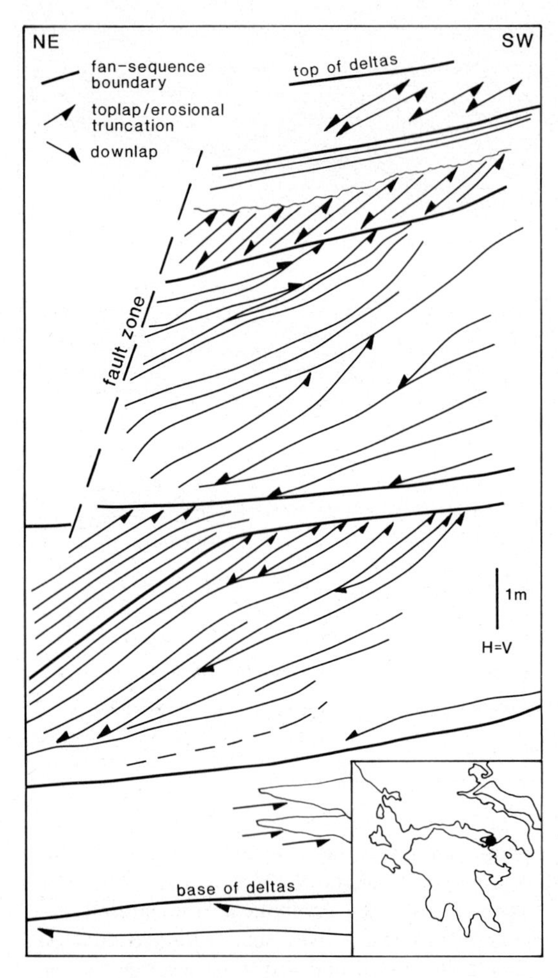

Fig. 9. Simplified line drawing of a section through fan deltas on the hanging wall of the Megara Basin, Greece (inset shows location). Note the complex internal geometry and the difference in scale compared to the footwall-derived fan deltas illustrated in Fig. 6.

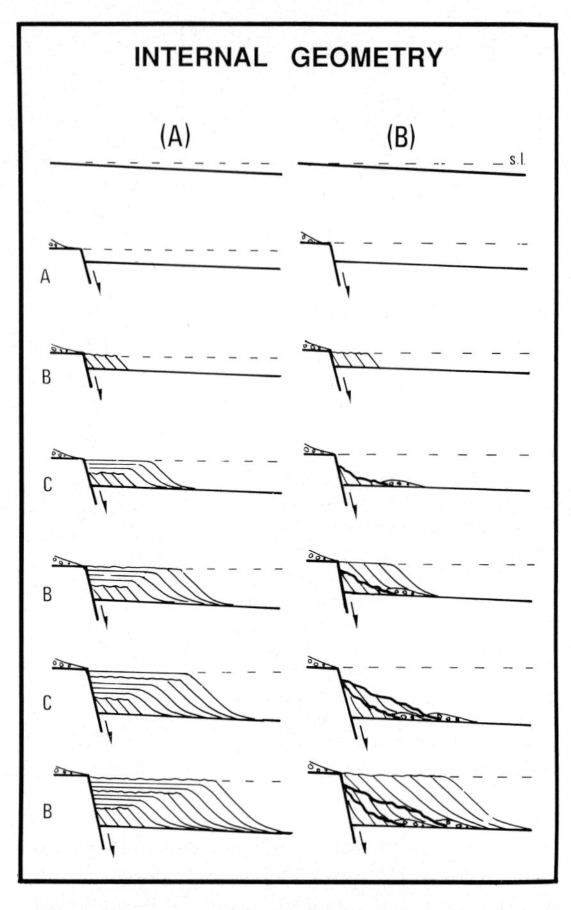

Fig. 10. Examples of the internal geometry of fan-delta bodies from the Crati Basin, Italy. (A) Simple foreset units composed of alternating sigmoidal and oblique clinoforms: this resulted from an initial high-magnitude increment of dip-slip fault motion, A, followed by episodic moderate magnitude slip events (coseismic deformation, C) alternating with periods of no slip, B. (B) Complex groups of foreset units separated by major delta-front slides resulted from high-magnitude slip events which cause failure of the delta front. Thick black lines represent slide planes.

control on the architecture of coarse-grained deltas is tilting, but evidence of variation in the degree of tilting is required to indicate that the tilting occurred during sedimentation. As already stated, evidence of tilting is best evaluated with respect to the datum provided by the transition zone in coarse-grained deltas. Examples of tilting effects on the internal geometry of coarse-grained deltas are seen in the Gulf of Suez and Crati Basin. In the Gebel Nazzazat area of the Gulf of Suez, topset beds in footwall-derived fan deltas show a progressive decrease in the amount of tilt from 45° at the base of the sequence to 22° at the top (Gawthorpe *et al.*, 1990) (Fig. 11). Thus, during deposition, the deltas underwent about 23° of rotation about a horizontal axis.

CONCLUSIONS

Coarse-grained deltas deposited in rift basins are strongly influenced by contemporary tectonic activity. These tectonic effects not only influence the location of coarse-grained deltas in the rift environment, but also have a marked control on their external form,

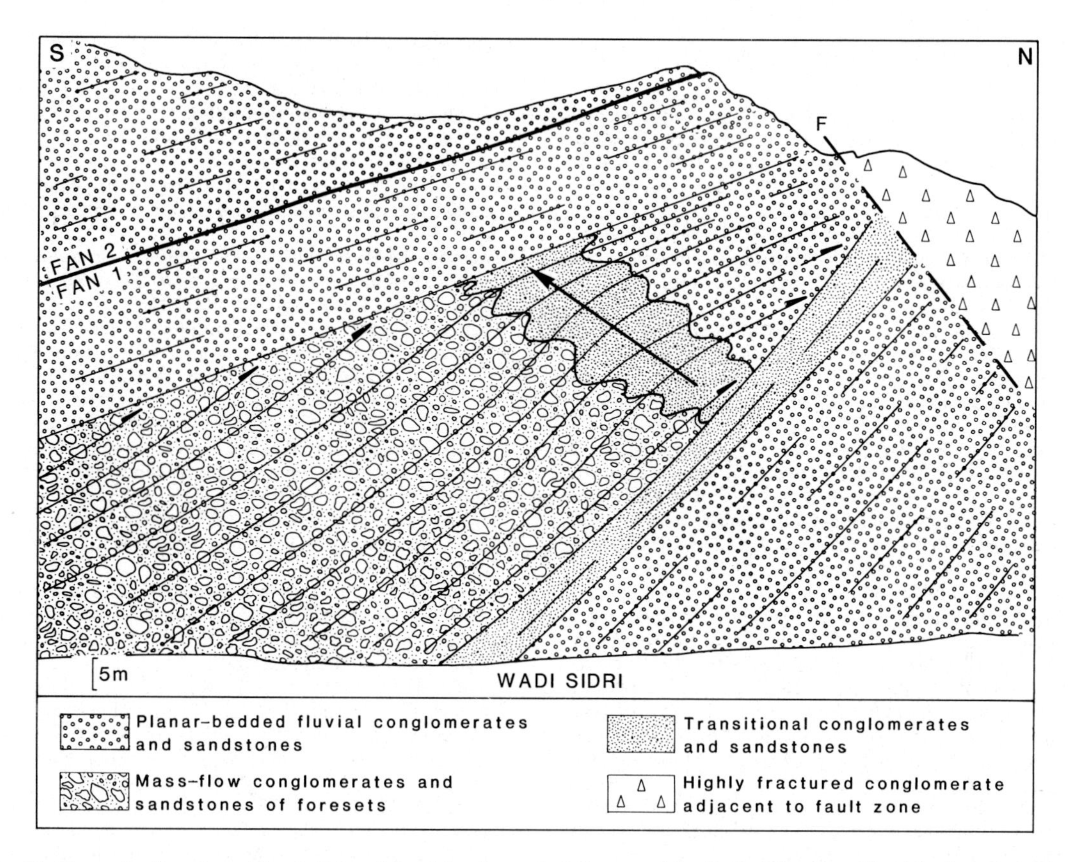

Fig. 11. Example of syndepositional tilting of fan deltas from the Miocene of the Gulf of Suez (see Fig. 1 for location). Note the decrease in the tilt of the fluvial conglomerates and sandstones that form the topset beds (deposited subhorizontally) from 45° at the base of the section to approximately 25° at the top of the cliff. Also note the evolution from sigmoidal to oblique clinoform geometry. The fault active during sedimentation lies just to the north of the fan delta illustrated.

architecture and internal geometry of such deltas. When attempting to determine the external controls on delta evolution from the ancient sedimentary record, it is important to identify a number of indicators of active tectonism before other possible mechanisms are discounted. For example, the presence of a Gilbert-type delta adjacent to a steep basin margin does not necessarily indicate a synchronous tectonic control. Similar stratigraphic relationships may be produced by eustatic variations in base level, whereby a fall in base level causes erosion and ria formation, and coarse-grained delta deposition follows during subsequent base-level rise. Alternatively, the basin margin may be fault controlled, but the phase of delta deposition post-dates tectonic activity and is therefore passively infilling an inherited bathymetry. Of the tectonic effects discussed, perhaps the most reliable is stratigraphic variations in tilting. This is most apparent close to the border fault in footwall-derived systems. In hanging-wall and axial coarse-grained delta systems, the lower slopes associated with these deposits makes identification of tectonic tilting difficult to recognize and it may be mistaken for, or overprinted by compactional effects.

ACKNOWLEDGEMENTS

We thank Drs G.G. Ori and R. Steel for their critical reading of the manuscript. Jan Alexander, Richard Collier, Mike Leeder and John Hurst are thanked for discussions on tectonic controls on sedimentation.

REFERENCES

Alexander, J. & Leeder, M.R. (1987) Active tectonic control on alluvial architecture. In: *Recent Developments in Fluvial Sedimentology* (Ed. by F.G. Ethridge, R.M. Flores and M.D. Harvey). Spec. Publ. Soc. econ. Paleont. Mineral., Tulsa, 39, pp. 243–252.

Aydin, A. & Page, B.M. (1984) Diverse Pliocene-Quaternary tectonics in a transform environment, San Francisco Bay region, California. *Bull. geol. Soc. Am.* **95**, 1030–1317.

Bridge, J.S. & Leeder, M.R. (1979) A simulation model of alluvial stratigraphy. *Sedimentology* **26**, 617–644.

Burchette, T.P. (1988) Tectonic control on carbonate platform facies distribution and sequence development: Miocene, Gulf of Suez. *Sedim. Geol.* **59**, 179–204.

Cheadle, M.J., McGeary, S., Warner, M.R. & Matthews, D.H. (1987) Extensional structures on the western UK continental shelf: a review of evidence from deep seismic profiling. In: *Continental Extensional Tectonics* (Ed. by M.P. Coward, J.F. Dewey and P.L. Hancock). Spec. Publ. geol. Soc. 28, pp. 445–465.

Colella, A. (1988a) Pliocene–Holocene fan deltas and braid deltas in the Crati Basin, southern Italy: a consequence of varying tectonic conditions. In: *Fan Deltas: Sedimentology and Tectonic Settings* (Ed. by W. Nemec and R.J. Steel), pp. 50–74. Blackie and Son, London.

Colella, A. (1988b) Quaternary Gilbert-type fan deltas of the Messina Strait. In: *Fan Deltas – Excursion Guidebook* (Ed. by A. Colella), pp. 139–152. Università della Calabria, Consenza, Italy.

Colletta, B., Quellec, P.Le, Letouzey, J. & Moretti, I. (1988) Longitudinal evolution of the Suez rift structure (Egypt). *Tectonophysics* **153**, 221–233.

Collier, R.E.L. (1990) Eustatic and tectonic controls upon Quaternary coastal sedimentation in the Corinth Basin, Greece. *J. geol. Soc.* **147**, 301–314.

Crossley, R. (1984) Controls of sedimentation in the Malawi Rift Valley, central Africa. *Sedim. Geol.* **40**, 33–50.

Crowell, J.C. (1975) The San Gabriel Fault and Ridge Basin. In: *San Andreas Fault in Southern California* (Ed. by J.C. Crowell). Spec. Rept. California Division of Mines and Geology 19, pp. 208–233.

Crowell, J.C. (1982) The Violin Breccia, Ridge Basin. In: *Geological History of the Ridge Basin, Southern California* (Ed. by J.C. Crowell and M.H. Link). Soc. econ. Paleont. Miner. Pacific Sect., pp. 89–98.

Denney, C.S. (1965) Fans and pediments. *Am. J. Sci.* **265**, 81–105.

Dunne, L.A. & Hempton, M.R. (1984) Deltaic sedimentation in the Lake Hazar pull-apart basin, south-eastern Turkey. *Sedimentology* **31**, 401–412.

Etheridge, F.G. & Wescott, W.A. (1984) Tectonic setting, recognition and hydrocarbon reservoir potential of fan-delta deposits. In: *Sedimentology of Gravels and Conglomerates* (Ed. by E.H. Koster and R.J. Steel). Mem. Can. Soc. Petrol. Geol. 10, pp. 217–235.

Ferentinos, G., Papatheodorou, G. & Collins, M.B. (1988) Sediment transport processes on an active submarine fault escarpment: Gulf of Corinth, Greece. *Mar. Geol.* **83**, 43–61.

Garfunkel, Z. & Bartov, Y. (1977) The tectonics of the Suez Rift. *Bull. Geol. Surv. Israel* **71**, 44.

Gawthorpe, R.L., Hurst, J.M. & Sladen, C.P. (1990) Evolution of Miocene footwall-derived fan deltas, Gulf of Suez, Egypt: implications for exploration. *Bull. Am. Assoc. Petrol. Geol.* (in press).

Harding, T.P. (1984) Graben hydrocarbon occurrences and structural style. *Bull. Am. Assoc. Petrol. Geol.* **68**, 333–362.

Hempton, M.R. & Dewey, J.F. (1981) Structure and tectonics of the Lake Hazar pull-apart basin, SE Turkey. *Eos, Trans. Am. geophys. Un.* **62**, 1033.

Hooke, R., LeB. (1972) Geomorphic evidence for late-Wisconsin and Holocene tectonic deformation, Death Valley, California. *Bull. geol. Soc. Am.* **83**, 2073–2098.

Howell, D.G., Crouch, J.K., Greene, H.G., McCulloch, D.S. & Veder, J.G. (1980) Basin development along the late Mesozoic and Cainozoic California margin: a plate tectonic margin of subduction, oblique subduction and transform tectonics. In: *Sedimentation in Oblique-slip Mobile Zones* (Ed. by P.F. Ballance and H.G. Reading). Spec. Publ. Int. Assoc. Sedim. **4**, pp. 43–62.

Jackson, J.A. (1987) Active normal faulting and crustal extension. In: *Continental Extensional Tectonics* (Ed. by M.P. Coward, J.F. Dewey and P.L. Hancock). Spec. Publ. geol Soc. **28**, pp. 3–17.

Leeder, M.R. & Gawthorpe, R.L. (1987) Sedimentary models for extensional tilt-block/half-graben basins. In: *Continental Extensional Tectonics* (Ed. by M.P. Coward, J.F. Dewey and P.L. Hancock). Spec. Publ. geol. Soc. **28**, pp. 139–152.

Leeder, M.R., Ord, D.M. & Collier, R.E.Ll. (1988) Development of alluvial fans and fan deltas in neotectonic extensional settings: implications for the interpretation of basin fills. In: *Fan Deltas: Sedimentology and Tectonic Settings* (Ed. by W. Nemec and R.J. Steel), pp. 173–185. Blackie and Son, London.

Link, M.H. & Osborne, R.H. (1978) Lacustrine facies in the Pliocene Ridge Basin Group, Ridge Basin, California. In: *Modern and Ancient Lake Sediments* (Ed. by A. Matter and M.E. Tucker). Spec. Publ. Int. Assoc. Sedim. **2**, pp. 169–187.

Link, M.H., Roberts, M.T. & Newton, M.S. (1985) Walker Lake Basin, Nevada: an example of Tertiary (?) to Recent sedimentation in a basin adjacent to an active strike-slip fault. In: *Strike-slip Deformation, Basin Formation, and Sedimentation* (Ed. by K.T. Biddle and N. Christie-Blick). Spec. Publ. Soc. econ. Paleont. Miner., Tulsa, **37**, pp. 105–125.

Manspeizer, W. (1985) The Dead Sea Rift: impact of climate and tectonism on Pleistocene and Holocene sedimentation. In: *Strike-slip Deformation, Basin Formation, and Sedimentation* (Ed. by K.T. Biddle and N. Christie-Blick). Spec. Publ. Soc. econ. Paleont. Miner., Tulsa, **37**, pp. 143–158.

Mitchum, R.M., Vail, P.A. & Sangree, J.B. (1977) Seismic stratigraphy and global changes of sea level; part 6: stratigraphic interpretation of seismic reflection patterns in depositional sequences. In: *Seismic Stratigraphy – Application to Hydrocarbon Exploration* (Ed. by C.E. Payton). Mem. Am. Assoc. Petrol. Geol. **26**, pp. 117–133.

Nilsen, T.H. & McLaughlin, R.J. (1985) Comparison of tectonic framework and depositional patterns of the Hornelen strike-slip basin of Norway and the Ridge and Little Sulphur Creek strike-slip basins of California. In: *Strike-slip Deformation, Basin Formation, and Sedimentation* (Ed. by K.T. Biddle and N. Christie-Blick). Spec. Publ. Soc. econ. Paleont. Miner., Tulsa, **37**, pp. 79–103.

Ori, G.G. (1989) Geologic history of the extensional basin of the Gulf of Corinth (Phocene-Pleistocene), Greece. *Geology* **17**, 918–921.

Postma, G. & Cruickshank, C. (1988) Sedimentology of a late Weichselian to Holocene, terraced fan delta, Varangerfjord, northern Norway. In: *Fan Deltas: Sedimentology and Tectonic Settings* (Ed. by W. Nemec and R.J. Steel), pp. 144–157. Blackie & Son, London.

Rine, J.M., Hassouba, A., Shishkevish, L., Shafi, A., Azazi, G., Nashaat, H., Badawy, A. & El Sis, Z. (1988) Evolution of a Miocene fan delta: a giant oil field in the Gulf of Suez, Egypt. In: *Fan Deltas: Sedimentology and Tectonic Settings* (Ed. by W. Nemec and R.J. Steel), pp. 239–250. Blackie and Son, London.

Rosendahl, R.B., Reynolds, D.J., Lorber, P.M., Burgess, C.F., McGill, J., Scott, D., Lambiase, J.J. & Derksen, S.J. (1986) Structural expression of rifting: lessons from Lake Tanganyika, Africa. In: *Sedimentation in the African Rifts* (Ed. by L.E. Frostick, R.W. Rentant and I. Reid). Spec. Publ. geol. Soc. **25**, pp. 29–44.

Sellwood, B.W. & Netherwood, R.E. (1984) Facies evolution in the Gulf of Suez area: sedimentation history as an indicator of rift initiation and development. *Modern Geol.* **9**, 43–69.

Smale. J.L., Thunell, R.C. & Schamel, S. (1988) Sedimentological evidence for early Miocene fault reactivation in the Gulf of Suez. *Geology* **16**, 113–116.

Sneh, A. (1979) Late Pleistocene fan deltas along the Dead Sea Rift. *J. sedim. Petrol.* **49**, 541–551.

Steel, R.J. (1988) Coarsening-upward and skewed fan bodies: symptoms of strike-slip and transfer fault movement in sedimentary basins. In: *Fan Deltas: Sedimentology and Tectonic Settings* (Ed. by W. Nemec and R.J. Steel), pp. 75–83. Blackie and Son, London.

Steel. R.J. & Gloppen, T.G. (1980) Late Caledonian (Devonian) basin formation, western Norway: signs of strike-slip tectonics during infilling. In: *Sedimentation in Oblique-slip Mobile Zones* (Ed. by P.F. Ballance and H.G. Reading). Spec. Publ. Int. Assoc. Sedim. **4**, pp. 79–103.

Stein, R.S. & Barrientos, S.E. (1985) High-angle faulting in the intermountain seismic belt: geodetic investigation of the 1983 Borah Peak, Idaho, earthquake. *J. geophys. Res.* **90**, 11355–11366.

Surlyk, F. (1977) Stratigraphy and tectonics and palaeogeography of the Jurassic sediments of the areas north of Kong Oscars Fjord, E. Greenland. *Bull. Gronland. Geol. Unders.* **128**, 108 pp.

Surlyk, F. (1978) Jurassic basin evolution of E. Greenland. *Nature* **274**, 130–133.

Tchalenko, J.S., & Ambraseys, N.N. (1970) Structural analyses of the Dasht-e-Bayaz (Iran) earthquake fractures. *Bull. geol. Soc. Am.* **81**, 41–60.

Westaway, R.W.C., Gawthorpe, R.L. & Tozzi, M. (1989) Seismological and field observations of the 1984 Lazio-Abruzzo earthquakes: implications for the active tectonics of Italy. *Geophys J.* **98**, 489–514.

Wilcox, R.E., Harding, T.P. & Seely, D.R. (1973) Basic wrench tectonics. *Bull. Am. Assoc. Petrol. Geol.* **57**, 74–96.

Woodcock, N.H. & Fischer, M. (1986) Strike-slip duplexes. *J. struct. Geol.* **8**, 725–735.

Spec. Publs int. Ass. Sediment. (1990) 10, 129–151

Pleistocene fan deltas in southeastern Iberian peninsula: sedimentary controls and sea-level changes

T. BARDAJI*, C. J. DABRIO[†], J. L. GOY[‡], L. SOMOZA[§] and C. ZAZO[§]

* Departamento de Geología, Facultad de Ciencias, Universidad de Alcalá de Henares, Madrid, Spain
Departamento de Geodinámica, Facultad de Geológicas, Universidad Complutense, 28040-Madrid, Spain
[†] Departamento de Estratigrafía , Facultad de Geológicas, Universidad Complutense, 28040-Madrid, Spain
[‡] Departamento de Geodinámica, Facultad de Geológicas, Universidad Complutense, 28040-Madrid, Spain
[§] Departamento de Geología, Museo Nacional de Ciencias Naturales, C.S.I.C., Madrid, Spain; Departamento de Geodinámica, Facultad de Geológicas, Universidad Complutense, 28040-Madrid, Spain

ABSTRACT

Examples of Pleistocene fan-delta deposits occur in the Neogene Basin of Cope (southeast Spain). The interaction of active faulting, successive sea-level changes and a variable pattern of sedimentary environments and processes resulted in the deposition of prograding offlapping sequences of interlayered marine and terrestrial deposits. Careful studies of morphology, geometry of sedimentary bodies and clinoforms, sedimentary structures and sequences have been used to infer the precise contribution of the various controls of sedimentation. In particular, the breaker-zone facies has been used as a marker of palaeo sea-levels. Cycles of sea-level fluctuations were deduced plotting present-day topographic positions of coastal fan-delta deposits. Comparison of those results with known global, theoretical curves of sea-level change permitted identification of local cycles and deduction of the precise contribution of tectonic factors (fault movements) during and after Pleistocene times.

INTRODUCTION

The Cope Basin (Fig. 1), located in the morphostructural domain of the eastern Betic Ranges (SE Spain), is one of the many Neogene basins formed in a left-lateral shear zone trending NE–SW (Montenat *et al.*, 1987). This zone has been submitted to an approximately N–S compressive regime, which produced a perpendicular extension, at least since the Tortonian. The rotation of the stress field and consequently the shifting of the shortening direction (NW–SE, Tortonian; N–S Tortonian–Pliocene; NW–SE, late Pliocene–Holocene) have notably influenced the generation and filling of these sedimentary basins.

A system of oblique N 120° E and N 60° E fractures, probably inherited from previous stages, produced the opening of the Cope Basin during late Miocene. The sedimentary fill is mainly by shallow-marine deposits up to the middle–late Pliocene, when tectonic uplift triggered a regressive tendency only interrupted by small rises of sea level; a trend that continues at the present time.

In this basin good examples of Pleistocene fan-delta deposits crop out (Figs 1, 2). These well-exposed deposits allow the study of sedimentary facies, sequences and responses to the geologic controls which determine the occurrence and development of fan deltas.

Sedimentation occurred in a complex realm due to the coincidence of: (1) active tectonics along faults directed N 60° E which marked the boundaries of the basin creating the slope necessary for the fans to develop, and also faults directed N 120° E, which did not cause a strong subsidence but determined both the lateral extension and morphology of the bodies and the flow directions of the distributary channels of fans (Fig. 3); (2) successive eustatic sea-level changes, which determined the differential subaerial exposure with the consequent different

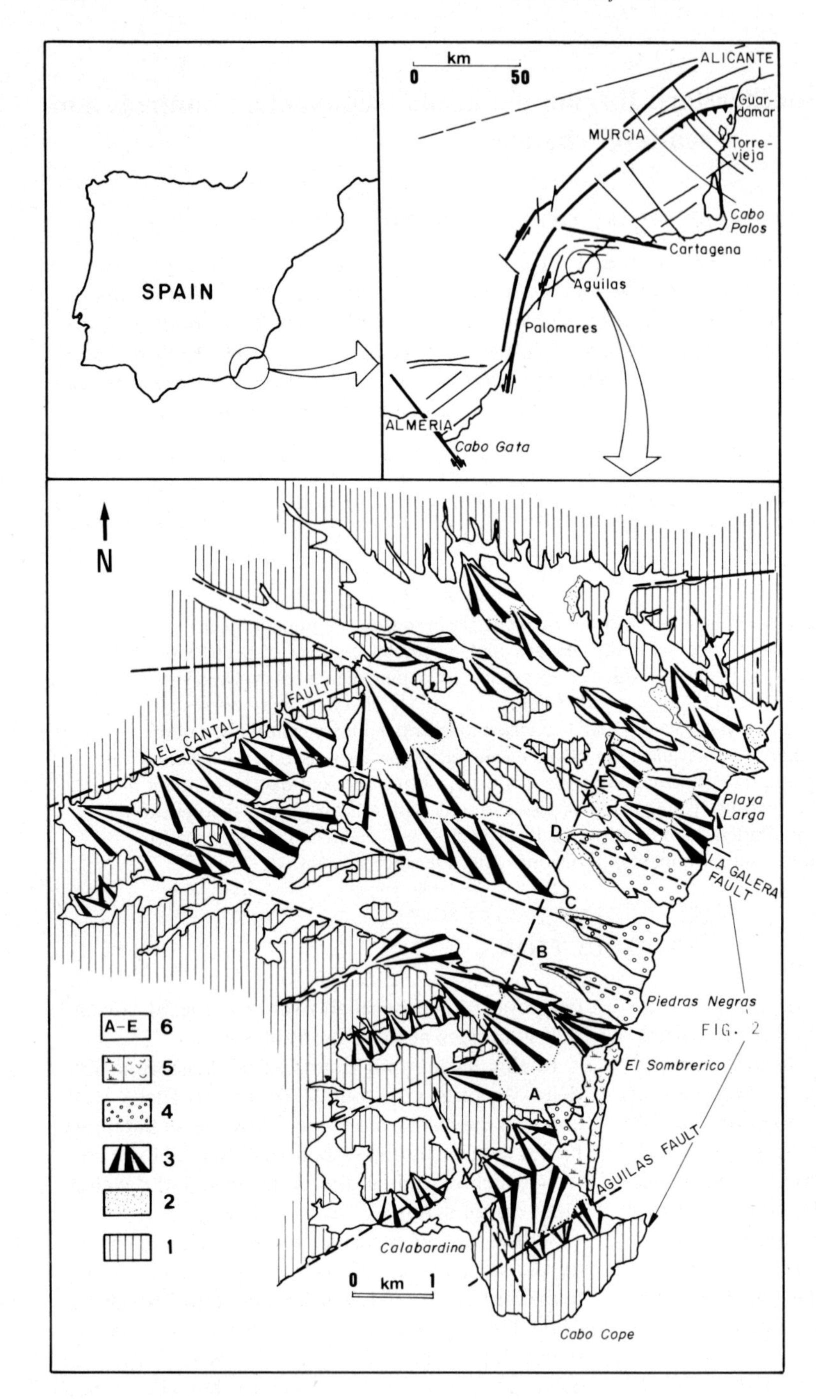

Fig. 1. Location map and geological map of the Cope Basin. Key: 1. metamorphic rocks of the Internal Betics; 2. fossiliferous sands and marls: 3. Pleistocene alluvial fans; 4. Pleistocene fan-delta deposits (prograding wedges of conglomerates and sands); 5. late Pleistocene oolitic and quartzitic aeolian dunes and lagoonal deposits; 6. fan-delta sedimentary bodies named from A (south) to E (north). Partly modified after Bardají *et al.* (1986).

Fig. 2. (A) Morphogeological map of the marine and terrestrial sequences in Cape Cope; A–E are fan-delta sedimentary bodies (see Figs 1 and 14). (B) Synthetic section of the marine and continental levels in Cope Basin. Modified after Bardají *et al.* (1986).

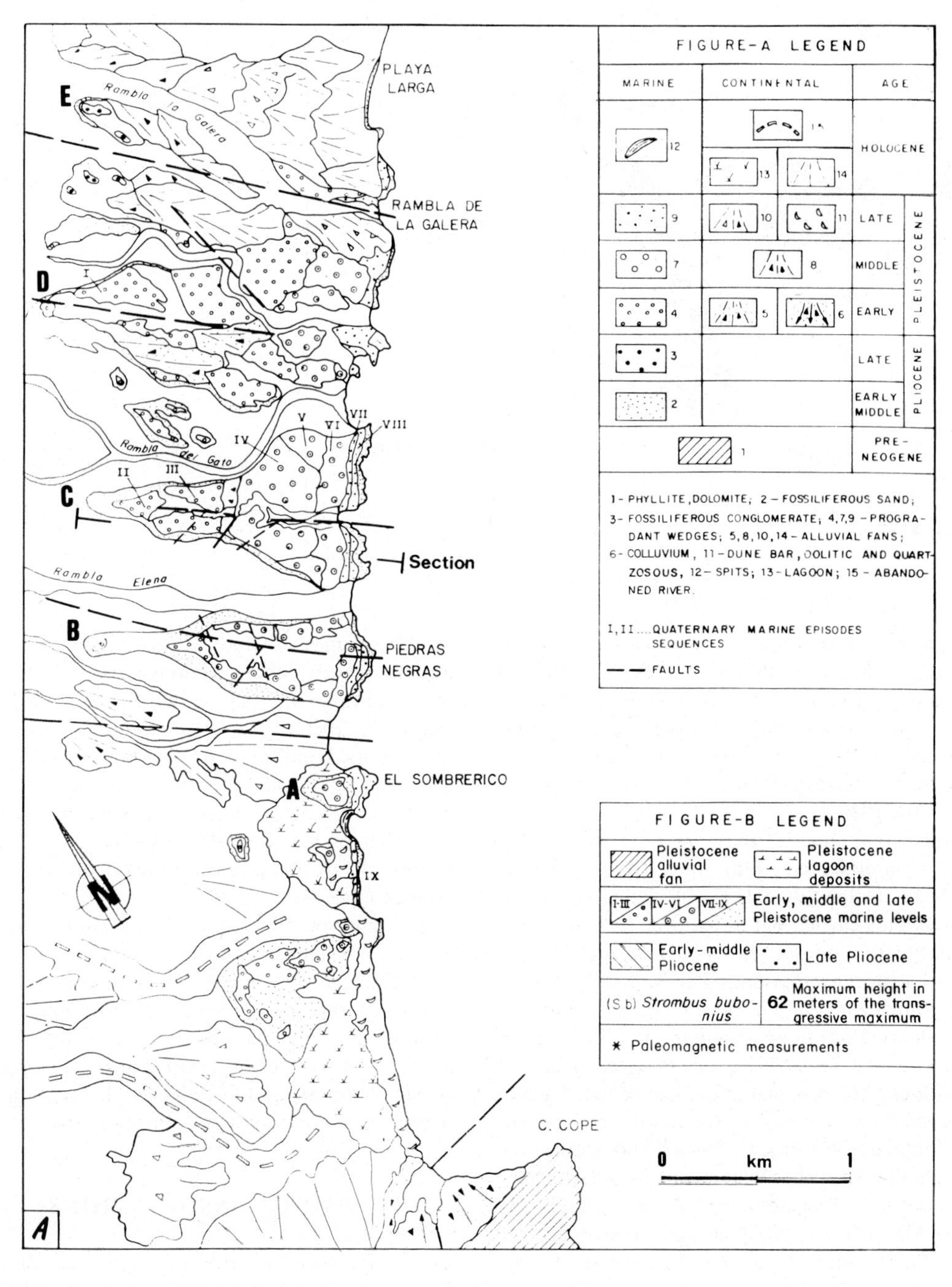
PLAYA LARGA
RAMBLA DE LA GALERA
Rambla la Galera
Rambla del Gato
Rambla Elena
Section
PIEDRAS NEGRAS
EL SOMBRERICO
C. COPE
E
D
C
B
A
I
II
III
IV
V
VI
VII
VIII
IX
N
0 km 1
FIGURE-A LEGEND
MARINE
CONTINENTAL
AGE
HOLOCENE
LATE
MIDDLE
EARLY
PLEISTOCENE
LATE
EARLY MIDDLE
PLIOCENE
PRE-NEOGENE
1- PHYLLITE, DOLOMITE; 2 - FOSSILIFEROUS SAND;
3- FOSSILIFEROUS CONGLOMERATE; 4,7,9 - PROGRADANT WEDGES; 5,8,10,14 - ALLUVIAL FANS;
6- COLLUVIUM, 11 - DUNE BAR, OOLITIC AND QUARTZOSOUS, 12- SPITS; 13 - LAGOON; 15 - ABANDONED RIVER.
I, II .... QUATERNARY MARINE EPISODES SEQUENCES
FAULTS
FIGURE-B LEGEND
Pleistocene alluvial fan
Pleistocene lagoon deposits
Early, middle and late Pleistocene marine levels
Early-middle Pliocene
Late Pliocene
(S b) Strombus bubonius
62 Maximum height in meters of the transgressive maximum
* Paleomagnetic measurements

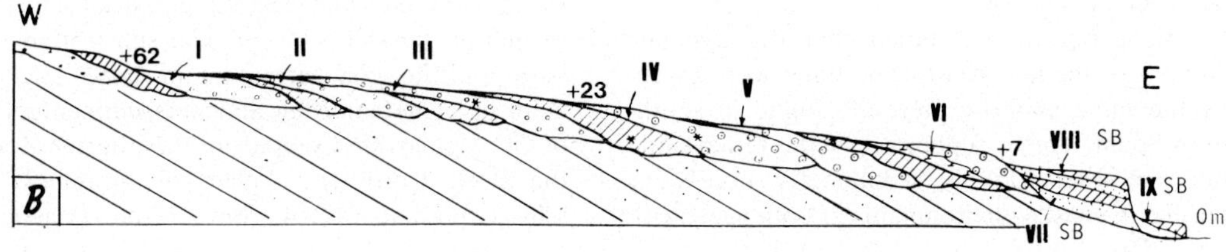
W
E
B
+62
+23
+7
I
II
III
IV
V
VI
VII SB
VIII SB
IX SB
0m

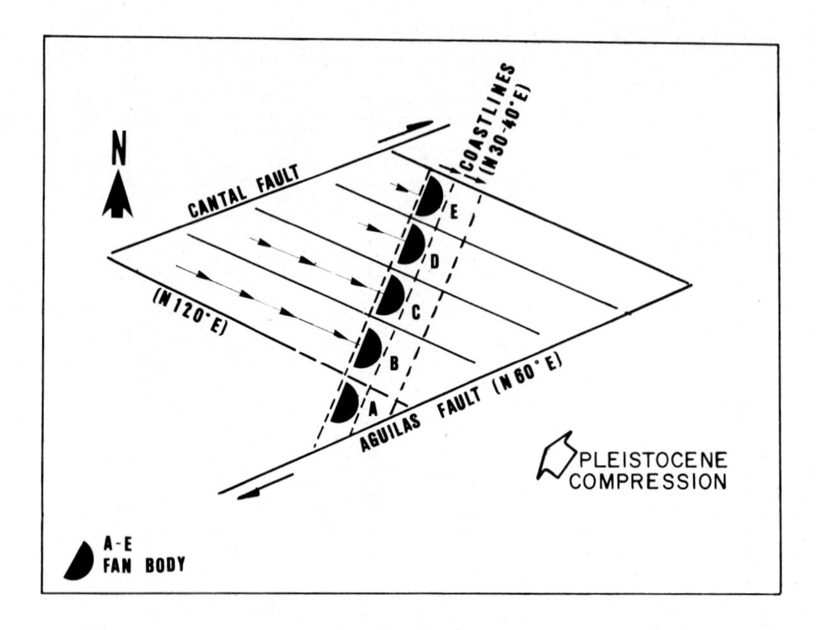

Fig. 3. Relationships of the systems of fracture in the Cope Basin and the shape of fan-delta sedimentary bodies. A–E represent the main bodies of marine deposits.

sedimentary behaviours; and (3) various sedimentary processes acting in both the fan deltas and the neighbouring environments. Mutual interference resulted in the deposition of prograding offlapping sequences, which have been referred to as 'sequences of marine and terrestrial levels' or 'marine terraces' (Goy *et al.*, 1986a).

In this area, nine of these marine Quaternary episodes interlayered with terrestrial deposits, have been distinguished (Fig. 2). Palaeomagnetic measurements carried out in the area indicate early Pleistocene (episodes I, II and III in Fig. 2) and middle Pleistocene (episodes IV, V and VI) ages. Three younger marine episodes (VII, VIII and IX, Fig. 2) bearing *Strombus bubonius* (i.e. Tyrrhenian in the sense of Issel, 1914) occur encased in the earlier ones. An approach to the age of these levels can be made by comparing them with one of the most complete sequences of Tyrrhenian episodes found in the Spanish Mediterranean, in Almería, where isotopic measurements (Zazo *et al.*, 1984; Hillaire-Marcel *et al.*, 1986) have given mean ages of 180 Ka (Tyrrhenian I), 128 Ka (Tyrrhenian II) and 95 Ka (Tyrrhenian III).

In this paper it is assumed that the dynamic conditions of the late Pleistocene times were essentially the same as those presently found in south-eastern Spain. This is supported by the remarkable parallelism observed in the sedimentary successions deposited in coastal environments of both ages, with similar regimes of wind as deduced by the general longshore drift, similar sedimentary environments, etc. (Dabrio *et al.*, 1984, 1985; Bardají *et al.*, 1986; Goy *et al.*, 1986b; Zazo *et al.*, 1989). With regard to the climatic conditions, it must be noticed that the occurrence of *Strombus bubonius* in the most recent episodes indicates a slightly warmer temperature than that of the present; nowadays, warm faunas bearing *S. bubonius* are restricted to the Gulf of Guinea (Zazo *et al.*, 1989).

The aim of this paper is to show the basic sedimentary features of these fan-delta deposits as related to the controls of sedimentation, because it is one of the places where tectonic and eustatic controls can be separated from each other. Some attention is paid to the comparison of the gravelly, coastal deposits of fan deltas with the sequences described in coarse-grained, linear coasts.

FACIES ASSOCIATIONS

The well-exposed Pleistocene deposits of the Cope Basin show interfingering of terrestrial and littoral to marine fan-delta facies. The substratum of the basin and the adjacent mountain ranges (sierras) is made up of metamorphic and metasedimentary rocks of the Alpujarride Complex of the Internal Zones of the Betic Cordillera: Palaeozoic graphitic mica-schists, quartzites and marbles, Permo-Triassic phyllites and Upper Triassic dolostones. Erosion of these source rocks supplied sediments to fan deltas which partly filled the shallow-marine basin.

Subaerial fan-delta deposits (alluvial fan)

These are the most extensive facies, forming bajadas along the fault bounding the Sierra de Almenara. The most abundant facies are reddish conglomerates with flat, rounded pebbles. These textural properties are related to the strongly schistose rocks forming the adjacent mountains. The high content of mica in the schists may have influenced the rheology of the masses of removed detritus. Weathering of iron-rich minerals, including phyllosilicates, in source areas is thought to cause the common red colour of the deposits.

Channellized conglomerate facies

These are bodies of conglomerate with irregular, erosional lower boundaries. The thickness of beds is variable but common values range between 5 m and 2 m. The most prominent internal sedimentary structure is crude horizontal stratification. Imbricated flat pebbles are common. This facies has been interpreted as channel-fill deposits having irregular shape both in the assumed longitudinal and transverse sections. Local cross-bedding associated with scours was interpreted as channel-fill cross-bedding. Rare large-scale planar cross-bedding found in some outcrops might be related to bars. Measured palaeoflow directions point (as it would be expected) roughly away from the hinterland. Poorer organization and somewhat coarser grain sizes are recorded in more proximal areas.

Disorganized conglomerate facies

Close to the hinterland there are masses of disorganized, matrix-supported conglomerates, with angular, irregular-shaped clasts, interpreted as debris-flow deposits. Median thickness is about 1–3 m.

Sandstone facies

They occur as beds of medium- to coarse-grained, litharenite (derived from metamorphic rocks). In proximal areas of the fans they usually form irregular, discontinuous beds which occur deeply incised by erosion prior to the overlying conglomerates. These beds become more continuous towards the distal parts. They are interpreted as channel-fill deposits covering the conglomerates.

Red mudstone facies

Massive, red sandy mudstones with some scattered clasts are characteristic of the distal alluvial-fan deposits. They are interpreted as the result of settling of fines after flooding. Any original bedding or interlamination has presumably been destroyed by pedogenesis.

Facies relationships

These facies occur interbedded. The channelized conglomerate facies accounts for most of the volume of sediments. It may be followed upwards either by sandy or, more commonly, by red mudstone facies. Masses of disorganized conglomerates (debris flow) of metric vertical scale may be found interbedded with channel-fill deposits near the hinterland. The amount and continuity of the finer-grained facies increase towards more distal parts. The red mudstone facies is often the only representative of the subaerial fan-delta facies in the most distal realms.

The association of facies and morphology of sedimentary bodies (well observed in air photographs) can be interpreted as alluvial-fan deposits with evidence for shifting, braided channels, mostly filled with coarse to sandy sediments, and a large extension of flood plain where finer, burrowed mudstones were deposited.

Marine fan-delta deposits (coastal and sublittoral)

Beach deposits

These consist mostly of conglomerates, sandstone and mudstone that form coarsening-upwards sequences, similar to those described by Dabrio *et al.* (1984, 1985). In ascending order, they comprise (Fig. 4):

1 a lower unit of parallel-laminated and wave-ripple, cross-laminated micaceous sandstones, commonly burrowed, representing the lower shoreface and transition zones;

2 a middle unit of trough cross-bedded sands and gravels corresponding to the shoreface zone under wave action;

3 an accumulation of the coarsest grain sizes available on the coastline including the large, heavy shells of the gastropod *Strombus bubonius*, de-

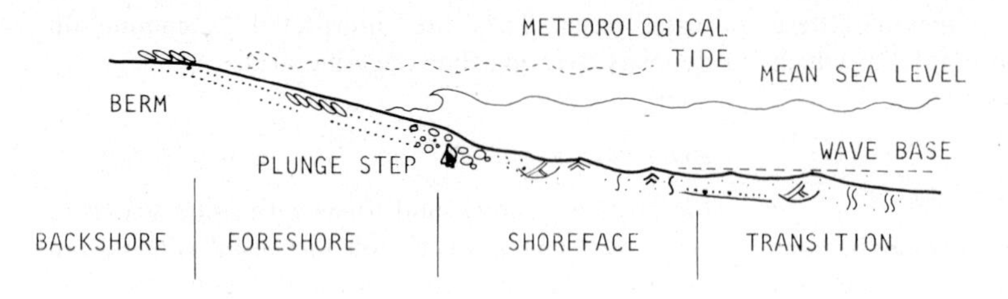

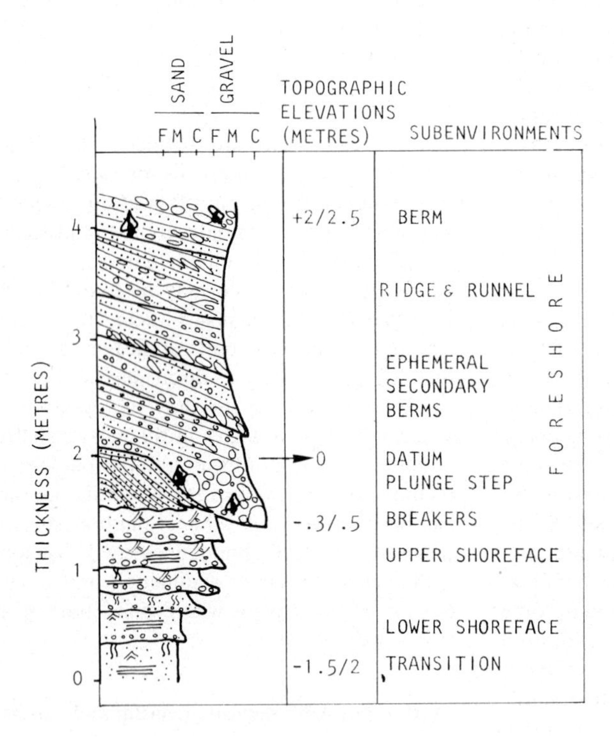

Fig. 4. Reconstruction of a typical Tyrrhenian beach and resulting sequence of deposits. Modified after Dabrio *et al.* (1985).

creasing upwards up to well-sorted fine gravels. These are deposits of the lower part of the foreshore, where the breaking waves accumulate the coarsest sediment available (Miller & Ziegler, 1968; Dabrio *et al.*, 1985). In low-energy coasts the breaker zone is characterized by a step at the base of the swash zone (Clifton *et al.*, 1971; Davis *et al.*, 1972) which probably marks the transition between upper and lower flow regime (Tanner, 1968). In accretionary beaches the step implies a sudden increase in the dip of the seaward-inclined laminae which, according to Clifton (1969) is formed by backswash. It is interesting to note that the breaker zone indicates very precisely the mean sea level during deposition in present tideless coasts. In fossil examples the step is preserved mostly in two ways depending on the predominant grain size: (a) in gravelly beaches it is marked by an accumulation of the coarsest grain sizes available; (b) in sandy beaches it is marked by a change from parallel lamination (the common deposits of the swash zone in the foreshore) to tabular cross-bedding (Dabrio *et al.*, 1985; Somoza *et al.*, 1987).

4 an upper interval of well-sorted, parallel-laminated gravel, gently inclined towards the sea, representing the upper part of the foreshore and berm.

The coastal deposits are the most characteristic facies of the exposed Pleistocene fan deltas of Cope Basin. These gravelly coastal deposits are identical to those fed by rivers or by longshore drift not directly related to fan deltas. As already noted, the most distinctive features are grain size (gravel or coarser), good roundness and, in general terms, high sorting, a particular vertical sequence of grain sizes and primary sedimentary structures, and the overall geometry of depositional units. These features are the result of a relatively steep foreshore: original deep slopes average 5° for beach units made up of coarse sand and 6–8° for those with fine gravel in Cope Basin. The slope becomes still more inclined towards the lower part of the assumed foreshore units, where the coarsest grain sizes accumulate. These values are similar to those obtained in the late Pleistocene and present beaches of the Gulf of Almería: 6° for gravelly foreshores with maximum grain size of about 10 cm and slopes up to 10° for those reaching maxima of about 15 cm (Dabrio *et al.*, 1984). Similar values are found in all the studied Pleistocene and present beaches of southeastern Spain as well.

Mass-flow deposits

These are conglomerates (Fig. 5) that occur at some places interbedded with sublittoral deposits. Two different lithologies could be distinguished:

1 Tabular beds of clast-supported coarsening-upwards conglomerates up to 2 m thick. They are thought to be the result of deposition of chaotic masses of clasts and sandy matrix following major unconfined floods of the fan delta.

2 Lenticular bodies of disorganized, heterometric conglomerates. The lower boundaries are erosional. They are interpreted as reworked subaerial or transitional marine deposits which filled subaqueous channels cut into the fan-delta front.

The interfingering of foreshore deposits and chaotic masses (Fig. 5) provides an interesting problem of interpretation. A most prominent feature is that beach deposits (in particular those marking the position of the shoreline) occur at variable height. This is thought to indicate that small-scale changes of sea level are involved. During highstands beach deposits are interbedded with non-channelized mass-flow deposits in the subaqueous areas. Contemporary alluvial sediments were deposited in the subaerial fan delta. A subsequent fall of sea level created erosional channelling over the former littoral deposits which were infilled by alluvial-fan deposits. Later, a sea-level rise produced reworking and 'solifluxion' of these alluvial-fan deposits which slid towards the lower foreshore.

CONTROLS OF SEDIMENTATION

Tectonics (fault movements)

Long-term subsidence occurred in the Cope Basin during Neogene times caused by normal faults directed N 60° E. This system of fractures was reactivated as dextral strike-slip faults during Quaternary times due to compression oriented N 150° E produced by the convergence of the European and African Plates in the western Mediterranean (Bousquet, 1979; Vegas & Banda, 1982).

The stress field created by the movement along dextral strike-slip faults triggered the activation of two secondary systems of fractures: N 120° E and N 30–40° E in the basin interior (Bardají *et al.*, 1987) (Fig. 1). These two systems define the orientation and geometry of the fan-delta bodies, clearly discernible in air photographs. Uplifting and drowning of blocks along the system N 120° E individualize areas of differential subsidence. As drowned blocks were preferentially used by floods, they predetermined both flow directions and potential areas of accumulation of fan-delta sediments. Thus, the system N 120° E controls the preferential direction of the feeding channels and also the elongation of the sedimentary bodies. Raised blocks were mainly out of reach of coarse sediments during floods and most of the deposition there consisted of red mudstone. The system N 30–40° E (at right angles to the former) determined the orientation of Pleistocene shorelines, indicating the possible limit to the subaerial part of the fan deltas. Both directions formed a square pattern of fault blocks. Those blocks experiencing positive subsidence were filled by fan-delta sediments, whereas raised blocks did not collect much fan-delta sediment (except for the red mudstone facies).

Another factor involved in the sedimentation of Pleistocene sequences, is the rate of uplifting/subsidence of the margin of the basin. It was (and still is) mostly influenced by tilting along the fault system directed N 30–40° E (parallel to the shorelines) which acted as a tilting line in response to the compressional stress. The continuous, gradual uplifting during the Pleistocene resulted in a relative fall of sea level which generated an offlapping sequence of successive units of fan-delta deposits (Fig. 6).

Subsidence was not very prominent as demonstrated by the offlapping geometry of the resulting sedimentary units. This fact is interesting because it

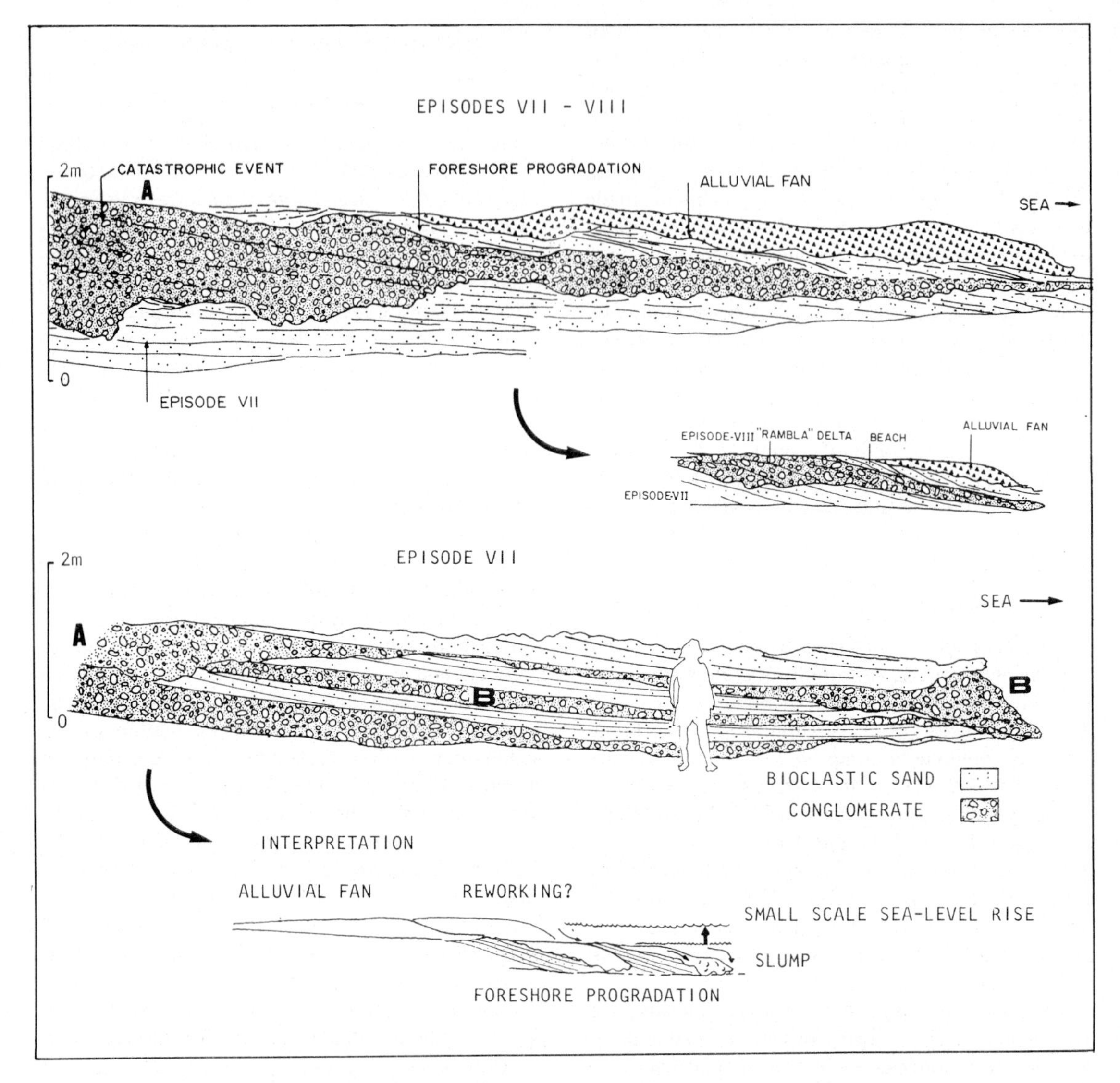

Fig. 5. Types of mass-flow deposits interlayered in beach sequences. (A) wedge-shaped, organized, heterometric conglomerate overlaid by foreshore deposits, interpreted as deposits of catastrophic flash floods in areas near the mouth of ephemeral rivers (*ramblas*). (B) disorganized, heterometric conglomerate, forming lenticular bodies with erosional lower boundaries, interpreted as reworking of subaerial or transitional deposits.

can be used as a criterion of differentiation between the effects of eustatic and tectonically influenced changes of sea level. In general, there is a clear relationship between tectonics and the generation of the Quaternary marine layers. When uplifting is continuous and gentle, sea level changes are recorded as separate ribbons of coastal deposits some metres apart from each other and at different heights. With reduced or no subsidence, successive sea-level changes generate sequences of stacked marine layers separated by erosional surfaces.

Eustatic changes of sea level

The gently inclined offlapping sequence of terrestrial and marine deposits found in Cope Basin is related to deposition under variable sea level.

We assume that during highstands most of the

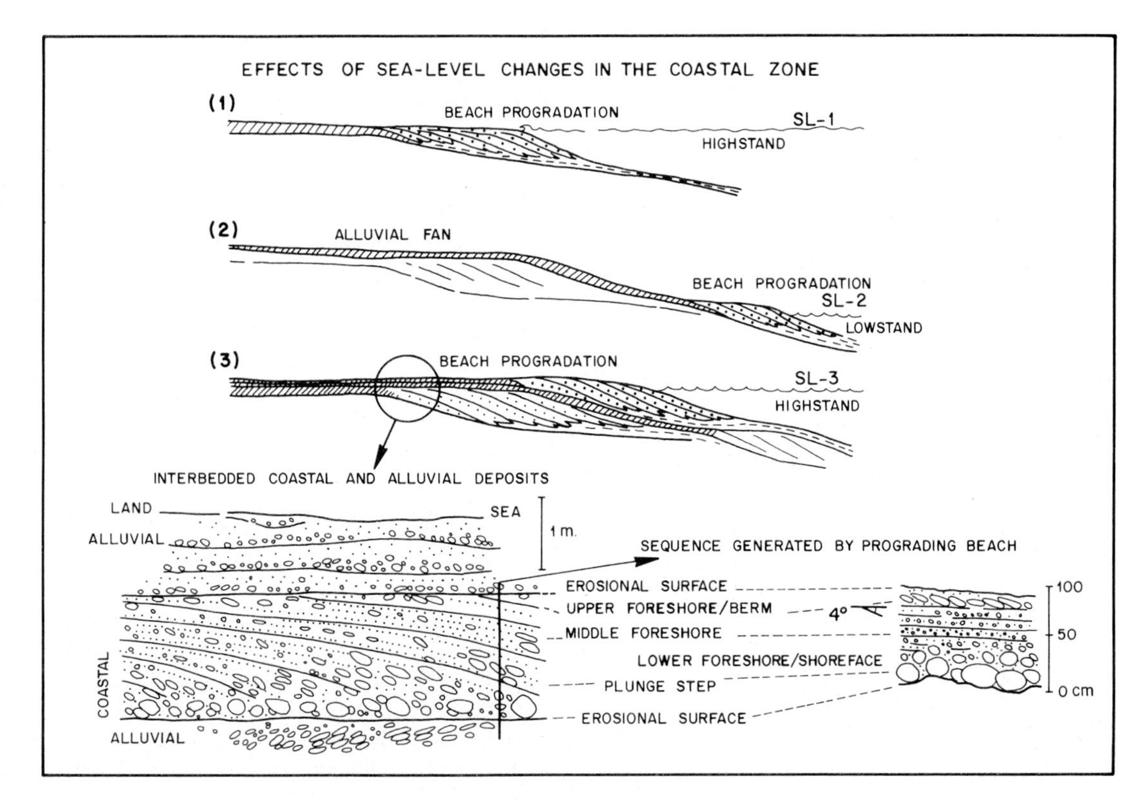

Fig. 6. Development of imbricated (offlapping) sequences of coastal and terrestrial units caused by relative changes of sea level in Cope Basin. Symbols: spots and circles — marine deposits; inclined lines — terrestrial deposits. Modified after Bardají *et al.* (1987).

coarse sediments of the fan deltas remained next to the hinterland and the areal extent of the subaerial fans was very restricted. Under these conditions, beaches on the periphery of the fan deltas actively prograded due to the high input of sediment. However, the preserved clinoforms indicate that sea level was stable or somewhat falling for any given unit (Fig. 6). In our opinion this is due more to the existence of higher relief rather than to uplift of the surrounding mountains or active subsidence of the basin. These features resemble present-day processes and allow us to differentiate between the effects of subsidence (tectonics) and eustatic sea-level changes.

During lowstands a large part of the basin margin and fan deltas is exposed. Weathering of the exposed deposits occurs because most of the new input concentrates into telescopic fans, fed by deeply incised channels (Fig. 7), and deposited at the toe of the fans formed during highstands. The typical deposit in exposed areas is massive red mudstone. Virtually no coarse sediment accumulated outside the active incised lobes. No exposures of the lowstand delta fronts are exposed today.

If, as supposed, sea-level falls are due to glacial causes, a colder, moister climate is to be envisaged during lowstands favouring fluvial processes instead of the catastrophic mass and debris flows. This agrees with field observations, because the dominant subaerial facies are alluvial with shifting, braided channels. Facies of disorganized conglomerate with angular clasts are restricted to proximal areas, although sometimes it is hard to say which phase they correspond to because the lack of physical continuity with areas where terrestrial and marine deposits interfinger with one another.

In conclusion, a succession of phases of highstand and lowstand can be observed, each producing their own facies and effects. On this basis, predictions can be made for other areas. However, only episodes

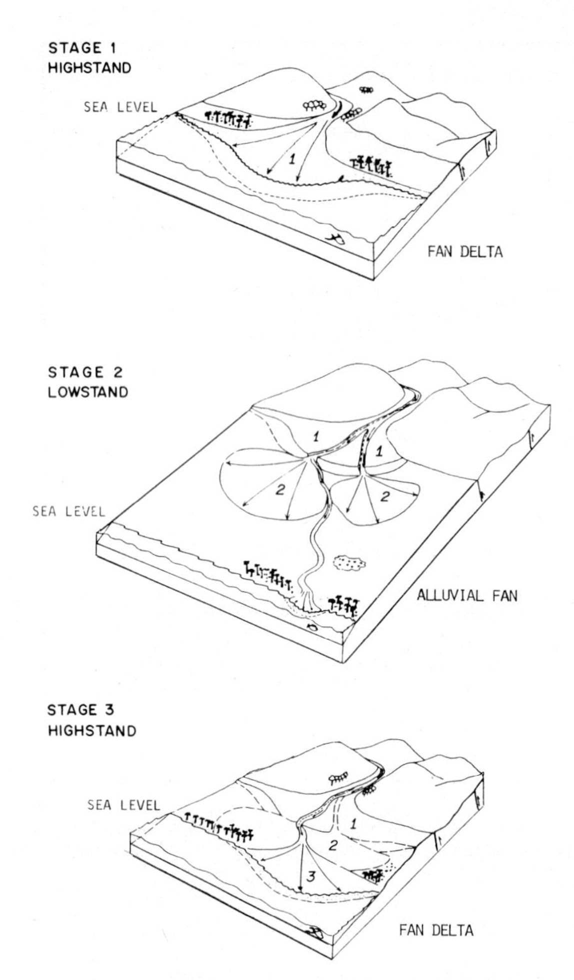

Fig. 7. Conceptual model of the evolution of fan deltas in response to sea-level fluctuations and uplift of the coastal areas.

when mean sea level was increased above the present datum could actually be observed and it must not be assumed that the studied sequence of episodes records every oscillation of relative sea level. Topographic heights must not be used as synonyms of stratigraphic position, because it is well known that the magnitude of relative oscillations may be very variable and, furthermore, younger tectonics (faulting) may change the original relative heights. Uplift may have preserved highstands parallel to present sea level.

Dynamics of fan deltas

In this paper, we assumed that the dynamic and climatic conditions of late Pleistocene times were essentially the same as those presently found in southeastern Spain. This is supported by the remarkable parallelism observed in the sedimentary successions deposited in coastal environments of both ages (Dabrio *et al.*, 1984; Bardají *et al.*, 1986; Goy *et al.*, 1986b).

The dynamic regime of the present-day fan deltas in southeastern Spain is controlled by episodic, but catastrophic, discharges which cause flooding of the fans. This happens because the *ramblas* (local name for wadi-like ephemeral streams subjected to episodic flash floods) are unable to keep pace with the huge volumes of water and sediment involved in the heavy rains generated by the seasonal change of the atmospheric circulation in the Mediterranean Sea. Maximal precipitation occurs in spring (April) and, above all, in autumn (MOPU, 1976): records of up to 300 mm of rain in one day are quite common in October, with a probable periodicity of 10 years. Those episodes of rain, very often of catastrophic nature, imply very large inputs of sediment to the fan delta. They are followed by rather long periods of inactivity when longer-term, but more continuous, coastal and shallow-marine reworking takes place accumulating sediment in beaches and barrier islands (Fig. 8).

Thus, the coarse sediments of the fan deltas and *ramblas* feed the coastal zone and the resulting beach deposit exhibits a large-scale prograding geometry. In the tideless southeastern littoral of Spain the major controls of coastal sedimentation are (Dabrio *et al.*, 1984): (1) exposure to the prevailing winds and storms; (2) the availability and size of sediment; and (3) recent regional tectonics (i.e. movements of fault blocks). A notable example of the variations caused by the change of these controlling features (both in time and space) is the area of the Gulf of Almería (Dabrio *et al.*, 1984) and the coast of El Campo de Dalías (Goy *et al.*, 1986b).

RECONSTRUCTION OF COASTAL MORPHOLOGY OF FAN DELTAS

The most characteristic facies of the exposed Pleistocene fan deltas of Cope Basin are the coastal deposits. Their distinctive features are grain size (gravel or coarser), good roundness and, in general terms, high sorting, a particular vertical sequence of grain sizes and primary sedimentary structures, and the overall geometry of depositional units.

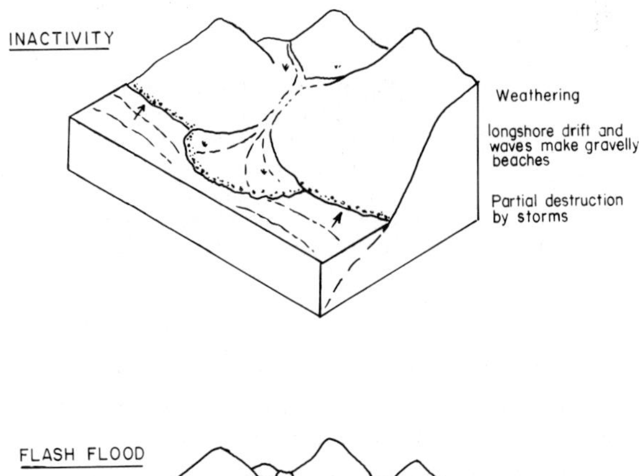

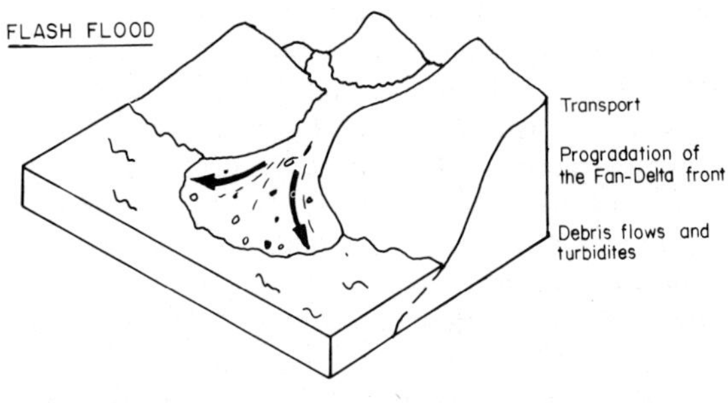

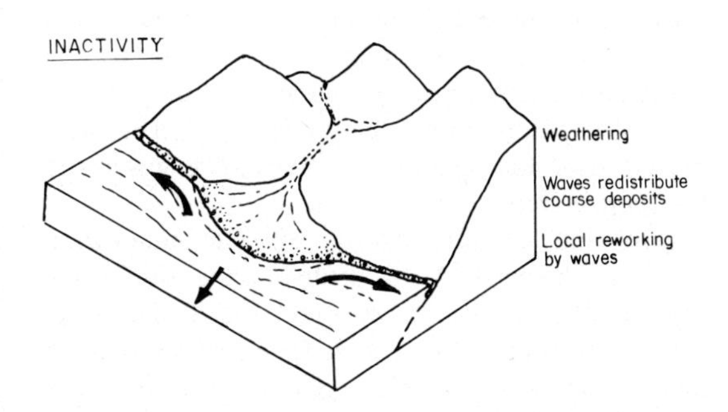

Fig. 8. Succession of events related to the sporadic, catastrophic activity of fan deltas.

All these features are closely comparable with the Pleistocene siliciclastic and carbonate shallow marine to coastal barrier islands of the Hergla region (Tunisia) (Mahmoudi *et al.*, 1987), although grain sizes are smaller. Massari and Parea (1988) described Messinian and Pleistocene prograding sequences of gravelly beaches in southern Italy that are almost identical to those of southeastern Spain. We conclude that coastal dynamics during late Pleistocene highstands were comparable to those of the Recent in the western Mediterranean area.

After considering a spectrum of beach conditions (Sonu, 1973; Short, 1979) two basic types of beaches have been delimited (Guza & Inman, 1975): reflective and dissipative. Reflective beaches reflect most of the energy of waves whereas in dissipative beaches most of the wave energy is dissipated in the surf zone (Wright *et al.*, 1979). The development of reflective beaches requires low waves (less than 1 m high) and coarse sediment (Md > 0·6 mm); consequently they tend to occur in sheltered coasts or in coasts of moderate to high energy as long as grain size is gravel (Short & Wright, 1983). Changes in grain size trigger modifications of the type of beach as shown by Short (1984).

In reflective beaches most of the sediment accumulates in the upper foreshore generating high berms and beach cusps. The foreshore is rather steep (about 8°) with a plunge step in the lower part, where the coarsest sediment concentrates. Towards the sea the shoreface is less inclined, but relatively deep, without bars or other morphological features. Consequently, the beach units, and still better the place of the change in slope of the foreshore (plunge step), indicate very accurately the mean sea level during deposition. Careful study of the topographic variations of the plunge step along sections normal to the shoreline (Fig. 9) may allow deduction of minor fluctuations of sea level (Bardají *et al.*, 1987).

Flow is essentially directed towards the land and the sea, with minor divergences due to the beach cusps. They are basically sheltered and accretional beaches (Bryant, 1983) where accumulation, lateral uniformity (but for the decorative effect of the beach cusps) and temporal stability dominate as long as wave height is smaller than 100 cm. The profile of beaches undergoes rapid changes because the volume of sediment involved is usually very small (Short & Wright, 1983).

This description fits perfectly the present coarse-grained beaches of southern Spain and it is very easily extrapolated to the Pleistocene raised beaches. Coarse sediment was retained in the beach face leaving behind a starved nearshore with little, fine-grained deposits.

The present southeastern Spanish coast is almost tideless. Tidal ranges of 8 cm were measured, but meteorological tides due to the daily, fair-weather wind setup reach up to 20–30 cm (Dabrio & Polo, 1981). Higher waves during storms were recorded as well (MOPU, 1976) which are thought to be responsible for flat, gently sloping (slope clearly lower than the one of normal foreshore laminae) erosional surfaces. This may explain the absence of terraces or

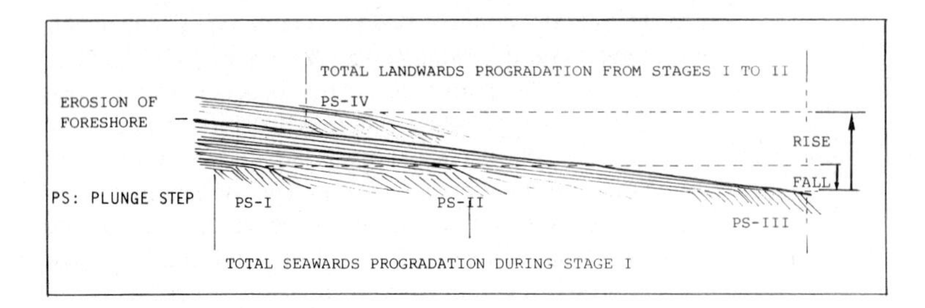

Fig. 9. Use of the planar cross-bedding generated by migration of the plunge step (PS) towards the sea in the reconstruction of progradations and retrogradations of beaches and the short-term tendencies of sea level. Modified after Bardají *et al.* (1987).

incipient low-tide bars usually found in present tidal sandy beaches. As a consequence, a microtidal or tideless Pleistocene sea is assumed.

COMPARISON WITH GRAVELLY COASTAL DEPOSITS

Additional data about prograding sequences of marine and terrestrial deposits caused by relative sea-level changes can be drawn from the comparison of the sequences described in Cope Basin and those of the Rambla de las Amoladeras in the basin of Níjar (Almería).

The *rambla* feeds the coast with coarse sediment. Grain sizes and relative abundance of lithologies are very much like the one described in the basin of Cope. Consequently, it can serve as a model for extrapolating features to be found in coastal deposits related to fan deltas.

The Rambla de las Amoladeras is a braided alluvial system where transport dominates over sedimentation. The lower reaches of the river heads towards the southwest under the strong control of subsidence along a left strike-slip fault running N 40–50° E (Goy & Zazo, 1983a). The ephemeral river acts as the longitudinal system draining the bajadas related to the mountains limiting the basin of Níjar (Sierra de Alhamilla and Serrata de Níjar) (Fig. 10A).

The *rambla* cannot form a fan delta, mostly due to the strong littoral drift induced by storm winds blowing from the west. Longshore drift sweeps the coast carrying sediment towards the southeast. The straight coastal morphology (Fig. 10B) has been affected since Pleistocene times by normal faults directed N 140°–160° E (Goy & Zazo, 1983b). Under these environmental constraints prograding coastal units of the linear type accumulated during Pleistocene and present times.

The exposed section is normal to the successive Pleistocene coastlines. Four offlapping episodes of progradation can be observed (three of them containing *S. bubonius*) separated by erosional surfaces (Fig. 11). The offlap is interpreted as a response to eustatic relative changes of sea level (Dabrio *et al.*, 1984). Major influence of fault-induced changes of subsidence was discounted for two reasons: (1) the repetitive nature of the process and (2) the tendency of the involved fractures to systematically sink the southwestern block of the fault (the one placed seaward from the fault). No criteria were found to invoke either reversion of movement of faults or the existence of flexures in the opposite direction.

During highstands, gravels carried to the coast accumulated into large-scale, cross-bedded units, with foresets dipping between 6° and 12°, generated by prograding coarse-grained foreshores. In the lower part a distinctive accumulation of the coarsest grain sizes visible in the section is found; it is interpreted to be the result of the seaward migration of the plunge step.

Lowstands caused subaerial exposure and erosion of these deposits. Most probably the siliciclastic input of the *rambla* was carried towards the newly formed coastal zones (nowadays several tens of metres below sea level) where marine dynamics moved it alongshore.

A vital difference with the sequences of Cope Basin is the almost complete absence of terrestrial deposits (corresponding to lowstands) between successive marine units (highstands): only calcretes and remains of aeolian dunes exist (Fig. 11). In our opinion, the difference is related to the distinct palaeogeographic settings involved. Floods of the braided system could reach most of the fan-delta areas. Only red mudstones were deposited in areas uplifted by faults (this is particularly true for subaerial areas of fan deltas of the highstand phase that were still more exposed during lowstands). However, in the linear coast of Níjar very little sediment was available to the raised beaches during lowstands because the feeding rivers (*ramblas*) tended to be

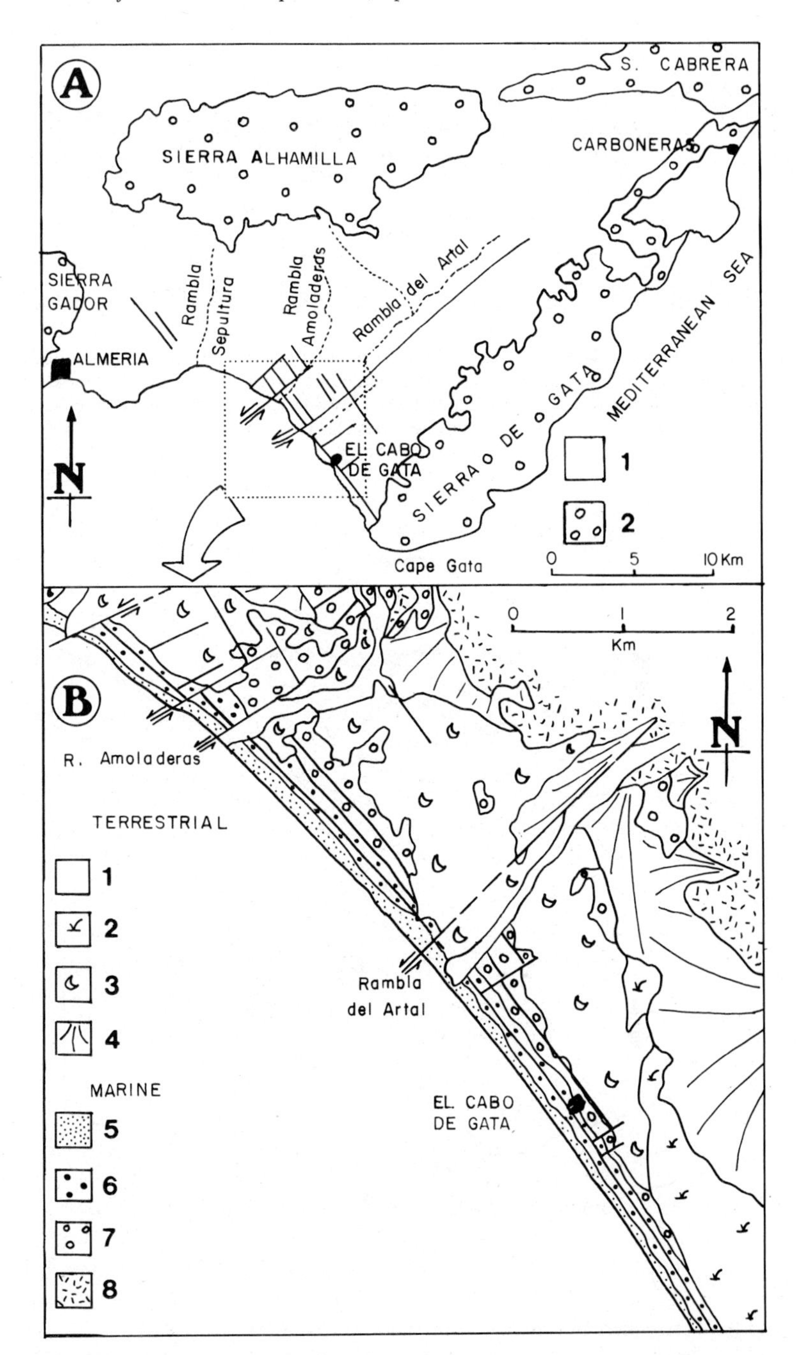

Fig. 10. (A) Map of the Almería-Nijar basin with the main reliefs, lines of fracture and typical courses of *ramblas* (ephemeral rivers). (1) Late Neogene and Quaternary deposits. (2) Substratum (mostly pre-Neogene). (B) geological map of a part of the eastern Gulf of Almería. Key: (1) *Rambla* riverbed; (2) coastal lagoon; (3) aeolian dunes; (4) alluvial fans; (5) present-day beaches; (6) marine deposits containing *Strombus bubonius*: prograding beaches (3 units); (7) marine deposits pre-*Strombus bubonius*; (8) Plio–Pleistocene marine deposits. Modified after Goy & Zazo (1983b).

confined, following lines of fault-induced subsidence. Under these circumstances, as relative falls of sea level were matched by descent of the palaeo *rambla* base level, the thalweg incised deeply into the former deposits becoming, in the process, unable to carry sediment to the relatively uplifted alluvial plain (Fig. 12). Consequently, these zones of non-deposition underwent weathering and encrusting by carbonates, serving as hard substratum for aeolian dunes.

The succession of Rambla de las Amoladeras

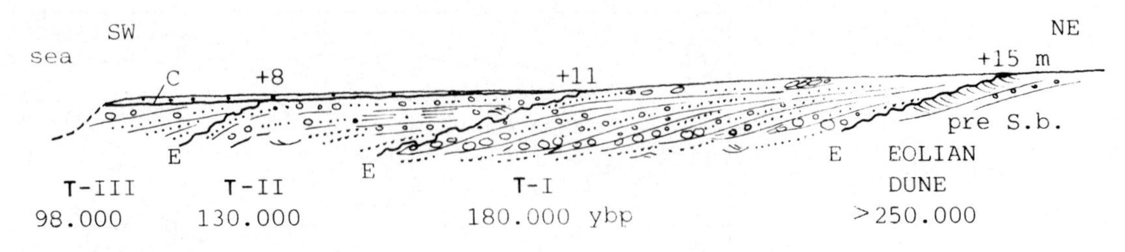

Fig. 11. Idealized sketch of Rambla de las Amoladeras section (real length about 300 m). There are four episodes of coastal progradation but only three (T-I–T-III) yielded *Strombus bubonius*. Progradation took place during highstands (figures indicative of minimum height reached by sea level in every highstand are included). These four episodes are separated by erosional surfaces cut down during lowstands. U/Th ages are indicated. Episodes T-I and T-II are partly covered by calcrete. Modified after Dabrio *et al.* (1984).

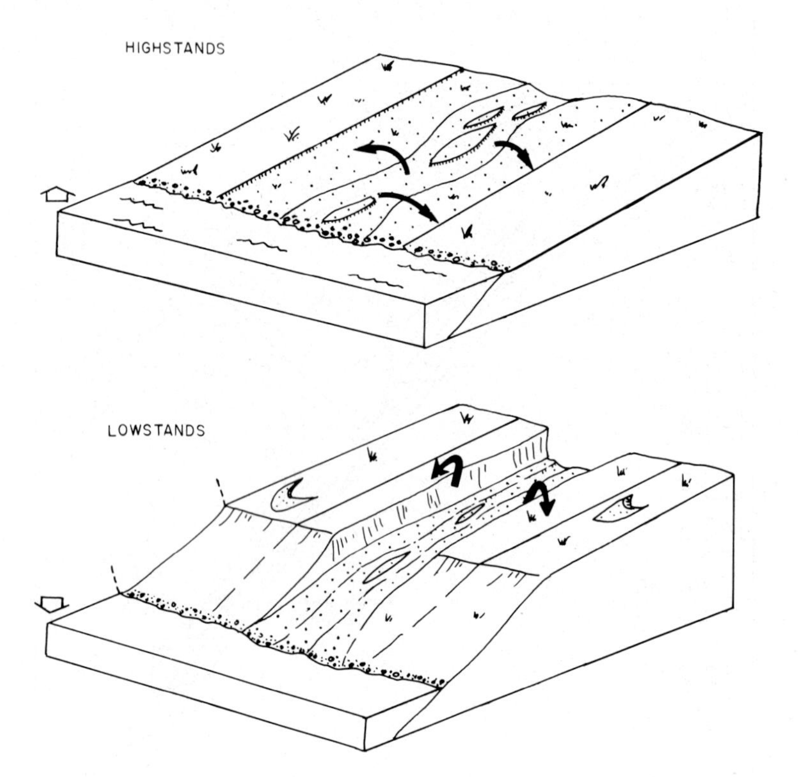

Fig. 12. Response of the encased course of Rambla de las Amoladeras to relative changes of sea level.

(Fig. 11) is also interesting from another point of view: it is the most complete sequence of 'Tyrrhenian' (defined in Issel's (1914) original sense as marine layers containing the indicative fossil *Strombus bubonius*) marine levels of the entire Mediterranean region. The age of the exposed deposits range between 250 Ka and 40 Ka (Th/U dating on *Strombus bubonius* shells, Hillaire-Marcel *et al.*, 1986).

It is interesting to note that it was in the section of Rambla de las Amoladeras where the Tyrrhenian marine episodes (T-I, T-II, T-III, and T-IV) were defined. These ideas were later extended to other coastal deposits along the Mediterranean littoral in Spain (Goy & Zazo, 1982; Zazo & Goy, 1989).

APPROXIMATE RECONSTRUCTION OF SEQUENCES OF ABSOLUTE SEA LEVEL

In this paper the reconstruction of absolute sea-level changes from the analysis of sedimentary sequences deposited in fan-delta environments has only been approximated. This is because of the lack of exact data and the complexity of the factors involved. Nevertheless such an approach is envisaged as a useful tool when calculating trends and coefficients of uplift and sinking in areas experiencing different crustal behaviour (Lajoie, 1986). The advantage offered by the analysis of Pleistocene fan-delta deposits in Cope Basin is that there are precise radio-metric datings of the transgressive maxima in neighbouring areas (Th/U isotopic dating: Butzer, 1975; Bernat *et al.*, 1982; Brückner, 1986; Hillaire-Marcel *et al.*, 1986) and amino acid measurements (Hearthy *et al.*, 1987). The available data and their processing by statistical methods and spectral analysis allow us to study the cycles inside the maxima of sea level during Pleistocene times in southeast Spain (Somoza *et al.*, 1987).

The proposed method for reconstruction of absolute sea levels from sea-level sequences of coastal fan-delta deposits is summarized as follows (Fig. 13):

1 Analysis and reconstruction of apparent level of the sea from the study of the break of slope of foreshore laminae (plunge step) as indicative of datum.

2 Preparation of dynamic theoretical models of curves of sea-level changes cycles based on summation of cycles of maxima, using spectral analysis of the Th/U datings histograms and the global models during the Quaternary.

3 Comparison 'loop reiterated' of the theoretical models and the observed sequences of apparent sea-level changes.

Analysis and reconstruction of apparent level of the sea

The existence of a slope break (plunge step) in the lower foreshore is very interesting because it enables valuable data to be inferred about small-scale changes of sea level. As observed in present-day microtidal or tideless beaches, the point where the seaward-inclined parallel lamination changes into the avalanche laminae of tabular cross-bedding indicates with a precision of centimetres the mean sea level (Dabrio & Polo, 1981; Bardají *et al.*, 1987). Any modification of sea level will produce lateral and vertical shifting of this particular point, assuming that it can be referred to a known datum (Fig. 9). According to this, even minor oscillations of sea

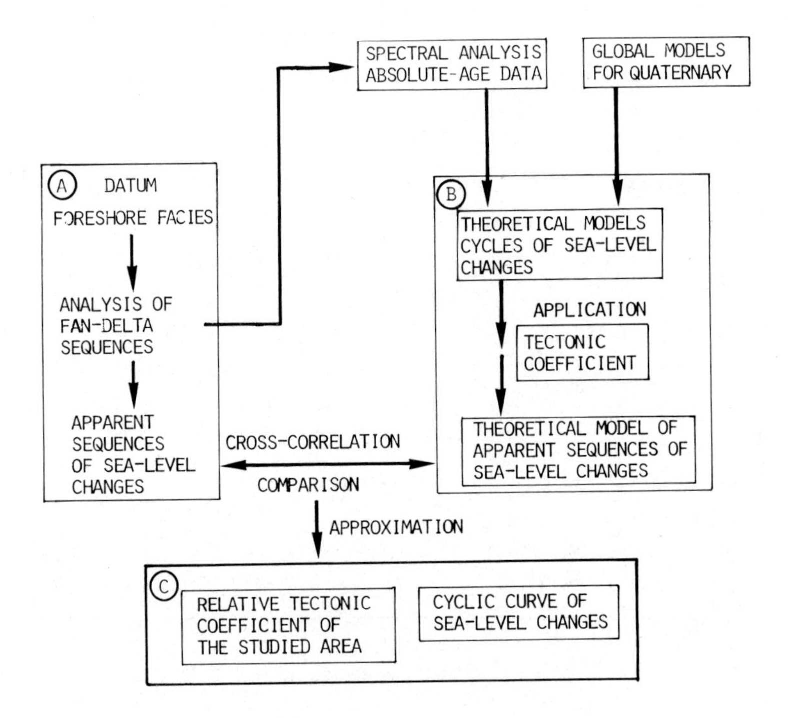

Fig. 13. Proposed methodology for the reconstruction of sea-level changes.

level can be measured and quantified just by measuring the displacement of the inflexion point (Somoza *et al.*, 1986–87). Complementary data can be deduced from the study of the vertical sequence and lateral evolution of facies.

The analysis of the sedimentary sequences laid down by fan deltas and the plunge steps in Cope Basin allow us to deduce relative fluctuations of sea level during the Pleistocene. The triangular morphology of fan-delta bodies is produced by successive migration of individual lobes during highstands which can be correlated (Fig. 14A). N 120° E-directed fractures control the differential elevation of each body of fan-delta deposits (Fig. 14B). The progressive uplift of the area generates the offlaping pattern of the sedimentary bodies corresponding to every stage of relative sea level.

Preparation of dynamic theoretical models

The present position of sea level is a result of a summation of all components of the vertical movements acting in the sedimentary basin. Theoretical models (Flemming, 1972; Schwarzacher & Schwarzacher, 1986) take into account all the (known) eustatic and tectonic components and show the relative positions of sea level as a function of the degree of tectonic activity (Fig. 15) (Mörner, 1987).

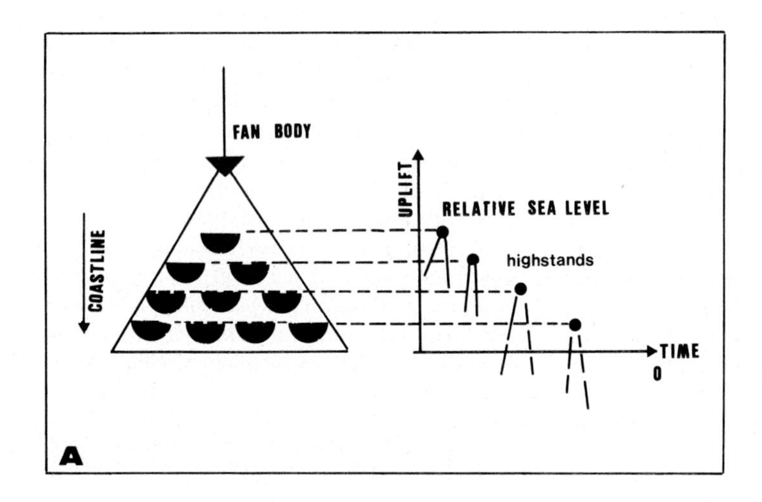

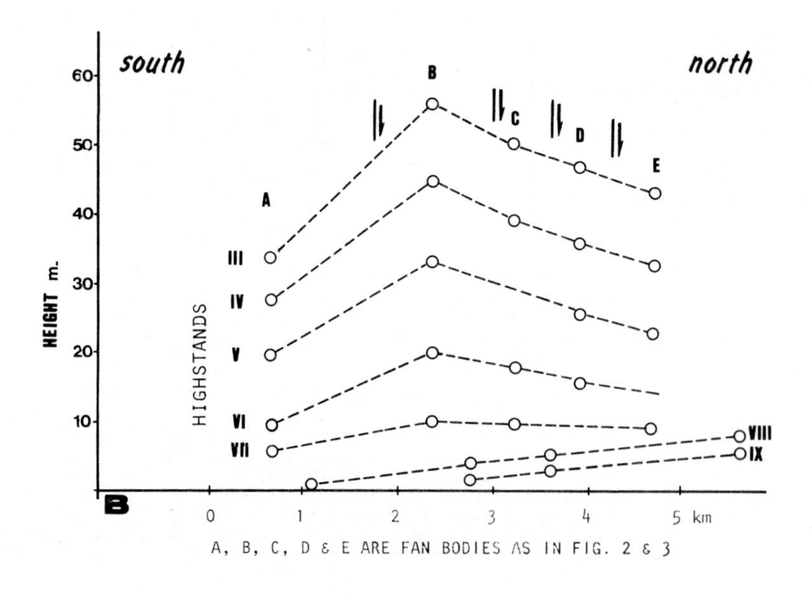

Fig. 14. (A) Sketch of the triangular-shaped geometry of the sedimentary bodies of 'fan deltas' generated as a consequence of the offlap of several generations of highstand deposits. Fluctuations of relative sea level are deduced from the marine–terrestrial sequence recorded in every sedimentary body. (B) Sea levels deduced from the sequences in every fan delta. Differences in height are produced by differential tectonic movements. Key: (o) height of break of slope of foreshore laminae (plunge step); A–E: areas with different degree of tectonic behaviour (see Fig. 2). Ages: III–VII middle Pleistocene marine episodes: cycle II; VIII–IX upper Pleistocene marine episodes; cycle III.

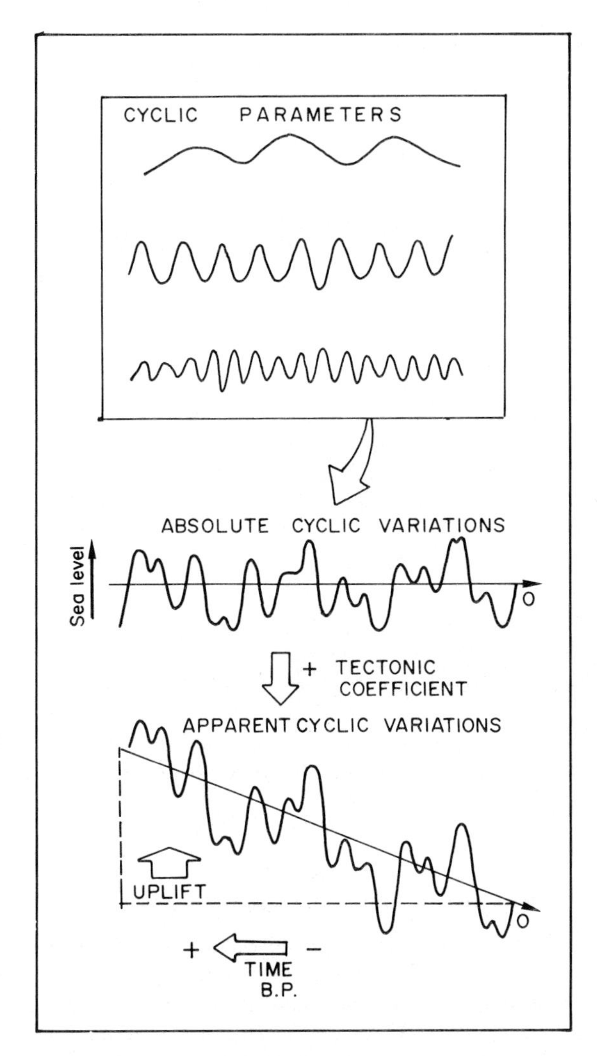

Fig. 15. Methodology for the construction of theoretical models of sea-level changes as a function of the eustatic-tectonic components. Modified after Schwarzacher and Schwarzacher (1986) and Mörner (1987).

These models help us to understand the geometric relationships and the resulting sedimentary sequences of fan-delta deposits as a result of sea-level fluctuations. In this paper these models are used to observe the theoretical relationship between the Pleistocene deposits and tectonic factors (Fig. 16).

A curve of absolute sea-level changes was prepared by putting together the cyclic parameters obtained from the spectral analysis of histograms of sea-level maxima based on the isotopic datings carried out by Hillaire-Marcel *et al.* (1986) in neighbouring areas (Somoza *et al.*, 1987). Diverse tectonic coefficients were added both in a linear and an exponential way to this curve of cyclic changes of sea level for varying degrees of tectonic uplift.

The result of the union of cyclic and linear parameters resulting from the tectonic movements is shown in Figure 16. This graph of apparent positions records only a possibility based upon two variables. Plenty of variations can be introduced in the numeric parameters of every variable, both in the tectonic coefficient and the amplitude and frequency of a given cycle. Computer-assisted curve design allows a better approximation of the models to the analysed sequences, because only curves matching more closely the real model are selected. This approximation yields information about the *cyclic parameters* affecting the sequences of sea-level changes in Pleistocene fan deltas as well as about the tectonic coefficients in each area. This last point is particularly important because of the active neotectonics affecting the studied area.

Comparison of the theoretical models and the observed sequences of apparent sea-level changes

Two main sedimentary stages can be deduced from the position of the relative curve in the theoretical models (Fig. 16).

First, highstands (h), correspond to the higher segments of the curve. The height reached by the curve for a given highstand depends on the amplitude of the cycle and also the tectonic coefficient. Maxima with diverse heights are visible: major cycles produce high maxima, whereas secondary or minor cycles produce lower maxima. There is a tendency for the maxima (episodes of highstand) to be smaller with time.

The second sedimentary stage is composed of sea-level falls or lowstands which primarily occur between major cycles. It should be noticed that the larger cycles are usually separated by stages of low relative sea level, although minor rises of relative sea level can take place during these periods.

The development of a rise of sea level implies several processes which are recorded in the sedimentary succession as diverse features (Fig. 17). In ideal, non-tectonic areas, a rise of sea level means transgression and coastal onlap ((a) in Fig. 17). During highstand, progradation of coastal deposits takes place ((b) in Fig. 17). Minor fluctuations of sea level can occur which are recorded as shifting of

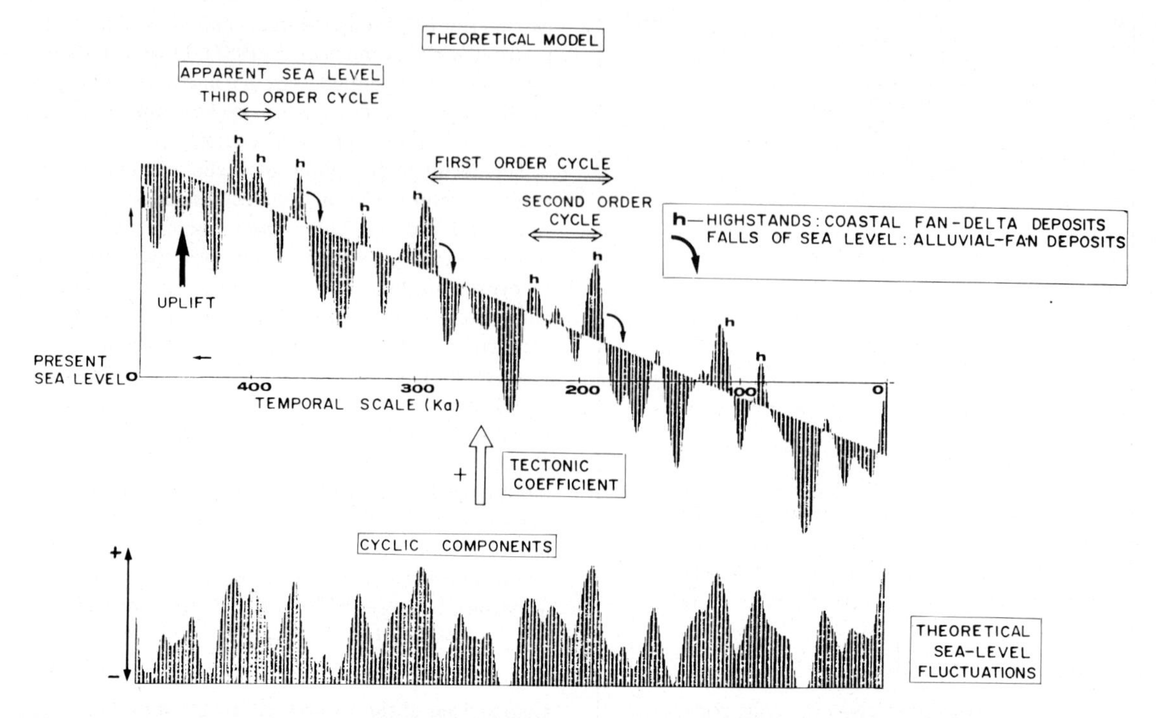

Fig. 16. Theoretical model of sea-level fluctuations and their relationship with the alternation of fan-delta and alluvial-fan deposits, applicable to the Pleistocene deposits of Cope Basin. Note the existence of major cycles of highstand separated by alluvial fan deposits. Minor cycles do not generate encased deposits or alluvial-fan deposits.

controlled facies ((c) in Fig. 17). Gentle lowering of relative sea level will probably increase progradation due to the exposure of new deposits to erosion ((d) in Fig. 17). Sea-level fall will eventually cause erosion ((e) in Fig. 17) and the coeval sediments will occur in the areas not exposed.

In Cope Basin, the occurrence of several episodes of major progradation may be related to the welding of several smaller episodes of highstand into a sequence of apparent continued, gentle fall of sea level (Fig. 6).

Lowstands result in erosional incision and development of confined alluvial fans on top of marine deposits of the former stage. It is interesting to note that there is a close correspondence between the type of alluvial deposits and sea-level movement. Muto (1987) has also correlated alluvial-fan deposits with the curves of late Pleistocene sea-level fluctuations. Alluvial fans deposited during lowstands can be distinguished from those of highstands. Incision during lowstands results in the confinement of deposits whereas open, mostly non-channelized alluvial fan deposits are found during highstands.

Cycles, tectonic degree and resulting sequences

After analysing the 'vertical' relationship between deposits generated during highstands and lowstands, it is necessary to study how they evolve through time, i.e. the resulting cycles. The comparison of the sequences of marine-terrestrial deposits in Cope Basin (Fig. 2) and the data from the theoretical model of apparent sea-level changes (Fig. 16) allows identification of three types of cycles (Fig. 18) which are related to the lateral association of deposits.

First-order cycles

Large maxima of highstand (recorded as prograding units of fan-delta sediments) are intercalated with long lowstand (marked by large accumulations of confined alluvial-fan deposits). Th/U dating indicates that these cycles lasted about 110–100 Ka. For instance, the Tyrrhenian levels T-I (dated as 180 Ka) and T-III (100–80 Ka) described by Hillaire-Marcel *et al.* (1986).

It is interesting to realize that in these cycles the

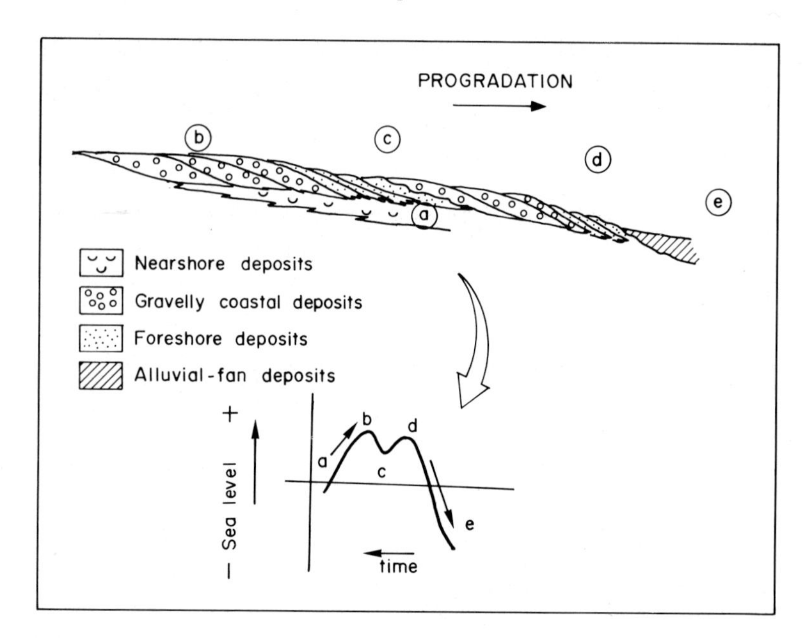

Fig. 17. Relationship of the observed sequence in coastal fan-delta deposits and sea-level changes during a cycle.

degree of incision produced by sea-level fall is directly proportional to the coefficient of tectonic uplift (i.e. the slope of the straight line defining the tectonic coefficient) (Fig. 18). This is observable in the Cope Basin, where the best-developed coastal deposits are those of the areas undergoing the most prominent uplift (sequences A and B in Fig. 14B).

Second-order cycles

These smaller-ranged cycles do not include alluvial-fan deposits but erosional surfaces indicative of lowstands (Fig. 18). The geometric relationships between deposits generated during the highstand depends primarily on the magnitude of tectonic uplift (TU) and, in lesser extent, on the amplitude of these cycles (AC) (Fig. 18).

If TU>AC, the second cycle is encased in the former. Some terrestrial deposits (indicative of lowstand) may occur in the interface between the two cycles.

If TU<AC, erosion, but no deposition, takes place between both cycles. If sea level rises more during the second cycle than during the first, fan-delta deposits of the highstand will occur stacked, separated by a scour surface of marine origin produced by erosion related to the onlapping foreshore.

Correlation of fan-delta deposits and isotopic datings indicate that the second-order cycles correspond to intervals of about 40 Ka. Examples of these cycles are the Tyrrhenian episodes T-II (130–125 Ka) and T-III (100–80 Ka described by Hillaire-Marcel *et al.*, 1986; Goy & Zazo, 1983b).

Third-order cycles

Smaller cycles can be distinguished inside the former. Small oscillations of sea level are necessary to generate interlayering of reddish chaotic conglomerates (terrestrial-derived debris flow) and foreshore deposits (Fig. 5) found for instance inside Tyrrhenian deposits (Episode VII). The period of these changes can be correlated to the small cycles found in the spectral analysis of Th/U datings performed in the area of southeastern Spanish Mediterranean (Somoza *et al.*, 1987). They are supposed to correspond to the intervals of 23–20 Ka and 10 Ka which probably are responsible for the occurrence of two maxima of sea level inside the cycles T-III (80 and 100 Ka) and T-II (120 and 130 Ka) (Kaufman, 1986).

CONCLUSIONS

In the basin of Cope, Pleistocene fan-delta sedimentation occurred in a complex realm due to the coincidence of: (1) active tectonics that did not cause strong subsidence; (2) successive eustatic sea-level changes; and (3) variable sedimentary processes

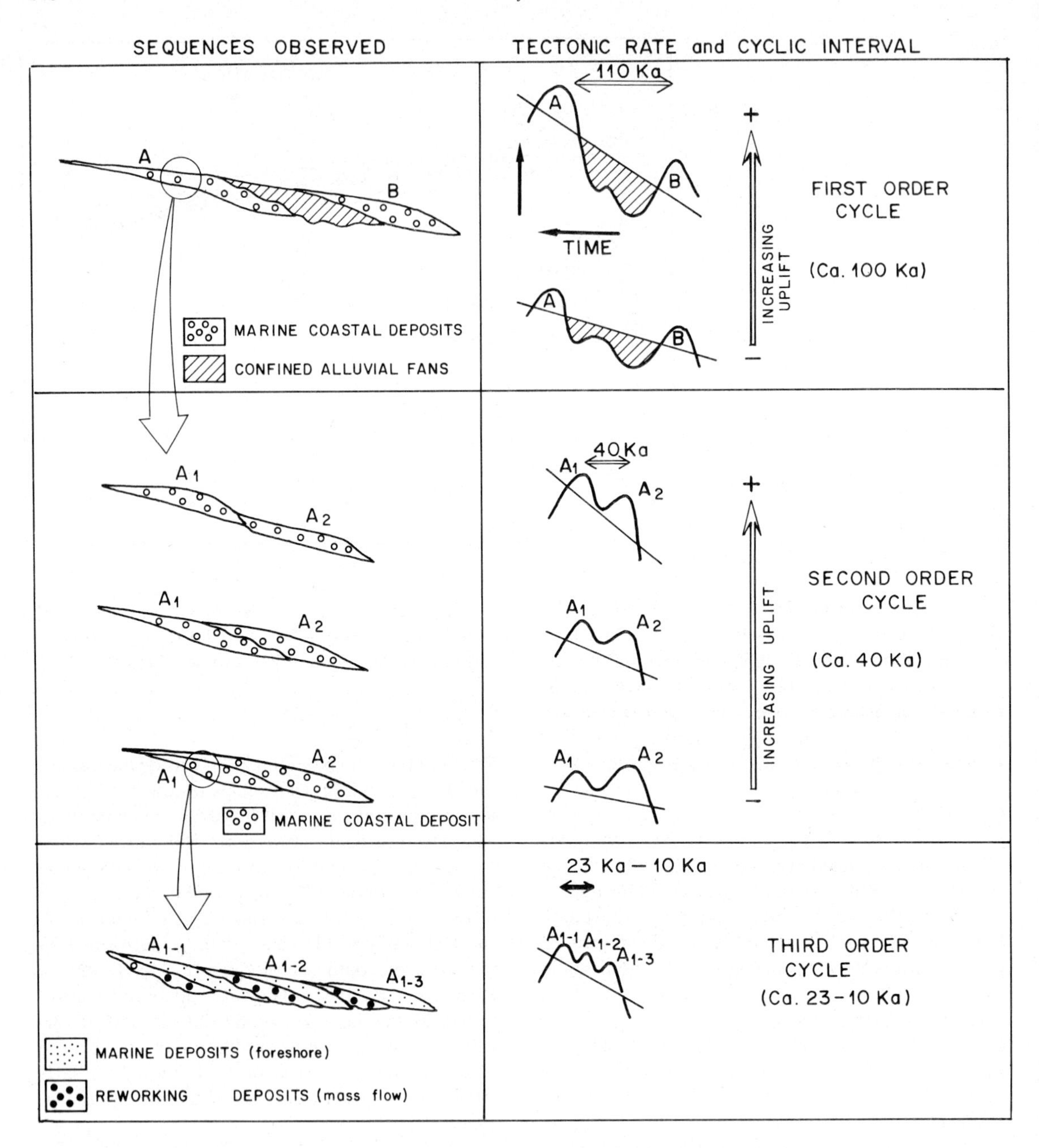

Fig. 18. Comparison between the observed sequences and cycles of sea-level changes.

acting in both the fan deltas and the neighbouring environments. Mutual interference resulted in the deposition of complex offlapping sequences which are commonly referred to as 'sequences of marine and terrestrial levels' or 'marine terraces'. The study of coastal morphology and detailed analysis of the sedimentary features provides useful tools for monitoring small-scale changes of sea level, inside larger eustatic oscillations, which are usually ignored.

The recognition of the several types of cycles in sedimentary sequences can be a useful tool for constructing curves of sea-level changes and calcu-

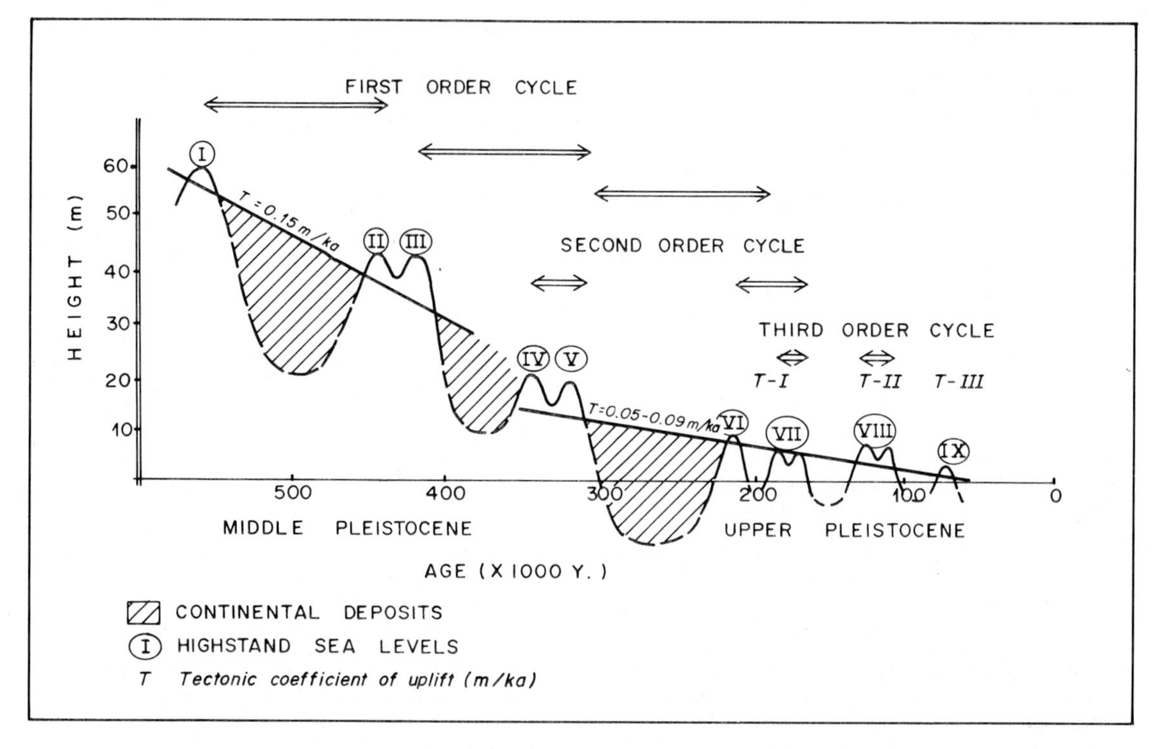

Fig. 19. Construction of the curve of sea-level changes based on measurements of the height reached by the breaker-zone facies, the interpretation of sequences and their relations with first, second and third-order cycles.

lating tectonic coefficients. Three categories of cycles were differentiated: (1) First-order cycles include the largest highstands with a period of about 100–110 Ka. They are separated from each other by terrestrial deposits. (2) Second-order cycles corresponding to highstands of smaller magnitude, which do not include terrestrial deposits but erosional surfaces indicative of lowstands. They occur with a periodicity of about 40 Ka. (3) Third-order cycles correspond to minor fluctuations of sea level distinguished inside larger cycles. They occur with a period of 23–20 Ka and 10 Ka.

The curve deduced for the study area (Fig. 19) is based on:

1 interpretation of first-order cycles from sediments of confined alluvial-fan deposits;
2 deduction of the height of highstand based on the analysis of lower-foreshore facies;
3 interpretation of second-order cycles based on the relationships of coastal fan-delta facies (erosion) and differences in height of the lower-foreshore facies;
4 interpretation of third-order cycles inside highstand episodes based on the analysis of features indicative of fall of sea level interlayered between prograding coastal deposits.

An approximation of the chronology of the events recorded in the prograding sequences of the Cope Basin can be carried out from study of the deduced curve of sea-level changes (Fig. 19). Episodes I–IV are younger than 700 Ka, as confirmed by palaeomagnetic datings (Mörner, pers. comm., 1989); they correspond to the middle Pleistocene. The curve was also used to deduce the linear coefficient of tectonic uplift because it is time-dependent (slope of the line) (Fig. 19). Two tendencies of the coefficient are observed corresponding to rates of uplift of 0.15 m/Ka during the middle Pleistocene and 0.05–0.09 m/Ka during the upper Pleistocene. These coefficients are rated as corresponding to moderately active areas (Lajoie, 1986), in particular for the middle Pleistocene times.

ACKNOWLEDGeMENTS

This paper is a part of a wider research programme with financial support of the Spanish CAICYT Project 2460/83 and DGICYT Project PB 88-0125. C.J.D. enjoyed financial support from 'Programación Científica del CSIC 630/070. The authors greatly appreciate the help of Dr M. Marzo and an anonymous referee who carefully reviewed the manuscript suggesting many improvements.

REFERENCES

Bardaji, T., Civis, J., Dabrio, C.J., Goy, J.L., Somoza, L. & Zazo, C. (1986) Geomorfología y estratigrafía de las secuencias marinas y continentales cuaternarias de la Cuenca de Cope (Murcia). In: *Estudios sobre geomorfología del sur de Espana. Com. Meas. Theory and Applic. in Geomorphology*. (Ed. by F. Lopez Bermudez and J. B. Thornes), pp. 11–16. Int. Geogr. Union. Universidades de Murcia y Bristol.

Bardaji, T., Dabrio, C.J., Goy, J.L., Somoza, L. & Zazo, C. (1987) Sedimentologic features related to Pleistocene sea level changes in the SE Spain. In: *Late Quaternary Sea-level Changes in Spain*. (Ed. by C. Zazo) Trab. Neog. Cuat. Museo Nal. Cienc. Nat. 10, 79–93.

Bernat, M., Echailler, J.C., Bousquet, J.C. (1982) Nouvelles datations Io-U sur des *Strombus* du dernier Interglaciaire en Mediterranée (La Marina, Espagne) et implications géologiques. *C. R. Acad. Sc. París* **295**, II, 1023–1026.

Bousquet, J.C. (1979) Quaternary strike-slip faults in the southeastern Spain. *Tectonophysics* **52**, 277–286.

Bryant, E. (1983) Sediment characteristics of some Nova Scotian beaches. *Maritime Sediments and Atlantic Geology* **19**, 127–142.

Brückner, H. (1986) Stratigraphy, evolution and age of Quaternary marine terraces in Morocco and Spain. *Z. Geomorph. N.F.* **62**, 83–101.

Butzer, K.W. (1975) Pleistocene littoral–sedimentary cycles of Mediterranean: a Mallorquin view. *After the Australopithecus, The Hague*, 25–71.

Clifton, H.E. (1969) Beach lamination — nature and origin. *Mar. Geol.* **7**, 553–559.

Clifton, H.E., Hunter, R.E. & Phillips, R.W. (1971) Depositional structures and processes in the non-barred high-energy nearshore. *J. sedim. Petrol.* **41**, 651–670.

Dabrio, C.J., Goy, J.L. & Zazo, C. (1984) Dinámica litoral y evolución costera en el Golfo de Almería desde el 'Tirreniense' a la actualidad. *I. Cong. Español de Geología, Segovia*. **I**, 507–522.

Dabrio, C.J., Goy, J.L. & Zazo, C. (1985) A model of conglomeratic beaches in tectonically-active areas (Late Pleistocene–Actual, Almería, Spain). *Abstr. Int. Ass. Sedim. 6th Europ. Reg. Mtg., Lleida*, Univ. de Barcelona, 104–105.

Dabrio, C.J. & Polo, M.D. (1981) Flow regime and bedforms in a ridge and runnel system, SE Spain. *Sedim. Geol.* **28**, 97–109.

Davis, R.A., Jr., Fox, W.T., Hayes, M.O. & Boothroyd, J.C. (1972) Comparison of ridge and runnel systems in tidal and non-tidal environments. *J. sedim. Petrol.* **42**, 413–421.

Flemming, N.C. (1972) Eustatic and tectonic factors in the relative displacement of the Aegean coast. In: *The Mediterranean Sea* (Ed. by D. J. Stanley), pp. 180–201. Dowden, Hutchinson and Ross Inc. Stroudsberg.

Goy, J.L. & Zazo, C. (1982) Niveles marinos cuaternarios y su relación con la tectónica en el litoral de Almería (España). *Bol. R. Soc. Esp. Hist. Nat. (Geol.)* **80**, 171–184.

Goy, J.L. & Zazo, C. (1983a) Los piedemontes cuaternarios de la región de Almería (España). Análisis morfológico y relación con la neotectónica. *Comunicaciones. IV Reunión del Grupo Español de Trabajo de Cuaternario*. Universidad de Santiago de Compostela, 397–419.

Goy, J.L. & Zazo, C. (1983b) Pleistocene tectonics and shorelines in Almería (Spain). *Bull. INQUA Neotectonics Com.* **6**, 9–13.

Goy, J.L., Zazo, C., Bardaji, T. & Somoza, L. (1986a) Las terrazas marinas del Cuaternario reciente en los litorales de Murcia y Almeria (España): el control de la neotectónica en la disposicíon y número de las mismas. *Est. Geol.* **42**, 439–443.

Goy, J.L., Zazo, C., Dabrio, C.J. & Hillaire-Marcel, C.L. (1986b) Evolution des systémes de lagoons-îles barriére du Tyrrhenian a l'actualite a Campo de Dalías (Almería, Espagne). *Edit. de l'Orsrom, Coll. Travaux et Documents* **197**, 169–171.

Guza, R.T. & Inman, D.L. (1975) Edge waves and beach cusps. *J. geophys. Res.* **80**, 2997–3012.

Hearthy, P.J., Hollin, J.T. & Dumas, B. (1987) Geochronology of Pleistocene littoral deposits on the Alicante and Almeria coasts of Spain. In: *Late Quaternary Sea Level Changes in Spain*. (Ed. by C. Zazo) Trab. Neog. Cuat. Museo Nal. Cienc. Nat. **10**, 95–105.

Hillaire-Marcel, CL., Carro, O., Causse, CH., Goy, J.L. & Zazo, Z. (1986) Th/U dating of *Strombus bubonius* bearing marine terraces in southeastern Spain. *Geology* **14**, 613–616.

Issel, A. (1914) Lembi fossiliferi quaternari e recente osservati nella Sardegna meridionali. *R. C. Acad. Lincei, 5a sèrie* **23**, 759–770.

Kaufman, A. (1986) The distribution of ^{230}Th/^{234}Th ages in corals and the number of high-sea stand. *Quat. Res.* **25**, 55–62.

Lajoie, K.P. (1986) Coastal tectonics. In: *Studies in Geophysics*. National Academy Press. Washington, D.C.

Mahmoudi, M., Purser, B.H. & Plaziat, J.C. (1987) Quaternary shallow marine carbonates in the eastern coast of Tunisia. Third day, The Hergla Region: Spatial relationships between Khnis and Rejiche Units; internal structures of littoral barrier ridge. In: *Int. Ass. Sedim. 8th Mgt.-Field Trip Guidebook, Excursion A-4*, Univ. de Tunis, 129–172.

Massari, F. & Parea, G.C. (1988) Progradational gravel beach sequences in a moderate- to high-energy, microtidal, marine environment. *Sedimentology* **35**, 881–913.

Miller, R.L. & Ziegler, J.M. (1968) A model relating dynamics and sediment pattern in equilibrium in the region of shoaling waves, breaker zone, and foreshore. *J. Geol.* **66**, 417–441.

Montenat, C., Ott D'Estevou, P. & Masse, P. (1987) Tectonic–sedimentary characters of the Betic Neogene Basins evolving in a crustal transcurrent shear zone (SE Spain). *Bull. Centre Rech. Explo.-Prod. Elf-Aquitaine* **11**, 1–22.

Mopu (1976) *Plano Indicativo de Usos del Dominio Litoral (PIDU)-Murcia*. MOPU, Madrid. 2 vols.

Mörner, N. (1987) Models of global sea-level changes. In: *Sea-level Changes* (Ed. by M. J. Tooley and I. Shennan). Blackwell Scientific Publications, Oxford.

Muto, T. (1987) Coastal fan processes controlled by sea-level changes: a Quaternary example from the Tenryugawa Fan system, Pacific coast of Central Japan. *J. Geol.* **95**, 716–724.

Schwarzacher, W. & Schwarzacher, W. (1986) The effect of sea-level fluctuations in subsiding basins. *Computer and Geosciences* **12** (2), 225–227.

Short, A.D. (1979) Wave power and beach-stages: a global model. *Proc. 16th Int. Conf. Coastal Eng. Hambourg*, 1145–1162.

Short, A.D. (1984) Temporal change in beach type resulting from a change in grain size. *Search* **15**, 228–230.

Short, A.D. & Wright, L.D. (1983) Physical variability of sandy beaches. In: *Sandy Beaches as Ecosystems*. pp. 133–144. Dr W. Junk Publishers, The Hague.

Somoza, L., Bardaji, T., Dabrio, C.J., Goy, J.L. & Zazo, C. (1986–87) Análisis de secuencias de islas barrera Pleistocenas en relación con variaciones del nivel del mar, Laguna de la Mata. *Acta Geológica Hispánica* **21–22**, 151–157.

Somoza, L., Zazo, C., Bardaji, T., Goy, J.L. & Dabrio, C.J. (1987) Recent Quaternary sea-level changes and tectonic movements in SE Spanish coast. In: *Late Quaternary Sea-level Changes in Spain* (Ed. by C. Zazo), Trab. Neog. Cuat. Museo Nal. Cienc. Nat. 10, pp. 49–78.

Sonu, C.J. (1973) Three-dimensional beach changes. *Jr. Geol.* **81**, 42–64.

Tanner, W.F. (1968) High-Froude phenomena. *Abstr. Geol. Soc. Am., Ann. Mtgs, New Orleans, La.*, 1967, Program, 220.

Vegas, R. & Banda, E. (1982) Tectonic framework and Alpine evolution of the Iberian Peninsula. *Earth Evol. Sc.* **4**, 320–343.

Wright, L.D., Chappell, J., Thom, B.C., Bradshaw, M.P. & Cowell, P. (1979) Morphodynamics of reflective and dissipative beach and inshore systems: Southeastern Australia. *Mar. Geol.* **32**, 105–140.

Zazo, C. & Goy, J.L. (1989) Sea-level changes in the Iberian Peninsula during the last 200 000 years. In: *Late Quaternary Correlations and Applications*. (Ed. by D. Scott, P. Pirazzoli and G. Honing) **256**, 27–39, Kluwer Academic Publishers.

Zazo, C., Goy, J.L. & Aguirre, E. (1984) Did *Strombus* survive the last interglacial in the western Mediterranean Sea? *Mediterranea* **3**, 131–137.

Zazo, C., Goy, J.L. & Meco, J. (1989) Highstands of sea level in the last 100 000 years in the littoral of Cadiz (Spain). *Meeting of I.G.C.P. Project 274*, September 1989.

Modern Alluvial Deltas

Spec. Publs int. Ass. Sediment. (1990) **10**, 155–168

Morphology and sedimentology of an emergent fjord-head Gilbert-type delta: Alta delta, Norway

G. D. CORNER, E. NORDAHL, K. MUNCH-ELLINGSEN *and* K. R. ROBERTSEN

Department of Geology, Institute of Biology and Geology, University of Tromsø, N-900 Tromsø, Norway

ABSTRACT

The Alta delta, in Finnmark, northern Norway (70° N, 23.5° E), is a Holocene, partially emergent, fan-shaped, sandy to sandy–gravelly Gilbert-type delta, set in a subarctic, fjord-head environment.

A study of the modern delta surface, subrecent subsurface and sections in older, raised terraces, shows that the delta comprises three major components: (1) a relatively thin (1–3 m), sandy–gravelly topset unit with erosional base, formed on an emerging, partly fluvial-, partly wave- and tide-dominated, intertidal delta plain; (2) a thick, sandy foreset unit comprising steeply dipping beds, deposited on a steep (5–35°), gravity-process-dominated delta slope; and (3) a muddy bottomset unit containing intermittent, turbidite sand beds, formed on a gently sloping (<3–4°) prodelta area.

Radiocarbon dates and a regional sea-level curve have been used to reconstruct palaeodepths, sedimentation rates and average rate of delta progradation (*c.* 0.4 m yr^{-1} over the last 6000 years).

INTRODUCTION

Gilbert-type deltas (Gilbert, 1885), which are characterized by their steep foreset slope and tripartite structure, appear to be common in three main marine settings: (1) tectonically active areas of high relief (Ethridge & Wescott, 1984; Postma, 1983, 1984; Postma & Roep, 1985; Colella *et al.*, 1987; Ori & Roveri, 1987; Colella, 1988a, b; Massari & Colella, 1988; Postma *et al.*, 1988); (2) glacier-marginal (glaciodeltaic) settings (McCabe *et al.*, 1987; McCabe & Eyles, 1988); and (3) fjord settings (Kostaschuk & McCann, 1987; Syvitski & Farrow, 1983; Prior & Bornhold, 1986, 1988; Postma & Cruickshank, 1988). Factors which these settings generally have in common (cf. Colella *et al.*, 1987; Colella, 1988b; Massari & Colella, 1988) are: (1) an abundant supply of relatively coarse (sandy, gravelly) sediment; (2) a relatively deep basin; and (3) a sheltered, low-energy, marine environment which causes relatively little reworking of the sediment.

Research on Gilbert-type delta systems has focused mainly on lithologic and stratigraphic aspects of ancient systems, or on processes, morphology and surficial sediments of modern systems. There have been relatively few attempts to integrate morphologic information from contemporary environments with stratigraphic data from raised features. The present work attempts to do this for an emergent, Holocene, fjord-head delta — the Alta delta in northern Norway.

The Alta delta (70° N, 23.5° E; Figs 1, 2) projects into the Alta fjord as a 5 km wide, unconfined, fan-shaped, distally thickening prism of predominantly sandy sediment. Postglacial isostatic uplift has partially raised the delta above sea level, forming terraces behind the modern, active delta surface. The delta is similar, in many respects, to other postglacial fjord-head deltaic deposits in northern Norway (Marklund, 1960; Corner, 1975; Calles, 1977; Fjalstad, 1986).

The present study concentrates on: (1) the morphology and sedimentology of the modern subaerial and subaqueous delta surface; (2) the lithology and structure of the delta subsurface investigated by coring and acoustic profiling; and (3) lithofacies exposed in raised-terrace sections. The aim is to provide an overview of the geometry, structure, lithofacies and evolution of the Alta delta, and to illustrate the relationship between the component parts of the recent and subrecent system. The availability of radiocarbon-dateable material and a well-

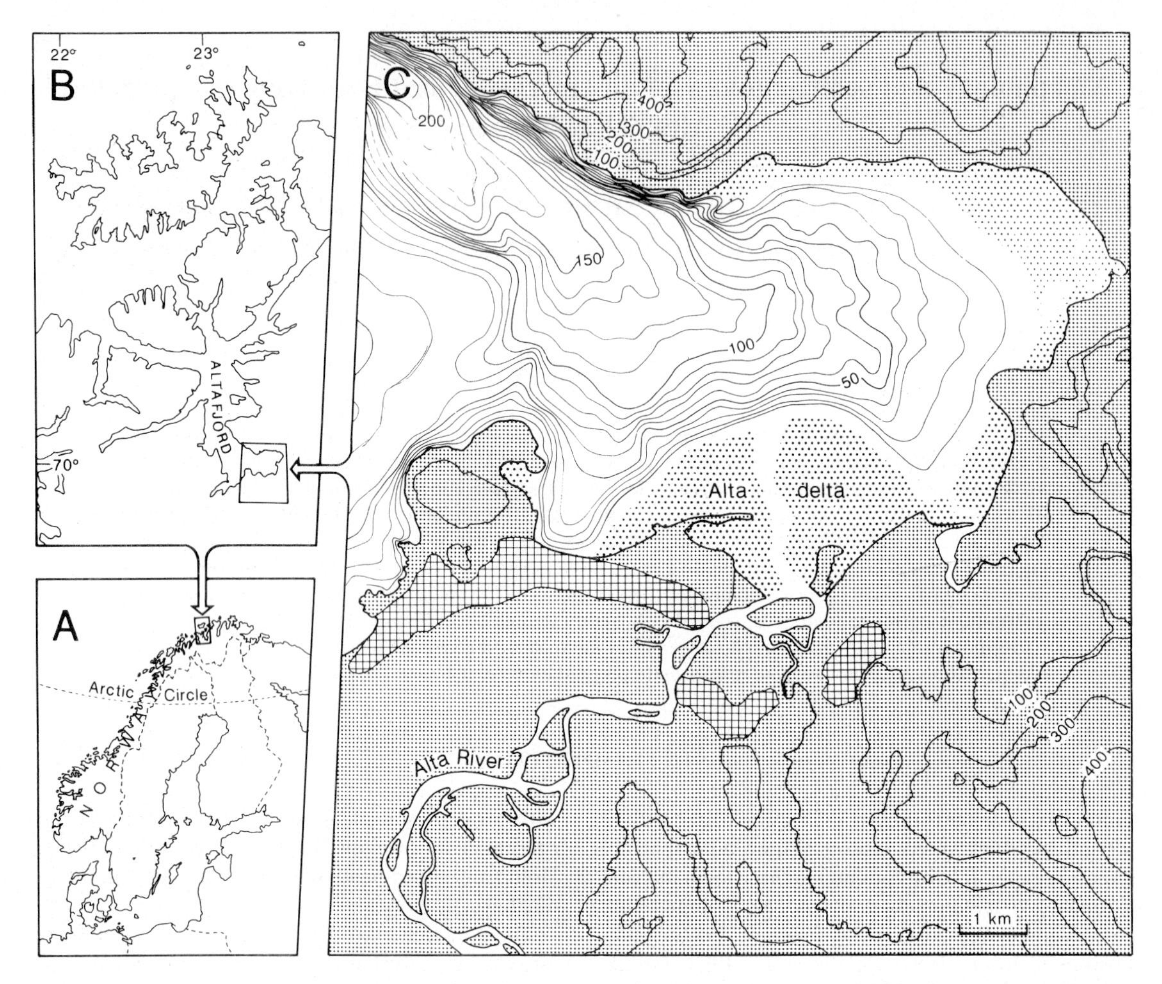

Fig. 1. Maps showing the geographic location (A), fjord-head setting (B) and physiography (C) of the Alta delta. Dotted areas: intertidal flats. Cross-hatched areas: marginal moraine formed *c.* 10 000 years ago. Bathymetry and altitude in metres.

documented history of sea-level change in the region have enabled progradation rates as well as depositional palaeodepths to be estimated.

Classification of deltaic systems is currently in a state of flux, as work on coarse-grained systems increasingly shows the great variety of deltaic types (Ethridge & Wescott, 1984; Massari & Colella, 1988; Prior & Bornhold, 1988; Wood & Ethridge, 1988; Chough *et al.*, 1988) and highlights the difficulty of choosing a suitable nomen-clature (Nemec & Steel, 1988; McPherson *et al.* 1987, 1988; Orton, 1988; Massari & Colella, 1988; Steel, 1988). Although the Alta delta shows obvious similarities with many coarse-grained systems described as fan deltas, it is not, genetically, a fan delta as currently defined (Holmes, 1965; McGowen, 1970; Nemec & Steel, 1988), since the Alta river is not part of an alluvial fan system. Neither is the Alta delta a typical braid delta (McPherson *et al.*, 1987, 1988), because braiding, both on the intertidal delta plain and along the river that feeds it, is poorly developed and secondary to a prominent main channel. Within the spectrum of deltaic systems, from low-relief, fine-grained river deltas to coarse-grained systems of low to high gradient (e.g. shelf-type fan deltas and Gilbert-type systems), the Alta delta may be described as a medium-grained (predominantly sandy), high-gradient (Gilbert-type) delta.

Fig. 2. Oblique aerial photograph of the Alta delta and hinterland looking southeast, taken at low tide, 27 July 1986. The western part of the delta has been partly destroyed by construction works and sand extraction. Photograph: Fjellanger-Widerøe A/S.

SETTING AND REGIME

The Alta delta is set in a glacial landscape of relatively high relief (up to 500–600 m a.s.l.), (Figs 1, 2). The delta is situated at the mouth of the Alta River, beyond a gap in a large end moraine formed during the retreat of the Fennoscandian ice-sheet about 10 000 years ago.

The Alta River, in its lower reaches, is meandering–braided (Schumm, 1981) with a gradient of *c.* 1.5 m km^{-1}. Discharge is highly seasonal and annually variable, being characterized by a brief period of peak discharge during snowmelt in May or June, when values normally reach 300–900 m^3 s^{-1} (Fig. 3). Sediment is derived primarily from the lower valley, which contains thick deposits of glacigenic and postglacial gravel, sand and (locally) mud (cf. Follestad, 1979). Falling base-level, caused by glacio-isostatic uplift, has ensured a high rate of sediment supply during the Holocene. Emergence, totalling about 30 m during the last 8000 years, has occurred at an average rate of *c.* 3.5 mm yr^{-1} (Fig. 13B). The present rate of emergence, estimated from regional geodetic surveys (Bakkelid, 1980; Sørensen *et al.*, 1987), is just over 1 mm yr^{-1}.

The climate is subarctic, with mean temperatures for the year, and the warmest (July) and

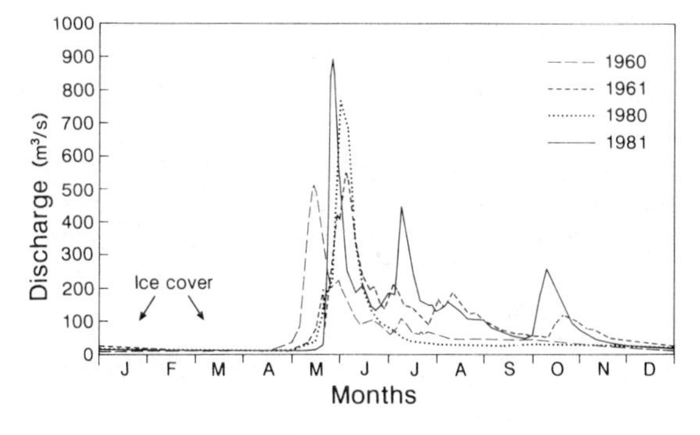

Fig. 3. Discharge curves from selected years for the lower part of the Alta River. Data source: Norwegian Water Resources and Electricity Board (NVE).

coldest (February) months, being 1.7°C, 14.3°C and −8.3°C, respectively (Norwegian Meteorological Institute). Mean annual precipitation is *c.* 400 mm. Sea-ice covers large parts of the delta during the winter months.

The receiving basin is sheltered and wave energy is low; prevailing onshore winds blow from a northwesterly to north–northwesterly sector during summer. Tides are semi-diurnal, mesotidal, with mean neap and spring ranges for the nearest tide-gauge station at Hammerfest, being 1 m and 2.5 m, respectively (Norges sjøkartverk).

Human activity, in the form of bank protection along the river, and sand-extraction and construction works on the delta, has had a relatively small, though increasingly important, effect upon the delta system during the last few decades.

MORPHOLOGY AND SEDIMENTS OF THE MODERN DELTA

The modern delta is enveloped by two major surfaces (Fig. 4): a predominantly intertidal *delta plain*, and a subaquatic, steep *delta slope*, which passes distally into a more gently sloping *prodelta* area lying deeper than *c*. 100 m. The boundary between the delta plain and the delta slope, the *delta rim*, is sharp and lies *c*. 2–5 m below mean sea level (*c*. 0.5–3 m below low-tide level). The delta plain is bounded proximally, at the high-tide limit, by an erosion scarp.

Delta plain

The delta plain (Fig. 5) can be divided, on the basis of morphogenetic criteria, into two major zones:

1 a central, fluvially-dominated zone (fluvio-tidal plain), having relief features oriented radially from the apex of the delta;

2 lateral wave- and tide-dominated areas (western and eastern tidal plains), having forms oriented predominantly obliquely to the central axis of the delta.

Fluvially-dominated zone (fluvio-tidal plain)

The central, fluvially dominated zone comprises a system of radially oriented channels and bars. The bars lie almost entirely within the intertidal zone.

The zone comprises a main, 2–5 m deep, fluvial channel, and a system of low-relief longitudinal bars and shallow channels. The main channel is bordered, on its western side, by a levee. To the east, the fluvio-tidal plain merges gradually into the tidal plain.

The radial orientation and splaying character of the bars in the central zone are interpreted as having formed predominantly during seasonal, snowmelt flood events. Apart from smaller bars which migrate annually down the main channel, the configuration of the fluvio-tidal zone has changed little during the 40 years or so (since 1946) that aerial photographs have been available, suggesting a relatively low level of fluvial activity during this period.

The fluvio-tidal plain is modified, outside flood season, by ice, tidal and, especially, wave processes, as evidenced by the widespread occurrence of wave-rippled surfaces and isolated, perennial, transverse bars (Fig. 5).

Wave- and tide-dominated areas (tidal plains)

Dominant features of the tidal plains are low-relief swells and moderate- (< 0.5 m) to high-relief (0.5–1.5 m) bars, alternating with very low-relief flats or partially enclosed basins (Figs 5, 6). The swells and

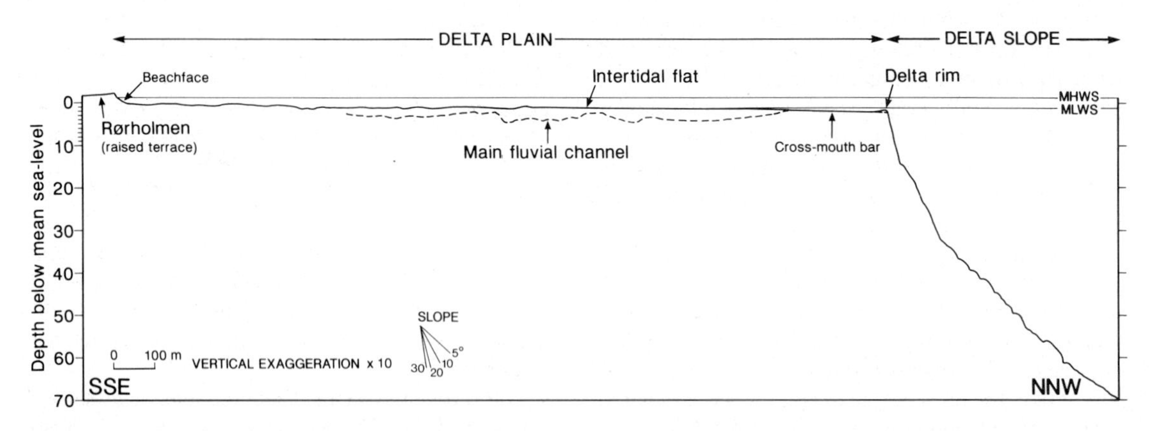

Fig. 4. Composite, axial profile across the Alta delta showing the morphology of the intertidal delta plain (levelled profile) and subtidal, main river channel and delta slope beyond (from echograms). The levelled profile traverses the fluvio-tidal plain and a short segment of the tidal plain nearest Rørholmen. Mean, high- and low-water, spring-tide levels (MHWS, MLWS) are indicated.

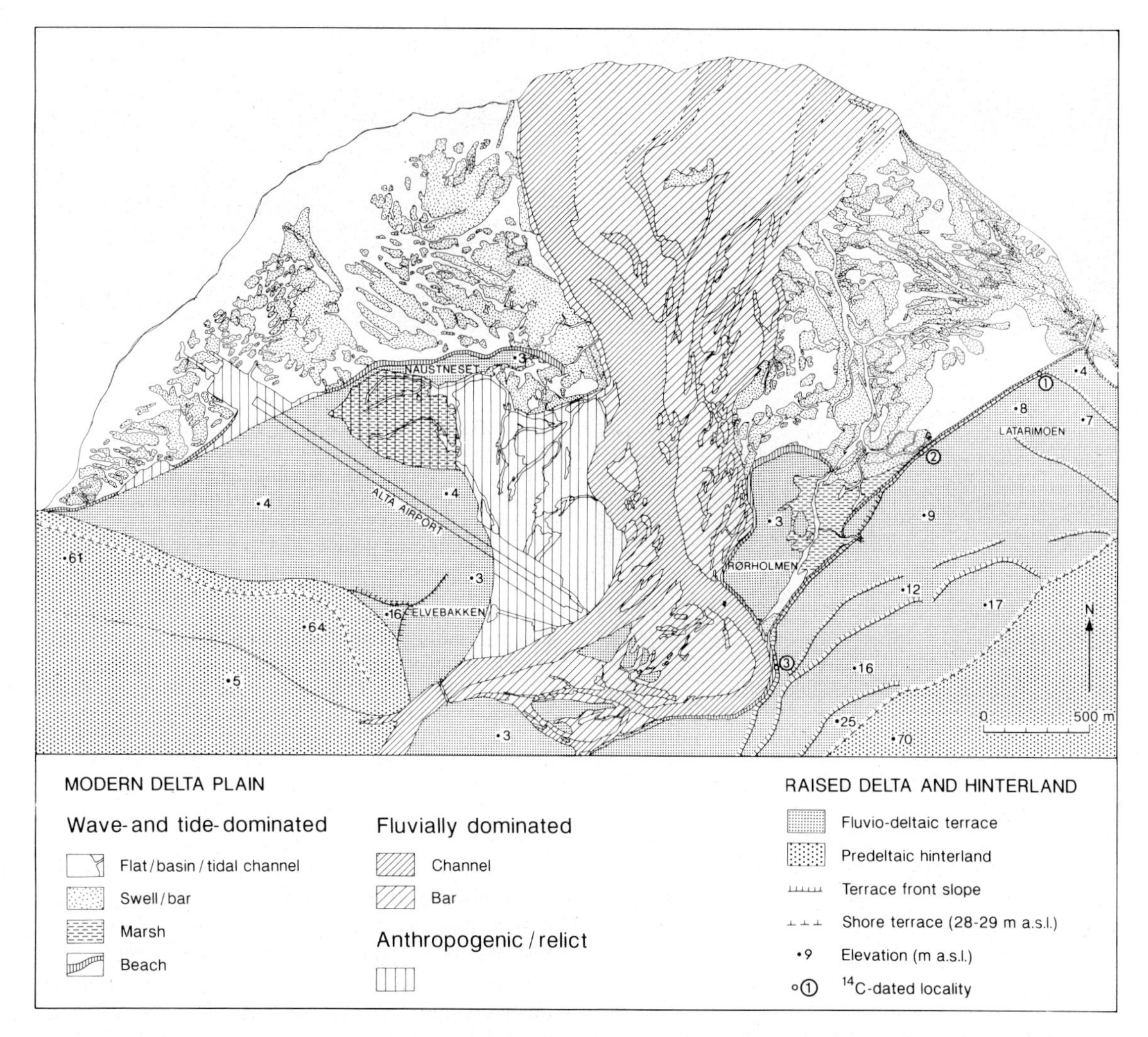

Fig. 5. Simplified morphogenetic map of the modern Alta delta plain and raised proximal terraces. Based on aerial photographs taken in 1984.

bars include both regular, elongate (longitudinal) and shorter (transverse) features, as well as more irregular forms. The irregular bars occur mostly on the eastern plain; their origin may be related to ice-push and ice-accretion processes, or they may comprise gravelly, palimpsest material exhumed or superimposed from an earlier, higher tidal plain by marine erosion. The regular swells and bars show a preferred orientation which is either parallel or transverse to a northwest–southeast direction (oblique to the axis of the delta). This orientation corresponds to the direction of the prevailing, northwesterly, onshore summer winds, and supports the interpretation that these are predominantly wave-formed features. The swells resemble those described from other relatively sheltered coasts (Nilsson, 1973; Luternauer, 1980). Some of the bars have migrated or extended a short distance towards the southeast during the last 40 years, but most have shown little change.

Shallow (0.5–1 m deep), stable, tidal channels are developed locally on the upper half of the intertidal zone, especially on the eastern plain where they form an integrated drainage system. Saltmarsh is developed locally on higher, protected, inner parts of the tidal plain. A beachface (Komar, 1976) bor-

Fig. 6. Photographs illustrating delta-plain morphology, sediments and processes. (A) High-relief, longitudinal bar with gravel–cobble lag overlying pebbly coarse sand. Middle part of the eastern tidal flat, looking northwest. Compass (right foreground) for scale. (B) Gravelly, sandy flats with scattered pebbles and gravel patches. Wave ripples and ice-block tool mark (arrow) in the foreground; disintegrating, drifted ice-blocks in the background. Transition zone between the fluvio-tidal plain and eastern tidal flat, looking east, 6 April 1988. (C) Sandy flat with grooves formed by pebbles attached to drifting algae (*Fucus vesiculosus*). Inner part of the eastern tidal flat, looking north. Trowel (left foreground) for scale.

ders the tidal plain proximally, at the interface between the high-tide limit and raised terraces behind (Fig. 8).

Sediments

Sediments on the delta plain (Fig. 7) are predominantly medium to coarse sands or bimodal sand–gravel mixtures. A small component of cobbles or silt is present in some sediments.

Three aspects of the character and distribution of the sedimentary facies may be emphasized:

1 The sediments show a general pattern of proximal to distal fining, with gravels dominating proximally along the main fluvial channel, and sands dominating distally around the delta rim. The finest, silty, sandy sediments occur at sheltered, inner locations on the tidal flats. This distribution is a predictable result of normal, hydrodynamic sorting.

2 Large areas of the middle and upper delta plain are characterized by an irregular distribution of gravelly sand and sandy gravel, which only partially corresponds with major morphological features. This *mixed* sand–gravel facies (Fig. 6) contains a variable proportion of clasts transported by drifting sea-ice and drifting algae (*Fucus vesiculosus*), (cf. Dionne, 1969; Martini, 1981; Gilbert, 1984).

3 Gravel lag layers, formed by wave-winnowing are fairly common on both bars (Fig. 6A) and flats on the tidal plain. In some cases, the gravel may have accumulated through ice — and algae-assisted transport in areas where wave energy is too low to rework other than the sand component. In other cases, the gravel may represent winnowed, palimpsest material.

Delta slope

The delta slope has a generally concave-upward, parabolic, longitudinal profile (Fig. 4), which decreases from *c*. 35° at the top to just under 5° at its foot. Profiles approaching parabolic shape are shown by many other, relatively high-gradient deltas (Luternauer, 1980; Bogen, 1983; Kostaschuk & McCann, 1983).

Sediments on the delta slope are predominantly

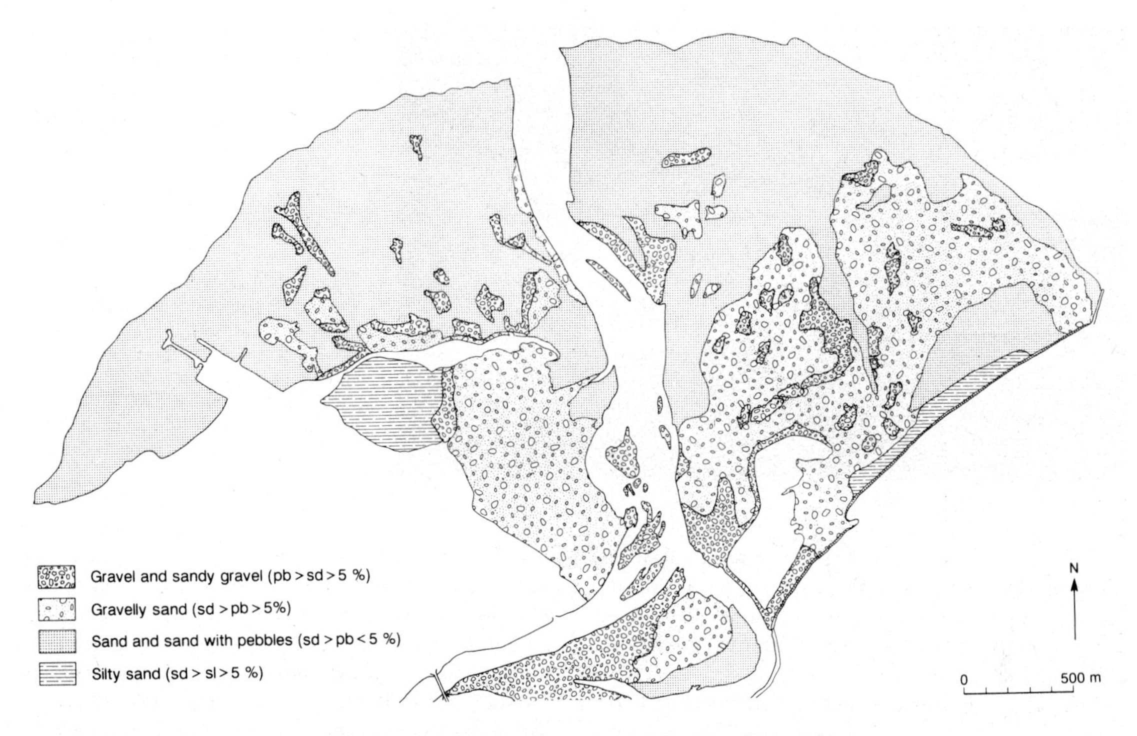

Fig. 7. Map showing the distribution of sediment grain-size facies on the Alta delta plain.

sands: fine to very fine sand to a depth of *c*. 30 m, silty sand to *c*. 70 m and sandy mud below *c*. 70 m.

Preliminary results from acoustic profiling reveal a system of downslope-oriented troughs and ridges, and series of smaller transverse ridges which give a characteristic hummocky topography in longitudinal profile (Fig. 4). First-order troughs are up to 10 m deep and 200 m wide and decrease in amplitude with depth. The exact nature of the relief features on the delta slope has yet to be elucidated, but they appear to include chutes, splays and transverse ridges similar to features reported from other, steep, gravity-process-dominated delta slopes (Prior *et al*., 1981; Kostaschuk & McCann, 1983, 1987; Prior & Bornhold, 1988).

Prodelta

The prodelta area occupies a narrow trough beyond the delta slope at depths below *c*. 100 m. The gradient is generally less than 3–4°. Surface sediments are predominantly bioturbated muds containing occasional pebbles, plant debris and shells deposited by drifting ice or other agents.

STRUCTURE AND LITHOFACIES OF THE DELTA SUBSURFACE

An investigation of the delta subsurface, based on seismic profiles, cores and exposures in raised terraces (cf. Fig. 8), indicates that the delta has a Gilbert-type, tripartite structure comprising topset, foreset and bottomset beds. These three components are described in the next section and compared with their modern analogues on the present delta plain, slope and prodelta, respectively.

Topsets

Topset beds observed in a 2 km, proximal-to-distal section in the raised 8–9 m eastern terrace (cf. Localities 1–3, Fig.5), form a 1–3 m thick unit of sand, pebbly sand and gravel, erosively overlying

Fig. 8. Raised-terrace section bounding the eastern tidal flats (Locality 2, Fig. 5). Sandy foreset beds are visible to the right and, above them (centre), topset beds (cf. Fig. 9). Sandy beach and saltmarsh at high-tide level in the foreground. Figure (circled) for scale.

foreset beds. There is considerable lateral variation in facies character and thickness; the unit thins locally to < 0.5 m beneath a tidal channel distally on the terrace and wedges out abruptly proximally (Fig. 13C).

The topset unit at an intermediate locality (Locality 2), comprises a lower unit of channel-fill, cross-bedded, pebbly coarse sands, overlain by an upper unit of alternating, thin, laterally persistent, horizontal to subhorizontal beds of gravel, and sets of cross-bedded, coarse sand (Figs 9, 10). The upper unit represents intermittently formed gravel lag deposits similar to those forming on the present delta plain, and accretionary bars and megaripples. The whole topset unit is interpreted as a shallowing-upward, delta-plain sequence passing from subtidal fluvial channel to intertidal plain deposits.

The thickness (1–3 m) of the topset unit in the terrace, is somewhat less than the amplitude of the modern delta plain (*c*. 3–5 m, Fig. 4). The difference may be due, at least in part, to regression and consequent erosion of the emerging plain during formation, resulting in topset units of reduced thickness. This appears to have occurred, for example, at Locality 4 (Fig. 13C), where a radiocarbon date from the base of the topset unit, suggests that its lower part (channel deposits) started forming when sea level was several metres higher than the present top of the terrace.

Foresets

Steeply dipping foreset beds beneath the distal, axial part of the modern delta plain (Fig. 11), appear to extend from near the base of the main river channel, to a depth of several tens of metres, where they overlie more gently dipping beds which may represent distal foresets, bottomsets or predeltaic glacigenic deposits. Borings and seismic refraction studies (Systad, 1979, 1980) on the western distal flank of the delta, suggest the presence there of several tens of metres of sandy foreset sediments, with a predominance of medium to very fine sand and some silty sand in the upper 10 m.

Foreset beds exposed normal to the strike, in the raised 8–9 m terrace section (Localities 1–3, Fig. 5), comprise regular, more or less parallel-bedded, steeply dipping (8–37°), thick to very thin bedded (millimetre to decimetre thick) sands, with occasional, very thin beds of mud (Figs 10, 12). Various bedding types occur: massive to indistinct lamination, parallel to low-angle cross-lamination, normal and inverse grading and deformed bedding. Scattered pebbles, intraclasts, plant debris, shells and shell fragments are common in some beds. Biogenic material also occurs concentrated in occasional thin beds or lenses.

Although the general impression of the foreset unit along the terrace section is one of regularity, there is considerable local variation in dip and lithology. However, average dip, local variation in dip and average grain size, appear to increase in a distal direction. This trend can be related to depositional palaeodepth, which decreased as the delta front prograded. The trend can be explained, with reference to the modern delta slope, as follows:

1 *Average dip*. Average foreset dip directly reflects the average gradient of the contemporaneous

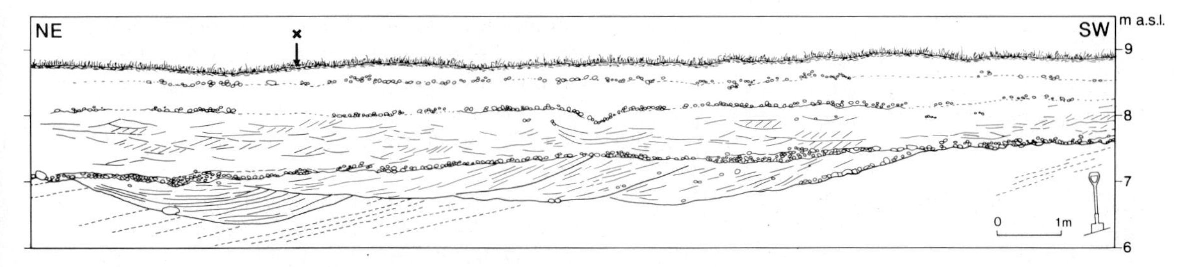

Fig. 9. Drawing constructed from photographs, showing topset beds overlying foreset beds (sloping dashed lines) in the raised terrace section at Locality 2 (Fig. 5; cf. Fig. 8). The topsets comprise channel-fill cross-bedded pebbly sands (below) and alternating gravel lag and cross-bedded sands above. An 'x' marks the position of the sedimentary log shown in Fig. 10.

front slope which, on the modern delta, decreases with increasing depth (Fig. 4). Average measured dips at Localities 1, 2 and 3 are 25°(14), 17°(8) and 12°(1), respectively (number of readings in parentheses). Comparing these values with slopes on the modern delta suggests corresponding depositional depths of < 5 m, 10–20 m and 20–30 m, respectively. An independant estimate of palaeodepth, obtained by correlating radiocarbon dated foreset beds from the three localities (Fig. 13C) with former sea level (Fig. 13B), gives corresponding values of 5 m, 12 m and 21 m. Estimates of palaeodepth obtained by the two methods compare well.

2 *Local variation in dip*. Local variation in foreset dip reflects the amplitude of relief features on the contemporaneous delta slope. On the present slope, relief increases upwards. Of the three localities, the largest range in dip values (19–37°) was registered at Locality 1, where deposition occurred highest on the delta slope.

3 *Grain size*. Grain size tends to becomes coarser towards the distal locality where foresets were deposited at lowest depth, in agreement with the pattern on the modern delta.

Bottomsets

Bottomset beds have not been directly observed in section, except in poor exposures just south of Locality 3, where silts occur between silty sands and rhythmically laminated silts and clays, interpreted as distal foresets and predeltaic, glaciomarine sediments, respectively.

Cores penetrating the upper 2–3 m of sediment at depths of 144 m and 197 m in the present, distal prodelta area, reveal predominantly muds with intermittent, thin beds of sand. The sand beds occasionally contain plant debris and shell fragments, are thicker in the proximal core, and are interpreted as turbidites. The muds contain scattered pebbles, shells and shell fragments deposited by drifting ice and other agents.

Seismic data and calculated sedimentation rates (see following section) suggest that the bottomset unit beneath the present distal prodelta area (*c.* 3–7 km from the delta rim) is of the order of 5–20 m thick.

PROGRADATIONAL HISTORY AND SEDIMENTATION RATES

The Alta fjord was deglaciated around 10 000 yrs ago, when the glacier retreated from the large end moraine shown on Fig. 1. Sea level then lay about 70–75 m above present sea level (Marthinussen, 1960; Follestad, 1979) and a marine basin developed behind the end moraine. Rapidly falling sea level and infilling of the basin caused the river mouth to prograde until it lay beyond the end moraine, near the present apex of the delta.

The highest terraces proximally (Fig. 5), lie slightly lower than the 28 m 'Tapes' (Middle Holocene) shoreline. This suggests, judging from the sea-level curve (Fig. 13B), that the delta first started prograding beyond the moraine after about 7000–8000 yrs ago.

Delta progradation after 7000 yrs ago, has been reconstructed using radiocarbon-dated samples from foreset beds at three localities (Localities 1–3, Fig. 13) and from topset beds at one locality (Locality 4, Fig. 13) in the raised 8–9 m terrace. The dated sediments have been correlated with corresponding delta-front positions and terrace levels, using the sea-level curve (Figs 13B, C).

The results show that the delta has prograded at

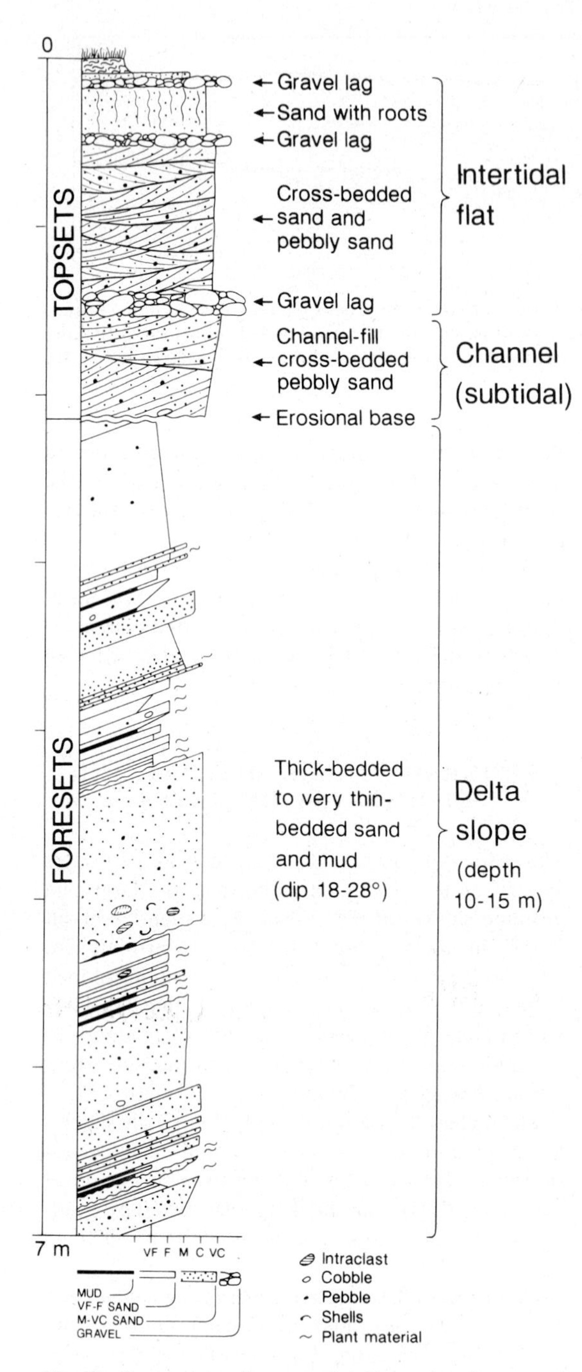

Fig. 10. Composite sedimentary log illustrating the general character and suggested depositional environment of the topset and foreset units in the raised terrace section at Locality 2 (Fig. 5; cf. Fig. 8). Interpreted depositional depth for the foreset beds is based primarily on radiocarbon dates and shore-level reconstruction (cf. Fig. 13).

an average rate of 0.4 m yr^{-1} during the last 6000 years. Progradation has decelerated as the slope advanced into deeper water on a progressively expanding front.

Estimates of average, net sedimentation rates on the delta slope, based on the rate of progradation and assuming a constant, equilibrium profile, give values of from 100 mm yr^{-1} at 15 m depth to 35 mm yr^{-1} at 50 m depth (*c*. 40 m and 300 m from the rim, respectively).

For the prodelta area, radiocarbon-dated samples from the 144 m and 197 m cores, gave ages of 1510 ± 80 and 3740 ± 170 yrs B.P., for sediments 206 cm and 293 cm below the surface, respectively. This gives average sedimentation rates of 1.4 and 0.7 mm yr^{-1} at distances of 3 km and 7 km from the delta rim, respectively.

EMERGENCE OF THE DELTA PLAIN

It is evident from Fig. 13, that terraces representing former delta plains, must previously have had considerably greater extent, and that most of the modern delta plain is primarily an erosional feature (fluvial and wave-cut platform), with a thin veneer of sediments (topsets). Similar conditions have been described by Postma and Cruickshank (1988) and Postma (1988). It follows, therefore, that considerable quantities of sediment eroded from these terraces, must have been redeposited, not only by fluvial processes operating within a confined zone, but also by marine processes acting on a broader front. This, together with river-channel migration, will have contributed to the delta's arcuate form.

Preservation of the delta plain and topset unit evidently takes place by gradual emergence and abandonment of broad areas of low-relief, upper tidal flat, with little accompanying progradation of the uppermost proximal facies of the delta plain, such as beachface facies, which remains attached to its terrace source. Emergence will result in renewed marine incision of the delta plain at some point on the rising surface near the high-tide mark, where a notch and beachface will develop. The rate of uplift and erosion will determine the extent to which the newly formed terrace will survive.

A recently raised terrace which exemplifies this is Rørholmen (Figs 4, 5), which at present has a well-developed gravelly beachface on its distal side, and is being eroded laterally by the river. A future candidate for abandonment is the relatively high-

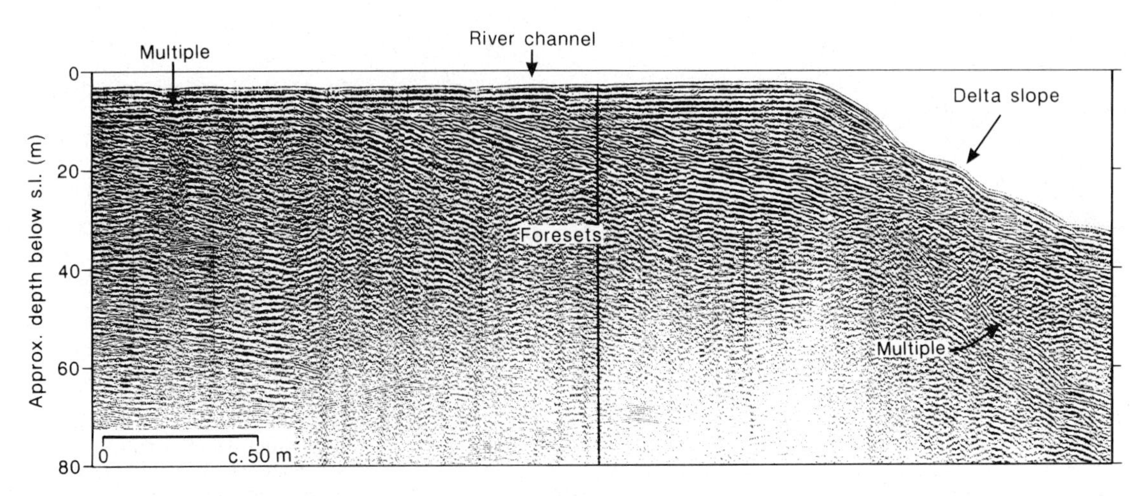

Fig. 11. Seismic profile axially across the front of the Alta delta. Steeply dipping reflectors, representing foreset beds, can be traced almost to the base of the river channel; horizontal, parallel reflectors immediately below the river bed are probably mostly multiples. Instrumentation: surface-towed boomer, 200 J.

Fig. 12. Photograph illustrating a sequence of relatively thin-bedded, sandy foreset beds exposed in the raised terrace section near Locality 2 (Fig. 5). These beds dip at a relatively moderate angle of *c.* 12°. Spade (1 m long) for scale.

lying, inner, eastern tidal plain, which is already notched on its seaward, northeastern side. Surviving terrace segments occur mostly on the southeastern side of the delta which has a sheltered position relative to prevailing winds and waves which cross the delta plain from the northeast.

CONCLUSIONS

The Alta delta system is a fan-shaped, partially emerged, medium-grained (predominantly sandy), Gilbert-type (high-gradient), fjord-head delta formed during the last 7000 years. Factors which have exerted a major control on its development are: (1) an abundant supply of relatively coarse sediment in the immediate hinterland; (2) periodic, high-energy fluvial discharge; (3) a fjord setting which has provided a deep depositional basin and sheltered, relatively low-energy, mesotidal marine environment; and (4) glacio-isostatically induced emergence which has promoted a high rate of sediment supply and influenced the evolution of the delta plain.

A study of the subaerial, subaqueous and subsurface components of the modern and subrecent delta, shows that the delta system comprises three main elements: (1) topset beds, which cap the modern delta plain and raised terraces and generally comprise a 1–3 m thick unit of sandy and gravelly sediments; (2) a thick (many tens of metres) foreset unit comprising steeply dipping beds of mostly sand and silty sand deposited on a steep (5–35°), gravity-process-dominated, subaqueous delta slope; and (3) a bottomset unit of moderate thickness (up to a few tens of metres) comprising predominantly muds, with interbedded turbidite sands, deposited on a gently sloping ($< c.$ 3–4°) prodelta area.

The modern, intertidal delta plain is composed of a system of channels, flats, swells and bars formed predominantly by fluvial processes in an axial zone, and wave- and tide-processes laterally. Drifting sea-

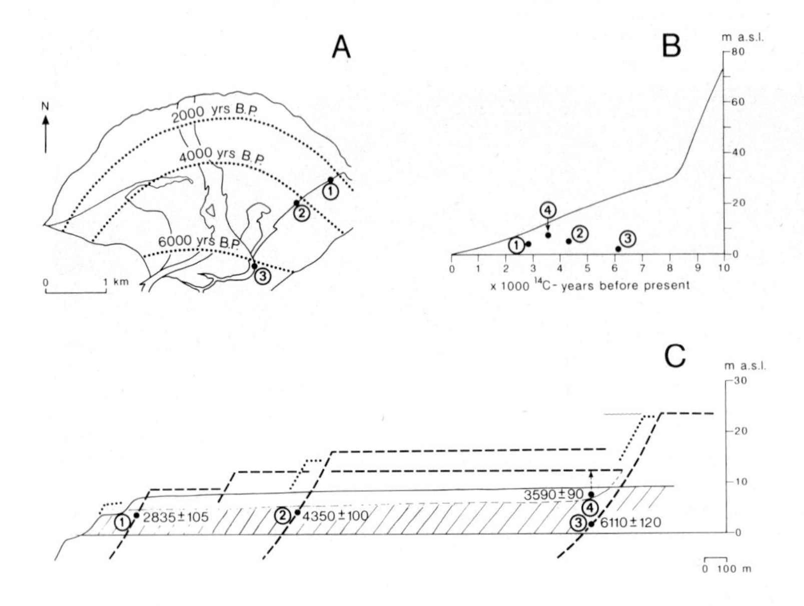

Fig. 13. Reconstruction of delta progradation in plan (A) and schematic longitudinal section (C), based on radiocarbon dates from the raised, eastern terrace section. Dated foreset (1–3) and topset (4) localities have been correlated with sea level using a generalized, sea-level displacement curve for the area (B). The sea-level curve is adapted from Møller (1987, 1989), based on regional data for a position at the 28 m 'Tapes' shoreline isobase. Heavy-dashed lines indicate reconstructed delta surfaces for the four dated localities (schematic; delta plains at mean sea level). Dotted lines indicate interpolated delta-rim positions at 2000, 4000 and 6000 years before present.

ice, algae and wave-winnowing of palimpsest material have played an important role in forming patchy, mixed sand–gravel and gravel lag facies on the emerging delta plain.

The delta has prograded at a decelerating rate, averaging 0.4 m yr^{-1} during the last 6000 years.

ACKNOWLEDGEMENTS

The work was financed by the University of Tromsø (UiTø) and the Norwegian Research Council for Science and the Humanities (NAVF). Offshore work was carried out using UiTø 's research vessels, F/F *Ottar* and F/F *Johan Ruud*. The radiocarbon datings were carried out at NAVF's Radiological Dating Laboratory at Trondheim and at the Tandem Accelerator Laboratory at Uppsala. Drafting work was done by Hilka Falkseth and photographic reproduction by Gunvor Granaas and Ole Sigmund Breimo. We are indebted to Yngve Kristoffersen and Inge Aarseth for their valuable contribution to the marine geophysical surveys, to Brian Bornhold and an anonymous reviewer for helpful reviews and to colleagues and students at the University of Tromsø.

REFERENCES

Bakkelid, S. (1980) Report to CRCM for the period 1974–1978 on activity within Fennoscandia on recent crustal movements, Norway. Unpublished report, Norges geografiske oppmåling, 9 pp.

Bogen, J. (1983) Morphology and sedimentology of deltas in fjord and fjord valley lakes. *Sedim. Geol.* **36**, 245–267.

Calles, B. (1977) Elvegårdselv och dess delta. En studie av fluvial transport och topografiska förändringer. *Naturgeografiska Institionen, Uppsala Universitet, Ungi Rapport.* **45**, 161 pp.

Chough, S.K., Choe, M.Y. & Hwang, I.G. (1988) The Miocene Doumsan fan delta, southeastern Korea: an example of composite (Gilbert-slope) system in back-arc margin. *Abstr. Int. Works. Fan Deltas, Calabria*, 11.

Colella, A. (1988a) Pliocene–Holocene fan deltas and braid deltas in the Crati Basin, southern Italy: a consequence of varying tectonic conditions. In: *Fan Deltas: Sedimentology and Tectonic Settings* (Ed. By W. Nemec and R.J. Steel), pp. 50–74. Blackie and Son, London.

Colella, A. (1988b) Fault-controlled marine Gilbert-type fan deltas. *Geology* **16**, 1031–1034.

Colella, A., de Boer, P.L. & Nio, S.D. (1987) Sedimentology of a marine intermontane Pleistocene Gilbert-type fan-delta complex in the Crati Basin, Calabria, southern Italy. *Sedimentology* **34**, 721–736.

Corner, R. (1975) The Tana valley terraces, a study of the morphology and sedimentology of the lower Tana valley terraces between Mansholmen and Maskejokka, Finnmark, Norway. *Department of Physical Geography, University of Uppsala, Ungi Report 38*, 293 pp.

Dionne, J.C. (1969) Tidal flat erosion by ice at La Pocatiere, St. Lawrence Estuary. *J. sedim. Petrol.* **39**, 1174–1181.

Ethridge, F.G. & Wescott, W.A. (1984) Tectonic setting, recognition and hydrocarbon reservoir potential of fan-delta deposits. In: *Sedimentology of Gravels and Conglomerates* (Ed. by E.H. Koster and R.J. Steel). Mem. Can. Soc. Petrol. Geol., 10, pp. 217–235.

FJALSTAD, A. (1986) Postglasial sedimentasjon i Breivika, Troms. Unpublished thesis, University of Tromsø, 166 pp.

FOLLESTAD, B.A. (1979) Alta. Description of the Quaternary geological map 1834 I — 1:50000. *Norges geol. Unders.* **349**, 1–41. (In Norwegian with English summary.)

GILBERT, G.K. (1885) The topographic features of lake shores. *U.S. geol. Surv. 5th Annual Rept.*, pp. 69–123.

GILBERT, R. (1984) The movement of gravel by the alga *Fucus vesiculosus* (L) on an arctic intertidal flat. *J. sedim. Petrol.* **54**, 463–468.

HOLMES, A. (1965) *Principles of Physical Geology*. Thomas Nelson, London, 1288 pp.

KOMAR, P.D. (1976) *Beach Processes and Sedimentation*. Prentice-Hall Inc., 429 pp.

KOSTASCHUK, R.A. & MCCANN, S.B. (1983) Observations on delta-forming processes in a fjord-head delta, British Columbia, Canada. *Sedim. Geol.* **36**, 269–288.

KOSTASCHUK, R.A. & MCCANN, S.B. (1987) Subaqueous morphology and slope processes in a fjord delta, Bella Coola, British Columbia. *Can. J. Earth Sci.* **52**, 52–59.

LUTERNAUER, J.L. (1980) Genesis of morphological features on the western delta front of the Fraser River, British Columbia — status of knowledge. In: *The Coastline of Canada* (Ed. by S.B. McCann). Paper Geol. Surv. Can. 80–10, pp. 381–396.

MARKLUND, H. (1960) *En studie av Rombaksälvens tidsvattenpåverkade delta i Rombaksbotn i norra Norge.* Unpublished report, Geografisk Institutionen, Uppsala Universitet, 31 pp.

MARTHINUSSEN, M. (1960) Coast- and fjord area of Finnmark, with remarks on some other districts. In: *Geology of Norway* (Ed. by O. Holtedahl). Norges geol. Unders. 208, pp. 416–429.

MARTINI, I.P. (1981) Ice effect on erosion and sedimentation on the Ontario shores of James Bay, Canada. *Z. Geomorph. N.F.* **25**, 1–16.

MASSARI, F. & COLELLA, A. (1988) Evolution and types of fan-delta systems in some major tectonic settings. In: *Fan Deltas: Sedimentology and Tectonic Settings* (Ed. by W. Nemec and R.J. Steel), pp. 103–122. Blackie and Son, London.

MCCABE, A.M. & EYLES, N. (1988) Sedimentology of an ice-contact glaciomarine delta, Carey Valley, Northern Ireland. *Sedim. Geol.* **59**, 1–14.

MCCABE, A.M., DARDIS, G.F. & HANVEY, P.M. (1987) Sedimentation at the margins of a late Pleistocene ice-lobe terminating in shallow marine environments, Dundalk Bay, eastern Ireland. *Sedimentology* **34**, 473–493.

MCGOWEN, J.H. (1970) Gum Hollow fan delta, Nueces Bay, Texas. *Bureau of Economic Geology, The University of Texas at Austin, Report of Investigation.* **69**, 91 pp.

MCPHERSON, J.G., SHANMUGAM, G. & MOIOLA, R.J. (1987) Fan-deltas and braid deltas: varieties of coarse-grained deltas. *Bull. geol. Soc. Am.* **99**, 331–340.

MCPHERSON, J.G., SHANMUGAM, G. & MOIOLA, R.J. (1988) Fan deltas and braid deltas: conceptual problems. In: *Fan Deltas: Sedimentology and Tectonic Settings* (Ed. by W. Nemec and R.J. Steel), pp. 14–22. Blackie and Son, London.

MØLLER, J.J. (1987) Shoreline relation and prehistoric settlement in northern Norway. *Norsk geogr. Tidsskr.* **41**, 45–60.

MØLLER, J.J. (1989) Geometric simulation and mapping of Holocene relative sea-level changes in northern Norway. *J. Coastal Research* **5**, 403–417.

NEMEC, W. & STEEL, R.J. (1988) What is a fan delta and how do we recognize it? In: *Fan Deltas: Sedimentology and Tectonic Settings* (Ed. by W. Nemec and R.J. Steel), pp. 3–13. Blackie and Son, London.

NILSSON, H.D. (1973) Sandbars along low energy beaches. Part 1. Multiple parallel sandbars of southeastern Cape Cod Bay. In: *Coastal Geomorphology* (Ed. by R.C. Coates), pp. 92–102. George Allen and Unwin, London.

NORGES SJØKARTVERK *Tide tables for the Norwegian coast.*

ORI, G.G. & ROVERI, M. (1987) Geometries of Gilbert-type deltas and large channels in the Meteora Conglomerate, Meso-Hellenic basin (Oligo-Miocene), central Greece. *Sedimentology* **34**, 845–859.

ORTON, G.J. (1988) A spectrum of Middle Ordovician fan deltas and braidplain delta, North Wales: a consequence of varying fluvial clastic input. In: *Fan Deltas: Sedimentology and Tectonic Settings* (Ed. by W. Nemec and R.J. Steel), pp. 23–49. Blackie and Son, London.

POSTMA, G. (1983) Water escape structures in the context of a depositional model of a mass flow dominated conglomeratic fan delta (Abrioja Formation, Pliocene, Almeria Basin, SE Spain). *Sedimentology*, **30**, 91–103.

POSTMA, G. (1984) Mass-flow conglomerates in a submarine canyon: Abrioja fan delta, Pliocene, southeast Spain. In: Sedimentology of Gravels and Conglomerates (Ed. by E.H. Koster and R.J. Steel). Mem. Can. Soc. Petrol. Geol., Calgary, 10, pp. 237–258.

POSTMA, G. (1988) Some examples of construction and destruction of Gilbert-type deltas. *Abstr. Int. Works. Fan Deltas, Calabria*, 51–52.

POSTMA, G. & CRUICKSHANK, C. (1988) Sedimentology of a late Weichselian to Holocene terraced fan delta, Varangerfjord, northern Norway. In: *Fan Deltas: Sedimentology and Tectonic Settings* (Ed. by W. Nemec and R.J. Steel), pp. 144–157. Blackie and Son, London.

POSTMA, G., BABIĆ, L., ZUPANIČ, J. & RØE, S.-L. (1988) Delta-front failure and associated bottomset deformation in a marine, gravelly, Gilbert-type fan delta. In: *Fan Deltas: Sedimentology and Tectonic Settings* (Ed. by W. Nemec and R.J. Steel), pp. 91–102. Blackie and Son, London.

POSTMA, G. & ROEP, T.P. (1985) Resedimented conglomerates in the bottomsets of Gilbert-type gravel deltas. *J. sedim. Petrol.* **55**, 874–885.

PRIOR, D.B. & BORNHOLD, B.D. (1986) Sediment transport on subaqueous fan-delta slopes, Britannia Beach, British Columbia. *Geo-Mar. Lett.* **5**, 217–224.

PRIOR, D.B. & BORNHOLD, B.D. (1988) Submarine morphology and processes of fjord fan deltas and related high-gradient systems: modern examples from British Columbia. In: *Fan-Deltas: Sedimentology and Tectonic Settings* (Ed. by W. Nemec and R.J. Steel), pp. 125–143. Blackie and Son, London.

PRIOR, D.B., WISEMAN, W.J. & BRYANT, W.R. (1981) Submarine chutes on the slopes of fjord deltas. *Nature* **290**, 326–328.

SCHUMM, S.A. (1981) Evolution and response of the fluvial

system, sedimentological implications. In: *Recent and ancient nonmarine depositional environments: Models for Exploration* (Ed. by F.G. Ethridge and R.M. Flores). Spec. Publ. Soc. econ. Paleont. Miner. Tulsa **31**, pp. 19–29.

Steel, R. (1988) Fan deltas: some recent advances. *Abstr. Int. Works. Fan Deltas, Calabria*, 57.

Systad, H. (1979) Planlagt trafikkhavn, Bukta. Seismiske grunnundersøkelser. Unpublished report, Norsk Teknisk Byggekontroll A/S.

Systad, H. (1980) Bukta, Alta: planlagt trafikkhavn. Grunnundersøkelser, prøvetaking, mudring, fylling, aktuelle kailøsninger. Unpublished report, no. 2, Norsk Teknisk Byggekontroll A/S.

Syvitski, J.P.M. & Farrow, G.E. (1983) Structures and processes in bayhead deltas: Knight and Bute inlet, British Columbia. *Sedim. Geol.* **36**, 217–244.

Sørensen, R., Bakkelid, S. & Torp, B. (1987) *Nasjonalatlas for Norge, kartblad 2.3.3, landhevning (land uplift)*. Statens Kartverk.

Wood, M.L. & Ethridge, F.G. (1988) Sedimentology and architecture of Gilbert- and month bar-type fan deltas, Paradox Basin, Colorado. In: *Fan Deltas: Sedimentology and Tectonic Settings* (Ed. by W. Nemec and R.J. Steel), pp. 251–263. Blackie and Son, London.

Spec. Publs int. Ass. Sediment. (1990) **10**, 169–181

Morphology and sedimentary processes on the subaqueous Noeick River delta, British Columbia, Canada

B. D. BORNHOLD* *and* D. B. PRIOR†

* *Geological Survey of Canada, Pacific Geoscience Centre, Sidney, British Columbia, Canada, V8L4B2*
† *Geological Survey of Canada, Atlantic Geoscience Centre, Dartmouth, Nova Scotia, Canada, B2Y4A2*

ABSTRACT

The Noeick River has deposited a large fan delta along the sidewall of South Bentinck Arm, a fjord on the central British Columbia coast. The intertidal and supratidal delta has an area of about 8×10^5 m^2; the underwater delta-front gradient averages 4–5° to water depths of 220 m, about 2 km from shore.

The Noeick River drains a high-relief basin of which 15% is covered by glaciers. In addition to large, annual floods related to snow- and icemelt, the river has experienced jokulhlaups, glacier outburst floods, the most recent of which were in October 1984 and August 1986. The floods caused major changes in the subaerial delta, including relocation of distributaries and the delivery of large quantities of coarse sediment, trees and other organic debris to the delta and adjacent inlet.

Side-scan sonar imagery and seismic profiling on the subaqueous delta revealed a suite of morphological features and sediment distribution patterns indicative of high-energy sediment-transport processes and little evidence of *in situ* slope failure. Seaward of the main subaerial distributary channel, radiating chutes and swales containing sand and gravel lead towards deeper water. Between 120 m and 170 m water depth, flutes, commonly aligned downslope, dominate the subaqueous delta. Further downslope, there is an extensive area of transverse symmetrical bedforms, 2–5 m high with wavelengths of 50–120 m, interpreted as antidunes. Sediments on the subaqueous delta front consist of interstratified organic-rich sandy muds and gravelly sands.

INTRODUCTION

From previous studies in British Columbia fjords, it is recognized that high-energy events play a major role in determining the morphology and facies distributions of fan deltas and other high-gradient depositional systems. For example, debris torrents accompanying rainfall-triggered floods along the east coast of Howe Sound transport tremendous concentrations of boulders, gravel and sand across the narrow deltaic platforms and through chutes incised into the delta foresets, depositing as coarse splays at the foot of the slope (Lister *et al.*, 1984; Prior & Bornhold, 1988). In Bute Inlet, delta-front slope failures, extreme rainfall events and seasonal snow and glacial icemelt lead to major sand-transporting turbidity currents which incise channels into the fjord floor and deposit sand as broad lobes more than 50 km from the river mouth (Prior *et al.*, 1986, 1987). On the Bear Bay (Bute Inlet) fan delta, cobbles were observed in incised sand- and gravel-filled chutes at 330 m water depth 1.4 km from the shoreline, presumably emplaced during high-energy events (Prior & Bornhold, this volume).

In this paper we describe the seafloor morphology and sedimentary facies on the Noeick River fan delta in South Bentinck Arm, British Columbia (Fig. 1). The Noeick River drainage system experiences not only the seasonal flood events described, related to rainfall, snow- and icemelt, but in recent years has also witnessed two major floods caused by jokulhlaups, the sudden draining of glacially dammed lakes.

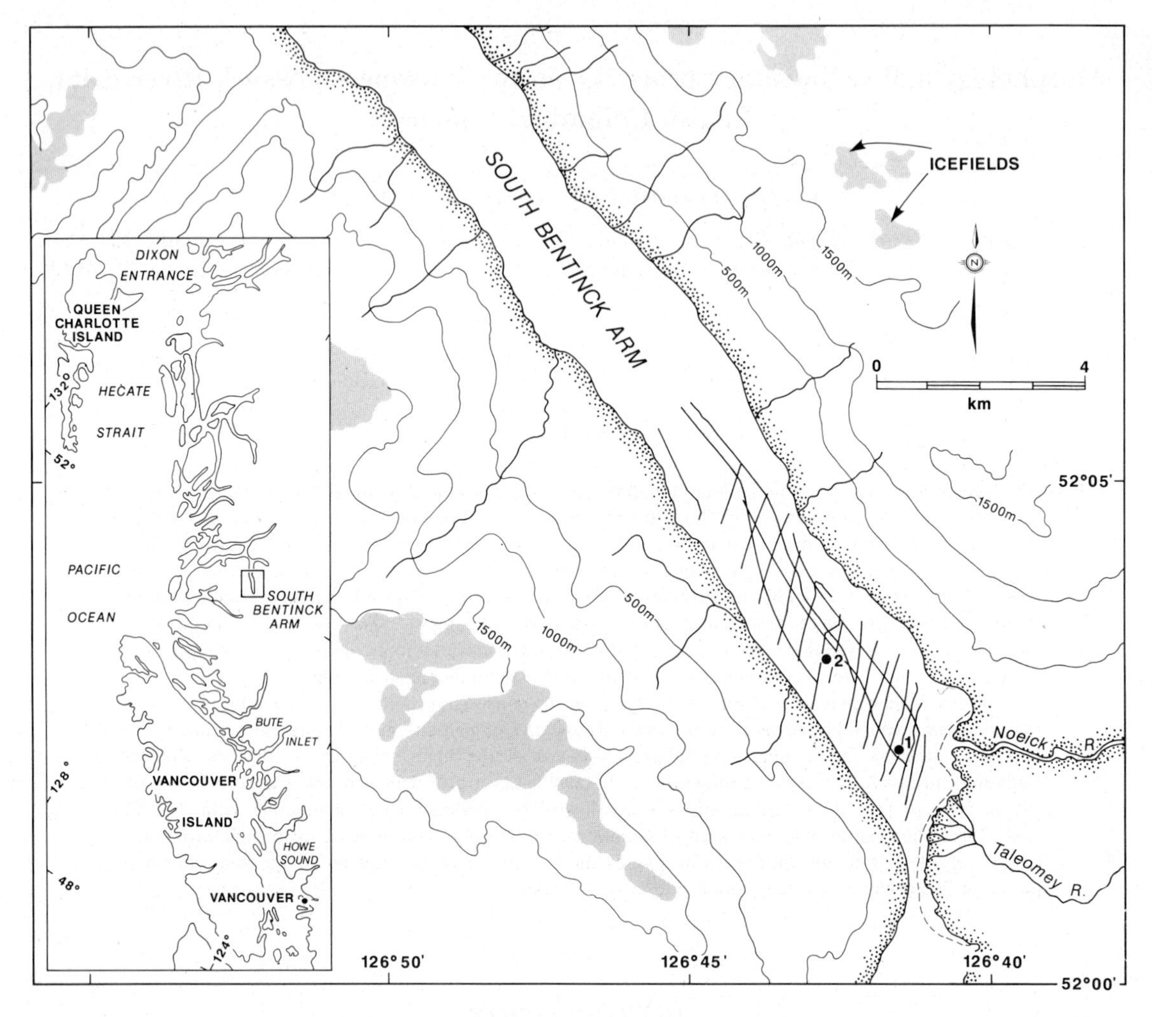

Fig. 1. Location map of South Bentinck Arm showing the positions of two piston cores and tracklines from the 1985 survey.

METHODS

The seafloor morphology was surveyed in 1985 and 1988 using a 100 kHz, scale-corrected, side-scan sonar system at a 200 m range setting. Bathymetry and high-resolution seismic profiling were obtained using a hull-mounted 3·5 kHz profiler, a deep-tow 'boomer' and a single-channel, air-gun seismic system using 16 cm^3 and 88 cm^3 air guns. A microwave positioning system was installed along the fjord for precise navigation.

Sampling was carried out using a Shipek grab sampler and a 1000 kg piston corer.

GEOGRAPHIC SETTING

The Noeick River drainage basin lies within the rugged Pacific Ranges of the Coast Mountains of British Columbia with local relief of more than 3000 m. The river drains an area of 562 km^2 of which 15% is covered by glaciers. Ape Lake, a glacially dammed lake which burst in 1984 and 1986, lies at 1400 m elevation near the headwaters of the river system. The mean annual flow of the river at its mouth is estimated to be about 210 $m^3\ s^{-1}$, with a maximum, mean daily discharge (100-year recurrence interval) of 460 $m^3\ s^{-1}$ (Jones *et al.*, 1985).

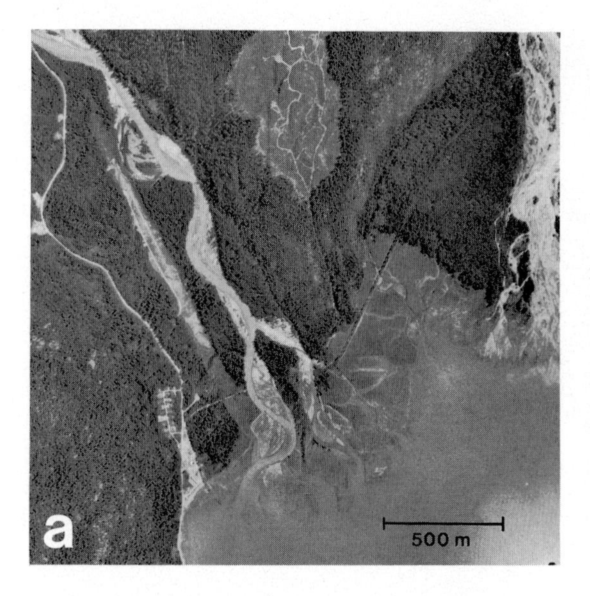

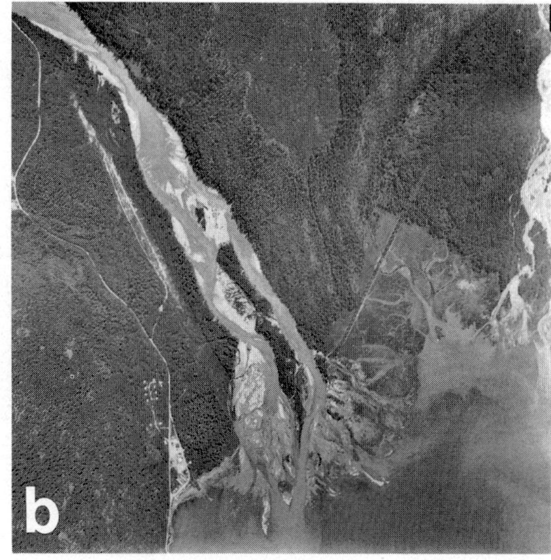

Fig. 2. Vertical air photographs of the Noeick River delta area showing major changes in fluvial and intertidal channel positions before (a) and after (b) the 1984 flood event.

Estimates of mean annual rainfall in the area range from 2100 mm to 3500 mm.

The lowermost few kilometres of the river occupy a broad (about 1 km) floodplain, terminating in an arcuate delta approximately 8×10^5 m^2 in area (Fig. 2). Within the tidally affected part of the delta (seawardmost 700–800 m), there are two principal distributary channels and several minor tidal channels (Fig. 3).

FLOOD EVENTS

On 20 October 1984, Ape Lake drained catastrophically through a subglacial tunnel; this major flood event was studied in detail by Jones *et al.* (1985), with later investigations carried out by Gilbert and Desloges (1987). The lake drained in less than 24 hours, with two pronounced discharge maxima. Peak discharges in the lower reaches of the Noeick River were between 900 and 1000 m^3 s^{-1}, compared to mean annual flood discharges of 210 m^3 s^{-1}. Approximately 46×10^6 m^3 of water were released from the lake during this event.

The flood destroyed forest access roads, washed out 200 000 newly planted trees, and severely damaged two bridges. Extensive erosion continued downstream as far as tidewater, causing changes in the location of some distributaries (Fig. 3). Huge quantities of coarse sediment, including boulders and cobbles, were delivered to the delta along with trees and organic debris. Sedimentological studies of other jokulhlaup-related deposits in Saskatchewan and North Dakota (Lord & Kehew, 1987) and in Iceland (Maizels, 1989) have been restricted primarily to subaerial depositional environments.

The subglacial tunnel resealed a few months after the flood and Ape Lake refilled by July 1985. As predicted by Jones *et al.* (1985), the lake drained again on 1 and 2 August 1986 (Gilbert & Desloges, 1987). Volumes of water released were apparently similar during this event, although only one discharge peak occurred.

SUBAQUEOUS SLOPE MORPHOLOGY

South Bentinck Arm is a typical British Columbia fjord, approximately 45 km long with an average width of 2 km (Fig. 1). In the innermost 8 km of the fjord, several rivers have built extensive deltas. The fjord floor northwest of the Noeick River delta deepens progressively to 285 m before rising to a sill (185 m depth) approximately 8 km from the fjord mouth.

In the study area, off the Noeick River delta, the inlet reaches 250 m depth at 4 km from the low-tide edge of the delta plain (Fig. 4). From 100 to 200 m,

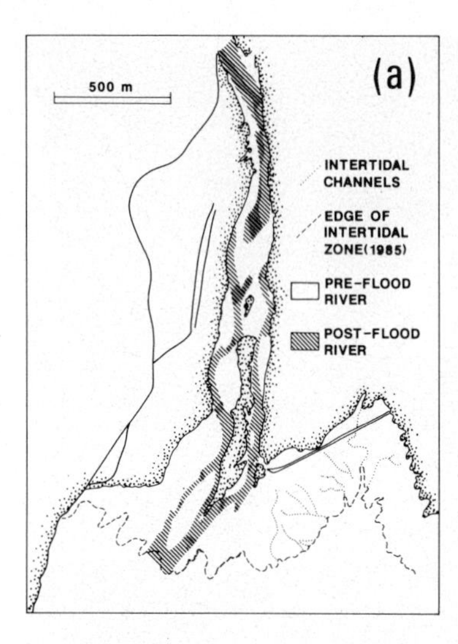

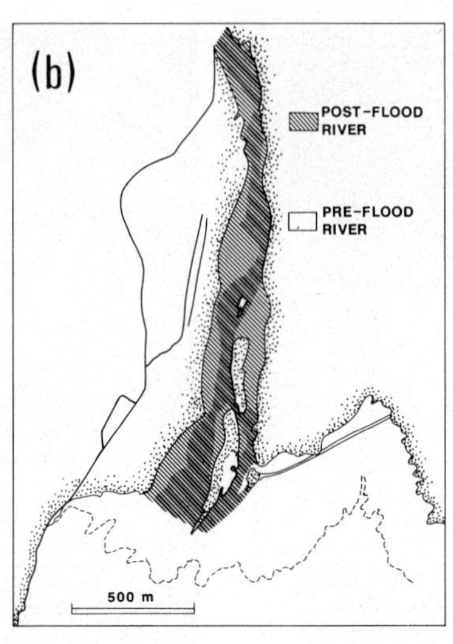

Fig. 3. (a) Channel configurations, pre- and post-1984 flood, in the lower Noeick River. (b) Lower Noeick River floodplain, pre- and post-1984 flood.

average seafloor gradients are 4.6°; locally the gradient is 6° to 6.5°, particularly on the shallower part of the subaqueous delta slope. From 200 m to the basin floor, slopes average less than 1.4°, diminishing with distance from the river mouth.

The seafloor on the subaqueous delta is characterized by a variety of distinctive morphologies which change both laterally and downslope (Fig. 5). Many of these have been found elsewhere on fjord fan deltas (Prior & Bornhold, 1988); some seabed features seen on the Noeick River delta, however, have not been reported from other modern fan deltas. We shall describe the seafloor off the Noeick River in terms of surficial zones established previously for the Bear Bay fan delta in Bute Inlet, British Columbia (pp. 75–91).

Upper delta slope

Because of navigational concerns and the concentrations of trees and other organic debris nearshore, the area immediately adjacent to the intertidal zone was not surveyed. Beyond this, from approximately 60 m water depth to about 110 m, the seafloor is similar to that of the 'transition zone' on the Bear Bay fan, though with much lower average gradients. This zone is characterized by seafloor morphologies indicative of both convergent and divergent flows.

Side-scan sonographs reveal a smoothly undulating surface, with 10–20 m relief, consisting of downslope-oriented ridges and broad, smooth swales (Fig. 6). The swales are usually acoustically highly reflective, compared to the intervening ridges (Fig. 6).

In contrast, the erosional chutes are from 10 to 40 m wide and separated by ridges up to 5 m high. In the upper part of this zone, the chutes have a distinctive anastomosing character (Fig. 6), but become clearly convergent downslope (Fig. 7). Side-scan sonographs reveal that between the major active chutes, systems of abandoned swales and chutes have been infilled with less reflective finer sediments (Fig. 7).

In the deeper part of this zone many of the converging chutes contain evenly spaced, arcuate, transverse depressions, concave-downslope, interpreted as flutes (Fig. 8). The spacing of these flutes in the smaller tributary chutes is commonly 10–15 m, increasing to 20 m or more in larger chutes. Amplitudes are less readily determined from side-scan sonographs or from seismic data but appear to be a few metres at most. Similar features were observed in Bute Inlet both acoustically and during submersible dives (pp. 75–91); in that system, the flutes have 2–2.5 m maximum relief, declining towards the sidewalls of the chute, and have an asymmetrical longitudinal geometry with the deepest part located nearer the upslope edge of the depression. Similar and larger features have been described from the Navy submarine fan by Normark *et al.* (1979), who have also interpreted them as flutes, and have been described from the ancient record by Winn and Dott (1979), Davies (1989) and Smith (1989).

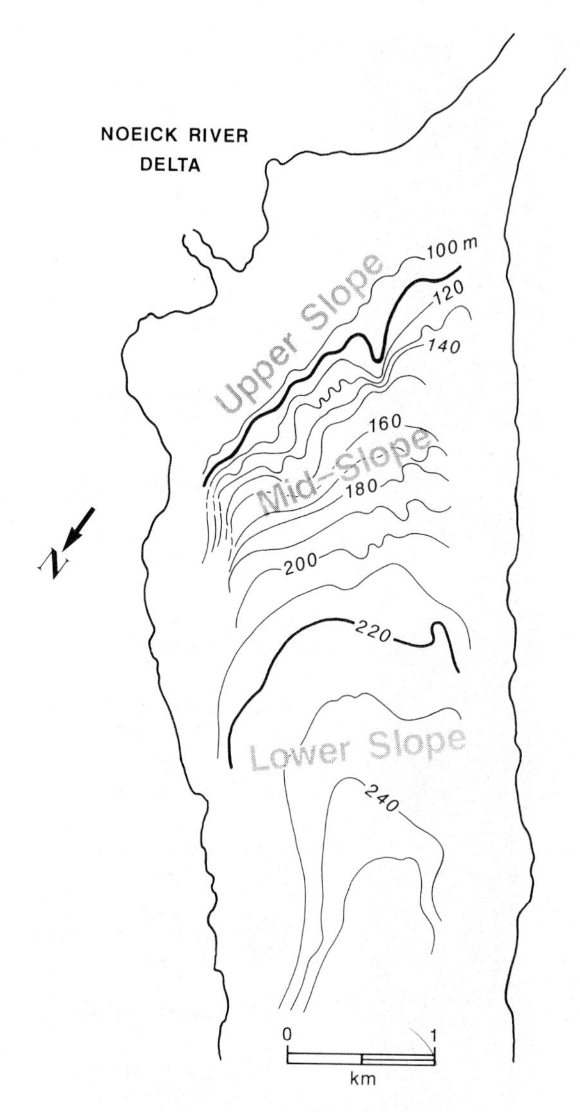

Fig. 4. Bathymetry (metres) of South Bentinck Arm off the Noeick River delta.

A major incised channel (130–160 m wide) is located on the southwestern side of the delta and appears to be fed via the upper fan, by sand and gravel from both the Noeick River and the adjacent Taleomey River (Figs 5, 9). The channel is cut 5–7 m deep and is bounded on the southwest for part of its length by the bedrock wall of the fjord (Fig. 5). Large flutes occur along the floor of the channel. These features have a spacing of about 100 m and a maximum downslope length, along the channel axis, of 40–50 m (Fig. 9).

Mid-slope

Between 120 m and 160 m water depth, the delta slope morphology consists of: (1) downslope-parallel, active and abandoned chutes, 30–50 m wide, containing trains of well-developed flutes with 30–50 m spacing; (2) a central zone of downslope-concave arcuate features interpreted either as flutes or possibly as shallow rotational failures due to delta-front loading; and (3) broad areas of undulating seafloor characterized by sinuous, transverse ridges interpreted as bedforms, with heights of 2–5 m, wavelengths of 50–100 m and crest lengths exceeding 150 m (Fig. 10). Further downslope, between 160 m and 200 m, the transverse bedforms dominate the seafloor and the chutes are absent. The bedforms become more laterally persistent with depth, attaining lengths in excess of 500 m, but retaining wavelengths of 50–120 m. The bedforms appear symmetrical on side-scan sonographs (Fig. 10) and on seismic profiles (Fig. 11); there is, however, a tendency for the upslope sides of the crests of the bedforms to appear smoother than the downslope sides which are characterized on sonographs by coarse-textured seafloor due to the presence of either gravels or accumulated organic debris.

In high-resolution seismic records, these bedforms appear as an undulating sequence of parallel-bedded strata which continue to a depth of 25–30 m below the seafloor (Fig. 11). Moving upward through this section, successive crests are displaced progressively upslope; total upslope migration of any one crest is 50–80 m. Similar, more deeply buried bedform sequences were also observed in the same area on the mid-fan.

Within this zone of large, laterally persistent bedforms are scattered a few very large flutes (Fig. 12). These concave-downslope features are up to 200 m wide with a maximum slope-parallel length of the depression of about 40 m.

Lower delta slope

The transverse bedforms described in the previous section diminish downslope (deeper than about 220 m water depth) in both amplitude and lateral persistence. High-resolution seismic profiles show that the sedimentary interval with bedforms overlies parallel-bedded, flat-lying fjord basin sediments and becomes progressively thinner with distance from the delta front (Fig. 12). The downslope edge of the bedform sequence is in about 250 m water

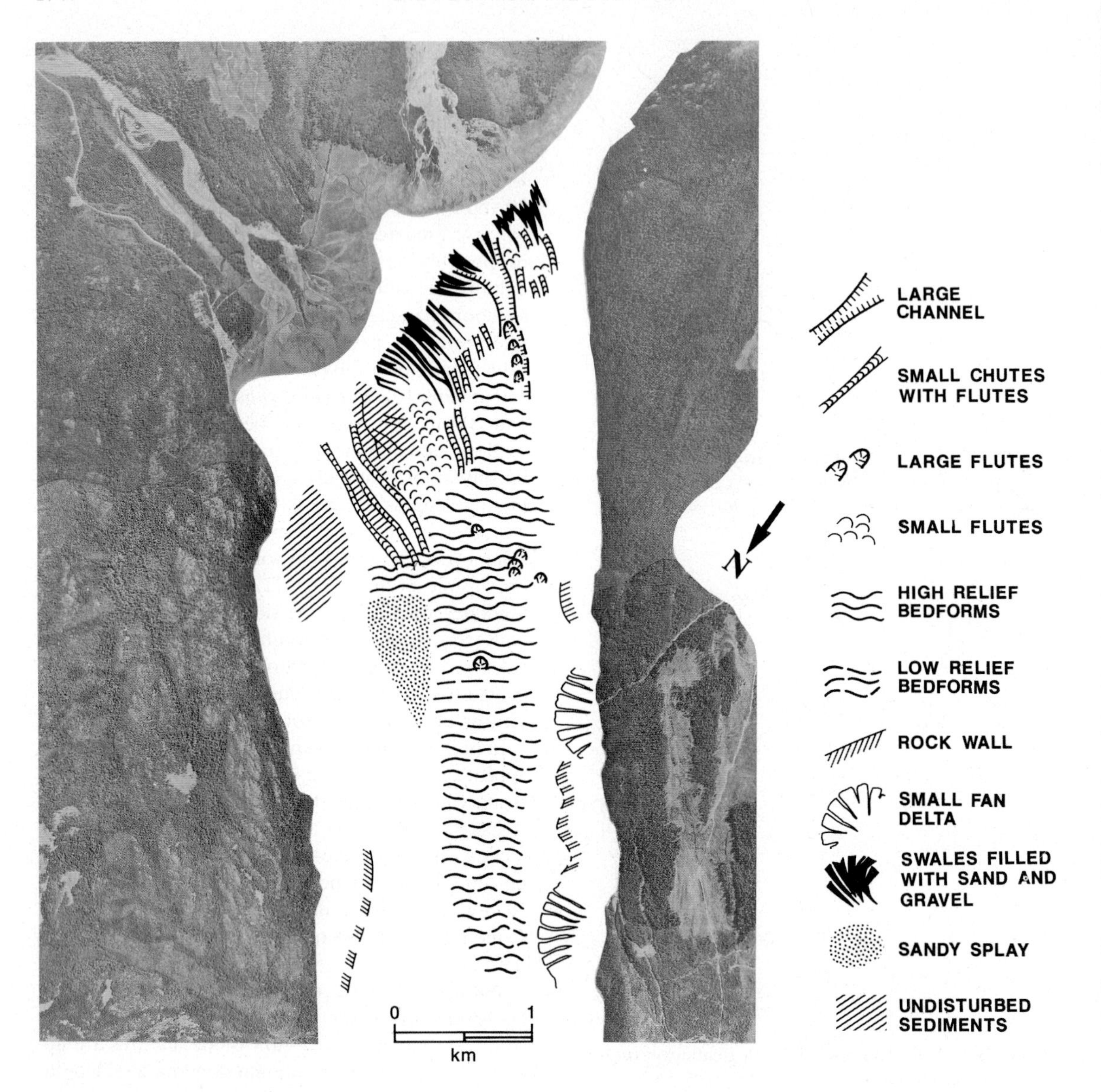

Fig. 5. Distribution of seafloor morphologies on the subaqueous Noeick River delta.

depth, approximately 4.5 km from the low-tide edge of the delta plain.

SEDIMENTS

A suite of grab samples and cores was collected on the subaqueous part of the delta. In general, surface sediments are muds and sandy muds with rare sandy sediments on the mid- and upper delta slope. Piston cores from the mid- and lower slope reveal graded and ungraded beds of mostly medium to coarse sand, 5–15 cm thick, with minor gravel and abundant organic debris, alternating with massive sandy muds (Fig. 13). A core from the lower fan, VEC85 A-2, consists of nearly 1 m of massive gravelly mud, interpreted as a debris flow, underlying gravelly sands, massive sands and sandy muds (Fig. 13).

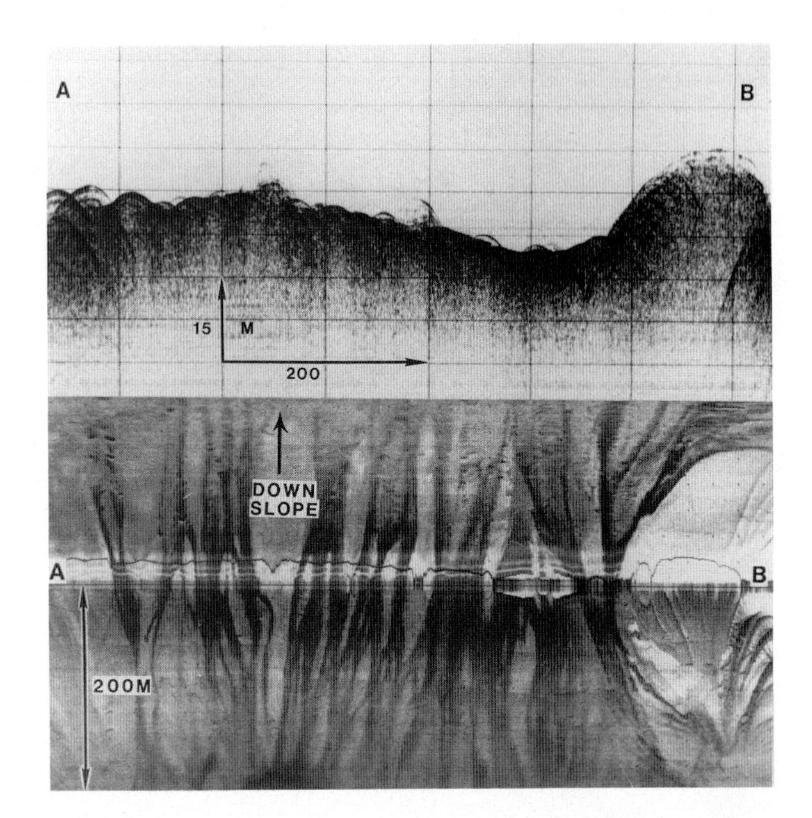

Fig. 6. Echosounding profile and side-scan sonograph across the upper slope.

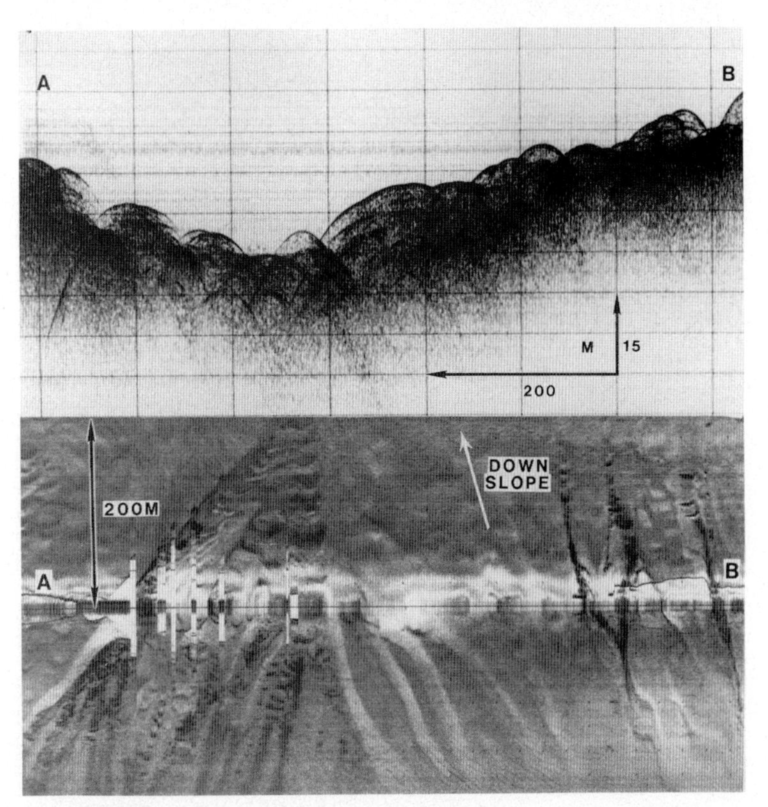

Fig. 7. Echosounding profile and side-scan sonograph across downslope-convergent chutes on the upper fan.

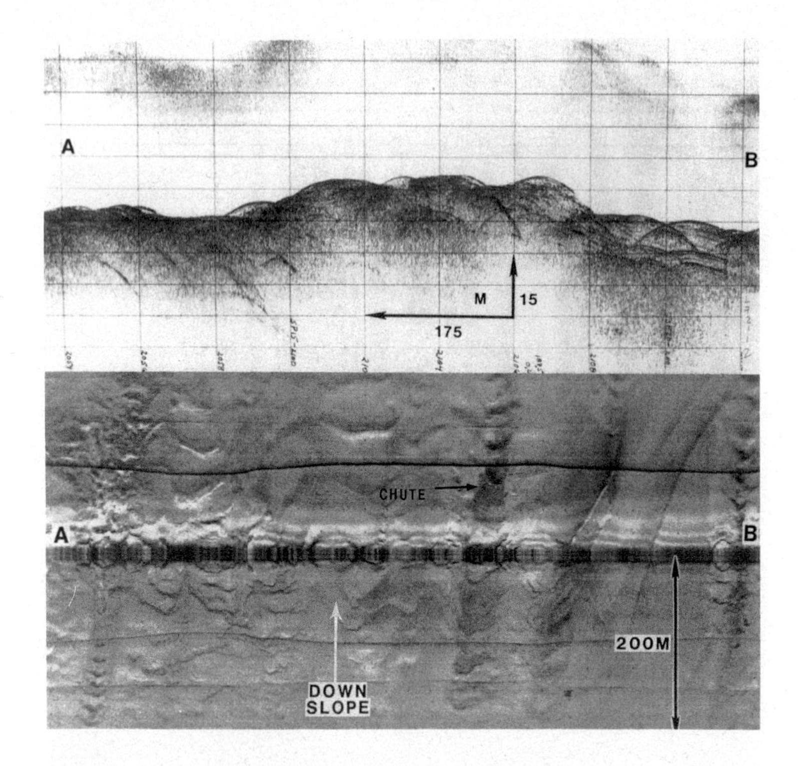

Fig. 8. Echosounding profile and side-scan sonograph across a prominent chute, containing flute trains on the upper fan.

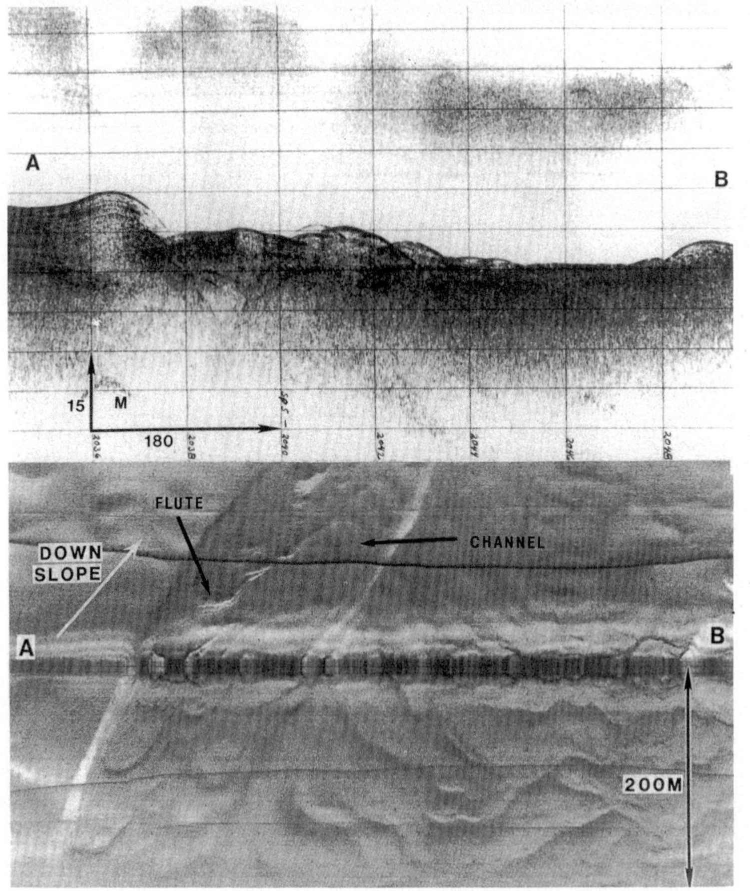

Fig. 9. Echosounding profile and side-scan sonograph of the major channel on the west side of the delta. The channel floor is characterized by a series of broadly spaced flute marks.

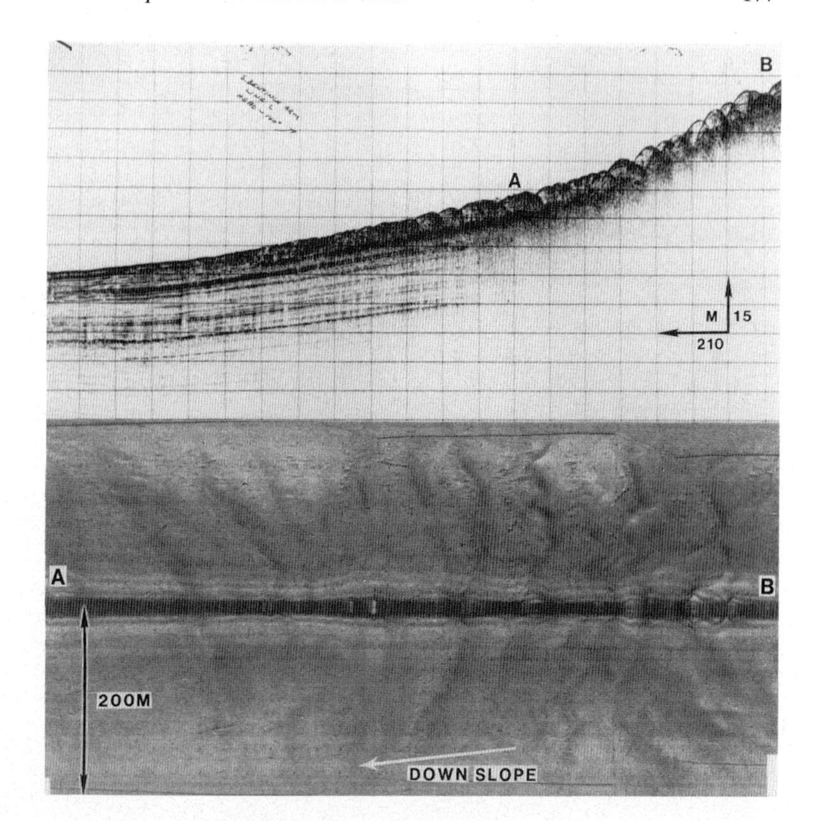

Fig. 10. Echosounding profile and side-scan sonograph across transverse, long-wavelength (50–100 m) bedforms, interpreted as antidunes on the mid-slope.

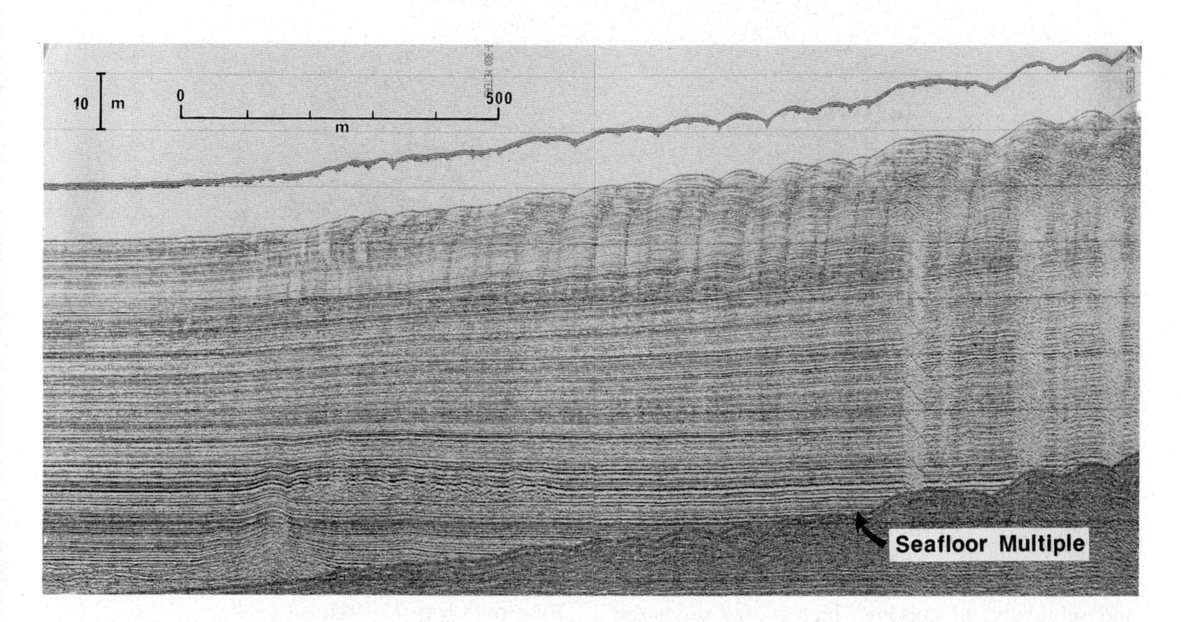

Fig. 11. Deep-tow 'boomer' seismic profile across the mid- and lower delta slope, showing the internal character of the transverse bedforms (inferred antidunes) and their upslope-migrating crests.

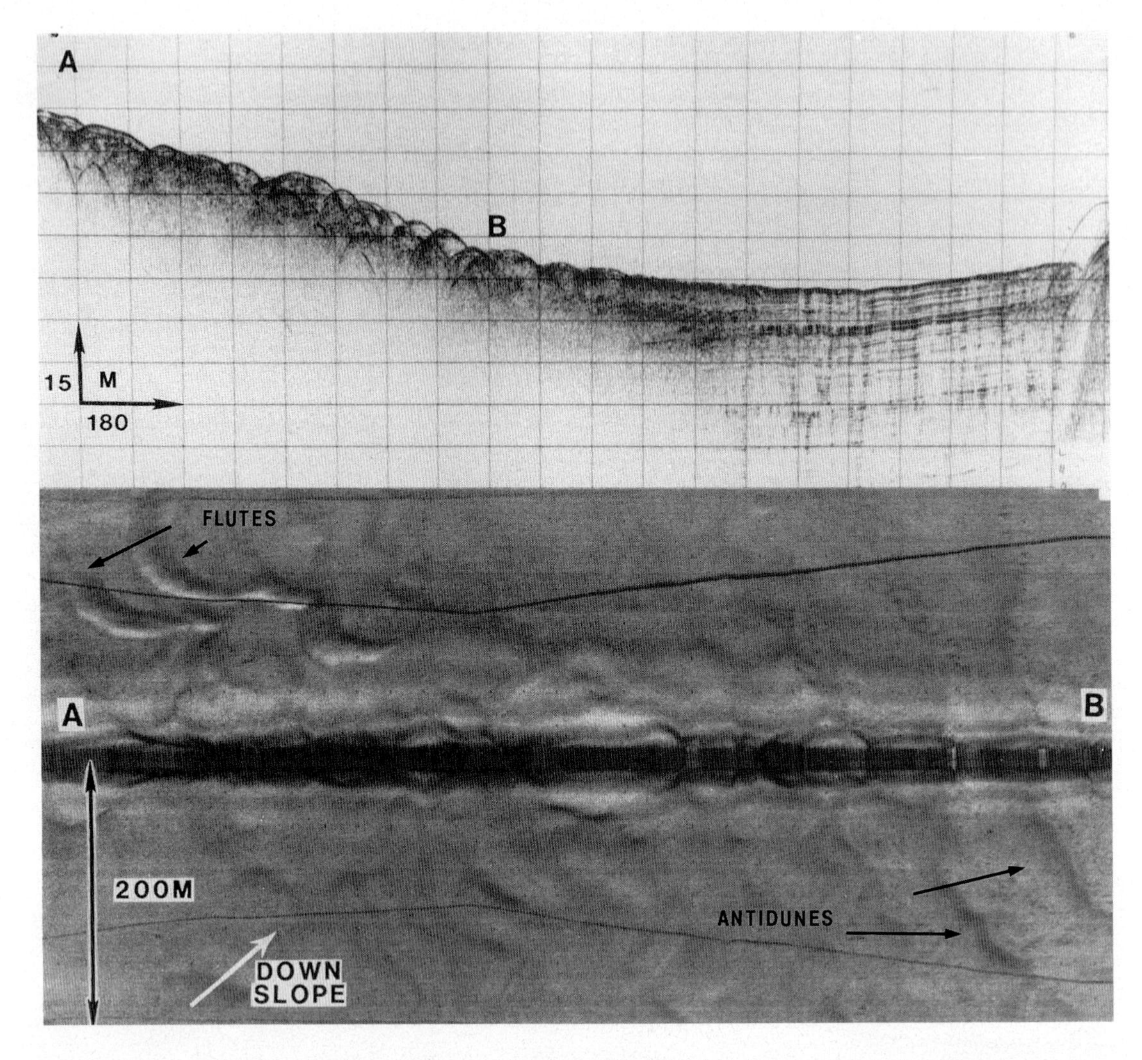

Fig. 12. Echosounding profile and side-scan sonograph of lower fan showing very large flutes (left) and inferred antidunes (right).

DISCUSSION

Many of the observed features of subaqueous morphology seen on the Noeick River delta are consistent with previous observations from Holocene high-energy fan deltas (Prior *et al.*, 1981; Prior & Bornhold, 1986, 1988, in press; Kostaschuk & McCann, 1983, 1987; Corner *et al.*, 1988, Corner *et al.*, this volume; Ferentinos *et al.*, 1988). In particular, the delta shows abundant evidence, both intertidally and subtidally, of episodic, high-energy discharge events: (1) major subaerial and intertidal channel migrations; (2) incised channels and chutes of varying sizes on the delta front; (3) flutes, both confined within incised chutes as well as on the 'open', lower delta slope; and (4) abundant coarse sediment on the lower delta slope. It is noteworthy that there is little indication of widespread *in situ* slope failure on the subaqueous delta; virtually all of the major morphological elements can be explained by dynamic sediment-flow phenomena such as high- and low-density turbidity currents and debris flows. A similar absence of slope failures was observed in Bear Bay, Bute Inlet (pp. 75–91).

The sinuous, transverse, upslope-migrating bedforms which characterize the mid- and lower

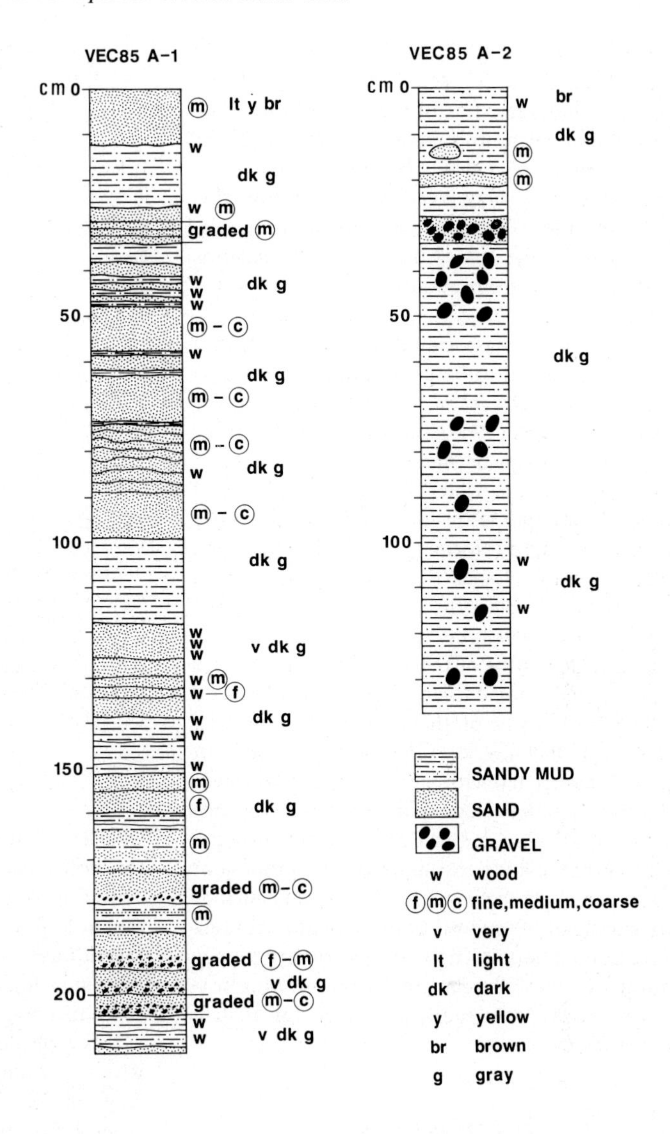

Fig. 13. Graphic logs of two piston cores taken on the Noeick River delta. Locations are shown in Fig. 1.

delta front have not been described previously from modern or ancient fan deltas. Strikingly similar features have, however, been described from both deep sea fans (e.g. Normark *et al.*, 1980) and more recently the Tiber River prodelta (Trincardi & Normark, 1988). On the Monterey deep-sea fan, upslope-migrating bedforms were interpreted as climbing antidunes created by turbidity currents with high sediment loads (Normark *et al.*, 1980). On the Tiber River delta front similar bedforms were interpreted to be the result of northward flowing shelf-depth currents (Trincardi & Normark, 1988). In South Bentinck Arm, only downslope hyperpycnal flow can be considered a viable mechanism to account for the observed transverse bedforms; perhaps a similar process could be invoked for the features on the Tiber delta.

Flow thicknesses and sediment concentrations of the turbidity currents responsible for the observed sediment waves in South Bentinck Arm can be estimated using the methods of Normark *et al.* (1980) based on the theoretical analysis of Allen (1970). Given the occurrence of coarse sand and gravels on the lower delta we infer that flow velocities must periodically exceed at least 50 cm s^{-1}. Since in Bute Inlet sand-transporting turbidity currents originating on the fjord-head delta have been measured at between 50 cm^{-1} and 350 cm^{-1} (Prior *et al.*, 1987), we estimate that velocities of similar turbidity currents on the smaller, lower gradient Noeick River delta would commonly fall between 50 and 200 cm s^{-1}, and on occasion perhaps even higher. Within this range of velocities, sediment concentrations necessary to account for the observed antidune wavelengths (50–120 m) range from about 2 g l^{-1} to 80 g l^{-1} (Fig. 14).

If Froude numbers associated with such turbidity currents are assumed to be approximately unity, as deduced by Normark *et al.* (1980) based on the results of Middleton (1966) and Ellison and Turner (1959), estimates can be made of the thickness of the turbidity currents which formed the observed antidunes; flow depth is approximately one-sixth the bedform wavelength (Normark *et al.*, 1980). Thus, flow depths on the Noeick River delta ranged from about 8 m for wavelengths of 50 m, to 20 m for bedforms with wavelengths of 120 m.

Although the widths of the turbidity currents are unknown, some must exceed 500 m, the observed crest length of many of the antidunes. For such flows, with a thickness of 12 m and a velocity of 100 cm s^{-1}, associated with antidunes of 75 m wavelength, the minimum sediment flux would be approximately 10^5 kg s^{-1}. Since the durations of the turbidity current events are unknown, total sediment transport cannot be estimated. Clearly, any further understanding of the genesis of these bedforms must await measurements of flow velocities, frequency and duration, and sediment concentration. Monitoring programmes are currently in progress in British Columbia fjords.

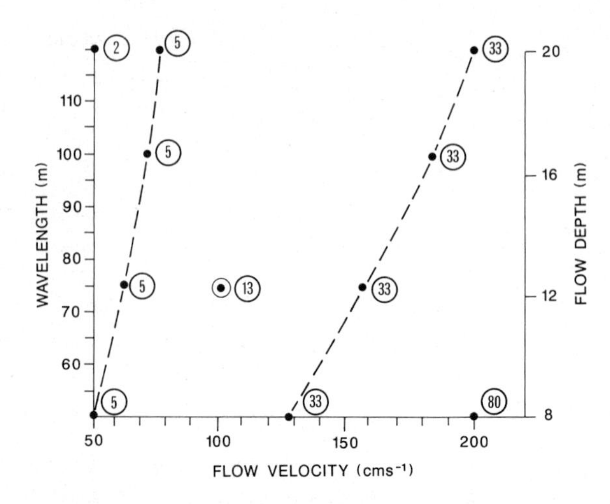

Fig. 14. Graph of all combinations of inferred flow velocity (50–200 cm s^{-1}), antidune wavelength (50–120 m) and inferred flow thickness. The figure shows that the range of possible sediment concentrations is from 2 to 80 g l^{-1} (circled values) and that the flow thicknesses vary from 8 to 20 m. The graph is based on theoretical analyses of Allen (1970) and Normark *et al.* (1980). The circled point represents bedforms of average observed wavelength (75 m).

CONCLUSIONS

Seafloor morphologies, sediment-distribution patterns, and the character of shallow seismic units all reflect the dominant role of episodic, energetic sediment underflows in the development of the Noeick River delta. We conclude, based on detailed monitoring studies in Bute Inlet and submarine morphology and sediment distribution on many fan deltas in British Columbia that these high-energy events are closely related to major increases in river discharge which deliver large quantities of coarse sediment rapidly to the subaqueous delta. As a result, these events may be expected to recur on the Noeick delta with an average periodicity of: (1) weeks to months for rainfall-related floods; (2) yearly for both snow- and icemelt; and (3) decades or centuries for jokulhlaups. Since nothing is known of the underflow dynamics associated with this range of magnitudes and frequencies, it is impossible to relate particular features or suites of features to particular types of flood.

It is especially important to note that, in common with other fan deltas in British Columbia (e.g. Bear Bay, pp. 75–91, there is almost no evidence on the Noeick River delta of significant *in situ* slope failure. River-derived underflow processes appear to be dominant in shaping the subaqueous delta.

ACKNOWLEDGEMENTS

We would like to express our appreciation to Ivan Frydecky, Gail Jewsbury, Trudie Forbes, Greg Liebzeit, Bill Hill, Graham Standen, Bill Wiseman and Karla Harris for their assistance in the field.

We also thank the officers and crew of the CSS VECTOR for their fine cooperation during the two surveys.

G.D. Corner, F.G. Ethridge, R. Higgs and F.

Surlyk critically read the paper and made many valuable suggestions for its improvement.

This is contribution number 17190 Geological Survey of Canada.

REFERENCES

ALLEN, J.R.L. (1970) *Physical Processes of Sedimentation*. Unwin University Books, London, 248 pp.

CORNER, G.D., NORDAHL, E., MUNCH-ELLINGSEN, K. & ROBERTSEN, K.R. (1988) Sedimentology of the Holocene Alta delta, northern Norway. *Abstr., Int. Works. Fan Deltas, Calabria*, 19–20.

DAVIES, P. (1989) The general characteristics, distinguishing features, and significance of large flute-shaped scours and their fills in the deep-water sequences of the Namurian Ross Formation, County Clare, Ireland. *Abstr. British Sedimentology Research Group*. Annual Meeting, Cambridge, U.K.

ELLISON, T.H. & TURNER, J.S. (1959) Turbulent entrainment in stratified flows. *J. Fluid Mech.* **6**, 423–428.

FERENTINOS, G., PAPATHEODOROU, G. & COLLINS, M.B. (1988) Sediment transport processes on an active submarine fault escarpment, Gulf of Corinth, Greece. *Mar. Geol.* **83**, 43–61.

GILBERT, R. & DESLOGES, J.R. (1987) Sediments of ice-dammed, self-draining Ape Lake, British Columbia. *Can. J. Earth Sci.* **24**, 1735–1747.

JONES, D.P., RICKER, K.E., DESLOGES, J.R. & MAXWELL, M. (1985) Glacier outburst flood on the Noeick River: the draining of Ape Lake, British Columbia, October 20, 1984. *Geol. Surv. Canada, Open File Report 1139*, 92 pp.

KOSTASCHUK, R.A. & MCCANN, S.B. (1983) Observations on delta-forming processes in a fjord-head delta, British Columbia, Canada. *Sediment Geol.* **36**, 269–288.

KOSTASCHUK, R.A. & MCCANN, S.B. (1987) Subaqueous morphology and slope processes in a fjord delta, Bella Coola, British Columbia. *Can. J. Earth Sci.* **24**, 52–59.

LISTER, D.R., KERR, J.W.G., MORGAN, G.C. & VANDINE, D.F. (1984) Debris torrents along Howe Sound, British Columbia. *Proc. IV Int. Symp. on Landslides, Toronto*, 649–654.

LORD, M.L. & KEHEW, A.E. (1987) Sedimentology and paleohydrology of glacial-lake outburst deposits in southeastern Saskatchewan and northwestern North Dakota. *Bull. geol. Soc. Amer.* **99**, 633–673.

MAIZELS, J. (1989) Sedimentology, palaeoflow dynamics and flood history of jokulhlaup deposits: palaeohydrology of Holocene sediment sequences in southern Iceland sandur deposits. *J. sedim. Petrol.* **59**, 204–223.

MIDDLETON, G.V. (1966) Experiments on density and turbidity currents: I. Motion of the head. *Can. J. Earth Sci.* **3**, 523–546.

NORMARK, W.R., HESS, G.R., STOW, D.A.V. & BOWEN, A.J. (1980) Sediment waves on the Monterey Fan levee: a preliminary physical interpretation. *Mar. Geol.* **37**, 1–18.

NORMARK, W.R., PIPER, D.J.W. & HESS, G.R. (1979) Distributary channels, sand lobes and mesotopography of Navy submarine fan, California Borderland, with application to ancient fan sediments. *Sedimentology* **26**, 749–774.

PRIOR, D.B. & BORNHOLD, B.D. (1986) Sediment transport on subaqueous fan-delta slopes, Britannia Beach, British Columbia. *Geo-Mar. Lett.* **5**, 217–224.

PRIOR, D.B. & BORNHOLD, B.D. (1988) Submarine morphology and processes of fjord fan deltas and related high-gradient systems: modern examples from British Columbia. In: *Fan Deltas: Sedimentology and Tectonic Settings* (Ed. by W. Nemec and R.J. Steel), pp. 125–143. Blackie and Son, London.

PRIOR, D.B. & BORNHOLD, B.D. (1989) Submarine sedimentation on a developing Holocene fan delta. *Sedimentology* **36**, 1053–1076.

PRIOR, D.B. & BORNHOLD, B.D. (this volume) The under water development of Holocene fan deltas.

PRIOR, D.B., BORNHOLD, B.D. & JOHNS, M.W. (1986) Active sand transport along a fjord-bottom channel, Bute Inlet, British Columbia. *Geology* **14**, 581–584.

PRIOR, D.B., BORNHOLD, B.D., WISEMAN, W.J., JR. & LOWE, D.R. (1987) Turbidity current activity in a British Columbia fjord. *Science* **237**, 1330–1333.

PRIOR, D.B., WISEMAN, W.J., JR. & GILBERT, R. (1981) Submarine slope processes on a fan delta, Howe Sound, British Columbia. *Geo-Mar. Lett.* **1**, 85–90.

SMITH, R.D.A. (1989) Giant mudstone-draped scours in turbidite systems. *Abstr. British Sedimentology Research Group*, Annual Meeting, Cambridge, U.K.

TRINCARDI, F. & NORMARK, W.R. (1988) Sediment waves on the Tiber prodelta slope: interaction of deltaic sedimentation and currents along the shelf. *Geo-Mar. Lett.* **8**, 149–157.

WINN, R.D., JR. & DOTT, R.H. (1979) Deep-water fan-channel conglomerates of Late Cretaceous age, southern Chile. *Sedimentology* **26**, 203–228.

Ancient Alluvial Deltas — Effects of Tectonics

Spec. Publs int. Ass. Sediment. (1900) **10**, 185–198

Fan-delta sequences in the Pleistocene and Holocene Burdur Basin, Turkey: the role of basin-margin configuration in sediment entrapment and differential facies development

N. KAZANCI

Department of Geological Engineering, University of Ankara, 06100 Beşevler, Ankara, Turkey

ABSTRACT

The Pleistocene and Holocene Burdur Basin of Turkey is a still active graben, about 12 km wide and 75 km long, which is presently occupied by Burdur and Yaraşlı lakes. Active extension, combined with lake-level oscillations and climatic changes, has produced a composite/cyclic basin-fill. The sedimentary facies and depositional characteristics vary along the basin margins, partly reflecting the effects of local morphological confinement by structural features. A late Pleistocene bouldery to sandy, Gilbert-type fan-delta complex, located in the Soğanlı area of the basin, is up to 70 m thick, consists of repetitive alluvial-fan and fan-delta sequences, and is separated into two segments by an intrabasinal high. The proximal segment of the complex includes bouldery, short-headed fan deltas which appear to have formed where the escarpment was steep and the substratum made of well-consolidated rocks such as limestone. Fan deltas in the distal segment (beyond the local high) are formed by high-angle, mass-flow-dominated, Gilbert-type delta-front sets, which intercalate with thick, lacustrine mud layers.

INTRODUCTION

A fan delta is the delta of an alluvial fan; its facies develop under the control of many factors such as climate, water-level fluctuation, synsedimentary tectonism, basin geometry, sediment input and basin-margin configuration (Colella, 1988; Nemec and Steel, 1988a; Steel, 1988). Among these controlling factors, basin-margin configuration is thought to be most important, strongly affecting fan-delta geometry (Kostaschuck, 1985; Leeder & Gawthorpe, 1987; Leeder *et al.*, 1988).

In this paper, a bouldery fan-delta complex in the Pleistocene–Holocene Burdur Basin of Turkey is discussed (Fig. 1). The fan-delta complex consists of four superimposed Gilbert-type fan-delta wedges and four alluvial-fan sequences. Basin-margin topography, as well as hydrodynamic and climatic conditions are important factors in fan-delta formation and depositional history.

PLEISTOCENE–HOLOCENE BURDUR BASIN

The Pleistocene–Holocene Burdur Basin of south-western Anatolia, one of the young lacustrine basins of Turkey, is an active graben superimposed on a late Tertiary intermontane basin. This fault-bounded graben is about 12 km wide and 75 km long, and its two large segments are occupied by Burdur and Yaraşlı lakes (Fig. 1). The lacustrine basin has been controlled since its origin in the Pliocene or early Pleistocene by two active extensional, normal faults trending northeast–southwest. The Burdur fault (Fig. 1) is presently active and generated a series of destructive earthquakes in 1963, 1964 and 1971 (Karaman, 1986). The water level of present-day Burdur Lake is rising, and it has increased by as

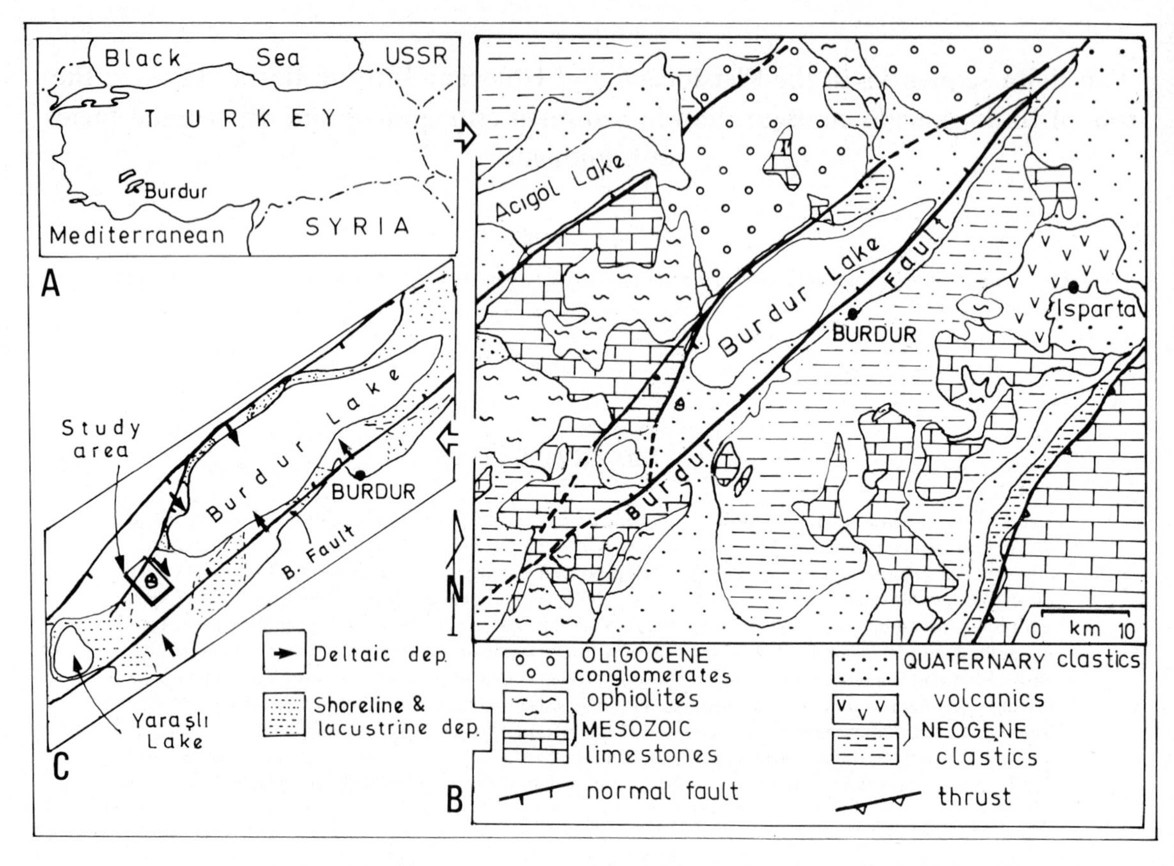

Fig. 1. Location (A), generalized geological map (B) of the Burdur region, and the study area in graben (C).

much as 1.5 m during the last 30 years, drowning large farmland areas, particularly along the northern margin of the lake. The fault to the northwest has created a steep basin-margin morphology, cutting and uplifting Mesozoic limestones and ophiolites, and Oligocene molassic conglomerates (Fig. 1B).

West-central Anatolia, including the Burdur area, is in a neotectonic transition zone between the west Anatolian extensional province and the compressional central Anatolian ova province (Şengör *et al.*, 1985). Ovas are large, approximately equant basins of Neogene age. Consequently, bedrock in this region has been influenced by both compressional and extensional neotectonic activity, creating a highly irregular topography. However, the Pleistocene–Holocene Burdur Basin was basically created in an extensional regime, as indicated by tectonic units, volcanism, regional stratigraphy and morphotectonics of the transition zone (Graciansky, 1968; Erol, 1972, 1978; McKenzie, 1972; Ozgül, 1976; Sungur, 1978; Koçyiğit, 1981, 1983; Şengör *et al.*, 1985; Karaman, 1986; Kozan *et al.*, 1988).

The Burdur Basin has a clastic infill, which is well exposed along the coast of present-day Burdur Lake (Fig. 1C) and which includes both alluvial-dominated and lacustrine-dominated facies (Erol, 1973; Kazancı & Erol, 1987). Thick deltaic sediments were mostly deposited along the basin's northwestern margin, and their depositional systems include Gilbert-type fan-delta complexes formed by the stacking of alluvial-fan and fan-delta lobes (Kazancı, 1988). Along the basin's southeastern margin, shoreline deposits are widespread but typically masked by younger alluvial sediments.

PLEISTOCENE BASIN-FILL IN SOĞANLI AREA

The basin-fill succession in the study area (Fig. 1C) is a cyclic-composite clastic wedge, composed of a bouldery proximal segment and a distal sandy to silty segment. An intrabasinal fault block forms a local structural high which separates the two segments (Figs 2, 3). In the field, the bouldery and sandy segments are in topographic asymmetry, the bouldery segment being higher. Together with the fault block, they form a feature of positive relief. The succession appears to have prograded into the graben where it has divided into two parts (Figs 2, 3, 4A). Sediments of the succession were derived from the northwestern margin of the basin. According to radiocarbon dating of *Dreissena* fossil shells by Erol *et al.* (1986), the succession was deposited within 35–37 Ka BP.

The proximal part of the succession, which directly abuts the graben-margin fault, is a wedge-like gravel deposit, at least 550 m long and up to 70 m thick. It consists of four fan-delta sequences that prograded against the intrabasinal structural high, and four intervening braided stream units (alluvial fans). Individual rockfall blocks and four fossiliferous (*Dreissena* spp., small gastropods) lacustrine mud layers are associated (Fig. 3). This dominantly coarse gravelly deposit, composed of well-rounded limestone clasts, thus appears to have been trapped by the palaeotopographic relief created between the graben margin and the adjacent fault block (Figs 2, 3).

The distal sandy segment of the basin-fill succession developed beyond the intrabasinal high and extended far into the palaeolake area of the graben centre (Fig. 2). This sandy segment is up to 38 m thick and is comprised of three fan-delta sequences interlayered with lacustrine muds (Fig. 3). It interfingers with prodelta silts that in turn grade into lacustrine muds. The prodelta/lake facies boundary of each sequence is transitional and generally unclear in the field. The succession has an overall dip of 4–5°.

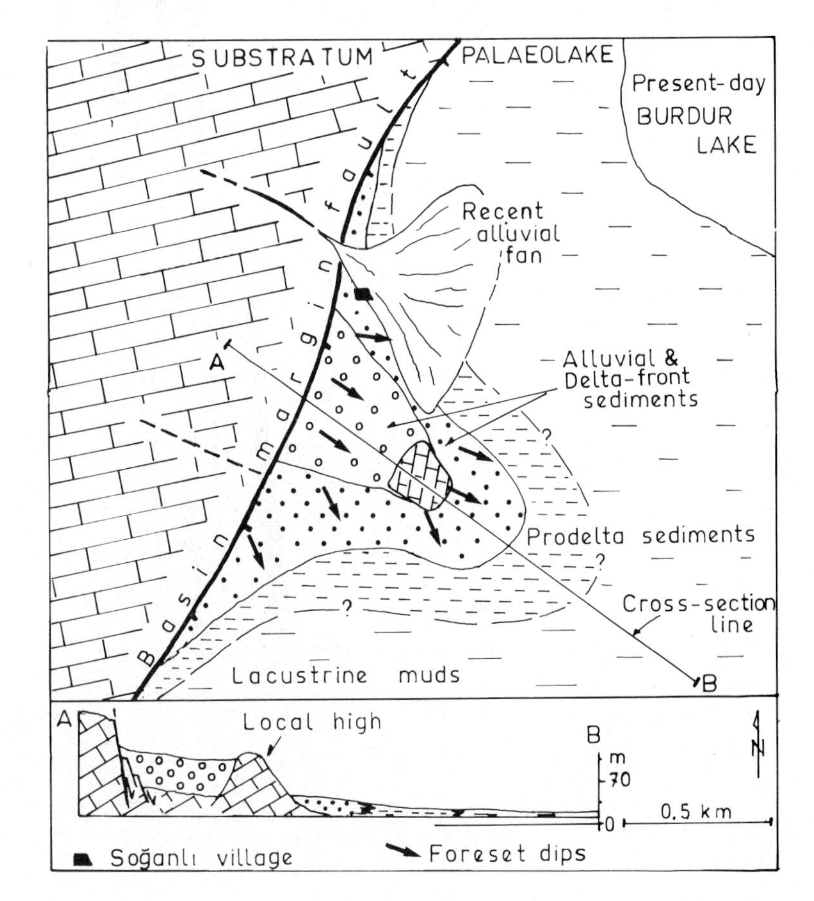

Fig. 2. Facies distribution and topographic setting of the basin-fill of the Soğanlı area. AB line indicates location of cross-section shown in Fig. 3.

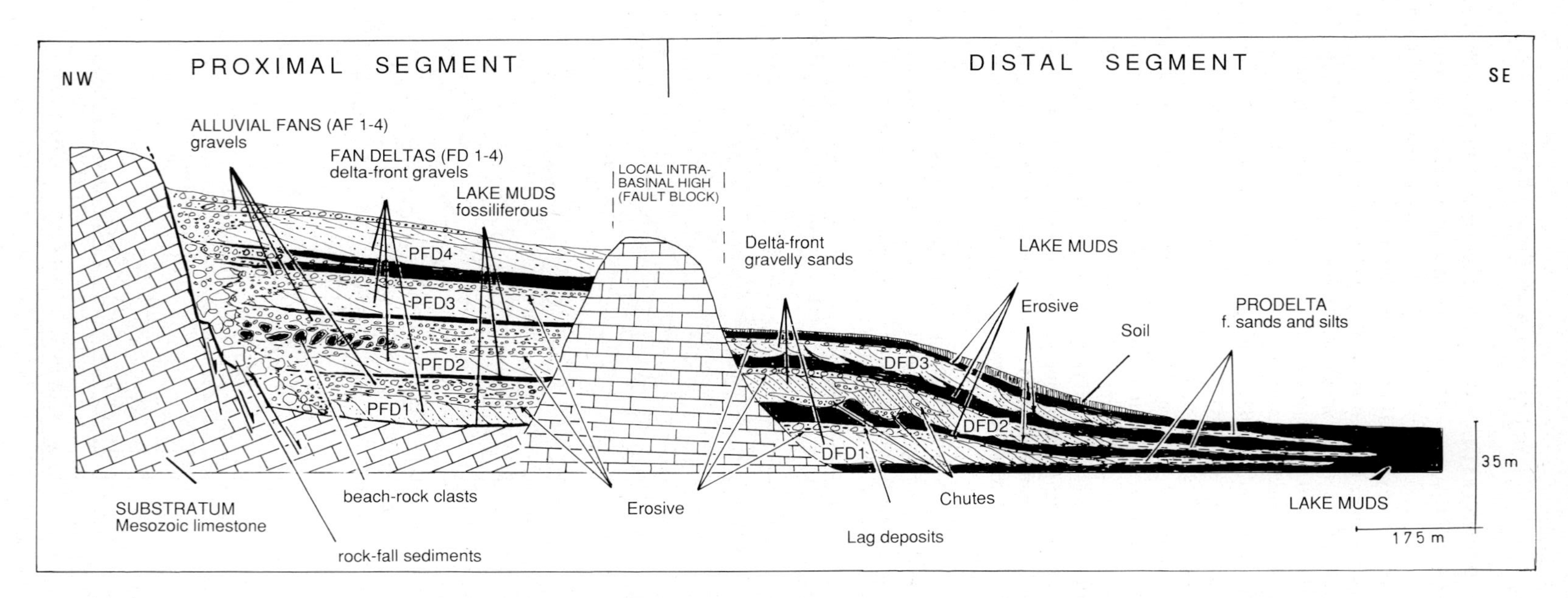

Fig. 3. Generalized cross-section of the Soğanlı fan-delta complex through the AB line of Fig. 2. The thicknesses of the lag deposits are exaggerated.

The Pleistocene basin-fill in the Soğanlı area contains repetitive/cyclic deposition, which is highlighted by the fossiliferous lacustrine muds. Each of the cyclic units of the succession begins with lacustrine mud, which is succeeded in turn by a deltaic and an alluvial sequence. In the distal segment, alluvial sequences are not observed; however, erosional surfaces and lag deposits are present (Fig. 3). The last cyclic unit is represented only by the basal lacustrine mud layer in the distal segment.

Fan-delta sequences

Fan-delta sequences of the Soğanlı study area are wedged between alluvial and lacustrine sediments. Their internal characteristics are quite different in the proximal and distal segments.

Fan-delta sequences in the proximal segment

Fan-delta units in the proximal segment form four fan-delta-front sequences, each overlying one of the fossiliferous lacustrine mud layers (Fig. 3). Prodelta facies were not developed, and moreover, subaerial delta-plain facies are not distinguished easily from alluvial sediments (Figs 4E, F). The proximal fan-delta sequences are numbered from base to top PFD1 to PFD4 (Fig. 3).

The basal sequence (PFD1), exposed in a limited area, forms a progradational delta-front set that is 8 m thick and is thicker towards the intrabasinal high. The set overlies the lacustrine muds and its upper surface is eroded and covered by alluvial sediments (Figs 3, 4F). The steeply dipping (up to 37°) planar foresets of the unit consist of moderate-to coarse-grained beds of sand, alternating with silty beds. Randomly scattered small pebbles do not have a preferred orientation or imbrication. Foreset beds do not exceed 10 cm in thickness, and commonly are 2–5 cm thick. Wavy bedding-surfaces and ripple cross-laminations are visible (Fig. 4F). Individual beds are normally graded; however the unit shows a general coarsening and thickening upward.

The second (PFD2, 7 m thick) and the third (PFD3, 11 m thick) fan-delta units are similar; both consist of gravel beds, thin sand layers, and some large rockfall blocks. The gravels, comprising 95% of the units, are commonly clast supported and composed of sand- to cobble-sized fragments. Clasts are rounded and of Mesozoic limestone. Sand layers have lens-like geometries (2–25 cm thick and 1–6 m long). The average grain size of the units decreases basinward although unit thicknesses increase in the same direction.

The deltaic units form progradational delta-front sets with a depositional foreset inclination of 5–12°. Foreset beds, 25–75 cm thick, are commonly normally graded; at some localities, bedding surfaces are not visible because of clast heterogeneity. Matrix is almost absent. Clast orientation and b-axis imbrication (dipping upslope) are characteristic of coarse pebbly layers. Foresets become tangential and grain size gets finer in the lower parts of the units. The upper bounding surfaces of the sequences are truncation surfaces, which are overlain by alluvial sediments creating scour-and-fill features.

The youngest delta sequence (PFD4, 13 m thick) is relatively fine-grained. It is composed of pebble-supported gravel associated with abundant fossil shells (*Dreissena* spp., and small gastropods) and lacustrine muds. Matrix is absent and individual beds are 3–25 cm thick. Upward-fining size gradation is typical in thicker beds. Bed bases are sharp, planar and erosive to non-erosive. This unit is dominated by stacked sets of planar cross-strata in which foreset-dip angles range between 3° and 20° (Fig. 4F). Alluvial-fan sediments cover the unit.

Deposition of the proximal fan-delta sequences involved progradation of coarse clastic-wedges, which are interpreted as mouth bars. Relatively small alluvial fans/cones created the mouth bars, which prograded into lake water probably 10–15 m deep. Similar depositional processes have been described in recent environments (e.g. Dunne & Hempton, 1984; Colella, 1988). Increasing foreset thickness of the units distally may indicate that lake water was dammed behind the intrabasinal high. After filling the local depression, the youngest unit (PFD4) was probably deposited by stacking of cross-strata in a relatively wide area, resulting in a wide range of foreset orientations. Beach deposits that may have formed during stable intervals were cannibalized by later alluvial processes when the lake level lowered.

Fan-delta sequences in the distal segment

The distal segment of the fan-delta complex, consisting of delta slope, prodelta and lacustrine facies, surrounds the proximal segment and intrabasinal high, and its facies locally abuts on bedrock. In the field, the facies distribution and depositional foreset dips show a half-fan pattern, probably a function of palaeogeographic basin-margin configuration

Fig. 4. Photographs showing details of the proximal segment sequences:
(A) Areal setting of the structural high and the substratum.
(B) Rockfall blocks in alluvial sediments; note that they abut on the substratum.
(C) Beachrock clasts in AF2 unit; they are underlain and overlain by fluvial sediments.
(D) Varve-like lacustrine muds interbedded with debris-flow deposits. The man is looking at a synsedimentary fault.
(E) Stacked cross-sets (fossil-rich) of the PFD4 unit, overlain by alluvial sediments (AF4 unit).
(F) High-angle foresets of PFD1 unit overlying lake muds. Note the erosive top (cut-and-fill) and wavy bedding surfaces.

(Fig. 2). In this segment, there are three wedge-shaped fan-delta units, deposited as succesive, cyclic delta-front sequences, separated by thick, lacustrine mud layers (Fig. 3). The fan-delta units are referred to here as DFD1, 2 and 3. The basal contact of each unit and particularly the DFD2 unit, is erosive into the underlying lacustrine mud unit. Moreover, the upper boundary of each unit is delineated by a boulder lag deposit (Figs 3, 5A–C). Thicknesses of units 1, 2 and 3 are 8 m, 7 m and 5 m, respectively.

The distal fan-delta wedges consist of an assemblage of relatively fine-grained, texturally immature clastic deposits that include randomly scattered pebbles and cobbles up to 20 cm in diameter. Some layers in DFD1 and DFD2 are matrix supported; this matrix consists of red-coloured clay and silt. However clast-supported beds also contain as much as 10% muddy matrix. The texture is poorly sorted, but clasts are moderate to well rounded.

Gilbert-type foresets with depositional dips up to 35° are the basic features of the units, and they form a set in each unit. The foresets are typically planar, becoming tangential and flatter basinward, where the beds are also thinner and generally better sorted, similar to the upper delta-slope conglomerates of Espirutu Santo Formation described by Postma and Roep (1985). The bed thickness averages 25–35 cm. In the DFD3 unit, foresets are relatively thin and interbedded with prodelta silts (Fig. 5C). The beds are usually non-graded and no internal cross-stratification has been observed. Soft-sediment deformation and small-scale normal faults are common in all units (Fig. 5).

The deltaic units, especially DFD2, include some chaotic gravel lenses typified by brownish-coloured matrix, boulder-sized clasts and a poorly sorted texture within the foreset beds. These gravel bodies are up to 30 m long and 3 m thick in transverse cross-section, and have concave-upward erosive bases (Figs 3, 5B, C). Similar features have been interpreted elsewhere as chute or trough deposits, indicating gravity processes or instability of delta fronts (e.g. Prior *et al.*, 1981; Postma, 1984; Colella *et al.*, 1987; Nemec *et al.*, 1988; Postma *et al.*, 1988).

The internal characteristics of the distal fan-delta units suggest that they include abundant mass-flow deposits. The presence of matrix, immature textures, outsize clasts, chutes and the erosive base of each unit indicate that subaqueous avalanching or gravity flows were the dominant depositional processes. However, slip-face sedimentation was represented by minor graded beds. Mass flow is a critical process in large-scale, fan-delta formation, and good examples of such mechanisms have been reported in the literature (Nemec *et al.*, 1984; Postma, 1984; Postma & Roep, 1985; Ori & Roveri, 1987; Choe & Chough, 1988). These examples are generally very thick deltaic deposits compared to the Soğanlı succession. In general, however, it is accepted that mass-flow deposits and related structures such as chutes, slump scars and soft-sediment deformation occur in tectonically active areas, and reflect slope instability (Colella *et al.*, 1987; Colella, 1988; Nemec *et al.*, 1988; Postma *et al.*, 1988).

In the study area, apart from mass-flow-dominated foresets and chutes, soft-sediment deformations and synsedimentary faults are common, reflecting significant tectonic activity combined with effects of basin-margin configuration during deposition. Coarse clasts on eroded tops of deltaic units represent the topsets or lag deposits of alluvial stages. Finer-grained sediments of these layers must have been cannibalized during lake transgressions and they are covered by thick lacustrine muds.

The fan-delta sequences of the proximal and distal segments were deposited by different mechanisms, but in the same system. The proximal units were formed by the progradation of alluvial fans into structurally created topographic lows, and their development did not directly affect distal deposition.

Alluvial-fan sequences

Alluvial-fan sequences occur in the proximal segment of the fan-delta complex and they form individual, stacked gravel wedges that extend from the basin margin to the intrabasinal high. They will be referred to here as AF1–4, according to their relative stratigraphic position from base to top (Fig. 3). Each one rests on a fan-delta unit, and is overlain by a lacustrine mud layer, except for the AF4 unit. Their thicknesses vary but generally decrease basinward. Because their internal facies characteristics are fairly similar, they are discussed as a single unit. The three basic facies of the units are rockfall, debris flow and braided-stream deposits.

Rockfall deposits

Rockfall sediments consist of large blocks up to 5 m in diameter, set in weakly consolidated, very coarse gravel, angular boulders and cobbles. They mostly occur near and on the escarpment as unstratified deposits interfingering with braided-stream sediments (Figs 3, 4B). Minor vadose carbonate cement with ferrinoxide is developed within the clasts. Some individual blocks up to 75 cm in diameter are observed in braided-stream deposits and also in lacustrine mud sequences, making it often difficult to distinguish the rockfall sediments from stream deposits; as a result, deposits with angular clasts larger than 40 cm were interpreted as rockfall

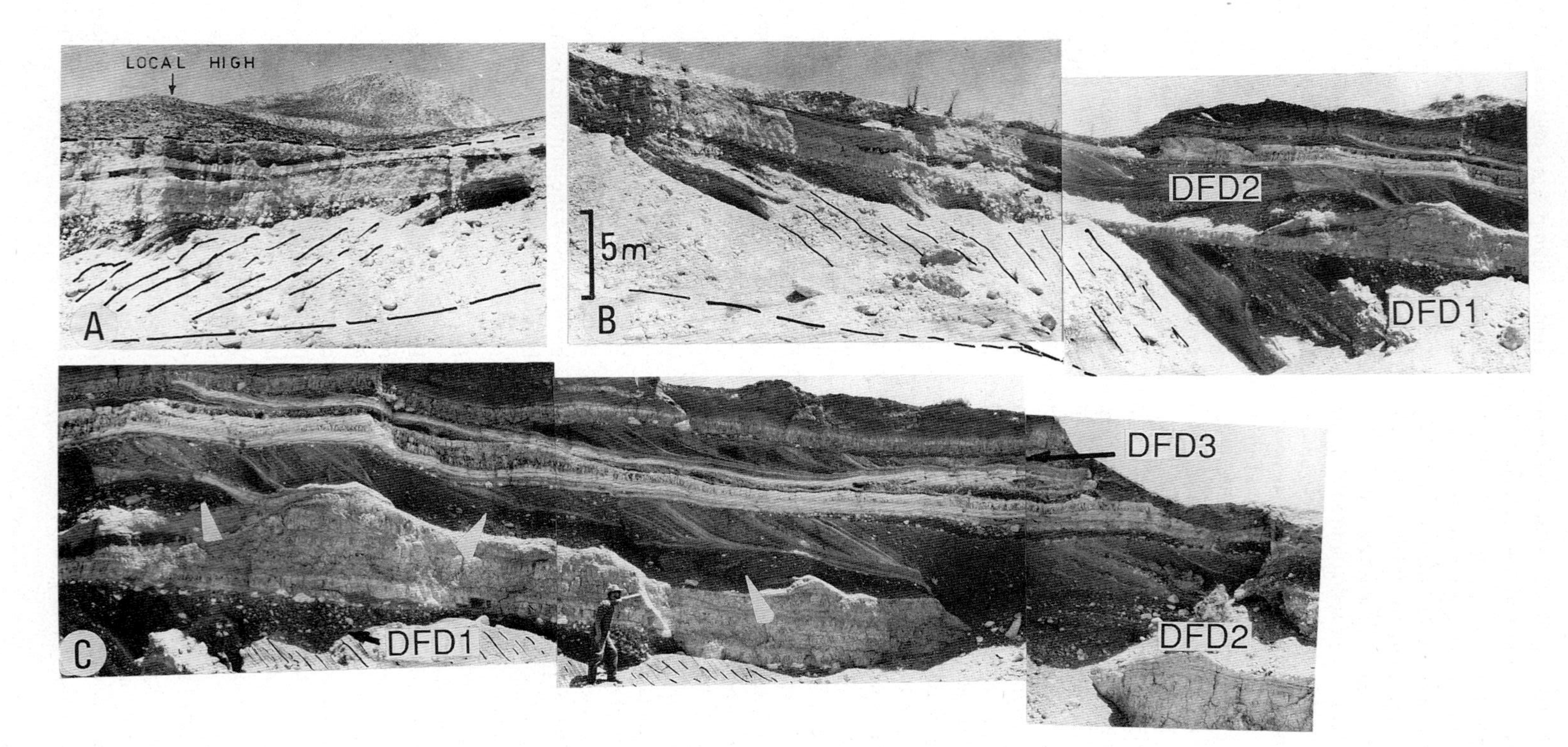

Fig. 5. Photographs showing details of the distal segment sequences. (A, B) Lateral continuity and a general view of distal segment sequences. The lag sediments become thicker towards the left on the DFD1 unit. (C) Close view of deltaic sets and irregular surfaces of lacustrine muds. Arrows (white) show chutes. The man is looking at a thickness of mud where a small synsedimentary fault cuts the mud and chutes.

deposits. Although their areal position appears to suggest deposition as talus or colluvium, the huge blocks in the deposits support the rockfall interpretation.

The occurrence of rockfalls in the fan-delta complexes and in lacustrine muds is not usual; some outsize blocks have been observed in fjords and short-headed deltas (Flores, 1975; Prior & Bornhold, 1986). Some rockfall clasts in the study area may have been detached and fell or rolled downslope during seismically active periods.

Braided-stream deposits

These gravels make up approximately 70% of the alluvial-fan sequences in the AF1–4 units. Each unit has lateral continuity and rests on an older delta-front sequence (Figs 3, 4D, F). This facies consists of poorly- to well-sorted limestone gravel beds. Near the valley margin, rockfall blocks are included. The facies is clast supported and is divided into two subfacies: weakly stratified/massive bouldery gravel, and pebbly well-stratified gravel. The former is usually on the stratified subfacies and its thickness increases marginward. A crude stratification is often picked out by differences in average clast size or by a high concentration of larger clasts. Moderate- to well-rounded cobbles and boulders (with a maximum clast size of 40 cm) are basic components of this subfacies. Contact imbrication with the b-axis dipping upstream is usual. Beds are both ungraded and graded. Some relatively fine-grained beds, especially lenses, such as fine-pebble and coarse-sand lenses, are present, and these are locally overlain by very coarse-grained beds with openwork tops. The average clast size decreases basinward.

The second subfacies, well-stratified gravel, typically lies below the coarse, unstratified subfacies, and rests on a fan-delta unit. Distinctive features of the subfacies are relatively finer-grained fragments, with an average grain size of 8 cm (maximum 20 cm) and a well-stratified, sheet-like bed geometry. Sheet-like gravel beds are 45–115 m long and 25–125 cm thick, with erosive, irregular basal surfaces. Some plane-stratified and cross-stratified beds are included. Cross-sets are markedly homogeneous and locally grade laterally into plane-stratified beds.

The internal characteristics of the two subfacies just described are consistent with a fluvial origin. In the first subfacies, the absence of cross-stratification and relative lack of sand indicate that the bedload was not subject to flow separation and that depth of flow was not great (Rust, 1978; Bluck, 1979). This subfacies is interpreted to represent longitudinal channel bars deposited by gravelly braided streams. Rapid textural variations in the subfacies may be explained by the alternation of high- and low-discharge events (Bluck, 1979; Nemec & Steel, 1984; Marzo & Anadón, 1988). The second subfacies, finer-grained and well-stratified gravel, lies directly on deltaic bodies, and, therefore the sheet-like gravels of the subfacies probably represent delta topsets, deposited by braided streams on alluvial plains. Their sheet-like geometry may have resulted from the lateral shifting of braided channels (Miall, 1977; Rust, 1978). Graded beds of this subfacies indicate a gradual filling of active channels by migrating of gravel bars (e.g. Miall, 1977; Marzo & Anadón, 1988).

Debris-flow deposits

The AF2 and AF3 sequences include a total of two matrix-supported gravel layers. Each one consists of clasts up to 60 cm in diameter in a sandy to muddy matrix. Clast percentages by volume range from 20 to 60%. The matrix-supported layer of the AF2 unit is composed of abundant beachrock clasts with platy shapes and a sandy-silt matrix (Figs 3, 4C). Beachrock clasts are fossiliferous and their a-axes are imbricated upcurrent (Fig. 4C). The matrix-supported layers in the AF3 unit are overlain by lacustrine muds and have a muddy matrix with red colour (Figs 3, 4D). Matrix content locally can reach up to 80%. Clasts are inversely graded and aligned parallel to the bounding surfaces. The basal boundaries of the two layers are non-erosive.

The internal characteristics of these layers, such as non-erosive bases, inverse grading and a-axis orientation, are consistent with a debris-flow origin (e.g. Enos, 1977; Nemec & Steel, 1984). Beachrock clasts indicate that lake level was high and stable for a long time, probably during a wet or cold climatic period. Then lake level was drastically decreased during an arid or warm period, destroying the beach sediments.

Alluvial-fan sequences of the Soğanlı fan-delta complex occur only in the proximal segment, and their equivalent deposits in the distal segment are the lag deposits separating the lacustrine and deltaic sets. Basin-margin topographic relief may not have been adequate for alluvial development in the distal segment (Fig. 6).

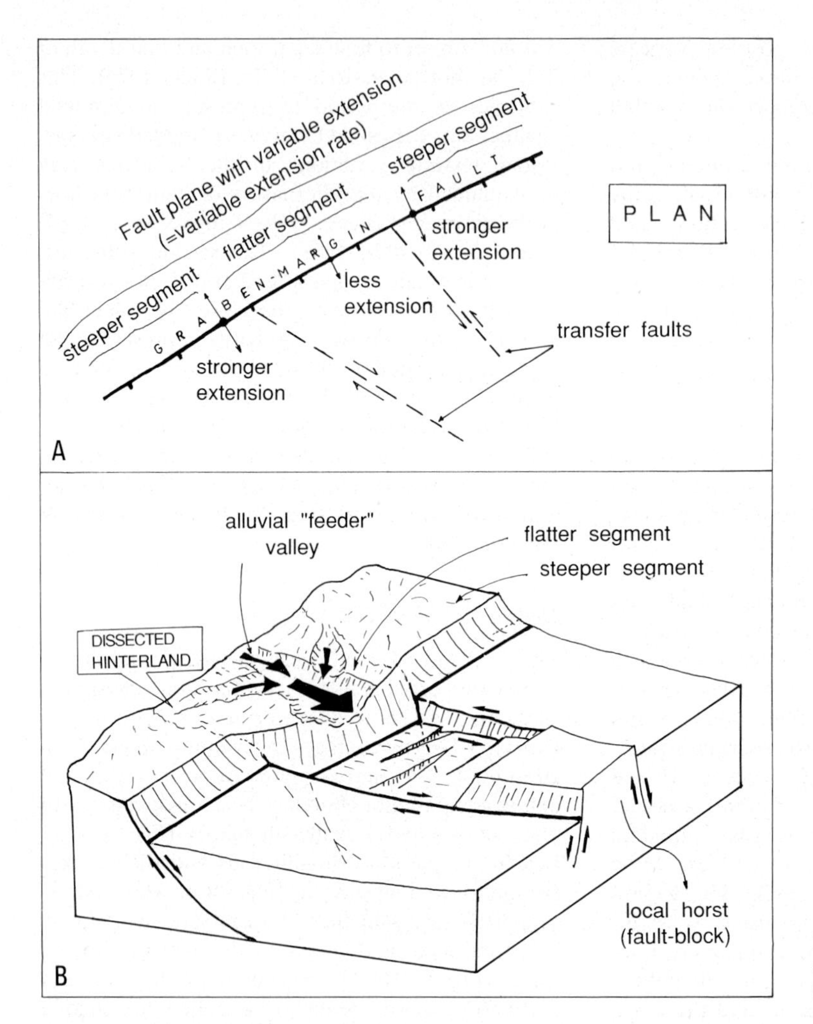

Fig. 6. Possible basin-margin configuration and tectonic model of Soğanlı area of Burdur Basin.

Prodelta and lacustrine sequences

Prodelta facies were developed in the distal segment of the succession, intercalating with and grading into lacustrine muds (Figs 2, 3). Overall, prodelta deposits are reddish-coloured, fine-grained sands and clayey silts. Parallel laminations, cross-laminations, soft-sediment deformations and small-scale faults are common. The distal, prodelta, prodelta–lake facies boundaries are transitional and generally unclear in the field.

The Pleistocene lacustrine muds occur both as transgressive horizons between alluvial and deltaic sequences, and as end-members of the fan-delta complex in the Soğanlı area (Fig. 3). The mud units of the proximal segment are relatively thin (10–350 cm) and of grey colour. One of them, within the AF3 and PFD4 units, displays a varve-like alternation of dark and light colours (brownish and grey), probably reflecting climatic changes during deposition; this unit also includes a small gravel lens, interpreted to be a debris flow (Figs 3, 4D). In the distal segment, mud units are much thicker (2–6 m) and their upper surfaces are deeply and irregularly eroded (Fig. 5A–C). The thickness difference between the proximal and distal units might be created by basin-margin topography and/or palaeotopographic asymmetry, which controlled the sedimentation (Fig. 7).

Most of the lacustrine deposits are now under lake water and their stratigraphic thickness is unknown. These are clays with grey colour, featuring massive bedding or weakly parallel lamination. Soft-sediment deformation and synsedimentary faults have locally disturbed the lacustrine sediments.

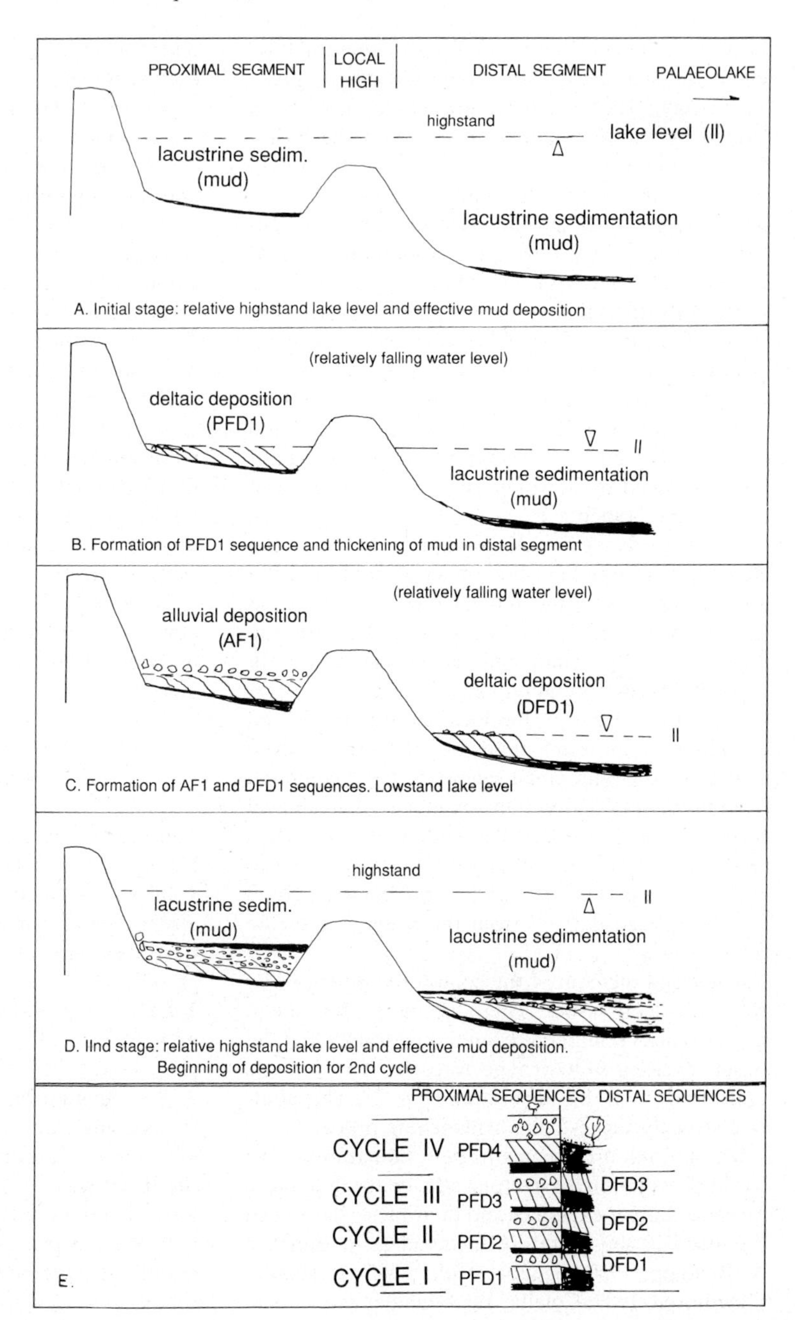

Fig. 7. Depositional model of a cycle in the proximal and distal segments (not to scale). The model is based on similarity of vertical facies organization between the proximal and distal segments.

DISCUSSION

Basin-margin configuration

Differential facies development in the study area indicates that basin-margin configuration played the primary role in controlling the sedimentation of the fluvial and deltaic facies. Vertical and lateral facies development of the Pleistocene–Holocene basin-fill of the Soğanlı area suggests that the proximal segment of the fan-delta complex was deposited in a local depression or a topographic low. Figure 6 illustrates a possible model creating a structural low, where alluvial and deltaic deposition took place repetitively. The structural low could have been formed by transfer faults together with the basin-margin fault (Figs 3, 6). There is no strict field evidence for the transfer faults apart from deep active valleys which cut the substratum perpendicularly; such a valley is now the head-point of a recent alluvial fan (Figs 2, 4A). The origin of transfer faults is based on the variable extension rate of a basin-margin fault plane (Fig. 6A). Because of stronger extension in steeper segments, a part of the substratum (fault block) transferred about 600 m into the graben, and a structural low was opened (Fig. 6B). Outside the structural low, the basin floor was influenced by downfaulting and inclined down towards the graben axis.

Alternatively, such a depression could have been created by a large landslide or an olistolite but the areal distribution of the alluvial and deltaic facies in the proximal segment and no relation between the deposits of the proximal and distal segments do not support this interpretation (Figs 2, 7E).

Deposition began in the local structural low and elsewhere at the same time, but the low received coarser clastics (proximal segment of the complex) from the dissected hinterland by means of an alluvial feeder valley (Fig 6B). The sediments of such a feeder valley were not observable, probably due to recent erosion. The AF4 unit of the complex could also have been derived from the sediments of the feeder valley.

The distal segment of the complex was formed by relatively finer clastics. There was no feeder valley, and avalanched material from the bedrock filled the basin, creating Gilbert-type foresets. The foresets abut directly on the substratum (Fig. 2). This abutting strongly suggests the avalanching process.

The Soğanlı area seems to be a miniature model for the evolution of tectonically controlled basin margins. Larger examples and their facies have been discussed in the literature in detail (e.g. Ballance & Reading, 1980; Şengör *et al.*, 1985; Leeder & Gawthorpe, 1987; Colella, 1988; Leeder *et al.*, 1988; Massari & Colella, 1988).

Deposition and differential facies development

The Soğanlı succession shows three main sedimentary characteristics: (1) repetitive deposition of alluvial-fan, fan-delta and lacustrine mud sequences; (2) mass-flow-dominated fan-delta units in the distal segment; and (3) differential facies development in both segments.

Several studies have suggested that very thick fan-delta sequences could be repeated, when tectonism was the principal controlling factor during deposition and/or basin formation (e.g. Gawthorpe *et al.*, 1988; Nemec & Steel, 1988b; Zupanič *et al.*, 1988). Intense climatic changes are also able to create cyclic deposition of fan deltas, but in that case fan-delta facies are often localized (e.g. Lamy au Rousseau, 1988). The infill of the Pleistocene–Holocene Burdur Basin is not so thick (up to 110 m); however, it provides an excellent record of repetitive deposition resulting from graben-related tectonism combined with climatic changes (Kazancı, 1988). This study indicates that marginal faults in the narrow Burdur graben tended to be active in an alternate fashion rather than simultaneously (see-saw effect), thus creating cyclic deposition.

In the Soğanlı succession, differential facies development resulted from both see-saw effects of the marginal faults and unusual topography of the basin margin.

CONCLUSIONS

The Soğanlı succession (Fig. 3) suggests the following history of facies development (Fig. 7).

1 Initial stage: lake transgressed onto the structurally complex basin margin, probably due to differential subsidence along the northwestern side of the graben.

2 Lacustrine mud was deposited both in the proximal and distal segments, during lake highstand (Fig. 7A).

3 A fan delta prograded into the local low, resulting in deposition of a clastic wedge (PFD1 unit) on the lake mud. Lacustrine sedimentation continued in the distal segment, as the lake water was probably fairly deep because of the basin-floor asymmetry between the proximal and distal segments. Consequently, lacustrine mud layers were much thicker in the distal segment (Figs 5, 7B).

4 A lowstand in lake level resulted in formation of a subaerial proximal area and an alluvial fan, including rockfall blocks (AF1 unit). On the other hand, in the distal segment, a high-angle, mass-flow-dominated fan delta, prograded into the lake (DFD1 unit), overlying the lake muds (Fig. 7C). The underlying lacustrine muds were highly disturbed by avalanching and slope failure during deposition.

5 Lake level relatively increased, drowning all alluvial and deltaic sequences, forming a highstand again. This was actually the beginning of formation of a new cycle (the 2nd cycle) in the succession (Fig. 7D). That highstand position of the lake appears to have lasted for a long time, because beach sediments, clasts of which are observed in the AF2 unit (Fig. 3) of the proximal segment, were deposited and cemented.

In all these stages, lake level changed relatively, probably due to extension-created subsidence. Renewed subsidence started the new cycle.

Because of the superimposed fan-delta, alluvial-fan and lacustrine mud sequences, relative lake-level fluctuations apparently repeated the same pattern, creating at least four sedimentary cycles. After the formation of PFD4 unit, the lake level drastically decreased and the study area became subaerial, with soil formation developed on the lacustrine mud of the distal segment (Figs 3, 7E).

ACKNOWLEDGEMENTS

I would like to thank the following people who contributed in various ways to the study: W. Nemec for constructive criticism, especially about tectonic model; O. Tekeli and R. Breuninger for fruitful discussions and critical review of an earlier draft of this manuscript; and A. Colella for financial support to be able to present the study in the Fan-Delta Workshop in Calabria. I am also indebted to the referees T.H. Nilsen and R.J. Steel for critically reading this manuscript and providing helpful suggestions. This work was supported by research funds from Ankara University.

REFERENCES

Ballance, P.F. & Reading, H.G. (Eds.) (1980) Sedimentation in oblique-slip mobile zones. Spec. Publs. int. Assoc. Sediment. 4, 265 pp.

Bluck, B.J. (1979) Structure of coarse-grained braided stream alluvium. *Trans. R. Soc. Edin.* **70**, 181–221.

Choe, M.Y. & Chough, S.K. (1988) The Hunghae Formation, SE Korea: Miocene debris aprons in a back-arc intraslope basin. *Sedimentology* **35**, 239–256.

Colella, A. (1988) Pleistocene–Holocene fan deltas and braid deltas in the Crati Basin, southern Italy: a consequence of varying tectonic conditions. In: *Fan Deltas: Sedimentology and Tectonic Settings* (Ed. by W. Nemec and R.J. Steel), pp. 50–74, Blackie and Son, London.

Colella, A., De Boer, P.L. & Nio, S.D. (1987) Sedimentology of a marine intermontane Pleistocene Gilbert-type fan-delta complex in the Crati Basin, Calabria, southern Italy. *Sedimentology* **34**, 721–736.

Dunne, L.A. & Hempton, M.R. (1984) Deltaic sedimentation in the Lake Hazar pull-apart basin, southeastern Turkey. *Sedimentology* **31**, 401–412.

Enos, P. (1977) Flow regimes in debris flows. *Sedimentology* **24**, 133–142.

Erol, P. (1972) Konya-Tuzgölü ve Burdur havzalarındaki plüvyal göllerin çekilmelerinin jeomorfolojik delilleri (Geomorphological evidence of the recessional phases of the pluvial lakes in the Konya-Tuzgölü and Burdur basins in Anatolia). *Coğrafya Araştırmaları Dergisi* **3–4**, 13–52.

Erol, O. (1973) Burdur Havzası Kuvaterner depoları (Quaternary deposits of Burdur Lake Basin). In: *Cumh. 50 yılı Yerbilimleri Kongresi* (Ed. by G. Sagıroğlu), pp. 386–391. Ankara.

Erol, O. (1978) Quaternary history of the lake basins of central and southern Anatolia. In: *The Environmental History of the Near and Middle East Since the Last Ice Age* (Ed. by W.C. Brice), pp. 111–139. Academic Press, London.

Erol, O., Şenel, S. & Kış, M. (1986) Preliminary results of C14 datings of the coastal deposits of the Pleistocene pluvial Burdur Lake. *Turk. Bull. Archaeometry* **15**, 81–88.

Flores, R.M. (1975) Short-headed stream delta: model for Pennsylvanian Haymond Formation, west Texas. *Bull. Am. Assoc. Petrol. Geol.* **59**, 2288–2301.

Gawthorpe, R.L., Hurst, J.M. & Sladen, C.P. (1988) Geometry and evolution of Miocene footwall-derived fan deltas, Ras Budran area, Gulf of Suez (east). *Abstr. Int. Works. Fan Deltas, Calabria*, 26–27.

Graciansky, P.C. (1968) Teke yarımadası (Likya) Toroslarının üst üste gelmiş ünitelerinin stratigrafisi ve Dinaro-Toros-lardaki yeri. *Maden Tetk. Ar. Enst. Derg. (Turkey)* **71**, 73–92.

Karaman, E. (1986) Burdur ili çevresindeki yerleşim alanlarının depremselliği. *Mühéndislik Jeolojisi Bült. (Turkey)* **8**, 23–30.

Kazanci, N. (1988) Repetitive deposition of alluvial-fan and fan-delta wedges at a fault-controlled margin of the Pleistocene–Holocene Burdur Lake graben, southwestern Anatolia, Turkey. In: *Fan Deltas: Sedimentology and Tectonic Settings* (Ed. by W. Nemec and R.J. Steel), pp. 186–196. Blackie and Son, London.

Kazanci, N. & Erol, O. (1987) Sedimentary characteristics of a fan-delta complex from Burdur Basin, Turkey. *Z. Geomorph.* **31**, 261–275.

Koçyiğit, A. (1981) Isparta Büklümü'nde (Batı Toroslar) Toros karbonat platformunun evrimi (Evolution of Taurus carbonate platforms in Isparta Bend (western Taurus). *Jeol. Kur. Bült. (Turkey)* **24**, 15–23.

Koçyiğit, A. (1983) Hoyran Gölü (Isparta Büklümü) dolayının tektoniği (Tectonics of the Hoyran Lake (Isparta Bend) region). *Jeol. Kur. Bült. (Turkey)* **26**, 1–10.

Kostaschuk, R.A. (1985) River-mouth processes in a fjord-delta, British Columbia, Canada. *Mar. Geol.* **69**, 1–23.

Kozan, A.T., Bozbey, E., Oğdüm, F. & Tüfekçi, K. (1988) Burdur Havzasının oluşumu ve gelişimi, G.B. Anadolu (Formation and evolution of the Burdur Basin,

SW Anatolia). *Abstr., 12th Congress on Geomorphology*, Ankara, 15–16.

LAMY AU ROUSSEAU, R. (1988) Lacustrine level fluctuations as revealed by proglacial delta migrations (Würm, Jura, France). *Abstr. Int. Works. Fan Deltas, Calabria*, 32–33.

LEEDER, M.R. & GAWTHORPE, R.L. (1987) Sedimentary models for extensional tilt-block/half-graben basins. In: *Extensional Tectonics* (Ed. by P.L. Hancock, M.P. Coward and J.F. Dewey). Spec. Publ. geol. Soc. London 28, pp. 139–152.

LEEDER, M.R., ORD, D.M. & COLLIER, R. (1988) Development of alluvial fans and fan deltas in neotectonic extensional settings; implications for the interpretation of basin-fills. In: *Fan Deltas: Sedimentology and Tectonic Settings* (Ed. by W. Nemec and R.J. Steel), pp. 173–185. Blackie and Son, London.

MARZO, M. & ANADÓN, P. (1988) Anatomy of a conglomeratic fan-delta complex: the Eocene Montserrat Conglomerate, Ebro Basin, northeastern Spain. In: *Fan Deltas: Sedimentology and Tectonic Settings* (Ed. by W. Nemec and R.J. Steel), pp. 318–340. Blackie and Son, London.

MASSARI, F. & COLELLA, A. (1988) Evolution and types of fan-delta systems in some major tectonic settings. In: *Fan Deltas: Sedimentology and Tectonic Settings* (Ed. by W. Nemec and R.J. Steel), pp. 103–122. Blackie and Son, London.

MCKENZIE, D.P. (1972) Active tectonics of the Mediterranean region. *Geophys. J.R. Astron. Soc.* **30**, 109–185.

MIALL, A.D. (1977) A review of a braided-stream depositional environment. *Earth Sci. Rev.* **13**, 1–62.

NEMEC, W. & STEEL, R.J. (1984) Alluvial and coastal conglomerates: their significant features and some comments on gravelly mass-flow deposits. In: *Sedimentology of Gravels and Conglomerates* (Ed. by E.H. Koster and R.J. Steel). Mem. Can. Soc. Petrol. Geol. 10, pp. 1–31.

NEMEC, W. & STEEL, R.J. (1988) What is a fan delta and how do we recognize it?. In: *Fan Deltas: Sedimentology and Tectonic Settings* (Ed. by W. Nemec and R.J. Steel), pp. 3–13. Blackie and Son, London.

NEMEC, W. & STEEL, R.J. (Eds.) (1988a) *Fan Deltas: Sedimentology and Tectonic Settings*. Blackie and Son, London, 444 pp.

NEMEC, W., STEEL, R.J., GJELBERG, J., COLLINSON, J.D., PRESTHOLM, E. & ØXNEVAD, I.E. (1988) Anatomy of collapsed and re-established delta front in Lower Cretaceous of eastern Spitsbergen: gravitational sliding and sedimentation processes. *Bull. Am. Assoc. Petrol. Geol.* **72**, 454–476.

NEMEC, W., STEEL, R.J., POREBSKI, S.J. & SPINNAGR, A. (1984) Domba Conglomerate, Devonian, Norway; process and lateral variability in a mass flow-dominated, lacustrine fan delta. In: *Sedimentology of Gravels and Conglomerates* (Ed. by E.H. Koster and R.J. Steel), Mem. Can. Soc. Petrol. Geol. 10, pp. 295–320.

ORI, G.G. & ROVERI, M. (1987) Geometries of Gilbert-type deltas and large channels in the Meteora Conglomerate, meso-Hellenic basin (Oligo-Miocene), central Greece. *Sedimentology* **34**, 845–859.

ÖZGUL, N. (1976) Torosların bazı temel jeoloji özellikleri (Some geological aspects of the Taurus orogenic belts). *Jeol. Kur. Bült. (Turkey)* **19**, 65–78.

POSTMA, G. (1984) Slumps and their deposits in delta front and slopes. *Geology* **12**, 27–30.

POSTMA, G., BABIĆ, L., ZUPANIČ, J. & RØE, S.L. (1988) Delta-front failure and associated bottomset deformation in a marine, gravelly, Gilbert-type fan delta. In: *Fan Deltas: Sedimentology and Tectonic Settings* (Ed. by W. Nemec and R.J. Steel), pp. 91–102. Blackie and Son, London.

POSTMA, G. & ROEP, T.B. (1985) Resedimented conglomerates in the bottomsets of Gilbert-type gravel deltas. *J. Sedim. Petrol.* **55**, 874–885.

PRIOR, D.B. & BORNHOLD, B.D. (1986) Sediment transport on subaqueous fan-delta slopes, Britannica Beach, British Columbia. *Geo-Mar. Lett.* **5**, 217–224.

PRIOR, D.B., WISEMAN, W.J., JR. & BRYANT, W.R. (1981) Submarine chutes on the slopes of fjord deltas. *Nature* **290**, 326–328.

RUST, B.R. (1978) Depositional models for braided alluvium. In: *Fluvial Sedimentology* (Ed. by A. Miall). Mem. Can. Soc. Petrol. Geol. 5, pp. 605–625.

ŞENGÖR, A.M.C., GÖRÜR, N. & ŞAROĞLU, F. (1985) Strike-slip faulting and related basin formation in zones of tectonic escapes: Turkey as a case study. In: *Strike-Slip Deformation, Basin Formation and Sedimentation* (Ed. by K.F. Biddle and N. Christie-Blick). Spec. Pub. Soc. Econ. Paleont. Mineral., Tulsa 37, pp. 227–264.

STEEL, R.J. (1988) Fan deltas: some recent advances. *Abstr. Int. Works. Fan Deltas, Calabria*, 57.

SUNGUR, K.A. (1978) Burdur ve Acıgöl depresyonları ile Tefenni ovasının fiziki coğrafyası (Physical geography of the Burdur and Acıgöl depressions and Tefenni plain). *Ist. Üniv. Coğr. Enst. Yay.* **95**, 126 pp.

ZUPANIČ, J., BABIĆ, L., RØE, S.L. & POSTMA, G. (1988) Construction and destruction of Gilbert-type deltas during transgression: an example from Eocene Promina Formation, Yugoslavia. *Abstr. Int. Works. Fan Deltas, Calabria*, 58.

Spec. Publs int. Ass. Sediment. (1990) **10**, 199–222

Stacked Gilbert-type deltas in the marine pull-apart basin of Abarán, late Serravallian–early Tortonian, southeastern Spain

H. C. VAN DER STRAATEN

Instituut voor Aardwetenschappen, Vrije Universiteit, 1081 HV Amsterdam The Netherlands

ABSTRACT

The Abarán Basin is a marine pull-apart basin that developed inside a releasing overstep along a right-slip fault zone. Due to progressive opening of the basin, extension in a direction parallel to the major strike-slip faults prevailed throughout the sedimentary history. Periods of filling were followed by faulting episodes leading to periodic deepening. Tectonic subsidence controlled the development of three depositional sequences bounded by angular unconformities. Each sequence reflects a major axial system of predominantly sand-sized deltaic sediments and a minor lateral system comprising a series of gravelly fault- and reef-taluses that rim the basin.

A Gilbert-type delta complex formed the axial system of the oldest sequence. It consists of deltas stacked in an imbricate way which are separated from each other by inclined truncational surfaces interpreted as major slide scars. The threefold division in top-, fore- and bottomsets typical for Gilbert-type deltas, is recognized. Foresets, dipping about 20° W and SW, asymptotically pass into more gently inclined bottomsets. Locally, subhorizontal, shallow-marine topsets, modified by wave action, truncate the underlying foresets.

The extensional regime during the development of the delta complex is illustrated by a prominent deepening and the activity of normal faults. Progressively deeper water is deduced from the height of the delta foresets increasing from 30 m to more than 150 m. NNW-trending normal faults express extension in a direction parallel to the major E–W strike-slip fault zone and fit in the structural concept of the pull-apart basin. They dip away from the basin centre and migrate progressively basinwards. Normal faulting is considered to have controlled the imbricate stacking of the deltas.

INTRODUCTION

Gilbert-type deltas commonly develop in laterally confined and protected basins within an area of pronounced topographic relief (see Colella, 1988). These conditions are met in the Abarán Basin interpreted as a shallow-marine pull-apart basin that formed an intramontane basin almost entirely enclosed by emerged palaeohighs. During the initial depositional episode, the basin was filled by an axial system interpreted as a Gilbert-type delta complex.

The paper concerns not only delta sedimentation, but is largely focused on the position of the delta complex within the sedimentary and structural framework of the pull-apart basin. A model is presented that explains how normal faulting and related deepening during the opening of the pull-apart basin, controlled the internal architecture of the delta complex.

GEOLOGICAL AND STRUCTURAL SETTING

The Abarán Basin is situated in the External Zone of the Betic Cordilleras, the Alpine mountain belt in southeast Spain (Fig. 1). The development of the External Zone took place during two compressional stages in early and middle Miocene. Although major deformation had been completed in the Tortonian, strike-slip faulting caused rapidly subsiding basins to develop (Baena Perez & Jerez Mir, 1982; Benavente

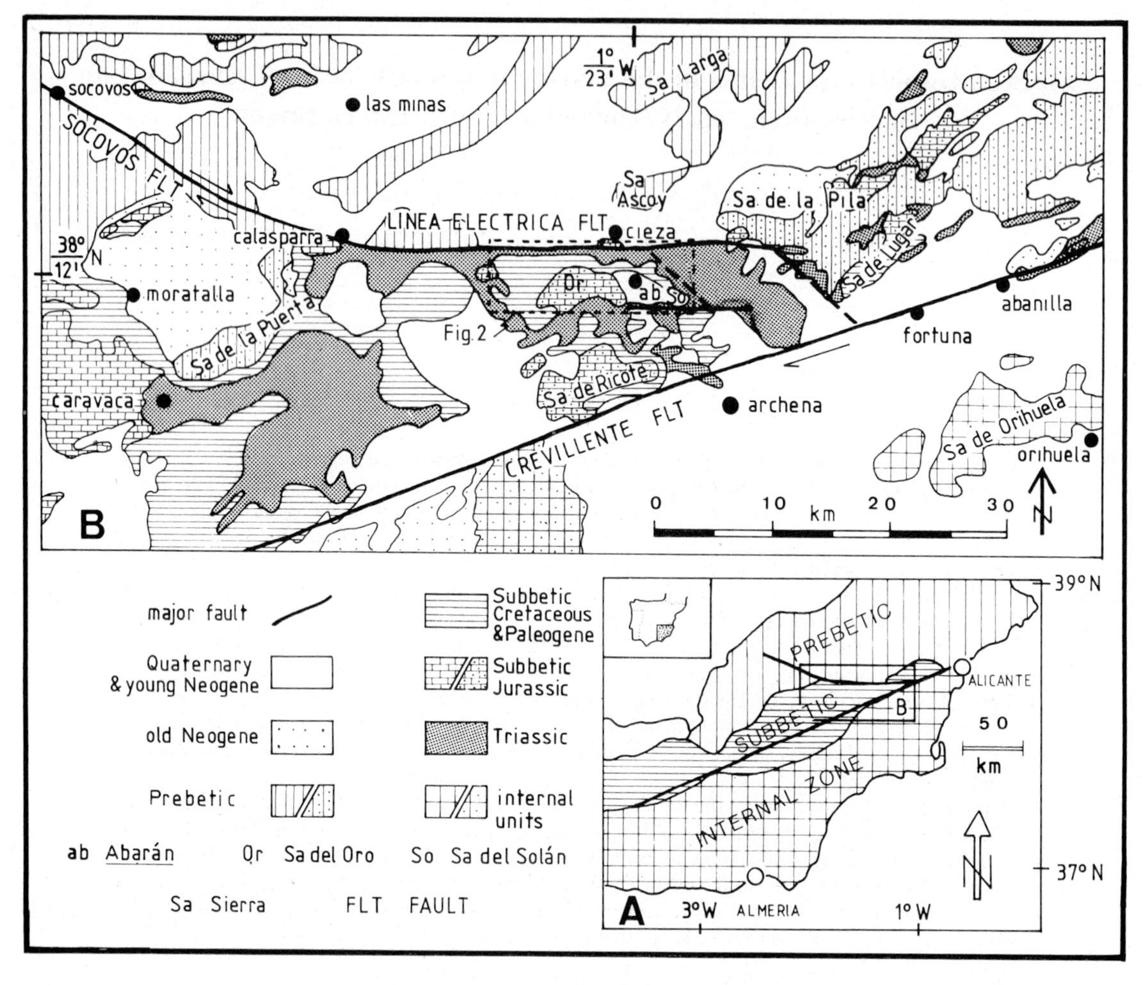

Fig. 1. Geological setting. (A) Tectonic sketch map of the central and eastern Betic Cordilleras showing the Internal, the Subbetic and the Prebetic zone. (B) Geological map of the area of the Abarán Basin (simplified after Baena Perez & Jerez Mir, 1982). The map shows the distribution of the Prebetic and the Subbetic Units, and the Triassic rocks (not differentiated in Prebetic and Subbetic). 'Old' and 'Young' Neogene refers to middle Serravallian and older, and late Serravallian and younger, respectively.

Herrera & Sanz de Galdeano, 1985; Montenat *et al.*, 1987). The relief created by the second compressional phase — very similar to the present-day topography — was flooded in late Serravallian times. Hence, sedimentation took place in intramontane basins confined by emerged structural highs (Montenat, 1973; Baena Perez & Jerez Mir, 1982). These 'post-orogenic' marine sediments are of late Serravallian to early Tortonian age and include the fill of the Abaràn Basin. They unconformably overlie middle Miocene and older rocks that were deformed by the second compressional phase and that in fact form the 'basement' of the Abarán Basin fill.

The Abarán Basin developed along an E–W trending, right-slip fault zone (Fig. 2) that runs along the Subbetic–Prebetic boundary (two structural units in the External Zone) between Calasparra and Cieza and continues further eastward to the area due south of the Sierra de la Pila. The 'Falla de la Linea Electrica' of Jerez Mir (1979) delimits the fault zone to the north (Linea Electrica Fault in Fig. 1). Fault zone characteristics, shown in Fig. 2, are diagnostic for right-lateral strike-slip

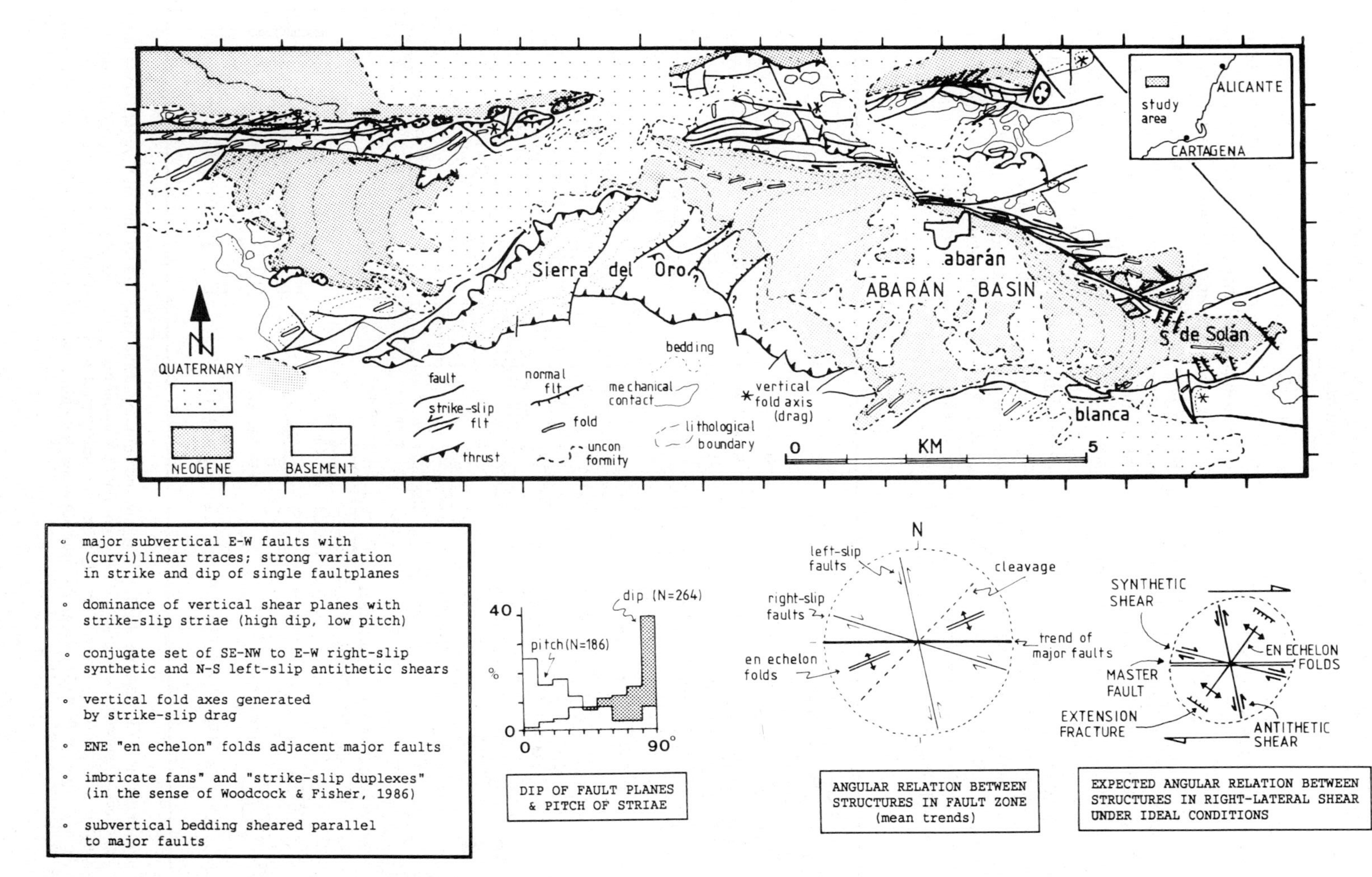

Fig. 2. Right-slip fault zone. Structural map of the right-slip fault zone in the Subbetic basement bounded in the north by the 'Linea Electrica fault' (see Fig. 1). The geology of the thrust slice forming the Sierra del Oro is based on IGME (1974). The diagram in the lower right corner that shows the expected structures in right-lateral shear under ideal conditions, is after Christie-Blick and Biddle (1985). See Fig. 1 for location.

faulting (e.g. Wilcox *et al.* 1973; Christie-Blick & Biddle, 1985).

THE ABARÁN BASIN

Basin infill

Sedimentation in the Abarán Basin took place under marine conditions during late Serravallian to early Tortonian times. Since the entire infill was restricted to biozone N15 (10.2–9.5 Ma BP; Haq *et al.*, 1987), basin stratigraphy goes beyond the resolution of foraminifera biostratigraphy. The stratigraphic thickness in the east amounts to *c.* 280 m (Fig. 3A). In the western part of the basin, the minimum thickness of the stratigraphic column is estimated between 250 m and 350 m (Fig. 3D). The actual stratigraphic thickness remains unknown due to synsedimentary tilting and overlap.

The sedimentary record of the Abarán Basin reveals an overall upward-fining and deepening trend, in which three sedimentary units can be recognized based on abrupt changes in lithology. Successive units show finer and more distal facies. Each unit is bounded by angular unconformities locally accompanied by chaotic, extremely coarse slide- or debris-flow deposits that probably represent seismites (cf. Kleverlaan, 1987). The units are thought to reflect distinct depositional episodes separated from each other by pulses of tectonic subsidence. The deepening of the basin is demonstrated by synsedimentary deformation and fossil content (Table 1). The sedimentary units bounded by unconformities will be called depositional sequences. In stratigraphic order, the lower sequence, the

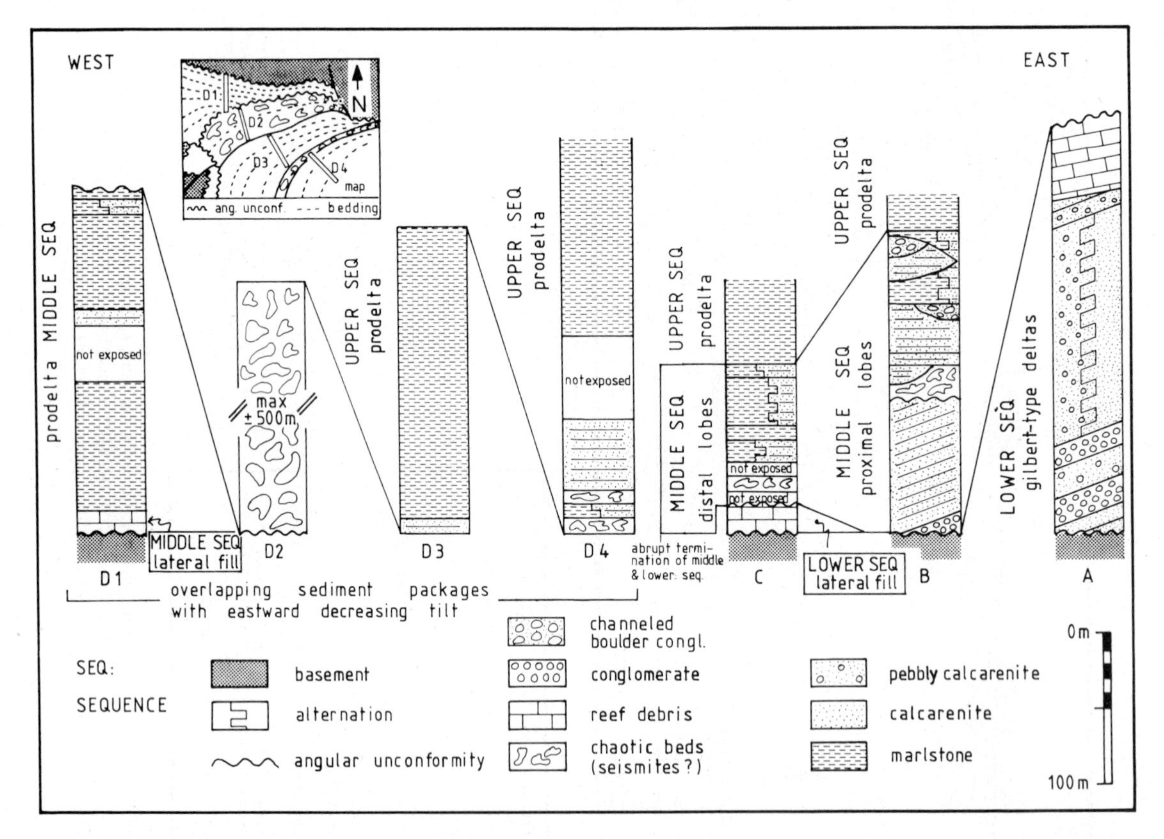

Fig. 3. Sections through basin fill. Sections A, B, C, D show the lithology of the axial and the lateral basin fill of the three depositional sequences. Composite section D (1–4) represents a continuous series, whereby in eastward direction progressively younger beds overlie basement. The phenomena is interpreted as overlap formed by synsedimentary tilt. The small map in the upper part of the figure shows the four segments that built up section D (see Fig. 4 for scale). The traces of sections A, B, C, D are shown in Fig. 4.

Table 1. Basin-fill. The fill of the Abarán Basin has been divided into three depositional sequences, i.e. sedimentary units bounded by angular unconformities that reflect distinct depositional episodes in basin history. Each sequence shows a major, sand-sized (or finer-grained) deltaic axial system and a minor gravel-sized lateral system comprising predominantly reef- and fault-talus deposits. Younger sequences reveal finer and more distal facies; fauna suggests progressively deeper water and a decreasing input of reworked fossils

Sequence		Sedimentary system	Transport	Interpretation	Dominant fauna
Upper	Axial	Silty marlstone with subordinate thin, fine calcarenite beds	?	Prodelta: hemipelagic mud and distal turbidites	Pelagic and deep-water, small benthonic foraminifera [1]
	Lateral	Series of channelled conglomerates	Transverse	Fault-talus	—
Middle	Axial				Pelagic and deep-water, small benthonic foraminifera and transported shallow-water foraminifera [2]
	West	Silty marlstone with subordinate thin, fine calcarenite beds	W	Prodelta: hemipelagic mud and distal turbidites	
		Distal calcarenite lobes (alternation marlstone and fine calcarenites)	WNW?	Prodelta lobes in front of submarine distributary (channel or chute in deltafront ??)	
	East	Proximal calcarenite lobes (coarse-grained and thick-bedded)	NW	Proximal: sandy high-density turbidites; distal: classical turbidite	
	Lateral	Apron of reef debris and series of channelled boulder conglomerates	Transverse	Forereef- and fault-talus	Transported reef debris (algae, bryozoans, bivalves) and pelagic foraminifera
Lower	Axial	Delta foresets largely made up of massive (pebbly) calcarenites	(S)W	*Gilbert-type delta complex*	Pelagic and shallow-water small benthonic foraminifera [3]
	Lateral	Algal boundstones, apron of reef debris and boulder conglomerates	Transverse	Fringing algal reefs and forereef- and fault-talus	Algae and bryozoans *in situ*, shell beds and transported reef debris

[1] Deep: *Bulimina* sp., *Gyroidina* sp., *Sphaeroidina* sp., *Melonis* sp., *Pullenia* sp., *Globobulimina* sp.,
Very poor to very rich pelagic fauna
Rare reworked foraminifera

[2] Shallow: *Elphidium* sp., *Florilus* sp., *Nonion* sp., miliolids
Deep: *Bulimina* sp., *Gyroidina* sp., *Melonis* sp., *Pullenia* sp., *Vulvilineria* sp.
Pelagic fauna of normal richness
Transported sponge probably indicative for water of *c*. 100–200 m deep
Common reworked Cretaceous to Palaeogene foraminifera

[3] Shallow: *Ammonia* sp., *Elphidium* sp., miliolids, bryozoans, charophyta, gastropods
Rare (reworked ?) deep-water foraminifera
Poor pelagic fauna
Rich in reworked Cretaceous to Palaeogene foraminifera

middle sequence and the upper sequence are recognized (Table 1; Figs 3, 4).

In each sequence, the facies distribution is controlled by *major axial deltaic systems* and *minor lateral talus systems* (Table 1). The axial system of the lower sequence comprises calcarenites and conglomerates forming sediment bodies with large delta foresets. The system is interpreted as a complex of submarine deposits of Gilbert-type deltas that prograded W and SW. The axial system of the middle sequence consists of stacked, calcarenite high-density turbidite lobes that pass westward into hemipelagic marlstones with distal turbidite interbeds. The sediment bodies are interpreted as lobes in front of submarine distributaries that laterally pass into finer-grained prodelta deposits. The lobes show a sediment transport towards the W and NW, following the basin axis. The upper sequence is dominated by silty marlstones with thin-bedded and fine-grained distal turbidites interpreted as an axial system of prodelta deposits. The basin has been subordinately side-fed by a lateral system consisting of a series of reef- and fault-taluses, forming a narrow conglomerate belt along the fault-controlled basin margins.

Basin structure

The fill of the Abarán Basin is far less affected by the E–W right-slip fault zone than the basement and unconformably overlies structures generated by the strike-slip faulting (Figs 2, 5, 6). Basin sedimentation therefore post-dates the initiation of the fault zone. Yet, Tortonian deposits are truncated and offset by the strike-slip faults and are locally incorporated in the fault zone. This documents post-Tortonian, right-lateral movement along the fault zone (Fig. 2). Since the basin was bounded by active E–W faults that document right-slip before and after basin development, synsedimentary right-slip is suggested.

The Abarán Basin has a rhomboidal geometry determined by the configuration of faults along the basin margins active during sedimentation (Fig. 6). Sedimentation was restricted to this fault-bounded depression enclosed by emerged basement highs forming the source area of the basin-fill. The present outline of the basin does not represent an erosional or tectonic fragment, but largely reflects the original basin geometry. Arguments for this assumption are found in: (1) synsedimentary faulting documented by deformation and talus deposits along the faulted margins; (2) basement highs indicated by onlap (Fig. 5) and talus deposits; (3) major, axial sediment transport parallel to the bounding faults. Given the angular relation and the synsedimentary nature of the bounding faults (Fig. 6), the basin is interpreted as a releasing overstep formed by two, right-hand-stepping, major, E–W right-slip faults connected by SE–NW oblique- and normal-slip faults (Figs 6, 7).

Older beds are faulted, folded and tilted more than younger beds and deformation is commonly restricted to deposits below intrabasinal unconformities (see Figs 5, 6). Synsedimentary faulting along the basin margins includes both vertical and lateral components, leading to basin subsidence and the growth of the basin in an E–W direction. Episodic subsidence is reflected in the three depositional sequences. NW-trending normal faults in the Gilbert-type delta complex document actual down-faulting of the basin floor. Since these faults truncate older deposits, but do not affect younger ones, they must have been active during sedimentation. Along the northern and northeasten fault margin, the basin-fill shows a progressive decrease of tectonic tilt in younger deposits, forming 'progressive unconformities' (in the sense of Riba, 1976; Fig. 5). The structure is associated with packet sliding and is considered to document a downward movement of the basin relative to the adjacent basement highs.

Right-slip along the E–W basin margins is indicated by *en echelon* folds (Figs 2, 6) and strike-slip striae on fault planes. Because faults and folds are restricted to older basin-fill and cut off by intrabasinal unconformities, right-slip took place during basin development. The western basin-fill is considered to be dragged in a right-lateral way. Drag is inferred from beds that swing round to parallel the orientation of the E–W fault zone. The prominent out-fanning of the beds, i.e. their southward divergence, reveals that older beds were dragged more than younger beds (Figs 3, 6).

Growth of the basin in E–W direction in response to the right-slip is thought to be reflected in the structure of the western part of the basin. Here, the central basin-fill forms a monotonously eastward-dipping series which leads to an apparent cumulative stratigraphic thickness of more than 2000 m (Figs 5, 7C). Along the southwestern and northern basin margin, successive younger beds unconformably overlie basement in an eastward direction. Consequently, younger sediments accumulated not only upon, but largely next to the older ones. This over-

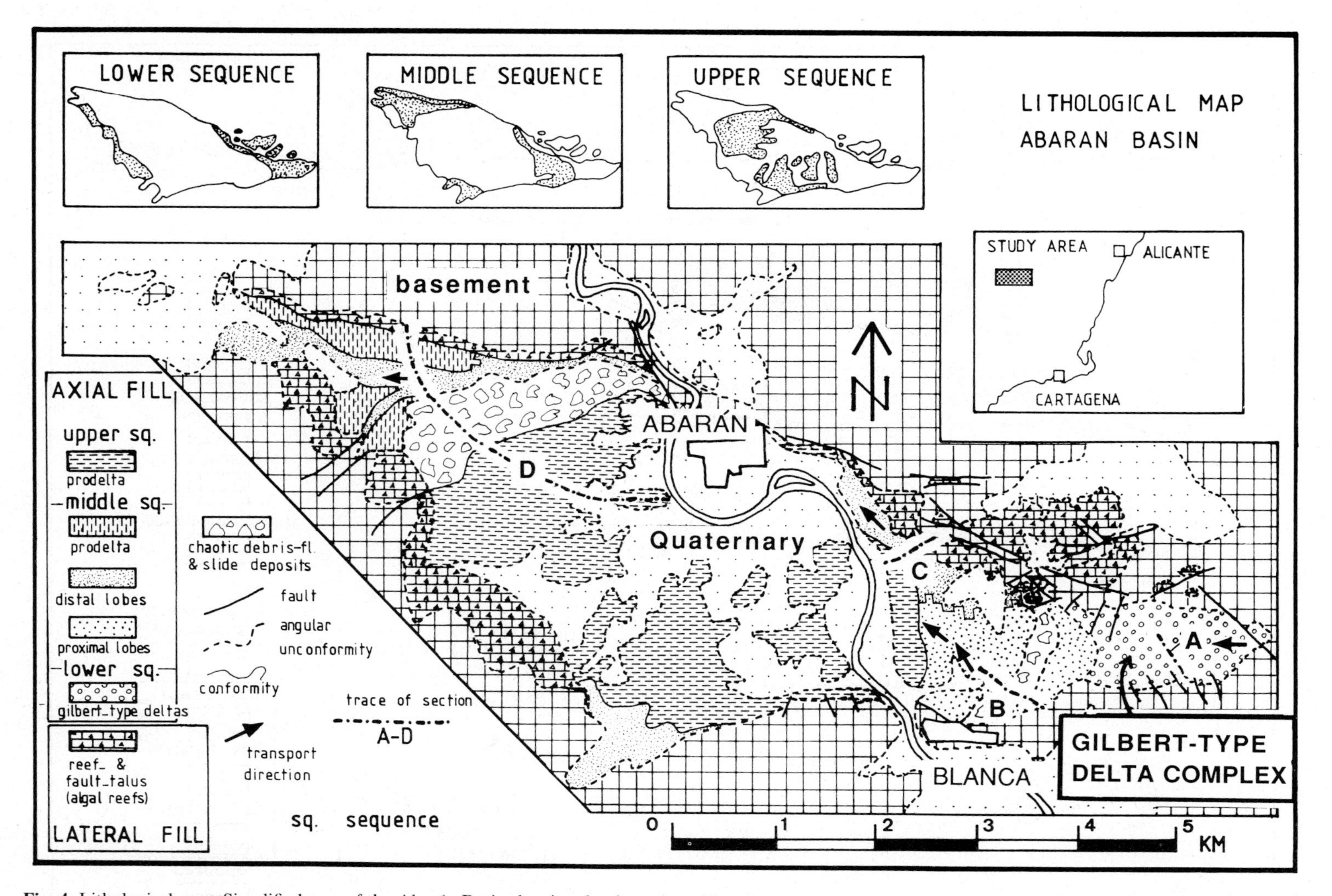

Fig. 4. Lithological map. Simplified map of the Abarán Basin showing the three depositional sequences and the distribution of the different lithologies within each sequence (see also Table 1). Note that the 'chaotic debris flow and slide deposits' are associated with angular unconformities bounding the depositional sequences. A, B, C, D mark the traces of sections shown in Fig. 3.

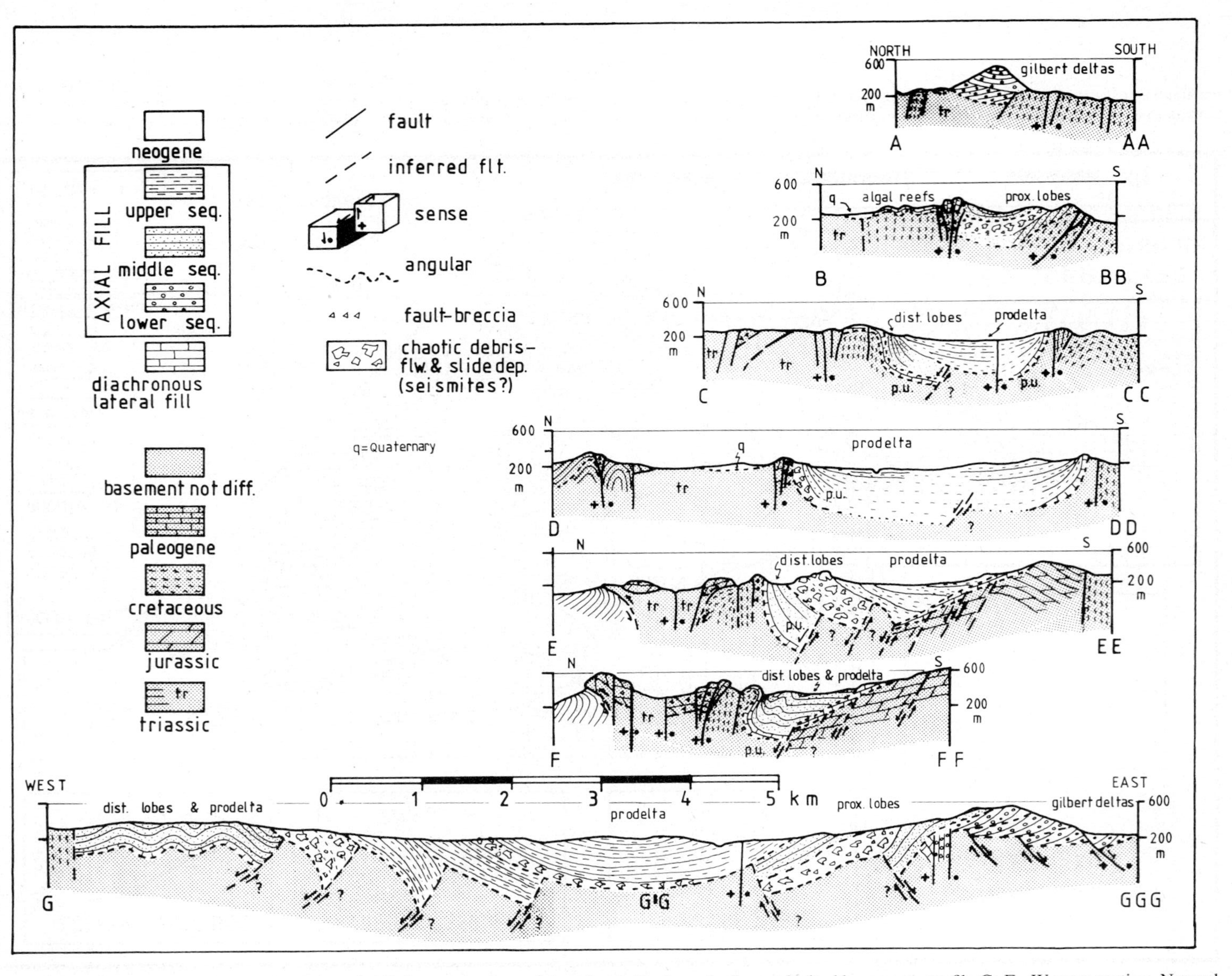

Fig. 5. Cross-sections. Profiles A (east) to F (west): N–S cross-sections of the Abarán Basin and adjacent faulted basement; profile G: E–W cross-section. Normal faults under the western basin-fill are inferred from synsedimentary normal faults that cut the basement of the Sierra del Oro and lateral fill along the SW margin. The presented way of faulting is, however, speculative. Note that if the beds do not terminate against basement, the thickness of the basin-fill would be

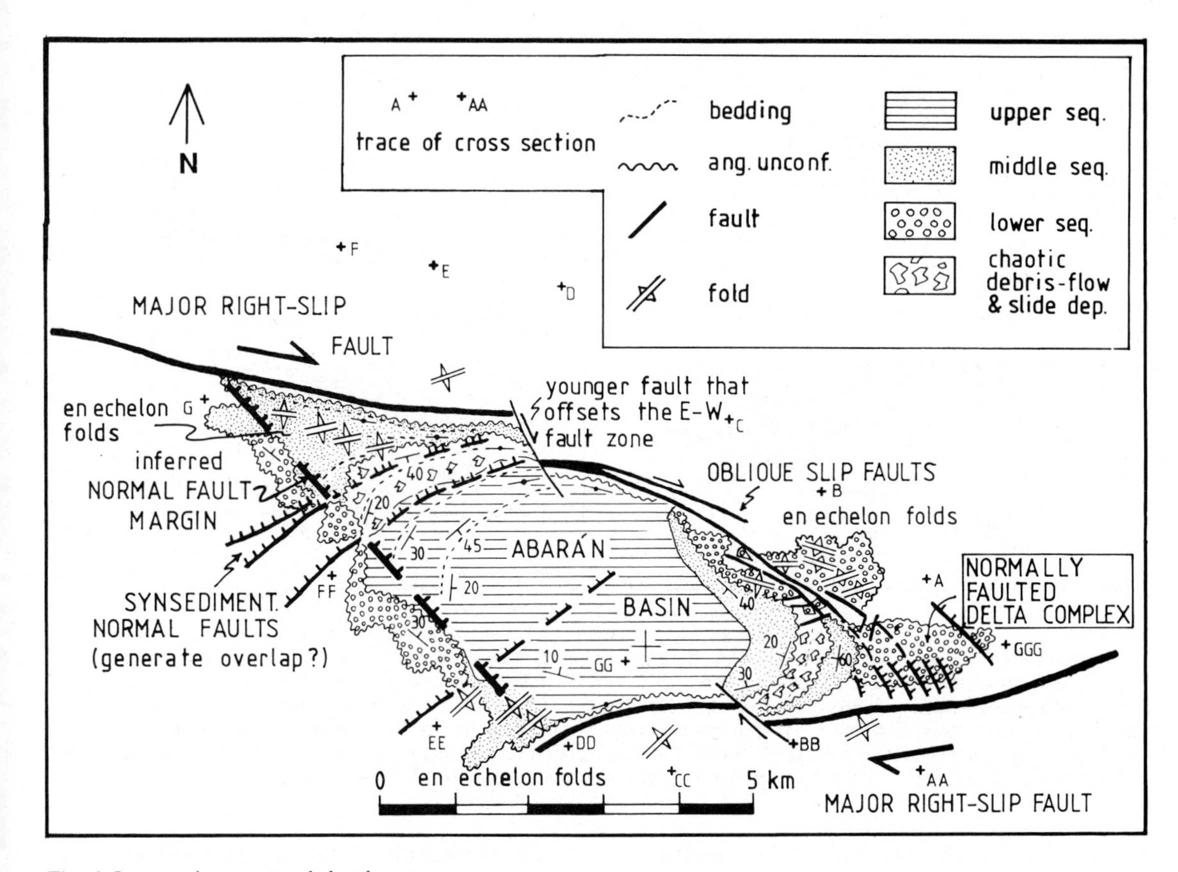

Fig. 6. Interpretive structural sketch map.

lap explains the anomalously large cumulative thickness which does not reflect the true thickness of the basin-fill. The dip of *c.* 40° of prodelta beds and the general westward sediment transport, show that the bedding configuration does not represent downlapping, prograding sediments, but suggest that the overlap is of tectonic origin. Synsedimentary normal faults in the Sierra del Oro dipping NW (IGME, 1974), laterally extend beneath the basin-fill and are considered to be related to the overlap (Figs 5, 6).

These structural elements lead to the interpretation illustrated in Fig. 7C: due to the right slip along the E–W fault zone, the basin is extended in an E–W direction by normal faulting, whereby sediments were progressively tilted and moved away from the basin centre, creating overlapping beds. The distribution of the three sequences (Fig. 4) may support the lateral extension of the basin. It evokes the impression that the older deposits were torn apart and moved away from the centre of the basin, causing the younger deposits to accumulate laterally next to, rather than upon the older deposits. Overlap, E–W extension and lateral migration of the older sediments away from the basin centre is inferred in the western part of the basin from rather circumstantial evidence. The phenomena can actually be observed in the Gilbert-type delta complex.

THE GILBERT-TYPE DELTA COMPLEX (LOWER SEQUENCE)

In the eastern part of the Abarán Basin, the axial system of the lower sequence comprises predominantly (pebbly) calcarenites and associated conglomerates that are exposed in the Sierra de Solán (Figs 3A, 4; Table 1). The deposits are *c.* 280 m thick and are characterized by large foresets, i.e. inclined beds that were deposited under a significant sedimentary angle (Figs 8, 10). Apart from the foresets, subhorizontal bottomsets and (rare) topsets have been recognized, each with a characteristic

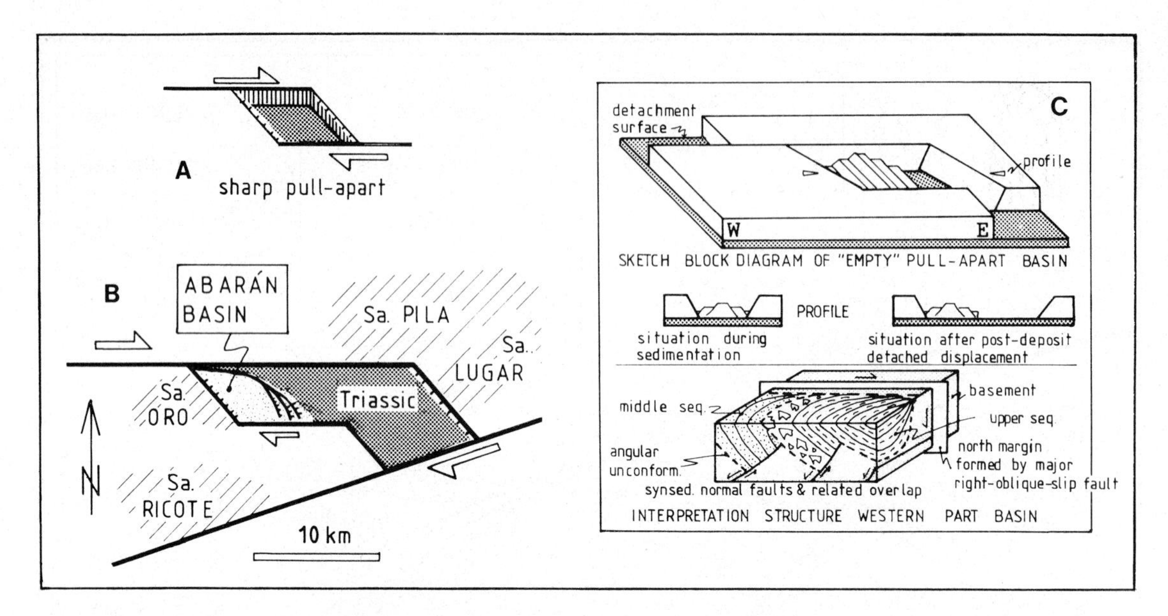

Fig. 7. Structural cartoon of the Abarán Basin as rigid pull-apart. (A) A sharp pull-apart on a right-slip fault. (B) A simple model illustrating the detached displacement of the entire Abarán Basin as part of a larger lithological block. The displacement separated the basin from its source area (Sa. de la Pila and Sa. de Lugar) and tore apart an angular gap floored with plastic Triassic lithologies that acted as décollement. (C) Interpretive cartoon of the Abarán Basin as a rigid, detached pull-apart basin in which E–W extension generates normal faults. In the western part of the basin, synsedimentary normal faults underlie the basin-fill. Faulting, related tilt and tectonic transport away from the basin centre generate overlapping beds and consequently an anomalously large cumulative thickness.

facies (Fig. 9). The deposits are interpreted as Gilbert-type deltas, whereby the term 'delta' refers to any outbuilding of a shoreline directly related to a sediment input (definition of McPherson *et al.*, 1987). Based on the marine deposits of the Abarán Basin, no distinction could be made between fan delta or river delta.

The interpreted internal geometry of the Gilbert-type delta is shown Fig. 9. The deltas are characterized by large *foresets (Facies I)* up to 150 m high dipping *c.* 20° towards the west and southwest. They asymptotically approach the base of the delta forming gently dipping *bottomsets (Facies II)*. Locally, the foresets are truncated and overlain by subhorizontal *topsets (Facies III*, Fig. 11). A fourth element that characterizes the deltas, are channel-shaped *conglomerate bodies (Facies IV*, Fig. 12) that cut deeply into the delta front. The Gilbert-type deltas prograded westward and southwestward (see Fig. 10).

Delta units and their boundaries

The delta deposits reveal large-scale truncational surfaces forming minor unconformities that separate three major sediment bodies stacked in an imbricate way (Fig. 10). These sediment bodies will be called 'delta units' and are considered to represent individual Gilbert-type deltas. The boundaries of the delta units are formed by westward-inclined, smooth and planar surfaces that truncate the underlying unit and that dip at a slightly lower angle than the delta foresets (Fig. 10). The foresets of the overlying unit downlap onto the boundary surface. The angular relation with the basal unconformity and recognized bottom- and topsets indicate that the significant surface dip is not created by tectonic tilt, but is largely a primary feature.

Because of its primary dip, its lateral extension, its smooth and planar nature and the absence of lag deposits, the surface was probably not formed by scour or abrasion. This leaves little alternative but to interpret the surface as scars produced by sediment failure, probably major slides. Similar boundary surfaces in delta complexes have been described by Postma and Roep (1985) and Colella (1988). Soft-sediment deformation beneath the boundary surface which would support the slide-scar interpretation,

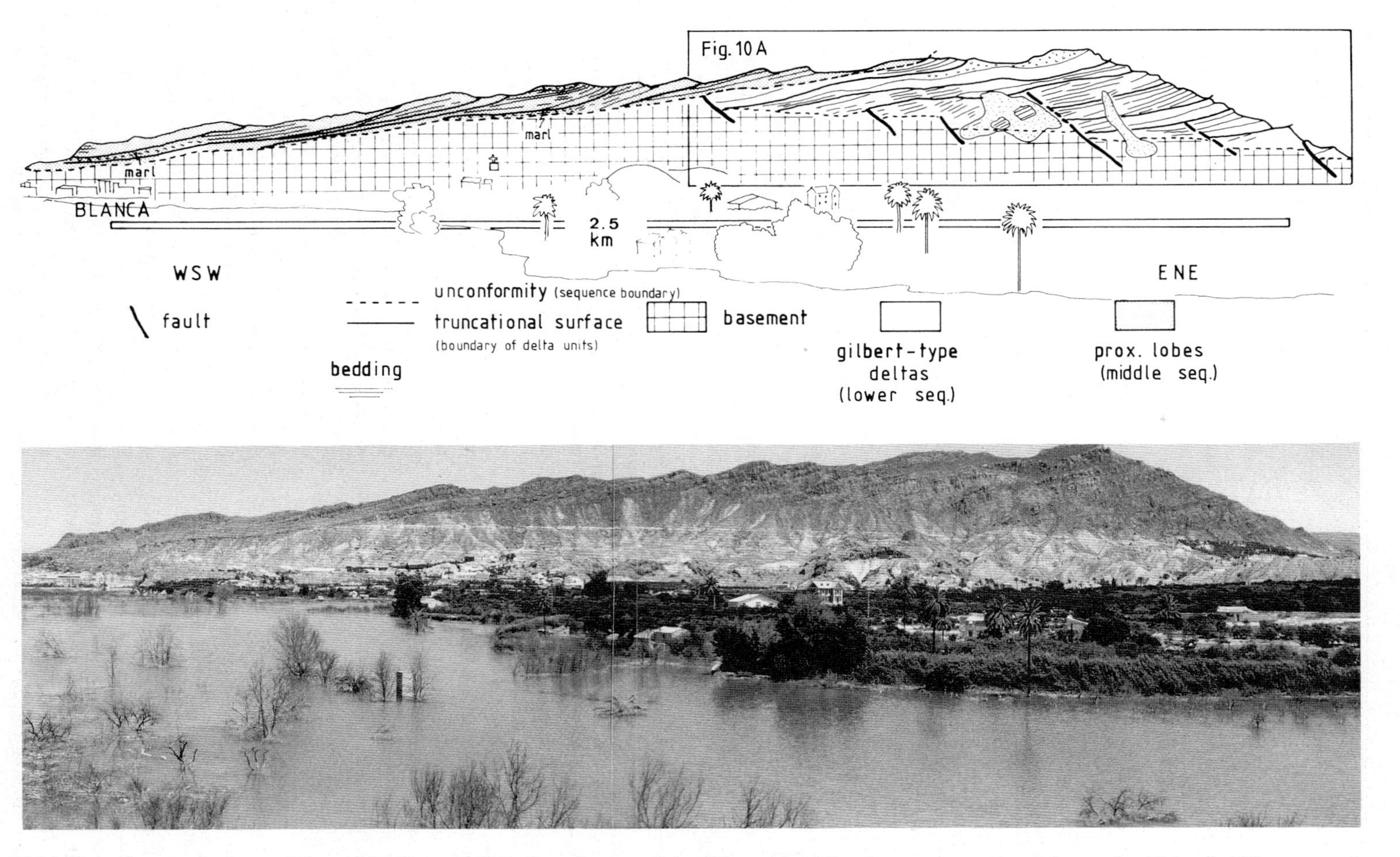

Fig. 8. Delta foresets. In the south face of the Sierra de Solán large foresets of the Gilbert-type deltas (lower sequence) and the proximal lobes (middle sequence) are exposed. View is towards the north (top of Sierra N25°E). Shown exposure extends from Blanca to the most eastern side of the Sierra de Solán, a distance of about 2.5 km. An interpretation of the marked section (in box) is presented in Fig. 10A.

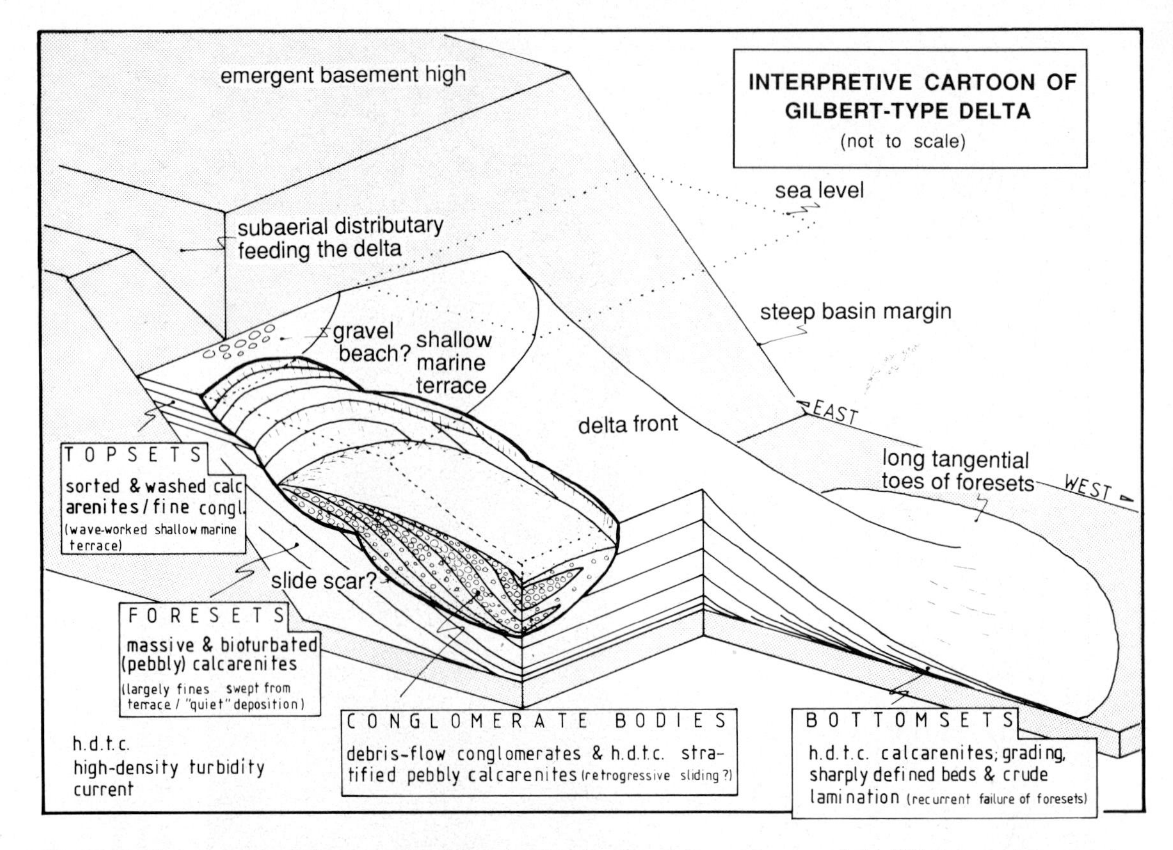

Fig. 9. Architecture of the Gilbert-type delta. Interpretive drawing of the architecture of the Gilbert-type delta, showing the *foresets (Facies I), bottomsets (Facies II), topsets (Facies III)* and the channel-shaped *conglomerate bodies (Facies IV)*. The coastal and subaerial part of the drawing is speculative. Drawing is not to scale.

is generally not encountered, probably due to the impossibility to recognize sediment distortion in these largely structureless and massive delta deposits. Better stratified and well-sorted topsets just beneath one of the surfaces (Figs 10B, C), however, show signs of liquefaction (Fig. 11), which may be related to the slide event.

Facies

Facies I: foresets

Description. Foresets are largely made up of structureless (pebbly) calcarenites showing no bedding or internal organization. Generally, the deposits are poorly sorted, fine- to medium-grained and rich in lime-mud. They frequently are intensely bioturbated. At one location, vertical (escape?) burrows have been encountered of 1–1.5 m long. The structureless deposits locally show thick, poorly defined bedding ranging from 0.5 to 2 m, or distinct bioturbation horizons with a vertical separation suggesting bed thicknesses of 3–4 m (Fig. 13B). The pebbly calcarenites contain scattered, floating clasts that range from granule- to pebble-size with outsize clasts up to 10 cm in diameter. With decreasing clast density, pebbly calcarenites pass into massive calcarenites containing a few isolated pebbles.

Interpretation. Foresets are documented by the observed primary dip of the calcarenites (Fig. 8). Bioturbation combined with a generally high lime-mud content, indicates that the largest part of the calcarenites represent deposits of small-sized flows mixed with silt and mud deposited out of suspension. The occasional occurrence of thicker, non-burrowed

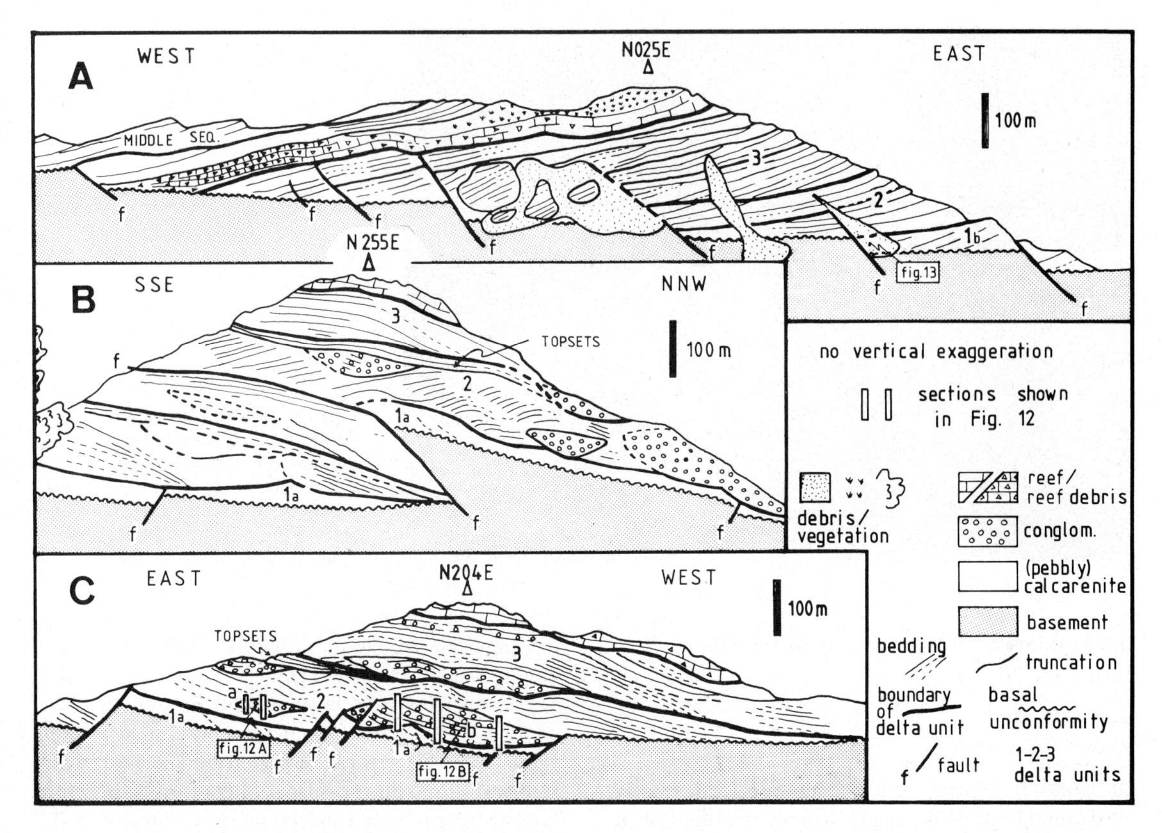

Fig. 10. Gilbert-type delta complex. The Sierra de Solán observed from different angles showing the imbricate stacking of the Gilbert-type deltas. In A, B and C, the view is towards N, WSW and SSW, respectively. A is drawn from a photo presented in Fig. 8. Delta foresets dip SW (unit 2) and W (unit 3). Delta-unit 3 is capped by massive algal and bryozoal reefs that pass downslope into bedded reef-talus deposits (drawing A). Note that the syndepositional normal faults (drawing A) dip towards the east and become younger in the westward direction. Drawings A and C show that the basal unconformity of the older delta units is slightly tilted westwards in respect to the younger unit (angles of *c.* 3°). B and C show elongated channel-shaped conglomerate bodies; sections of body 'a' and 'b' (marked with open bars) are presented in Figs 12A and B, respectively.

beds and long, vertical escape burrows may reflect accumulation of a larger volume of sediment during a single event.

Facies II: bottomsets

Description. The foresets asymptotically pass downslope into more gently-dipping to subhorizontal bottomsets. The bottomsets generally comprise bedded calcarenites that are poorly sorted, and show crude and vague parallel or cross-lamination. Bedding is characterized by sharp, planar or undulating boundaries. Some beds are normally or inversely graded; the grading may be restricted to only part of the bed. Others show poorly-defined, medium to thick beds that pinch out laterally over a short distance (less than 10 m). Apart from the bedded calcarenites the bottomsets may include massive calcarenites similar to Facies I.

Interpretation. The position relative to the foresets and the gentle sedimentary dip, demonstrate that the *Facies II* beds represent bottomsets. Lamination documents traction flows, normal grading suggests deposition under waning flow conditions, while inverse grading indicates grain interaction in a highly concentrated flow (Middleton & Hampton 1976; Lowe, 1982; Nemec & Steel, 1984). This combination of depositional processes characterizes sandy high-density turbidity currents (Lowe, 1982), so that

Fig. 11. *Topsets (Facies III)* of delta-unit 2. Low-angle cross-laminated calcarenites and fine-gravel interbeds. Note the vertical gravel-filled structures at the right side of the picture, probably generated by water escape.

the bottomsets are considered to be deposited by such flows.

Facies III: topsets

Description. Locally, the foresets are truncated by horizontal beds consisting of sharply stratified, well-sorted calcarenites and fine conglomerates. The calcarenites are medium- to coarse-grained and are characterized by parallel, low-angle planar and high-angle trough lamination (Fig. 11). Lamination is generally far more sharply defined than the vague and crude lamination in other delta facies. One outcrop shows a cross-lamination characterized by undulating parallel laminae, many curved internal truncations, symmetrical crests and offshooting and draping foresets (Fig. 13 C).

The fine conglomerates generally consist of thin, tabular and continuous layers of well-sorted granules to small pebbles that are clast supported and closely packed. Most show remarkably sharp and straight boundaries and some contain well-rounded and bored pebbles. Small and shallow channels of *c.* 10–30 cm deep are common, which are generally filled by (cross-stratified) calcarenites and floored by one-pebble-thick lag deposits (Fig. 11). The deposits include shallow-marine fossils. However, in contrast to the other delta facies, they are free of intergranular lime mud and do not contain pelagic foraminifera. Locally, the lithology is strongly bioturbated or shows sandy, matrix-supported disorganized conglomerates and associated water-escape structures (Fig. 11).

The topsets of the lowermost delta unit (unit 1a) contain thin layers of plant remains. Fossil rootlets arise from this layer and penetrate the underlying sediment up to 30 cm. The rootlets form subvertical linear structures that are commonly defined by decolourization of the sediment. In many cases, however, the peaty matter of the rootlet itself is still preserved. They bifurcate and thin in a downward direction. These layers are associated with clay (rich in plant debris), thin gypsum layers (*c.* 2 cm) and distinct horizons of gastropod moulds. The gastropods are marine and resemble *Cerithium* sp. (pers. comm., T. Geel).

Interpretation. The subhorizontal beds can be interpreted as either topsets of the delta unit, or bottomsets of the following delta. A marine depositional environment is illustrated by well-rounded, occasionally bored clasts, bioturbation and marine (reworked?) fossils. The layers of plant debris underlain by fossil rootlets are interpreted as paleosols. The paleosols and the associated gypsum layers indicate a restricted, very shallow-marine environment that occasionally became emerge. Given the very shallow-marine setting, the beds concerned are interpreted as topsets.

Good sorting and the lack of intergranular mud reveal that the depositional processes were effective in washing and sorting the sediment. The cross-

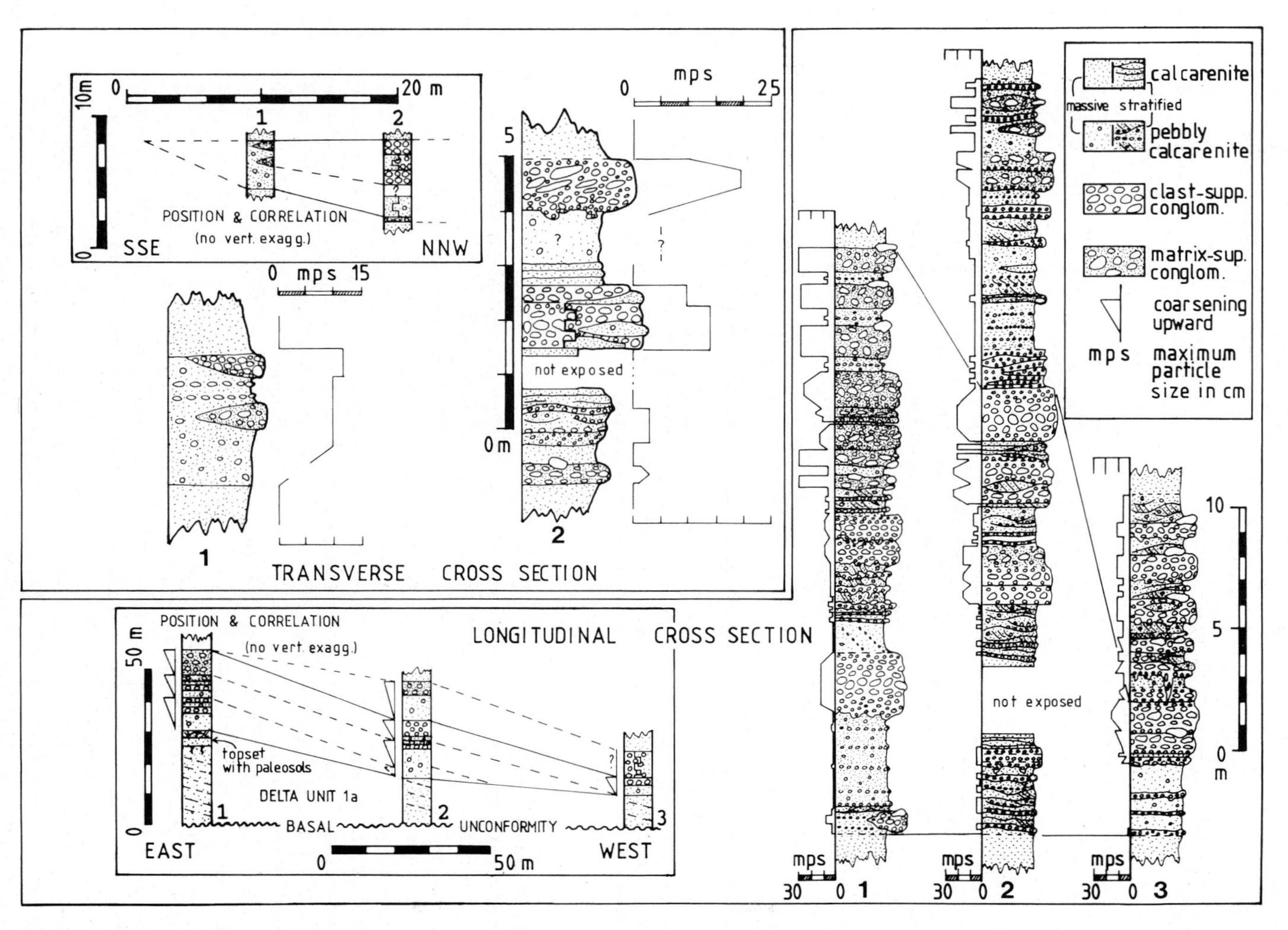

Fig. 12. Cross-sections of conglomerate bodies. (A) A flow-transverse cross-section of one of the conglomerate bodies shown in Fig. 10C (body 'a'). (B) A flow-parallel cross-section of one of the conglomerate bodies shown in Fig. 10C (body 'b'). Note internal foresetting, coarsening-upward and proximal-distal trend.

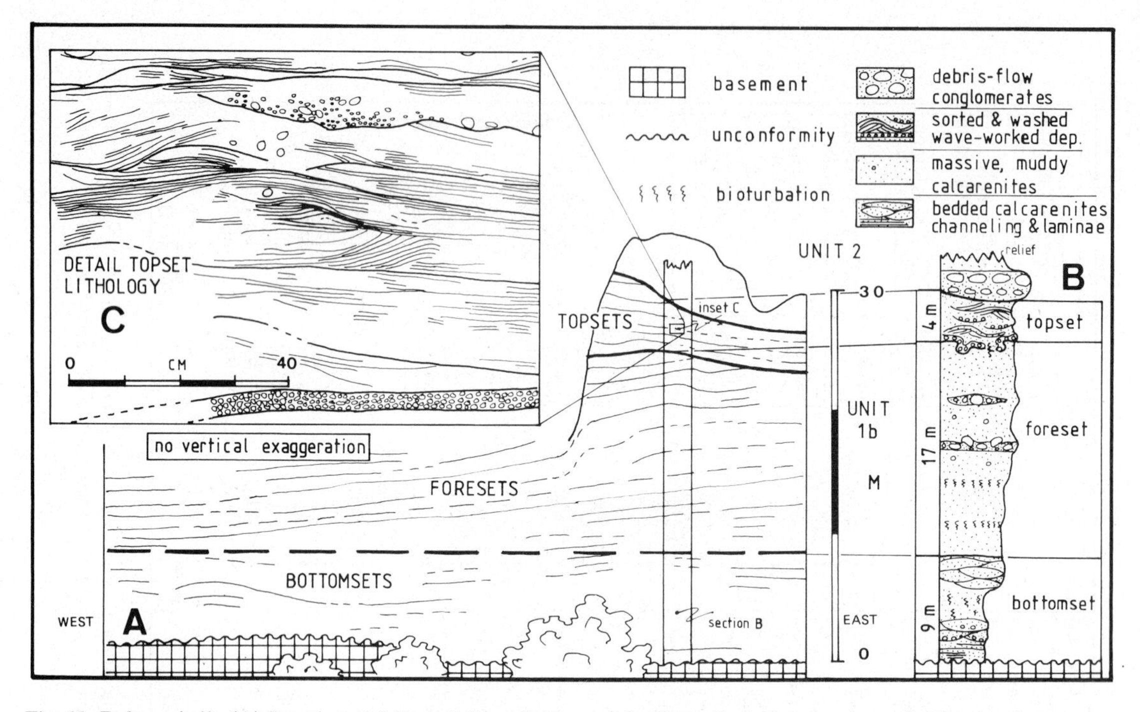

Fig. 13. Delta-unit 1b. (A) Drawing of delta-unit 1b at the base of the Sierra de Solán, showing a division into bottomsets, foresets and topsets. The undulating of the beds is caused by the receding relief of the outcrop. View towards N. See Fig. 10A for location. (B) Detailed section of the outcrop shown in A, the lower 34 m of the delta complex. (C) Detail of topsets comprising cross-laminated calcarenites and thin tabular layers of fine gravel (Facies III).

lamination, as shown in Fig. 13C, is thought to be generated by wave activity. Sharply bounded and lateral, persistent, tabular gravel beds are interpreted as lag deposits also generated by wave activity; they probably represent storm-related erosional surfaces. Small channels and cross-lamination are formed either by waves or shallow-marine (wave-induced?) currents. Hence, the topsets accumulated in a shallow-marine environment and reflect wave action.

The locally observed, disorganized matrix-supported conglomerates are interpreted as debris-flow deposits. Debris-flow deposits and associated water-escape structures are thought to reflect liquefaction. Sorted, washed sand-sized sediment deposited subhorizontally, do not liquefy easily under 'normal' depositional conditions. The phenomena may be explained by assuming an alluvial fan sloping into standing water (see Nemec *et al.*, 1984). On the other hand, liquefaction may be induced by a seismic shock related to the activity of normal faults that controlled the stacking of the delta units.

Facies IV: conglomerate bodies

Description. Conglomerates are largely restricted to the channel-shaped bodies that cut down deeply into the underlying delta front (Fig. 10; schematic in Fig. 9). Their geometry and E–W orientation are apparent from a channel-shape (concave lower boundary and planar upper boundary) in a transverse NNW–SSE cross-section, and a tabular- to wedge-shape in a longitudinal E–W cross-section. One conglomerate body of 30–35 m thick has been examined in more detail and is shown in Fig. 12 B (body 'b' in Fig. 10). Internally, it shows foresets with a maximum apparent dip of *c.* 20° W that asymptotically pass into beds that dip *c.* 5°. The fill of the depression largely consists of coarse conglomerates, but starts with finer-grained, stratified pebbly calcarenites. Although tractional structures and small-scale truncations are present, the sediments do not record strong erosion. The conglomerates show a distinct downslope-fining and upward-coarsening trend. The prominent overall coarsening-upward trend reflects the upward transition from the finer-grained toe to the coarser-grained upper part of the foresets. A transverse section of another conglomerate body is presented in Fig. 12A (body 'a' in Fig. 10). Similar to the previous body, it shows a coarsening-upward trend. This cross-section suggests tongue-shaped conglomerate beds interbedded in pebbly calcarenites.

The conglomerates are characterized by clast-support, poor sorting, inverse-to-normal grading and thick tabular beds with sharp bases that are non-erosive or slightly erosive. However, matrix support and inverse or normal grading are common. In several graded beds a non-graded, disorganized part is present, whereas others are disorganized throughout. Clasts vary from granule- to boulder-size. The largest clasts are outsized boulders up to 50 cm, which are commonly located on top of the bed and occasionally at the base. Some conglomerates display clasts that protrude above the bed.

Conglomerates are closely associated with stratified pebbly calcarenites, frequently forming bipartite beds. The pebbly calcarenites generally are poorly sorted and show parallel or cross-stratification. Their common 'streaky' appearance is caused by diffuse alternation of granule- and sand-sized layers a few centimetres thick. Locally small and shallow channels are present. Bed thickness (10–30 cm) is defined by either set height or the presence of very thin horizontal layers in which fine gravel is concentrated. The pebbly calcarenites contain conglomerate interbeds of 5–30 cm thick. They are better sorted and show a closer packing than the thick-bedded conglomerates described previously. Clastsize in the pebbly calcarenites and interbedded fine conglomerates ranges from granule- to pebble-size (max. 10 cm). Both conglomerates and stratified pebbly calcarenites frequently show water-escape structures.

Interpretation. Similar conglomerates associated with stratified pebbly sandstones or calcarenites have been interpreted as the deposits of subaqueous mass flows that either represent gravelly high-density turbidity currents or debris flows (e.g. Lowe, 1982; Kleinspehn *et al.*, 1984; Postma *et al.*, 1988). The presence of grading illustrates that the clasts within the flow were free to move relative to each other, suggesting low cohesion or no cohesion at all. However, the general muddy matrix (lime-mud) and poor clast organization and the occurrence of matrix support or clasts projected above the bed, do indicate a considerable viscosity. The close association with conglomerates and signs of traction and scouring, indicate that the stratified pebbly calcarenites were deposited by high-density turbidity currents. So, the conglomeratic fill is interpreted as the deposit of debris flows that partly evolved into high-density turbidity currents.

Channels or chutes dissecting the delta front are common. They may represent scours eroded by stream or mass flows or may result from sediment failure (e.g. Prior & Coleman 1980; Postma, 1984; Massari & Colella, 1988; Postma & Cruickshank, 1988; Prior & Bornhold, 1988). Given the consistent deepening trend in the delta complex and the extensional regime expressed in basin deformation, erosion resulting from a sea-level lowstand is unlikely. Moreover, the presence of marine topsets and the general lack of subaerial deposits seem to exclude any significant sea-level fall. The depression may represent a submarine channel or chute scoured by stream or mass flows. In that case, however, a highly erosive event must have generated channels of more than 30 m deep. If fill and formation are related, a coarse lag deposit and an overall fining-upward trend are expected, reflecting waning capacity to transport material. This is not the case. However, according to Prior and Bornhold (1988) delta fronts may show 'empty' scours that function as passive conduits filled much later, so that formation and fill are not related.

An alternative explanation is that the depression was created by sediment failure. Retrogressive upslope retreat of successive slides may form narrow chutes or channels carrying a mass of slurried sediment (Prior & Coleman, 1980; Allen, 1985; Prior & Bornhold, 1988). Under subaqueous conditions these slides may easily evolve into debris flows, which in turn may induce high-density currents. Postma (1984) and Massari (1984) present case studies of slide scars in ancient fan deltas filled by mass-flow conglomerates showing large-scale foresets that are considered to originate from retrogressive sliding extending up to the beach. The non-erosive fill and the absence of lag deposits suggests that the depression was formed by sliding rather than erosive scouring. Moreover, sliding matches the 'lazy w'-shaped base of one of the conglomerate bodies in a longitudinal section (body 'b' in Figs 10B, C), a shape that is expected to be formed by successive slides (cf. Allen, 1985). The lower-order coarsening upward sequences (Fig. 12) may reflect successive pulses of sliding evolving downslope into sluggish debris flows that were preceded by finer and smaller mass flows that moved faster.

Retrogressive sliding seems a probable explanation for the conglomerate bodies. The formation of a slide scar generated slope instability and induced another slide higher on the delta front. This secondary slide triggered, in turn, another slide. Thus sliding started a chain reaction of sediment failure extending in upslope direction, whereby finally coastal conglomerates became involved. The coastal sediments transformed during their downslope transport from slide masses into debris flows and high-density turbidity currents that followed and finally filled the depression created by earlier slides. Coarse debris was trapped in the depression, whereas the finer sediment largely moved further basinwards. Large-scale foresets reveal that successive debris flows filled the depression, which may reflect repeated sliding in the coastal realm.

Overall delta interpretation and summary

A threefold division in topsets, foresets and bottomsets may characterize not only fan deltas and river deltas, but also prograding reefs or carbonate shelf-and-slope systems. Carbonate slopes (cf. foreset) are fed by intrabasinal material formed on the shelf itself, whereas delta foresets are fed by extrabasinal detritus carried into the sea by subaerial distributaries. The calcarenites that largely built up the prograding sediment bodies in the Abarán Basin, are lithoclast packstones. The far largest part of the grains (up to 80%) represents carbonate detritus derived from a source area formed by the Sierra de la Pila and Sierra del Lugar (Fig. 1). Müller (1986) reports coeval fluvial and deltaic deposits in this area, due west of the Sierra del Lugar. So, the sediment bodies were obviously fed by a subaerial source outside the basin and therefore can be considered as river delta or fan delta. The threefold division makes it a Gilbert-type delta (cf. Elliot, 1978; Postma & Roep, 1985).

Since only submarine deltaic deposits are preserved and consequently little information about the subaerial part of the delta is available, the delta may have been fed by rivers, alluvial fans or other sediment sources. The structural setting of the basin, however, makes alluvial fans a probable source. The scarceness of stable constituents and the dominance of proximal and immature limestone debris indicate limited transport. The frequent occurrence of bored, well-rounded lime clasts, some encrusted by barnacles, document that a considerable part of the clasts stayed on the beach for some time.

A schematic model for the deltas is shown in Fig. 9. A load of carbonate detritus was dumped on a subhorizontal, shallow-marine terrace (topsets) which was washed and modified by waves (and currents?). The platform may originally have been submerged (cf. Prior & Bornhold, 1988) or may be flooded post-deposition (cf. Colella, 1988; Postma & Cruickshank, 1988). In the model, an originally submerged terrace is assumed, because there is no clear evidence for fluvial deposition. Mud and silt- to sand-sized material was swept from the terrace in suspension or bedload, and accumulated on the delta front, where it was transported further downslope by mass-flow processes. Strong bioturbation of the foresets suggests that relatively 'quiet' depositional conditions prevailed. During periods of high energy (high fluvial discharge?), pebble-sized and finer sediment was carried across the terrace, whereby coarser material remained on the beach. The delta front was at times too steep which led to slope failure. This generated high-density turbidity currents that accumulated at the toe of the foresets forming more gently inclined bottomsets (cf. Postma & Roep, 1985). Recurrently, channel-shaped depressions were formed cutting the delta front that were subsequently filled by beach-derived, gravelly debris flows. Retrogressive sliding is thought to represent a plausible process to create the depressions and to generate the debris flows.

Succession of delta units (see Fig. 10)

Successive delta units are stacked in an imbricate way, i.e. younger units partly sit upon older units and partly overlie basement (Figs 8, 10). In the delta-complex three major delta units are recognized, which are numbered in stratigraphic order. Units 1, 2 and 3 are *c.* 30, 100 and 150 m thick, respectively. For delta-units 1a, 1b and 2, their thickness is the vertical distance between the highest topsets and the point where the foresets rest upon basement. In unit 3 no topset facies has been encountered. Here unit thickness represents the vertical distance between the algal reefs capping the unit and where the reef-talus reaches the basement. Since the topset facies and algal reefs record very shallow-marine conditions, the thickness of a given unit directly reflects water depth. Based on lithological differences, the oldest delta unit is subdivided into 1a and 1b. Delta-unit 3 is internally intersected by several truncation surfaces, revealing that the unit is more complex than sketched in this description.

Delta-unit 1a

At the northern side of the Sierra de Solán, at the base of the delta complex, a delta unit of *c.* 30 m thick is recognized (Fig. 10 C). Foresets dip approximately to the west and are characterized by floating coarser components consisting of well-rounded or stretched, small gypsum pebbles, oysters and scallops, gastropods and coarse organic debris. At a height of approximately 25 m, the foresets are truncated and overlain by a 5 m thick package of subhorizontal topset beds, characterized by paleosols, clay interbeds, gypsum horizons and gastropod layers.

Delta-unit 1b

At the southern side of the Sierra de Solán, a basal delta unit of *c.* 30 m thick is observed that does not include fore-mentioned components (Figs 10A, 13). Here, subhorizontal bottomsets (*c.* 9 m thick) are characterized by sharply defined calcarenite beds with a crude lamination. Bottomsets are overlain by inclined foresets (*c.* 17 m thick) comprising massive fine-grained (pebbly) calcarenites that show a high lime-mud content and intense bioturbation. Foresets are truncated by subhorizontal topsets (*c.* 4 m thick) that consist of well-sorted and washed laminated calcarenites and thin, tabular layers of fine gravel (see Fig. 13).

Delta-unit 2

The topsets of delta-unit 1a are truncated and overlain by gently inclined bottomsets consisting of an alternation of well-defined, calcarenite beds (20–30 cm) and marlstones. Topsets of unit 1b are truncated by conglomeratic deposits (*Facies IV*?) that are overlain by massive calcarenites (*Facies I*?), both interpreted as sediments deposited at the base of delta unit 2 (Fig. 13). The foresets dip roughly SW and are locally cut by channel-shaped conglomerate bodies (Figs 10B, C). At the highest part of the delta unit, marked by a prominent step in the topography, the foresets are truncated and overlain by subhorizontal topsets (Figs 10, 11) that show signs of liquefaction. The maximum thickness of the delta unit has been estimated to range between *c.* 95 m and 110 m.

Delta-unit 3

Delta-unit 3 shows a height of approximately 150–175 m and foresets dipping westwards (Fig. 10). No topset facies has been encountered. Instead of topset facies, the unit is capped by oyster and scallop beds and algal and bryozoal reefs, interpreted as a shallow-marine deposit also. The boundstones exposed at the top of the Sierra de Solán, pass downslope into less massive reef-talus deposits (Fig. 10A). The occurrence of these reefs on top of the stacked delta units suggests a break in delta progradation.

SETTING OF THE DELTA COMPLEX IN THE PULL-APART BASIN

Sedimentation

At present, the Abarán delta complex is situated about 12 km west of its sediment source area. During sedimentation the basin was enclosed by the relief of the Sierra de la Pila, Sierra del Lugar and Sierra del Oro (see Fig. 7B). The basin continued to move post-deposition due to prolonged right-slip along the 'Linea Electrica fault' (Fig. 1). Arguments for the displacement of the basin are offset of Tortonian deposits and lithological mismatches. Coexisting deposits west of the Abarán Basin are cut and dragged over at least 1 km by right-lateral movement along the fault zone. Some of these deposits are incorporated in the fault zone (Fig. 2). Tortonian deposits north and south of the fault differ in composition and facies, and both document an emerged source area directly across the narrow fault zone. A likely explanation for this configuration is that the deposits north and south of the fault originally were located adjacent to extensive palaeohighs and subsequently were juxtaposed to each other.

The displacement is also reflected in the present-day position of the Gilbert-type delta complex. Gilbert-type deltas are expected adjacent to a steep coastline and an emerged relief that forms the source area of the delta. In our case, the expected source area due east, as indicated by the general westward sediment transport, is occupied by large areas of Triassic gypsiferous deposits that neither contain suitable source rock nor are likely to have formed a prominent topography. Total erosion of formerly existing relief seems unlikely, since all detritus giving topographic reliefs at that time are still in existence. So, in the present configuration the delta complex is deprived of its source area and occurs as an isolated sediment body far removed from the basement high that fed the deltas (Sierra de la Pila; Fig. 7B). Considering the thin-skinned deformation of the Subbetic Zone whereby the Triassic acted as décollement (Baena Perez & Jerez Mir, 1982), the fault-bounded outcrop of Triassic rocks (Figs 1, 7) evokes the impression that the area was stripped bare of all post-Triassic rock. This impression matches the lateral displacement of the Abarán Basin and suggests that the entire basin, as part of a larger detached block, slid over the plastic Triassic levels towards the west, leaving an angular gap (Fig. 7). The detached displacement tore apart and separated the basin-fill from its source area.

The original position of the Abarán Basin can be reconstructed by placing the basin sediments adjacent to basement highs that match clast composition. If the Abarán Basin is shifted to its original position, the basin is placed about 12 km towards the east, due south of the Sierra de la Pila. The lithologies of the Sierra de la Pila and Sierra del Lugar match the clast composition of the conglomerates of the Gilbert-type delta and of the reef- and fault-talus deposits (lateral fill) along the northern margin of the basin. In addition, the reef-talus deposits are placed adjacent to similar, coeval deposits. In its original position, the Gilbert-type delta is situated next to the palaeohigh of the Sierra del Lugar. Here, early Tortonian fluvial and deltaic deposits ('Delta von Rellano' of Müller, 1986) probably formed the proximal counterpart of the Abarán delta complex.

The reconstructed palaeogeography of the Abarán Basin are favourable for the formation of Gilbert-type deltas. Sedimentation took place in an active pull-apart basin within a shallow-marine, archipelago environment. The opening of the basin (Fig. 7) created a depositional depression enclosed by emerging structural highs. The margins of the basin were probably steep and controlled by normal- and oblique-slip faults active during sedimentation. The confinement of the basin by the surrounding palaeohighs protected the deltas against high-energy intrabasinal processes and the steep submarine bathymetry allowed the progradation of the large delta foresets. In addition, the emerging palaeohighs and the normal fault activity guaranteed a steady supply of coarse carbonate detritus feeding the deltas.

The sedimentation pattern in the Abarán Basin (Table 1) conforms to that of many other strike-slip basins. Examples are presented in Ballance and Reading (1980) and Biddle and Christie-Blick (1985). They reveal a major, generally sand-sized axial system that parallels the strike-slip faults. It enters the basin at one of its short sides controlled by normal faults. Along the strike-slip faults forming the long basin margins, a series of small-sized, gravelly fault-talus systems laterally fill the basin. This sedimentation pattern may find its origin in the topographic expression of the faults (see Steel & Gloppen, 1980) and may additionally be promoted by the fault sense. Offset along the strike-slip margins represses the development of large systems, because sediment sources move relative to depocentres. Consequently, sediment input is distributed along the entire length of the strike-slip fault margin as a series of small-sized sediment build-ups. In the case of the axial system, the sediment source faces the same deposition area throughout basin history.

Normal faults and imbricate stacking

The delta complex shows normal faults that affect older deltas and are overstepped by younger ones (Figs 8, 10), so that the faults must have been active during sedimentation. The normal faults trend NNW to NNE, reflecting an E–W extensional regime in the basin. They are interpreted as antithetic faults of an inferred major fault that bounds the basin in the east separating it from its source area (Fig. 7). The direction of extension is parallel to the E–W right-slip faults and is coupled to the opening of the releasing overstep (Fig. 7). Since the normal faults trend almost perpendicular to the E–W right-slip faults, the strike-slip component of the normal faults is not substantial.

The NNW-trending normal faults form the most southern extension of faults along the east margin that splay off from the major E–W strike-slip fault in the north (Fig. 6). The splays show a change in orientation from WNW in the north to NNW in the south. The orientation of the stress fields at the northern, northeastern and eastern margin of the Abarán Basin has been inferred from slip data (method of Michael, 1984). Since the results correspond with the general basin structure, distortion of the stress field due to rotation is thought to be insignificant. The stress field documents a gradual transition from right-oblique slip along the E–W fault zone to normal slip along the NNW faults intersecting the delta sequence (Table 2). The orien-

Table 2. Palaeostress around the Abarán Basin. In the area in and around the Abarán Basin 190 striated fault planes have been analysed. To establish the palaeostress a computer program is used which is based on a method of Michael (1984). First the stressfield orientation is calculated from all data in a given area. Data that strongly deviate from the calculated stress are aborted and a new stress is established using the remaining data. The process is repeated with decreasing tolerance of the angular deviation until a group of slip data is left that fit the calculated stress within a specific field of certainty. Angular deviation is the angle between the measured and the theoretic striations based on the calculated stress ellipsoid.

	Area 1[†]		Area 2	Area 3	Area 4
	E–W fault zone 8–12 km WNW of Abarán (see Fig. 2)		E–W faults along north margin	(W) NW faults along NE margin	NNW faults at E margin *Gilbert deltas*
$\sigma1$*	331/26	100/38	268/35	295/53	162/73
$\sigma2$*	165/64	233/41	84/55	126/37	338/17
$\sigma3$*	64/5	348/26	177/2	32/5	69/1
Mean angular deviation ± standard dev.	8.5° ± 6.3°	7.6° ± 5.6°	8.4° ± 4.3°	9.5° ± 4.8°	9.2° ± 5.5°
Number used slip for data	19	10	14	33	14
Total number data in area	35	16	34	82	39

* $\sigma1$, $\sigma2$, $\sigma3$: principal axes of maximum, intermediate and minimum compressional stress, respectively.
[†] For area 1 two stress fields are calculated; the right-hand column is calculated from the original data set minus the slip data used for the first calculated stress ($N = 19$)

tation of the principle stresses corresponds with the stress trajectories at a releasing overstep described by Gamond (1983) and Guiraud & Seguret (1985).

Figures 10 and 14A show the nature of the normal faults. They dip away from the basin and progress basinwards. Major offset along the normal faults is followed by the development of a new delta unit that partly overlies basement and partly sits on top of the downfaulted and slightly tilted older unit. The new unit oversteps the normal fault and shows a larger thickness, reflecting deeper water. Successive delta units show a thickness that increases from *c.* 30 m, through 100 m, to more than 150 m. The inferred large-scale slump scar, separating the new unit from the older one, is probably related to the faulting event. The close relation between the normal faults and the increasingly thicker delta units suggest that the deepening was largely tectonically controlled, in spite of the contribution of the Tortonian transgression to water depth. Figure 10 reveals that the slide scars cut through all previously deposited delta sediments, but die against the 'bottom' of the basin. This feature suggests that the slide surface laterally merges into the basal unconformity.

Figure 14B presents a speculative model of the relation between the normal faults and the stacked delta units. During an episode of E–W opening of the pull-apart basin, older delta units moved downward and away from the basin centre and were slightly tilted (*c.* 3°) due to normal faulting. Faulting triggered a major slide that truncated and largely removed the old delta. The downfaulted part was saved from destruction, because of its depressed position (below the slide scar). Blocking up by the footwall may additionally have helped to prevent this part sliding basinwards. Concurrently, the extensional regime caused the entire basin to subside. As a consequence, younger delta units accumulated in deeper water partly upon and partly next to the older units. Due to the basinward progradation of the normal faults, this repeated process resulted in a series of overlapping delta units.

Normal faults have been inferred in other strike-slip basins to explain overlapping sedimentary units (e.g. Crowell, 1974; Steel & Gloppen, 1980; Guiraud & Seguret, 1985). They conform the normal faults in the Abarán Basin in their progradation and their control on overlapping units (Fig. 14B). In contrast, they dip basinward and prograde away from the depositional centre (Fig. 14C). In the case of the Abarán Basin, the normal faults and the related stacked delta units, illustrate that the way of faulting

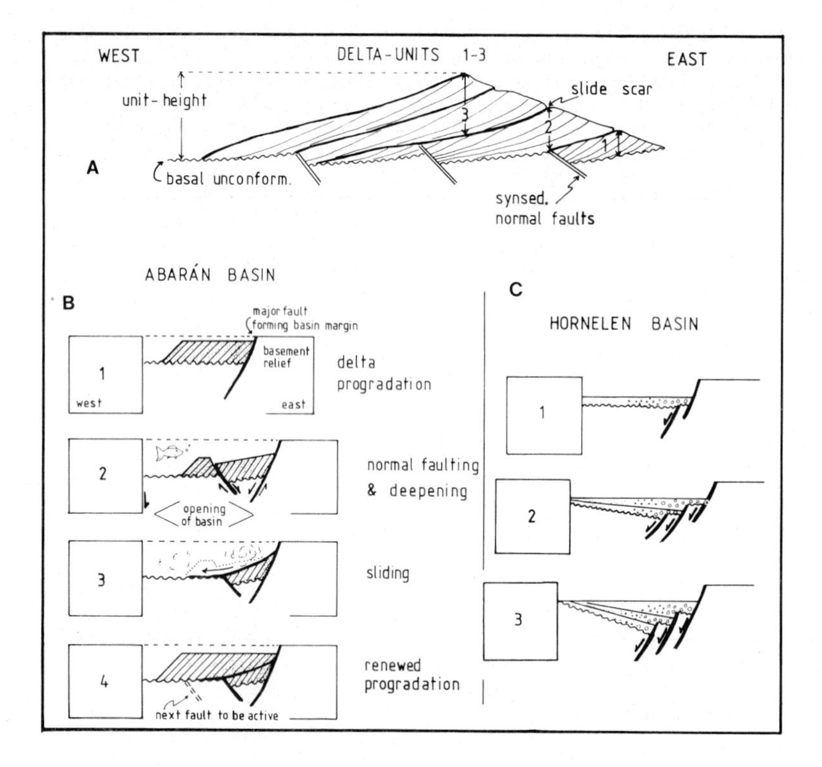

Fig. 14. Normal faulting and related stacking of the delta units. (A) Interpretation of sequence of stacked delta units shown in Fig. 10A. The synsedimentary normal faults dip away from the basin and progressively migrate basinwards. Slide scars, forming the delta-unit boundaries, truncate faults and older delta units. Younger units downlap the slide-scar surface and show a greater thickness than the down-faulted older delta, reflecting deeper water. (B) Cartoon illustrating how normal faulting generates the imbricate stacking of the delta units and the increasing height of the younger deltas. (C) Cartoon based on Steel and Gloppen (1980; their Fig. 9E) showing the inferred relation between normal faults and the overlapping sedimentary units in the Hornelen Basin.

is responsible for an anomalously large cumulative thickness and supports the structural interpretation of the western part of the basin (Figs 5, 7C). The inferred normal faults in the western part of the basin and those in the delta complex lead to a relative elevation in the centre of the basin bordered at both sides by graben-like depressions (Figs 5, 7C). The situation resembles the depressions at both sides of a pull-apart basin predicted by the theoretical model of Rodgers (1980).

CONCLUSIONS

The Abarán pull-apart basin is largely filled by major sand-sized axial systems and subordinately fed by lateral talus systems that formed a series of small-sized conglomerate bodies that rim the basin. The Gilbert-type delta complex represents the axial fill in an early stage of basin development. The deltas are stacked in an imbricate way and range in height from *c.* 30 m to *c.* 150 m at the base and the top of the delta complex, respectively. The increasing thickness of the delta units reflects an episodic deepening of the basin. The truncational surfaces that separate the deltas, are interpreted as large-scale slide scars related to the activity of synsedimentary normal faults. Slide scars generally truncate the normal faults and the new delta on top of the slide scar reflects considerably deeper water. The normal faults and the deepening trend are considered to express extension due to the opening of the pull-apart basin. The normal faults migrate basinward and dip away from the basin centre. Consequently, older deltas moved downward and away from the basin centre, so that younger deltas could accumulate partly upon and partly next to the older ones. Thus normal faulting controlled the imbricate stacking of the delta units and created overlapping sedimentary units.

ACKNOWLEDGEMENTS

The study is part of a PhD research project and has benefited from discussions with Tini Geel, Warner ten Kate, Thom Roep, Harm Rondeel and Wolfgang Schlager. The reconstruction of the source area of the carbonate detritus could be established thanks to the contributions of Tini Geel. Slip data have been analysed using a computer program by Peter Heyke and Fred Smid. Determination of microfauna was done by Jan Manuputty. The manuscript has been improved by readings of Thom Roep, Harm Rondeel, Wolfgang Schlager, Rob Gawthorpe and Cristino J. Dabrio. I thank them all most kindly.

REFERENCES

ALLEN, J.R.L. (1985) *Principles of Physical Sedimentology*. Allen and Unwin, London, 272 pp.

BAENA PEREZ, J. & JEREZ M.I.R., L. (1982) *Sintesis para un ensayo paleogeografico entre la Meseta y la Zona Betica*. Instituto Geologico y Minero de España, colleccion – informe, 256 pp.

BALLANCE, P.F. & READING, H.G. (Eds.) (1980) *Sedimentation in Oblique-slip mobile Zones*. Spec. Publ. Int. Assoc. Sedim. 4, 265 pp.

BENAVENTE HERRERA, J. & SANZ DE GALDEANO, C. (1985) Relacion de las direcciones de karstificacion y del termalismo con la fracturacion en las Cordilleras Beticas. *Estudios geol.* **41**, 177–188.

BIDDLE, K.T. & CHRISTIE-BLICK, N. (Eds.) (1985) *Strike-slip Deformation, Basin Formation, and Sedimentation*. Spec. Publ. Soc. econ. Paleont. Mineral., Tulsa 37, 386 pp.

CHRISTIE-BLICK, N. & BIDDLE, K.T. (1985). Deformation and basin formation along strike-slip faults. In: *Strike-slip Deformation, Basin Formation, and Sedimentation* (Ed. by K.T. Biddle and N. Christie-Blick). Spec. Publ. Soc. econ. Paleont. Mineral., Tulsa 37, pp. 1–34.

COLELLA, A. (1988) Pliocene–Holocene fan deltas and braid deltas in the Crati Basin, southern Italy: a consequence of varying tectonic conditions. In: *Fan Deltas: Sedimentology and Tectonic Settings* (Ed. by W. Nemec and R.J. Steel), pp. 186–196. Blackie and Son, London.

CROWELL, J.C. (1974) Sedimentation along the San Andreas Fault, California. In: *Modern and Ancient Geosynclinal Sedimentation* (Ed. by R.H. Dott and R.H. Shaver). Spec. Publ. Soc. econ. Paleont. Mineral., Tulsa 19, pp. 292–303.

ELLIOTT, T. (1978) Deltas. In: *Sedimentary Environments and Facies* (Ed. by H.G. Reading), pp. 97–142. Blackwell Scientific Publications, Oxford.

GAMOND, J.F. (1983) Displacement features associated with fault zones: a comparison between observed examples and experimental models. *J. struct. Geol.*, **5**, 33–45.

GUIRAUD, M. & SEGURET, M. (1985) A releasing solitary overstep model for the late Jurassic–early Cretaceous (Wealden) Soria Strike-slip Basin (northern Spain). In: *Strike-slip Deformation, Basin Formation, and Sedimentation* (Ed. by K.T. Biddle and N. Christie-Blick). Spec. Publ. Soc. econ. Paleont. Mineral., Tulsa 37, pp. 159–175.

HAQ, B.U., HARDENBOL, J. & VAIL, P.R. (1987). Chronology of fluctuating sea levels since the Triassic. *Science* **235**, 1156–1167.

IGME (1974) *Mapa Geologica de España, 1:50 000, Hoja Cieza (no. 891)*. Serv. Publ. Min. Industria, Madrid.

JEREZ MIR, F. (1979) Contribucion a una nueva sintesis de las Cordelleras Beticas. *Bol. Inst. geol. y min. España* **40**, 503–555.

KLEINSPEHN, K.L., STEEL, R.J., JOHANNESSEN, E. & NETLAND, A. (1984) Conglomeratic fan-delta sequences. Late Carboniferous–Early Permian, western Spitsbergen. In: *Sedimentology of Gravels and Conglomerates* (Ed. by E.H. Koster and R.J. Steel). Mem. Can. Soc. Petrol. Geol. 10, pp. 279–294.

KLEVERLAAN, K. (1987) Gordo Mega Bed: a possible seismite in a Tortonian submarine fan, Taberas Basin, province Almeria, southeast Spain. *Sedim. Geol.* **51**, 165–180.

LOWE, D.R. (1982) Sediment gravity flows: II. Depositional models with special reference to the deposits of high-density turbidity currents. *J. sedim. Petrol.* **52**, 279–297.

MASSARI, F. (1984) Resedimented conglomerates of a Miocene fan-delta complex, southern Alps, Italy. In: *Sedimentology of Gravels and Conglomerates* (Ed. by E.H. Koster and R.J. Steel). Mem. Can. Soc. Petrol. Geol. 10, pp. 259–277.

MASSARI, F. & COLELLA, A. (1988) Evolution and types of fan-delta systems in major tectonic settings. In: *Fan Deltas: Sedimentology and Tectonic Settings* (Ed. by W. Nemec and R.J. Steel), pp. 103–122, Blackie and Son, London.

MCPHERSON, J.G., SHANMUGAN, G. & MOIOLA, R.J. (1987). Fan deltas and braid deltas: varieties of coarse-grained deltas. *Bull. geol. Soc. Am.* **99**, 331–340.

MICHAEL, A.J. (1984) Determination of stress from slip data: faults and folds. *J. geophys. Res.* **89**, 11517–11526.

MIDDLETON, G.V. & HAMPTON, M.A. (1976) Subaqueous sediment transport and deposition by sediment gravity flows. In: *Marine Sediment Transport and Environmental Management* (Ed. by D.J. Stanley and D.J.P. Swift), pp. 197–218. Wiley, New York.

MONTENAT, C. (1973). *Les formations néogénes et quarternaires du Levant espagnol (Provinces d'Alicante et du Murcie)*. PhD thesis, University Orsay, Paris, pp. 1170.

MONTENAT, C., OTT D'ESTEVOU, P. & MASSE, P. (1987) Tectonic–sedimentary characters of the Betic Neogene Basins evolving in a crustal transcurrent shear zone (SE Spain). *Bull. Centres Rech. Explor. Prod. Elf-Aquitaine* **11**, 1–22.

MÜLLER, D.W. (1986) *Die Salinitätkrise im Messinian (spätes Miozän) der bekken von Fortuna und Sorbas (Südost Spanien)*. PhD thesis (NR 8056), Zurich, Schweiz.

NEMEC, W. & STEEL, R.J. (1984) Alluvial and coastal conglomerates: their significant features and some comments on gravelly mass-flow deposits. In: *Sedimentology of Gravels and Conglomerates* (Ed. by E.H. Koster and R.J. Steel). Mem. Can. Soc. Petrol. Geol. 10, pp. 1–31.

NEMEC, W., STEEL, R.J., PORĘBSKI, S.J. & SPINNANGR, A. (1984) Domba Conglomerate, Devonian, Norway: process and lateral variability in a mass flow-dominated, lacustrine fan-delta. In: *Sedimentology of Gravels and Conglomerates* (Ed. by E.H. Koster and R.J. Steel). Mem. Can. Soc. Petrol. Geol 10, pp. 295–320.

POSTMA, G. (1984). Mass-flow conglomerates in submarine canyon: Abrioja Fan-Delta, Pliocene, southeast Spain. In: *Sedimentology of Gravels and Conglomerates* (Ed.

by E.H. Koster and R.J. Steel). Mem. Can. Soc. Petrol. Geol. 10, pp. 237–258.

Postma, G. & Cruickshank, C. (1988) Sedimentology of a late Weichselian to Holocene terraced fan delta, Varangerfjord, northern Norway. In: *Fan Deltas: Sedimentology and Tectonic Settings* (Ed. by W. Nemec and R.J. Steel), pp. 144–157. Blackie and Son, London.

Postma, G., Nemec, W. & Kleinspehn, K.L. (1988) Large floating clasts in turbidites: a mechanism for their displacement. *Sedim. Geol.* **58**, 47–61.

Postma, G. & Roep, T.B. (1985) Resedimented conglomerates in the bottomsets of Gilbert-type gravel deltas. *J. sedim. Petrol.* **55**, 874–885.

Prior, D.B. & Bornhold, B.D. (1988) Submarine morphology and processes of fjord deltas and related high-gradient systems: modern examples from British Columbia. In: *Fan Deltas: Sedimentology and Tectonic Settings* (Ed. by W. Nemec and R.J. Steel), pp. 125–143. Blackie and Son, London.

Prior, D.B. & Coleman, J.M. (1980) Sonograph mosaics of submarine slope instabilities, Mississippi River Delta. *Mar. Geol.* **36**, 227–239.

Riba, O. (1976) Syntectonic unconformities of the Alto Cardener, Spanish Pyrenees: a genetic interpretation. *Sedim. Geol.* **15**, 213–233.

Rodgers, D.A. (1980) Analysis of pull-apart basin development produced by en echelon strike-slip faults. In: *Sedimentation in Oblique-slip Mobile Zones* (Ed. by P.F. Ballance and H.G. Reading). Spec. Publs. int. Assoc. Sedim. 4, pp. 27–41.

Steel, R.J. & Gloppen, T.G. (1980) Late Caledonian (Devonian) basin formation, western Norway: signs of strike-slip tectonics during infilling. In: *Sedimentation in Oblique-slip Mobile Zones* (Ed. by P.F. Ballance and H.G. Reading). Spec. Publs. int. Assoc. Sedim. 4, pp. 79–103.

Wilcox, R.E., Harding, T.P. & Seely, D.R. (1973) Basic wrench tectonics. *Bull. Am. Assoc. Petrol. Geol.* **57**, 74–96.

Woodcock, N.H. & Fisher, M. (1986) Strike-slip duplexes. *J. Struct. Geol.* **8**, 725–735.

Spec. Publs int. Ass. Sediment. (1990) **10**, 223–233

Transverse and longitudinal Gilbert-type deltas, Tertiary Coalmont Formation, North Park Basin, Colorado, USA

R. M. FLORES

US Geological Survey, Denver, Colorado 80225, USA

ABSTRACT

The transpressional, intermontane North Park Basin, Colorado, provides excellent examples of sand-rich Gilbert-type deltas that prograded into freshwater lacustrine environments. Gilbert-type deltaic sandstones are present in the coal-bearing Palaeocene and Eocene Coalmont Formation.

Two types of Gilbert-type sandstone facies are recognized based on their site of deposition, sedimentological properties and nature of the facies of contributing fluvial systems. The transverse Gilbert-type deltaic facies is formed along the basin margin and is composed of vertically stacked sandstone with foreset beds that are thin, slightly variable in thickness and steeply dipping. Multilateral, conglomeratic channel sandstones are formed landward of this deltaic sandstone facies. The longitudinal Gilbert-type deltaic facies is developed in the basin centre and consists of overlapped sandstone bodies with foreset beds that are thick, highly variable in thickness, gently dipping, and interbedded with silty interforesets. The longitudinal deltaic sandstone facies passes upslope into syndepositional growth faults and into multistorey, *en echelon* fluvial channel sandstones.

The transverse, Gilbert-type deltaic sandstone facies was deposited in a rapidly subsiding shallow basin fed by bedload of braided streams. The longitudinal, Gilbert-type deltaic sandstone facies was deposited in a slowly subsiding deep basin, fed by a mixed load or meandering streams. Here, slope-producing, syndepositional growth faults controlled accretion and thickness of the foreset beds.

INTRODUCTION

Gilbert's (1890) classic work on Pleistocene Lake Bonneville deposits, in which he described deltaic features of lake shores, has dominated concepts of modern and ancient deltaic models for the past century. A Gilbert-type deltaic deposit in Lake Bonneville consists of tripartite structures of bottomset, foreset and topset beds. The deposits form a coarsening-upward sequence of sandy, tangential to horizontal bottomset beds that grade upward into sandy and gravelly, steeply inclined foreset beds. The foreset beds grade upward into subhorizontal to flat-lying, gravelly topset beds. Deltaic deposits with these characteristics were formed by homopycnal flow of river waters into freshwater lakes, which resulted in rapid mixing and deposition of coarse sediments near river mouths. Although the classic Gilbert-type deltas developed in a freshwater lake, similar deltaic deposits have been reported by Stanley and Surdam (1978) in playa and saline–alkaline lakes and by Colella *et al.* (1987) in a marine environment. Nemec and Steel (1988) provided a comprehensive description of various Gilbert-type deltas in other environmental settings.

In the North Park Basin of north-central Colorado (Fig. 1), Gilbert-type deltaic deposits are found in the Palaeocene and Eocene Coalmont Formation (Flores, 1988). These deltaic deposits are rich in sand and exhibit diagnostic sedimentological features, depending on their site of deposition in the basin as well as the type of fluvial systems that fed the deltas. The purpose of this paper is: (1) to differentiate the sedimentological features of the various Gilbert-type deltaic deposits; (2) to characterize the styles of progradation of the Gilbert-type deltas; and (3) to understand the nature of the contributing fluvial systems.

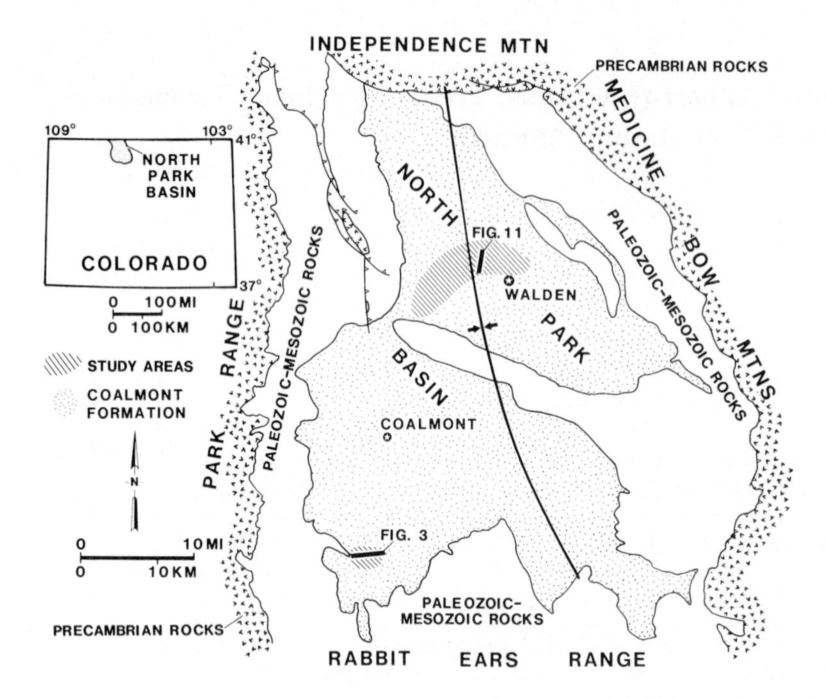

Fig. 1. Geologic map of the North Park Basin showing study areas in the southwest and north-central parts of the basin. Lines of cross-sections in Figs 3, 11.

GEOLOGIC AND STRATIGRAPHIC SETTINGS

The North Park Basin is interpreted by Maughan (1988) as a transpressional basin with several tectonic styles. Maughan suggested that north-central Colorado was subjected to several stages of compression and extension during the late Cretaceous to Eocene Laramide orogeny. Early Laramide compression along a northeast to southwest direction resulted in northwest-trending faults and folds. Late Laramide compression was directed east–west producing north-trending faults and folds. Subsequent Oligocene–Miocene tectonism consisted mainly of north–south compression and complimentary east–west extension, resulting in east-trending thrusts (for example, Independence Mountain thrust along the northern margin of the North Park Basin; Fig. 1) and folds, as well as formation of horst and graben by normal faults along north–south trends.

The North Park Basin began to develop during the Palaeocene with uplift of surrounding basement-cored areas and deposition of the synorogenic Coalmont Formation. The Coalmont Formation is as much as 2500 m thick (Hail, 1965, 1968) and consists of interbedded conglomerate, sandstone, siltstone, mudstone, carbonaceous shale and coal beds. The mudstone is the most common rock type. The formation can be divided into coal-bearing and non-coaly zones that interfinger with each other (Roberts & Flores, 1988a). The coal-bearing zone consists of coal beds as much as 25 m thick (Hendricks, 1977; Madden, 1977; Roberts & Flores, 1988b) and freshwater mollusc (viviparids) fossils are present (Joseph Hartman, oral commun., 1988). The non-coaly zone includes sandstone and siltstone beds and mudstone beds up to 135 m thick that contain freshwater fish scales and skeletal remains (Hail, 1965, 1968).

The study areas for the Gilbert-type deltaic deposits in the Coalmont Formation are located in the southwest margin and north-central part of the North Park Basin. Dip orientation of the foreset beds of the Gilbert-type deltaic deposits in the southwest margin shows an easterly (72–113°) direction, transverse to the length of the basin axis. Foreset beds of the Gilbert-type deltaic deposits in the north-central part exhibit a southerly (125–177°) dip direction or longitudinal to the length of the basin axis. Based on the site of deposition and dip orientations of the foreset beds, the Gilbert-type deltaic deposits can be classified into transverse (basin margin) and longitudinal (basin centre) Gilbert-type deltas. Figure 2 shows that the transverse Gilbert-type deltaic deposits are in the Palaeocene portion and the longitudinal Gilbert-type deltaic deposits are in the Eocene portion of the Coalmont Formation.

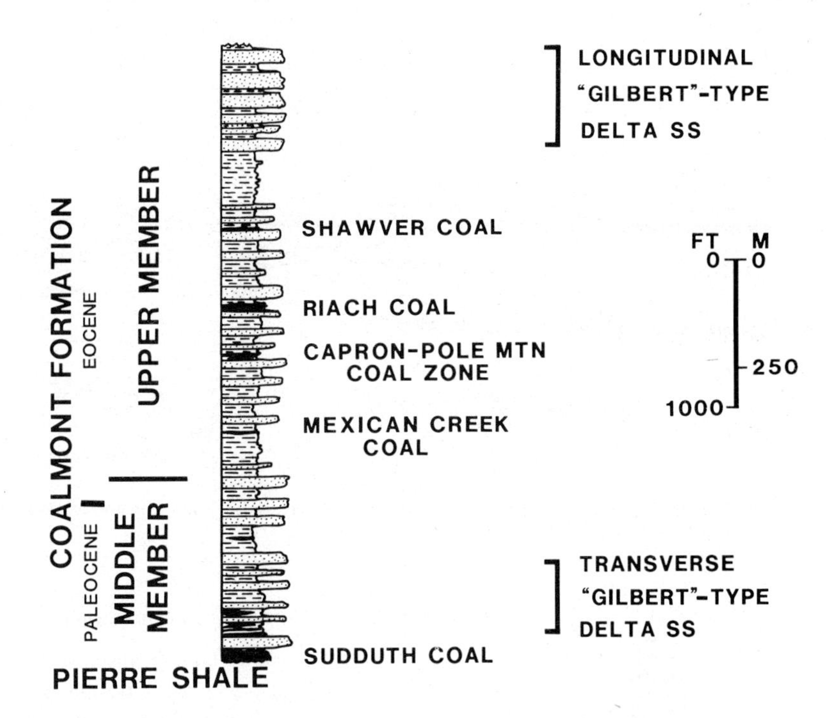

Fig. 2. Composite stratigraphic section indicating positions of the Gilbert-type deltaic deposits in the Coalmont Formation. SS = sandstone.

FACIES FRAMEWORK

Transverse Gilbert-type deltaic facies

The framework of the transverse Gilbert-type deltaic sandstone facies is shown in Fig. 3. Five vertically stacked, Gilbert-type deltaic sandstone facies with bottomset, foreset and topset beds pass eastward into mudstones that contain fish scales. The deltaic sandstones pass westward into multilateral sandstone facies with basal scours and beds of pebble conglomerate. The vertical facies profile (Fig. 4) of the sandstones and associated units displays basal, crudely imbricated conglomerates that grade upward into planar and trough cross-bedded, granule sandstones. This facies profile is repeated several times upwards with commonly developed trough cross-beds (small scale) and the uppermost part of the sandstone capped by a rooted, pebbly sandstone. Bodies of conglomeratic sandstone are interbedded with minor siltstone, mudstone and thin carbonaceous shale and coal (as much as 0.6 m thick) beds (Figs 3, 4).

A vertically stacked, transverse, Gilbert-type deltaic sandstone facies is shown in Fig. 5. Each Gilbert-type deltaic sandstone facies consists of tangential bottomset sandstone beds that pass updip into steeply inclined foreset sandstone beds which, in turn, are overlain by subhorizontal to flat-lying sandstone beds. Bottomset beds consist of laminae and beds as much as 2·5 cm thick of granule- to medium-grained sandstone that are normally graded (Fig. 6B). A graded lamination is commonly overlain by a ripple lamination. Thin laminae of comminuted plant fragments and coaly materials are interbedded with the graded-rippled couplets.

Foreset beds have an average dip of 20° and a total thickness as much as 4 m. Detailed internal structures of the downdip and updip parts of the foreset beds consist of granule and coarse sandstone laminae as much as 5 cm thick and are shown in Fig. 6. The downdip foreset beds are composed of ripple laminae that grade upward into dip-oriented parallel to subparallel laminae (Figs 6A, B) with locally formed rippled and graded laminae. The updip foreset beds, which are thicker than the downdip foreset beds, consist, from bottom to top, of graded, trough cross-bedded (small-scale), dip-oriented, parallel to subparallel laminated and ripple-laminated sandstones. This vertical sequence of internal structures is locally interrupted by large-scale trough cross-beds. Some trace fossils resembling faecal pellets and basally erosional, lenticular, trough cross-bedded sandstones are found in the foreset beds.

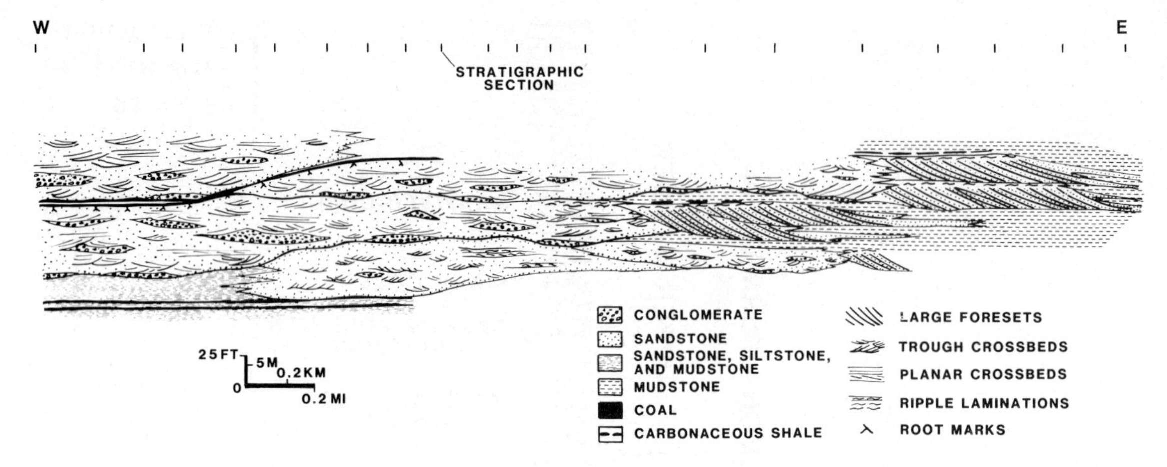

Fig. 3. Facies framework of the transverse, Gilbert-type deltaic deposits and associated sediments in the southwest part of the basin.

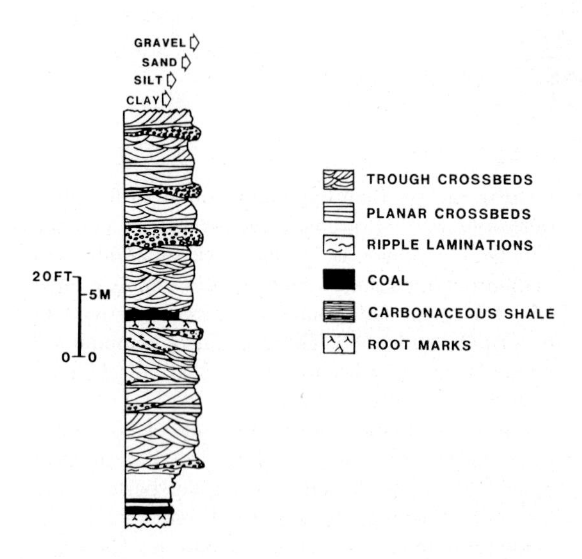

Fig. 4. Vertical facies profile of the conglomeratic sandstone facies and related sediments.

Topset beds are as much as 3 m thick and consist of coarse- to medium-grained sandstones that are subparallel and ripple laminated (Fig. 6). These beds are locally dissected by basally erosional sandstones that are up to 2 m thick with small-scale trough cross-beds in the lower part and ripple laminae in the upper part. The topset beds show common vertical burrows of straight, smooth-walled tubes as much as 15 cm long (Fig. 6). The topset sandstone beds may be locally capped by carbonaceous shales.

Interpretation

The transverse Gilbert-type deltaic sandstone facies was deposited in a freshwater lake. Westward, the Gilbert-type deltaic sandstone facies passes into conglomeratic channel sandstone facies characterized by mixtures of pebble and sand, as well as by rapid, vertical and lateral facies variations, suggesting mainly deposition of bedload during variable turbulent-flow conditions typical of braided streams (Smith, 1970). Variable flow conditions are indicated by the abundance of cut-and-fill structures, grouping of trough and planar cross-beds (bed accretion units of lower and upper flow regimes), and paucity of clay and silt. The braided fluvial system trailed the transverse Gilbert-type deltas along the southwest margin of the basin. The presence of thin coal beds interbedded with the braided-stream deposits suggests that some inactive, abandoned braidbelts well above sediment influx served as platforms for thin peat formation.

Vertical stacking of the transverse, Gilbert-type deltaic sandstone facies probably reflects multiple progradation events in a rapidly subsiding part of the basin. That the basin margin was covered by a shallow lake is indicated by the thickness of as much as 4 m of the foreset beds. Deltaic progradation was controlled by deposition of sands by turbidity currents along the delta face resulting in ripple, parallel, subparallel and graded laminae of the foreset beds. Density currents that flowed along submarine channels along the delta face are indicated by the presence of some channel sandstones in the

Fig. 5. Vertically stacked, transverse, Gilbert-type deltaic sandstone facies, each showing tripartite bottomset (B), foreset (F) and topset (T) beds.

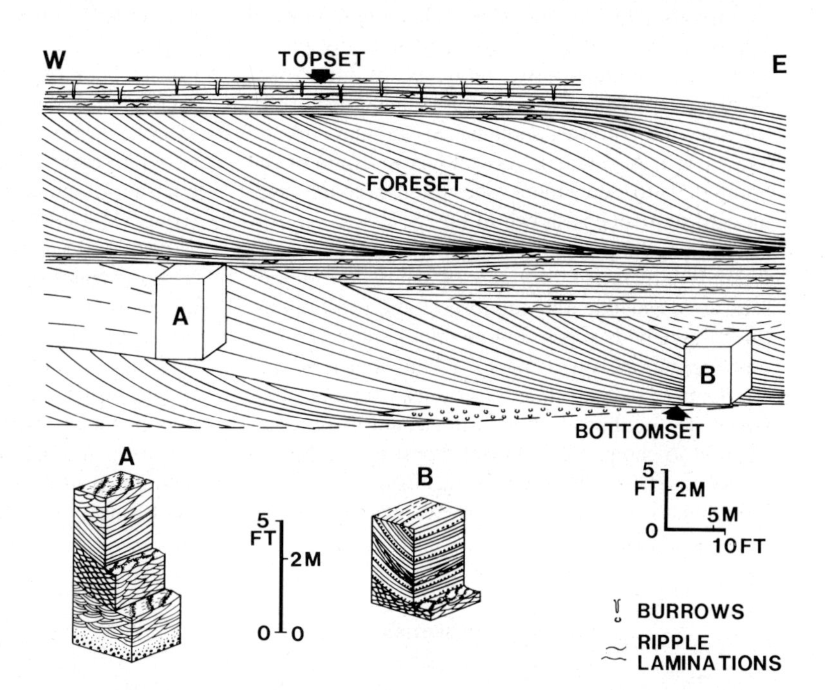

Fig. 6. Diagram of two stacked transverse, Gilbert-type deltaic sandstone facies showing detailed vertical sequence of internal structures in the updip and downdip parts of the foreset beds.

foreset beds. Waning periods of sedimentation on the delta face may have resulted in reworking of sediments by suspension feeders that transported sediments to the surface by their excrements or faecal pellets (Reineck & Singh, 1980). Bottom currents flowed across the delta face and basinward forming graded–rippled couplets or turbidites of the bottomset beds. Fluctuation of the lake level is probably indicated by alternating deposition by wave ripples and traction flows along avalanche slopes of the upper part of the delta face. Shoaling condition of the lake is suggested by development of scour-and-fill structures in the updip foreset beds.

Aggradation by fluvial channels in the delta top is indicated by topset channel sandstones. These channels were probably distributaries of the braided stream that fed the delta. Wave reworking of the topset beds is suggested by wave-ripple laminae. That the topset beds also were reworked by filter-feeding animals is indicated by the vertical burrows. Bioturbation may have followed abandonment of the delta, which also served as a sump for transported organic matter as reflected by the deposition of carbonaceous shale above the topset beds.

Longitudinal Gilbert-type deltaic facies

The framework of the longitudinal, Gilbert-type deltaic sandstone facies is shown in Fig. 7. The deltaic sandstone facies, which displays bottomset, foreset and topset beds, is overlapped, vertically stacked and interbedded with fish-scale-bearing mudstones in the northern part of the study area. In the southern part of the study area, the deltaic sandstone facies is overlain by *en echelon*, multistorey, erosionally based, medium- to fine-grained sandstone that is trough cross-bedded in the lower part, ripple-laminated in the upper part and overlain by subordinate silt and clay. This fining-upward sandstone is interbedded with and laterally passes into a coarsening-upward sequence of mudstone, siltstone and sandstone beds. The sandstone beds of the coarsening-upward sequence are tabular, rippled and rooted. A detailed, vertical facies profile of the fining-upward sandstones and related coarsening-upward sequence is shown in Fig. 8.

The longitudinal, Gilbert-type deltaic sandstone facies is shown in Figs 9 and 10. Here, the deltaic sandstones include only the foreset and topset beds; the foreset beds (up to 0·9 m thick) show variable dip orientations averaging 15° to the south and are overlapped to as much as 19 m thick. Foreset sandstones are medium to fine grained and as much as 11 m in thickness. The downdip part of the foreset beds consists of dip-oriented parallel- to subparallel-laminated, well- to poorly developed, ripple-laminated and normally graded sandstones. The ripple laminae are locally disrupted by remains of trace fossils resembling faecal pellets. The updip part of the foreset beds comprises alternating couplets of ripple-laminated and parallel- to subparallel-laminated sandstones (Fig. 11A). These couplets are normally graded, as shown in Fig. 11A, and are separated by siltstone interforesets that are ripple laminated. Lenses of granule sandstones occur in the parallel and subparallel laminae of the couplets (Fig. 11B). The sequence of internal structures of these couplets is similar to that of the Bouma (1962) turbidite sequence. The foreset beds pass downdip into tangential bottomset sandstone beds that are interbedded with siltstone beds (Fig. 10). The bottomset sandstone beds are fine to medium grained and up to 2 m thick. They are normally graded and ripple cross-laminated, similar to those of the bottomset beds of the transverse Gilbert-type deltaic sandstone facies. The bottomset beds commonly exhibit narrow U-shaped burrows with downward-bent (concave) sediment layers which are up to 4 cm long and 2 cm wide, and broad U-shaped burrows (concave, 4 cm long and 6 cm wide) without traces of disturbed sediment layers (Fig. 11C). In addition, V-shaped burrows (4 cm long and 3 cm wide) showing either downward-bent (chevron shape) sediment layers or absence of disturbed sediments are also present.

The topset beds of the longitudinal Gilbert-type deltaic facies, which are as much as 5 m thick, consist of multistoried, basally erosional, coarse- to medium-grained sandstones (Fig. 7). These topset sandstones are trough cross-bedded and where not

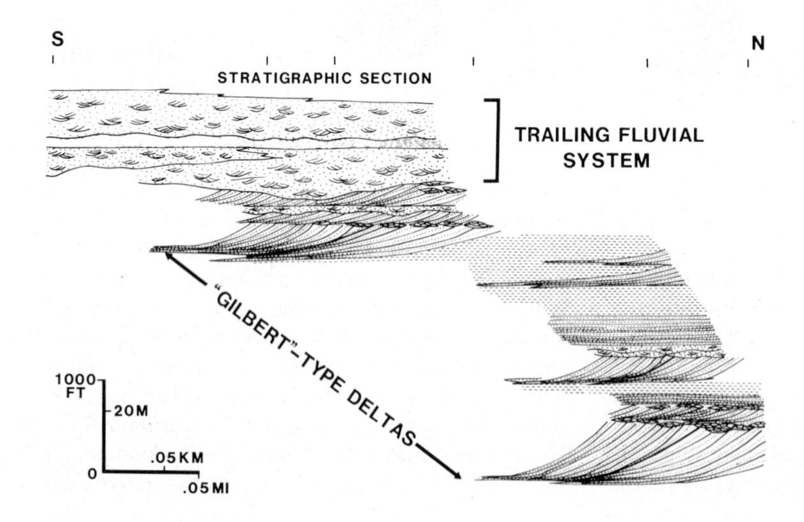

Fig. 7. Facies framework of the longitudinal, Gilbert-type deltaic deposits and associated sediments in the north-central part of the basin.

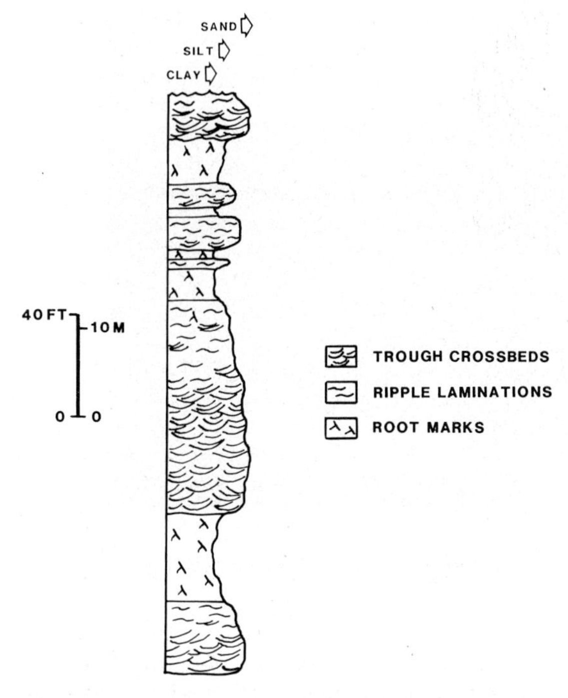

Fig. 8. Vertical facies profile of the fining-upward *en echelon* sandstone and related fine-grained sediments.

incised by these sandstones, the topset beds are composed of ripple-laminated sandstones.

Figure 9 shows a syndepositional fault developed in the proximal portion of the longitudinal, Gilbert-type deltaic sandstone facies. Sediment loading on the downdrop block of the normal fault resulted in contemporaneous movement and overthickening of the foreset beds. The break in slope produced by this syndepositional fault probably caused a difference in relief that initiated deposition of the foreset beds.

Interpretation

The longitudinal, Gilbert-type deltaic sandstone facies, which consists of foreset beds as much as 11 m thick and merges with fish-scale-bearing mudstones, suggests deposition in a deeper freshwater lake. The multistorey, *en echelon* channel sandstone facies that overlies the longitudinal, Gilbert-type deltaic sandstone facies indicates that the deltas were fed by laterally shifting sinuous channels. The fining-upward grain size and subordinate silt and clay in the upper part of the sandstone facies suggest deposition by mixed-load streams. Thus, these characteristics of the sandstone facies reflect deposits of a meandering fluvial system. The coarsening-upward sequence associated with the multistorey channel sandstone facies represents crevasse-splay deposits of floods that breached levees into the floodplain. The absence of coal and carbonaceous shale beds in this fluvial sequence, unlike those in the braided stream deposits that trailed the transverse Gilbert-type deltas, suggests that interchannel areas were probably either well drained and/or flooded by detrital influx, thus, inimical to peat formation. Flooding may also have been exacerbated by lake transgressions. However, study by Roberts and Flores (1988) of upstream deposits indicates peat formation in poorly drained interchannel

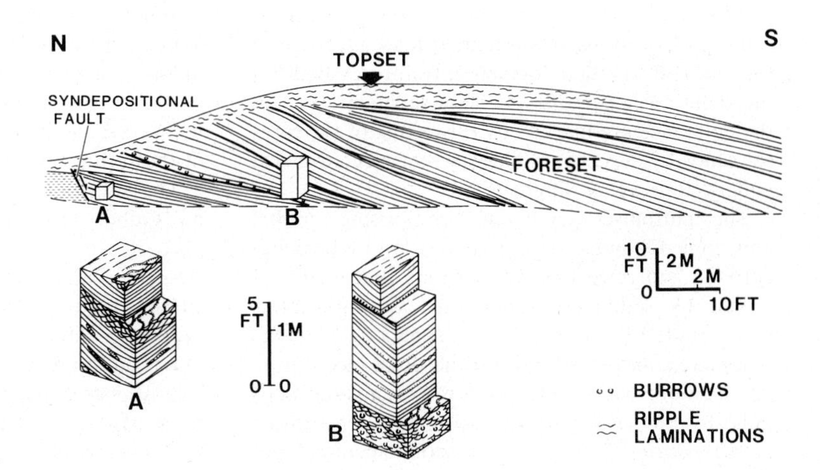

Fig. 9. Diagram of overlapped foreset beds of the longitudinal, Gilbert-type deltaic sandstone facies. Detailed internal structures in the updip and downdip parts of the foreset beds are also illustrated.

Fig. 10. Overlapping of the longitudinal, Gilbert-type deltaic facies indicated by variously dipping foreset (F) beds and underlying bottomset (B) beds.

areas in which as much as 25-m-thick coal formed. They suggest the origin of peat formation in these interchannel areas as raised bogs.

Overlapping of the longitudinal, Gilbert-type deltaic sandstone facies indicated by groups of foresets with different dip directions, reflects progradation of laterally shifting deltaic lobes into a slowly subsiding part of the basin. Progradation is controlled by deposition of bedload by turbidity currents along the delta face or avalanche slope during floods. Accretion of foreset beds on the delta face by turbidity currents is reflected by the graded, rippled and parallel–subparallel laminated couplets. The siltstone interforesets resulted from deposition of suspended load. Thus, rapid deposition of bedload followed by settling of suspended load gave rise to the normally graded foreset–interforeset beds of the delta face. Episodes of slow sedimentation on the delta face are marked by reworking of the foreset sands by suspension feeders that ejected faecal pellets (Reineck & Singh, 1980).

The bottomset sandstones comprising graded and rippled laminae suggest deposition by turbidity currents and reworking by currents that were not driven by sediment suspension. Siltstone interbeds resulted from settling of suspended sediments that were transported beyond the delta face during floods. Bioturbation of the bottomset sandstone beds indicates reworking of the sediments by benthic animals during periods of low sediment discharge. The variety of the morphology of burrows found in the bottomset sandstone beds reflects the diversity of benthic fauna that lived in the distal area of the longitudinal Gilbert-type delta. Some of the benthic fauna may have been suspension feeders that created U-shaped burrows (Seilacher, 1967). Similar U-shaped burrows with downward-bent sediments were identified as escape structures of freshwater pelecypods by Hanley and Flores (1987).

Occurrence of thick, basally erosional sandstones above the foreset sandstone beds indicates that the topset of the longitudinal Gilbert-type deltas consisted of well-developed fluvial channels that supplied bedload sediments to the delta face. These channels were probably distributaries of the mixed-load or meandering fluvial system that trailed the deltas. The distributaries may have fed shifting deltaic lobes that accompanied progradation. Abandonment of the deltaic lobes promoted wave reworking of the topset beds that developed ripple laminae.

The development of foreset beds of the longitudinal Gilbert-type deltas, particularly in the central part of the basin, far removed from the original coastline, may have been enhanced by syndepositional faulting. Normal faults probably produced breaks in slopes near to where foreset beds were deposited on a maximum angle of dip of about 34°. This dip of foreset beds is similar to that of the back edge of the Gilbert-delta foreset units marked by fault scarps in the Pliocene–Holocene Crati

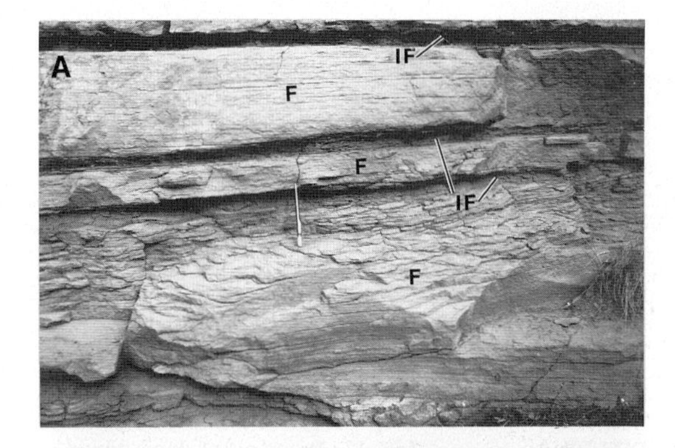

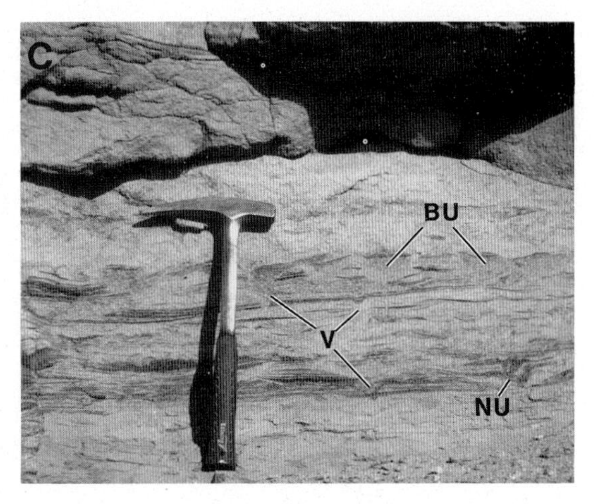

Fig. 11 (A) Vertical sequence of internal structures in the updip part of the longitudinal, Gilbert-type deltaic foreset (F) beds. IF is siltstone interforesets. (B) Granule-sandstone lenses (G) intercalated with parallel and subparallel laminations in the downdip part of the longitudinal, Gilbert-type deltaic foreset beds. Pencil is 13 cm long. (C) V-shaped (V), broad U-shaped (BU) and narrow U-shaped (NU) burrows (escape structures) in the longitudinal, Gilbert-type deltaic bottomset beds. Hammer is 0.3 m long.

Basin, Italy (Colella, 1988). This fault-scarp slope-producing process, which initiated accumulation of foreset beds, may be common in a mud-rich basin.

COMPARISON OF GILBERT-TYPE DELTAIC FACIES MODELS

Comparison of the sedimentological properties of the transverse and longitudinal, Gilbert-type deltaic sandstone facies and deposits of related fluvial systems shows major differences. Depositional models of these deltas and associated fluvial systems are shown in Fig. 12.

In the transverse, Gilbert-type deltaic sandstone facies, the topset beds contain burrows and are capped by carbonaceous shale beds. The foreset beds exhibit an average dip of 20° and thickness variations from 3 to 5 m. In addition, the updip part of the foreset beds shows alternating small-scale trough cross-beds, ripple laminae and dip-oriented parallel to subparallel laminae. The transverse, Gilbert-type deltaic sandstone facies is vertically stacked and the deposits of the associated fluvial system consist mainly of multilateral, conglomeratic, channel sandstones and coal beds interbedded with siltstone and mudstone beds.

In the longitudinal, Gilbert-type deltaic sandstone facies, foreset beds are burrowed, display an average dip of 15° and vary in thickness from 5 m to 19 m. The foreset beds are commonly composed of rippled and parallel- and subparallel-laminated couplets with siltstone interforesets. The bottomset beds commonly contain U-shaped and straight, vertical burrows. The longitudinal, Gilbert-type deltaic sandstones are overlapped and the deposits of the related fluvial system include multistorey, *en echelon* channel sandstones and tabular sandstones interbedded with siltstones and mudstones. A synsedimentary fault occurs proximal to overthickened foreset beds of the longitudinal Gilbert-type delta.

CONCLUSIONS

The intermontane North Park Basin in north-central Colorado contains two types of sand-rich Gilbert-type deltas that prograded into freshwater lakes.

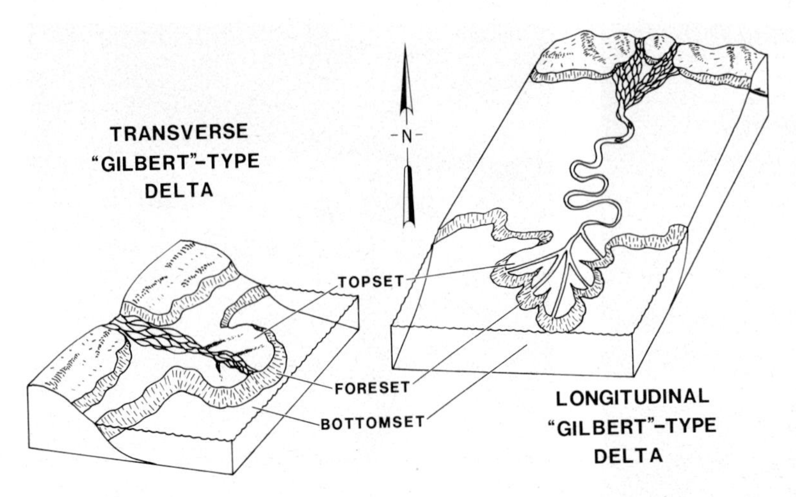

Fig. 12. Diagrams of depositional models of the transverse and longitudinal Gilbert-type deltas and contributing fluvial systems.

Homopycnal flow of river waters into the lake waters favoured thorough mixing and rapid deposition of sandy sediments near river mouths. The lack of density contrast between these freshwaters promoted deposition of Gilbert deltas with well-defined bottomset, foreset and topset beds. Despite these ideal hydrodynamics, differences exist between Gilbert-type deltas formed in various parts of the basin.

Along the basin margin, transverse Gilbert-type deltas fed by bedload or braided streams formed in shallow parts of lakes. That these deltas were developed in a rapidly subsiding part of the basin is indicated by vertical stacking of their deposits with well-preserved bottomset, foreset and topset beds. Subsidence may be related to Laramide thrust-loading of the basin margin. In addition, basin-margin subsidence may have formed slope breaks that permitted deposition of the Gilbert-type deltas with well-developed tripartite structures.

In the basin centre, longitudinal Gilbert-type deltas fed by mixed-load or meandering streams developed in relatively deep lakes. Overlapping of the foreset beds of these deltas suggests deposition and superposition by laterally shifting deltaic lobes in a slowly subsiding part of the basin. Here, syn-depositional growth faults may have created breaks in slopes that promoted deposition of Gilbert-type deltas with overthickened foreset beds. Mixed-load sediments of the streams which fed these deltas may have influenced deposition of suspended load that represents that siltstone interforesets, particularly during flood events.

Thus, during the early Tertiary, the North Park Basin, which was occupied by freshwater lakes much like its counterpart in the Pleistocene Lake Bonneville, served as an ideal setting for deposition of Gilbert-type deltas. In the North Park Lake Basin, the sedimentological characteristics of the sand-rich Gilbert-type deltas display diagnostic differences between depositional sites in the basin margin and basin centre. More importantly, the properties of the contributing fluvial systems played a major role in the dichotomy of the Gilbert-type deltas in the basin. Knowledge of these differences may assist in interpreting the sedimentary evolution of other intermontane freshwater-lake basins.

ACKNOWLEDGEMENTS

I am grateful to T. Nilsen and an anonymous referee for helpful comments on the manuscript.

REFERENCES

Bouma, A.H. (1962) *Sedimentology of Some Flysch Deposits*. Elsevier, Amsterdam, 168 pp.

Colella, A., De Boer, P.L. & Nio, S.D. (1987) Sedimentology of a marine intermontane Pleistocene Gilbert-type fan-delta complex in the Crati Basin, Calabria, southern Italy. *Sedimentology* **34**, 721–736.

Colella, A. (1988) Fault-controlled marine Gilbert-type fan deltas. *Geology* **16**, 1031–1034.

Flores, R.M. (1988) Transverse and longitudinal Tertiary Gilbert-type deltas, North Park Basin, Colorado, USA. *Abstr., Int. Works. Fan Deltas, Calabria*, 23.

Gilbert, G.K. (1890) Lake Bonneville. *U.S. geol. Survey Mon.* **1**, 438 pp.

Hail, W.J. (1965) Geology of northwestern North Park,

Colorado. *U.S. geol. Survey Bull.* **1188**, 133 pp.

HAIL, W.J. (1968) Geology of southwestern North Park and vicinity, Colorado. *U.S. geol. Survey Bull.* **1257**, 119 pp.

HANLEY, J.H. & FLORES, R.M. (1987) Taphonomy and palaeoecology of nonmarine Mollusca: Indicators of alluvial plain lacustrine sedimentation, upper part of the Tongue River Member, Fort Union Formation (Paleocene), northern Powder River Basin, Wyoming and Montana. *Palaios* **2**, 479–496.

HENDRICKS, M.L. (1977) Stratigraphy of the Coalmont Formation near Coalmont, Jackson County, Colorado. Unpublished MS thesis, Colorado School of Mines, 112 pp.

MADDEN, D.H. (1977) Exploratory drilling in the Coalmont coal field, Jackson County, Colorado. *U.S. geol. Survey Open-File Rpt.* **77–887**, 137 pp.

MAUGHAN, E.K. (1988) Geology and petroleum potential, Colorado Park Basin Province, north-central Colorado. *U.S. geol. Survey Open-File Rpt.* **88–450 E**, 36 pp.

NEMEC, W. & STEEL, R.J. (1988) *Fan Deltas: Sedimentology and Tectonic Settings* (Ed. by W. Nemec and R.J. Steel), pp. 3–13. Blackie and Son, London.

REINECK, H.E. & SINGH, I.B. (1980) *Depositional Sedimentary Environments*. Springer Verlag, New York, 549 pp.

ROBERTS, S.B. & FLORES, R.M. (1988a) Shifting coal depocentres in the Tertiary Coalmont Formation, North Park Basin, Colorado. *Abstr. Geol. Soc. Am.*, 169.

ROBERTS, S.B. & FLORES, R.M. (1988b) Fluvial pattern influenced by underlying coal-bed morphology, Coalmont Formation, North Park Basin. *Abstr. Soc. eco. Paleont. & Mineral., Tulsa, Ann. Midyr, Mtg.*, 46.

SEILACHER, A. (1967) Bathymetry of trace fossils. *Mar. Geol.* **5**, 413–428.

SMITH, N.D. (1970) The braided stream depositional environments: comparison of the Platte river with some Silurian clastic rocks, north-central Appalachians. *Bull. geol. Soc. Am.* **81**, 2993–3014.

STANLEY, K.O. & SURDAM, R.C. (1978) Sedimentation on the front of Eocene Gilbert-type deltas, Washakie Basin, Wyoming. *J. sedim. Petrol.* **48**, 557–573.

Spec. Publs int. Ass. Sediment. (1990) **10**, 235–254

The Miocene Chunbuk Formation, southeastern Korea: marine Gilbert-type fan-delta system

I. G. HWANG *and* S. K. CHOUGH

Department of Oceanography, Seoul National University, Seoul 151–742, Korea

ABSTRACT

The Miocene Chunbuk Formation represents part of the Doumsan fan-delta system and comprises various breccias and gravelstones deposited in the subaerial, transitional and subaqueous environments of a marine Gilbert-type fan-delta. Detailed facies analysis reveals that the system can be represented by 12 sedimentary facies, which are organized into five facies associations. Facies association I (interpreted as subaerial fan in origin) is characterized by massive/crudely stratified breccias (facies B1-a and B1-b) deposited by sheetfloods and debris flows. It grades laterally into facies association II (braided stream) in which massive/crudely stratified and cross-bedded gravelstones (facies G1-a and G4) show channel geometries. Facies association III (Gilbert-type topset) is represented by coarsening-upward sequences in which massive and laminated sandstones (facies S1 and S2) grade vertically into massive/crudely stratified and cross-bedded gravelstones (facies G1-a and G4). Facies association IV (Gilbert-type foreset) is characterized by steeply inclined (initial slopes of more than 20°) massive/crudely stratified and stratified gravelstones (facies G1-b, G2-a and G2-b) that were deposited by gravity slides or cohesionless debris flows. Facies association V (Gilbert-type toeset) comprises massive/crudely stratified gravelstones (facies G1-b) and massive muddy sandstones (facies MS1-b), in which slope failure resulted in extensive syndepositional deformation.

The development of this Gilbert-type system is most likely due to syndepositional tectonic subsidence. The abnormally thick foreset units (> 150 m in foreset height) are suggestive of rapid tectonic movement.

INTRODUCTION

A fan-delta system commonly comprises tripartite depositional elements of subaerial, transitional and subaqueous parts. The subaerial part is dominated by sheetfloods and debris flows, whereas sedimentary processes in the transitional part are largely modified by waves and currents (Wescott & Ethridge, 1980; Kleinspehn *et al.*, 1984). The subaqueous part is characterized by various sediment gravity flows such as rock falls, slides/slumps, debris flows and turbidity currents. In basins bounded by steep coastlines, the subaqueous part forms either Gilbert-type fore- and bottomsets (Gilbert, 1885; Postma, 1983, 1984; Postma & Roep, 1985; Colella *et al.*, 1987; Colella, 1988a; Postma *et al.*, 1988), or slope channels and chutes (Prior *et al.*, 1981; Prior & Bornhold, 1988) or submarine fans (Wescott & Ethridge, 1980; Ethridge & Wescott, 1984; Surlyk, 1984).

The Miocene Doumsan fan-delta system in the Pohang Basin, southeast Korea, contains an entire spectrum of fan-delta facies associations including subaerial fan, braided stream, Gilbert-type top-, fore- and toesets, prodelta, slope apron and basin plain (Chough *et al.*, 1989, 1990; Hwang, 1989). In this study, we focus on the depositional processes of the Gilbert-type fan-delta body (Chunbuk Formation), based on detailed analysis of facies, facies associations and sedimentary architecture.

GEOLOGIC SETTING

A number of small-scale Tertiary sedimentary basins occur along the eastern coast of the Korean Peninsula (Fig. 1). The Pohang Basin is the largest of these basins and has received much attention

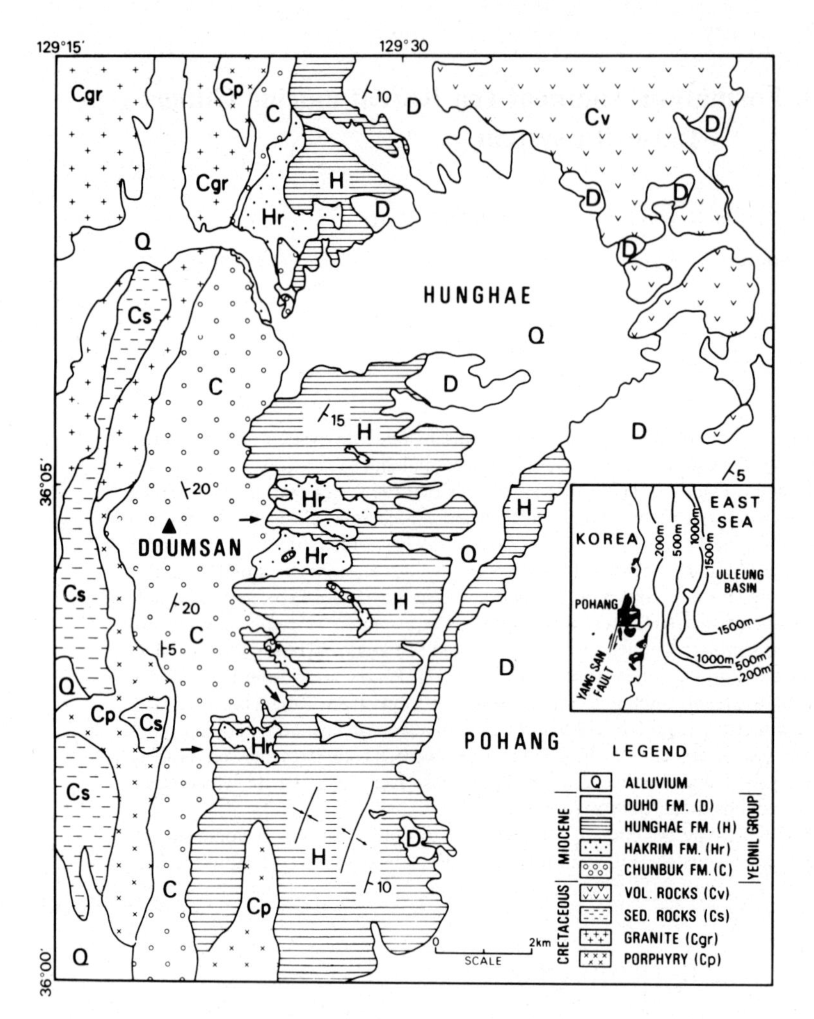

Fig. 1. Simplified geologic map showing distribution of Cretaceous and Neogene sequences in Pohang area, southeast Korea. Arrows indicate palaeoflow directions (inferred from gravel orientation, imbrication and cross-beds).

because of its hydrocarbon potential. The Tertiary sequence is far more extensive in the adjacent continental margin of the East Sea where it is more than 10 km thick on the continental shelf (Chough & Barg, 1987) and up to 4 km on the deep Ulleung Basin (Chough, 1983). The subsidence history of the Ulleung Basin margin based on Dolgorae-1 well (Barg, 1986; Chough & Barg, 1987) indicates that the basin experienced rapid subsidence (700 m/my) during Miocene time. The Ulleung Basin margin was uplifted in the late Miocene, during which the Pohang Basin was exposed on land.

The Yeonil Group sequence in the Pohang Basin occurs on the eastern part of the Yangsan Fault (Fig. 1) which has experienced right-lateral, strike-slip movement (Reedman & Um, 1975). The sequence dips gently eastward (5–25°) with local folding (Fig. 1) and consists of various lithologic units, including gravelstone, sandstone and mudstone (Um *et al.*, 1964). The entire sequence is more than 1 km thick and can be divided into four formations: Chunbuk (gravelstone), Hakrim (sandstone), Hunghae (sandstone/mudstone) and Duho (mudstone) formations (Fig. 1). The lowest Chunbuk Formation unconformably overlies the Cretaceous sedimentary and igneous rocks (Um *et al.*, 1964; Kim, 1965; Yoon, 1975; Yun, 1986) and is, in turn, conformably overlain by the Hakrim Formation. The Chunbuk Formation largely comprises gravelstones and is intercalated with sandstones and mudstones with variable thickness (150–600 m) (Um *et al.*, 1964; Kim, 1965; Yoon, 1975; Yun, 1986).

Previous studies on the Pohang Basin were mostly concentrated on the litho- and biostratigraphy and palaeoceanography (Um *et al.*, 1964; Kim, 1965; Yoon, 1975; You, 1983; Lee, 1984; Bong, 1985). These studies revealed that the Yeonil Group was of Miocene age, deposited under mixed conditions of cold and warm waters (Kim, 1965; Yoo, 1969; Lee, 1982). The climate of adjacent terrestrial areas was warm to temperate (Bong, 1982, 1985; Chun, 1982; Lee, 1984; Kim, 1987). For part of the sequence (i.e. Hunghae Formation; see Fig. 1), Choe (1986) Choe and Chough (1988) suggested that the sediments were deposited on the basin slope of fan deltas, where mass flows such as slide/slump, debris flow and turbidity current were dominant, forming debris aprons.

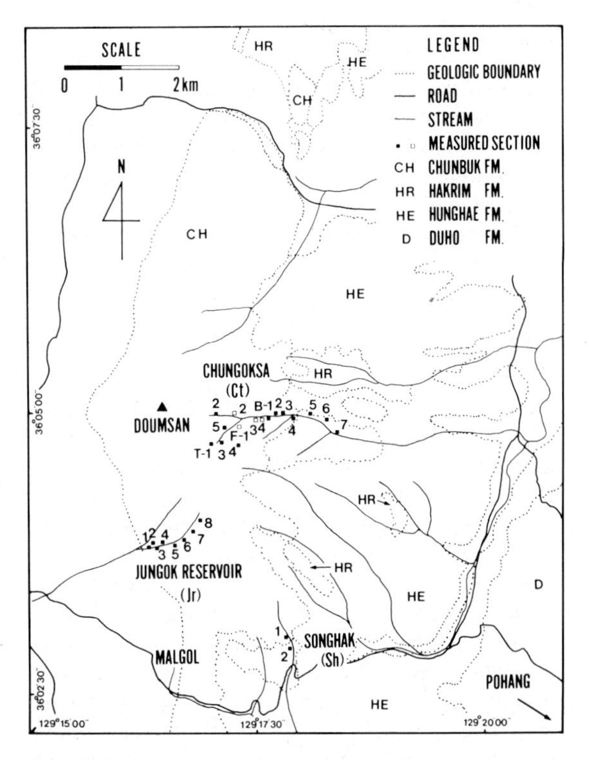

Fig. 2. Location map of measured sections.

SEDIMENTARY FACIES

The Chunbuk Formation is dominated by gravelstone and breccia with minor occurrence of sandstone and mudstone. Based on detailed measurements (1 : 10 scale) of entire outcrop sections (Fig. 2), 12 major sedimentary facies are established (Fig. 3). The classification is based on a two-tier system of grain size and sedimentary structures, supplemented by grain fabric and biogenic structures. Characteristics of individual facies, distinctive occurrence and their hydrodynamic interpretations are given in the following sections.

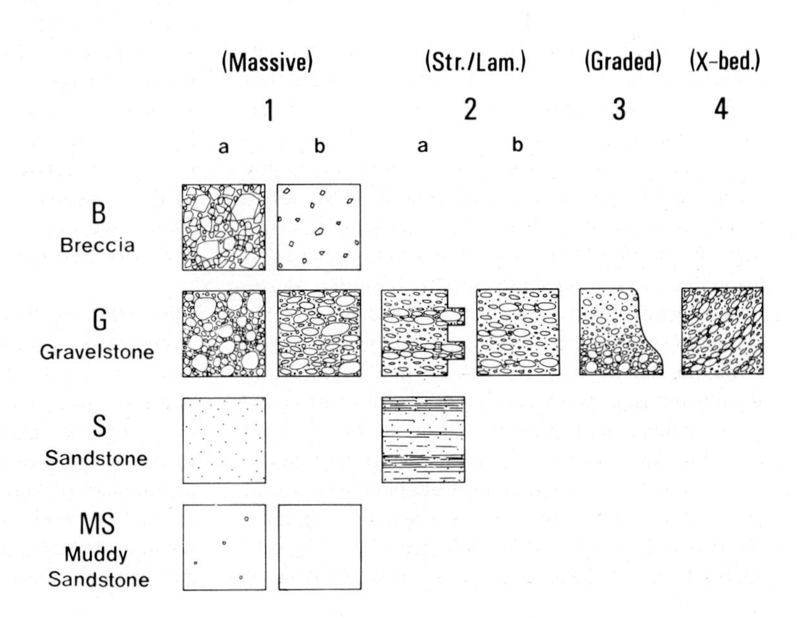

Fig. 3. Facies classification and symbolic summary of Chunbuk Formation.

Class B: breccia

The breccias comprise angular rock fragments that are cemented in a fine-grained matrix and occupy the lowermost part of the Chunbuk Formation. Clasts were mostly derived from the immediately underlying Cretaceous sedimentary, volcanic and granitic rocks (Fig. 1) and show a regional variation in composition.

Facies B1-a: massive/crudely stratified breccia (clast-supported)

Description. This facies is characterized by a clast-supported fabric albeit with a brownish muddy sand matrix (Fig. 4). Clasts range in size from granule to boulder (more than 1 m in long axis), and are arranged parallel to bedding. Each unit is laterally continuous and ranges in thickness from three or four grains to several metres. Some units are inversely graded. There are strong lateral variations in sedimentary characteristics at outcrop scale (< 100 m in width): (1) mud matrix decreases distally whereas sand matrix increases; (2) maximum clast size decreases abruptly; (3) the facies unit boundary is generally diffuse in the proximal part whereas it is rather sharp in the distal part; (4) crudely stratified units are more common in the distal part; and (5) cross-bedded units are partly intercalated in the distal part. This facies unit is interlayered with massive matrix-supported breccia (facies B1-b) rarely forming inversely graded units.

Interpretation. Although this facies is clast-supported, the abundant mud content indicates that cohesive strength played an important role. Debris flows in the subaerial environment range from slurry flows (mud flow or classical debris flow) to highly turbulent fluidal flow (e.g. Pierson, 1981; Nemec & Steel, 1984; Schultz, 1984). Orientations of clasts that are parallel to bedding indicate that clast collision occurred during transport. It is similar to the Facies Dcm (massive clast-supported diamictite) of Schultz (1984) deposited by pseudoplastic debris flows. In the distal part, lack of muds and crude stratifications as well as cross-bedded units are indicative of more fluidal flow than those in the proximal part (Fig. 4). Recently, the transformation of fluidal debris flows to water-laden flows has been recognized in alluvial fans and volcaniclastic sediments (Harrison & Fritz, 1982; Ballance, 1984; Smith, 1986). It may be due to increase in water content during the flow and consequent loss of yield strength. In its distal part, this facies is similar to the sheetflood deposits of Bull (1972), in which plane beds, antidunes and transverse ribs (Koster, 1978) are the equilibrium bedforms, forming massive, crudely stratified and cross-bedded sedimentary structures (Fig. 4).

Fig. 4. Massive/crudely stratified breccia (facies B1-a) formed in distal subaerial fan (section Jr-1). Breccias are clast-supported in poorly-sorted sand matrix. Scale in centimetres (in circle).

Facies B1-b: massive breccia (matrix-supported)

Description. This facies shows a massive, matrix-supported fabric. Clasts comprise granule to pebble-size breccias with small amounts of cobble to boulder-size breccias, which are widely dispersed in the brownish mud matrix and show random orientation. Some units are inversely graded. This facies is interlayered with massive clast-supported breccias (facies B1-a). In the distal part, however, this facies occurs as thin lenses (< 10 cm in thickness) and is intercalated with massive/crudely stratified (facies G1-a) and cross-bedded (facies G4) gravelstones.

Interpretation. The matrix-supported breccias are interpreted as subaerial debris-flow deposits, based on poor sorting, randomly oriented floating clasts and brownish mud matrix. In this facies, abundant mud content suggests that cohesive strength was dominant, whereas dispersive pressure and fluid turbulence played less important roles. Lack of grading as well as non-erosional lower bounding surface also suggest deposition from debris flows (or mud flows). Random clast orientation is commonly developed where the sediment contains abundant mud matrix (e.g. Cook, 1979) suggesting that clast collision was limited during flow (Lewis *et al.*, 1980; Gravenor, 1986). Alternatively, random orientation may also reflect either short transport (e.g. Lindsay, 1968), non-sheared (high-strength) plug flow or only weakly sheared (high-viscosity) flow (Lewis *et al.*, 1980; Nemec & Steel, 1984).

Class G: gravelstone

The gravelstone contains more than 30% of gravels in poorly to well-sorted yellowish-grey sand matrix. Clasts are rounded to subrounded and largely consist of Cretaceous sedimentary (58%), volcanic (25%) and granitic (17%) rocks.

Facies G1-a: massive/crudely stratified gravelstone [a(t), b(i)]

Description. This facies has a clast-supported and partly openwork fabric, with a poorly to well-sorted sand matrix (Fig. 5). Most clasts show random orientations, but some are oriented transverse to the flow direction [a(t), b(i)]. Individual facies units are generally thick (> 1 m) and are commonly interbedded with cross-bedded gravelstone (facies G4). This facies occurs in sections of Jr-2 to -8 and Ct-T-1 to -5 (Fig. 2), which form the lower and middle parts of the Chunbuk Formation.

Interpretation. The clast orientation and imbrication [a(t), b(i)] as well as the partly openwork fabric collectively indicate that the gravels were transported as bedload (Harms *et al.*, 1975; Harms *et al.*, 1982). The mechanics of gravel bedload transport and the origin of massive gravelstone are not well known. Smith (1970, 1972), Rust (1972) and Gustavson (1974) pointed out that migrating longitudinal bars tend to form poorly defined horizontal beds, possibly indicating transportation in planar sheets under high-flow energy. Because the longitudinal bar is of low amplitude, avalanching in the slip face is rare and large clast trains that lie oblique to bedding are common (Fig. 5, arrow). This facies is similar to the massive or crudely stratified gravelstone (Facies Gm) of Miall (1977, 1978) and Rust (1978).

Facies G1-b: massive/crudely stratified gravelstone [a(p), a(i)]

Description. This facies can be divided into two types in terms of clast orientation and bed attitude. The first type occurs in steeply inclined (> 20°) beds (sections of Ct-T-3,-4 and Ct-F-1,-2,-3,-4) forming Gilbert-type foreset sequences. Most clasts are arranged parallel to bedding and some are imbricated. Clasts range in size from granule to boulder, and are either matrix- or clast- supported in a poorly- to well-sorted sand matrix with clay content of less than 1%. Each unit ranges in thickness from three

Fig. 5. Massive/crudely stratified gravelstone (facies G1-a) formed in braided streams (section Jr-6). Gravels are randomly oriented; some parts are diffusely cross-bedded (arrow).

or four clasts to several metres. Large clasts which lie parallel to bedding show crude stratification. This facies is interlayered with stratified (facies G2-a and G2-b) and graded (facies G3) gravelstones.

The second type occurs in less steeply inclined beds (< 10°) than the above (sections of Ct-B-1 to -7 and Sh-1) and define the Gilbert-type toeset sequence. Clasts are generally oriented parallel to bedding, although some are randomly oriented. This facies is interlayered with massive muddy sandstone (facies MS1-b).

Interpretation. In the first type, lack of clay content suggests cohesionless flow, and relatively high clast content reflects that dispersive pressure played an important role in maintaining the clasts in a dispersed state. Furthermore, the well-developed clast orientation indicates that grain collisions were important during the last stages of flow (Allen, 1982; Gravenor, 1986). This facies is similar to the density-modified grain-flow deposits of Lowe (1982). It is also similar to deposits of sandy debris flows transitional to density-modified grain flows (Facies 4, non-graded, clast-supported, imbricated facies) of Surlyk (1984), and cohesionless debris-flow deposits of Postma (1986).

The second type contains less well-oriented clasts reflecting that dispersive pressure was less important than that of the first type. This facies is similar to the Facies 3 (non-graded, clast-supported conglomerate with random fabric) of Surlyk (1984) originating from high-density, non-cohesive, sandy debris flow. It is also similar to gravelly, high-density turbidity-current deposits of Lowe (1982).

Facies G2-a: stratified gravelstone

Description. This facies occurs in steeply inclined (> 20°) foreset beds with each stratified unit consisting of thin (one or two clasts thick) laterally discontinuous gravel layers. The layer boundaries are sharp, but laterally become diffuse and grade into large clast trains (Fig. 6). Some intercalated, gravelly sandstone layers contain thin, laterally discontinuous graded or inversely graded units. Clasts range in size from granule to boulder and are arranged parallel to the local slope; some are imbricated [a(p), a(i)]. They are either matrix- or clast-supported. The matrix comprises poorly to well-sorted sands with clay contents of less than 1%. Each facies unit is commonly interlayered with massive/crudely stratified (facies G1-b), stratified (facies G2-b) and graded (facies G3) gravelstones. This facies occurs in sections of Ct-T and Ct-F (middle part of the Chunbuk Formation) (Fig. 2).

Interpretation. The occurrence of this facies on steeply inclined initial slopes and the thin, laterally discontinuous graded and inversely graded gravelly-sandstone units indicate deposition from small-scale

Fig. 6. Stratified (facies G2-a and G2-b) and massive/crudely stratified (facies G1-b) gravelstones formed in Gilbert-type foreset (section Ct-T-5). Note thin laterally discontinuous flow units. Scale arrow is 20 cm long (in circle).

gravity slides. This facies is similar to 'flow slide' deposits of Colella *et al.* (1987), large-scale tangential foreset deposits (Facies IIIA) of Postma (1984), and deposits formed by 'sliding of cohesionless sediments' of Postma (1986), all of which occur on the Gilbert-type foresets. According to Postma (1986), their transport distance is likely to be short and steep slope angles will be required to overcome internal friction.

Facies G2-b: stratified gravelstone (pebble-train layers)

Description. This facies is characterized by laterally discontinuous, large-clast trains with diffuse upper and lower boundaries; it is similar to the stratified gravelstones of facies G2-a, but differs in that the layers are less distinct (Fig. 6). Clasts range in size from granules to outsized boulders. They are either matrix- or clast-supported in poorly- to well-sorted sands with clay content less than 1%. In Ct-T and -F sections (foreset), most clasts are oriented parallel to the local slope and some are imbricated [a(p), a(i)]. Sedimentary characteristics are transitional between the stratified gravelstone (facies G2-a) and massive/crudely stratified gravelstone (facies G1-b). In sections of Ct-B and Sh-1 (toeset), this facies shows a less well-developed clast orientation than those in the foreset beds and is interlayered with massive/crudely stratified gravelstone (facies G1-b) and massive muddy sandstone (facies MS1-b).

Interpretation. In Ct-T and Ct-F sections, the less well-developed stratification than that found in the stratified gravelstones (facies G2-a) may be due to an increase in fluid turbulence (Lowe, 1982). Deposition probably occurred by gravity slides transitional to sandy debris flows. In sections of Ct-B and Sh-1, depositional slope was less steep and clasts are less well-oriented than those in the foreset units, indicating that fluid turbulence played an important role. These flows are similar to density-modified grain flows that are transitional to high-density turbidity currents, sandy debris flows (Lowe, 1982; Surlyk, 1984) or cohesionless debris flows (Postma, 1986).

Facies G3: graded gravelstone

Description. This facies is characterized by graded (or inverse-to-normally graded) gravelstones. Clasts are either clast- or matrix-supported and most clasts are oriented parallel to bedding. Matrix comprises poorly- to well-sorted sands with clay content less than 1%. It occurs in sections of Ct-F-3, -4, and Ct-B-3, -4 (Fig. 2), forming the middle and upper parts of the Chunbuk Formation.

Interpretation. Various mechanisms have been suggested for graded (or inverse-to-normally graded) gravelstones with sand matrix: gravelly high-density turbidity currents of Walker and Mutti (1973), Aalto (1976), Walker (1978), Lowe (1982) and Massari (1984); density-modified grain flows of Middleton and Hampton (1976) and Lowe (1979); and cohesionless debris flows of Curry (1966), Winn and Dott (1977), Lowe (1982) and Postma (1986). In sections of Ct-F-3 and -4, steep slope ($> 20°$), well-oriented clasts and partly inversely graded units suggest that dispersive pressure played an important role during the last stage of the flow. In sections of Ct-B-3 and -4 (toeset), clasts are less well-oriented than those of foreset beds suggesting more turbulent flow.

Facies G4: cross-bedded gravelstone

Description. The facies G4 includes either planar, tangential or trough cross-bedded gravelstones. The foreset angle is variable (10° to 35°) and the foreset laminae consist of alternating units of gravel-stone, gravelly sandstone and sandstone as well as thin mudstone (Fig. 7). Gravel layers are commonly openwork and some are graded. Some cross-beds retain Gilbert-type sequences including top-, fore- and bottomsets. Upper bounding surfaces are generally planar, although mound-shaped geometry (gravel-bar geometry) also occurs. The trough cross-bedded units commonly show channel geometry with scoop-shaped, lower bounding surface. At the lower boundary, clasts are slightly larger than those in the cross-bedded units (channel lag deposits). Bed thickness ranges from tens of centimetres to several metres. This facies alternates with massive/crudely stratified gravelstones (facies G1-a) with diffuse, upper and lower boundaries and occurs in sections of Jr-2, -3, -4, -6, -7, -8, Ct-T-1, -2, -3 and -5 (Fig. 2), forming the lower and middle parts of the Chunbuk Formation.

Interpretation. This facies was most probably deposited by migration of various gravel bars and dunes, and by small-scale, secondary channel fills. Remnants of gravel bar geometry are formed in some

Fig. 7. Cross-bedded gravelstones (facies G4) formed in braided streams (section Jr-2). Scale bar is 1 m long.

planar and tangential cross-bedded units, whereas channel geometry is common in trough cross-bedded gravelstone. The thick cross-bedded gravelstone with small-scale Gilbert-type geometry indicates progradation of a microdelta.

Class S: sandstone

The class S contains more than 70% of sands with minor amounts of gravels and muds. It is commonly yellowish grey in colour and comprises grains of feldspar, quartz and rock fragment. The sand grains are generally angular to subrounded.

Facies S1: massive sandstone

Description. This facies is characterized by thick (generally > 1 m), massive and commonly bioturbated sandstone units. The sands are poorly to moderately-sorted, containing gravel pockets and lenses, plant debris and lignite fragments. Thin units of wave ripples (wave length, 0.12–0.15 m; wave height, 0.01–0.03 m) are partly intercalated. Part of the sandstone is cemented with calcite and contains brackish to shallow-marine mollusc fossils (Yoon, 1975, 1976a,b). Burrows with a diameter of about 0.01 m and length of several centimetres are common. This facies is interlayered with laminated sandstone (facies S2), and massive/crudely stratified and cross-bedded gravelstones (facies G1-a and G4). It occurs extensively in sections of Jr-5, -7, -8 and Ct-T-2, -5 (Fig. 2), forming the lower and middle parts of the Chunbuk Formation.

Interpretation. Massive sandstones can be formed either by rapid deposition from suspension, penecontemporaneous deformation or by bioturbation. Pebble trains indicate tractional process, whereas lignite fragments suggest slow suspension sedimentation. Lack of grading and fluid-escape structures also suggest that sedimentation rate was low. Brackish to shallow marine-type mollusc fossils, wave ripples and lignite fragments indicate that deposition occurred near river-mouth draining to shallow-marine environments (e.g. lagoons and salt marshes).

Facies S2: laminated sandstone

Description. This facies is composed of moderately- to well-sorted, very fine sands and silts, and is characterized by thin, lignite laminae (Fig. 8). Some units are bioturbated and contain abundant, mollusc fossil fragments, which occur as calcite-cemented sandstone blocks. This facies is interlayered with massive sandstone (facies S1), and massive/crudely stratified and cross-bedded gravelstones (facies G1-a and G4) with sharp upper and lower boundaries. It

Fig. 8. Laminated sandstone (facies S2) formed in transitional zone between subaerial and subaqueous parts (section Jr-8). Note thin lignite-fragment layers. Scale in centimetres.

occurs in sections of Jr-2 to -7 and Ct-T-2 (Fig. 2), forming the lower and middle parts of the Chunbuk Formation.

Interpretation. The fine-grained nature of the clastic grains (very fine sands and silts) and lignite fragments suggest that flow velocity was rather slow and that sediments were supplied periodically. Lack of both grading and fluid-escape structures also suggest slow suspension settling. Deposition may have occurred in (semi-) enclosed lagoons near river-mouth draining to shallow-marine environments where brackish to shallow-marine-type molluscs are abundant.

Class MS: muddy sandstone

The muddy sandstones are composed of sands (70% > sand > 50%) and muds (30% < mud < 50%) with small amounts of gravels.

Facies MS1-a: massive muddy sandstone

Description. This facies is generally thick and is either brownish or dark grey depending on the amounts of lignite fragments. Gravels range in size from granule to pebble and occur as thin, laterally discontinuous trains. Otherwise, this facies is extensively bioturbated with vertical burrows and rootlets. An individual facies unit is commonly scoured by massive/crudely stratified and cross-bedded gravelstones (facies G1-a and G4) with channel geometry. This facies occurs in sections of Jr-2, -3 and -4 (Fig. 2), which form the lower part of the Chunbuk Formation.

Interpretation. The existence of plant debris and the brownish to dark grey colour as well as the channellized gravelstones suggest flood-plain deposition. Muddy sands were deposited from suspension fallout due to overbank flooding. Thin, laterally discontinuous gravel trains are indicative of tractional processes during flood stage. Original sedimentary structures were obliterated by extensive bioturbation. It is similar to the Facies Fl of Miall (1977, 1978) and Rust (1978) deposited by vertical accretion on overbank area or flood plain.

Facies MS1-b: massive muddy sandstone

Description. This facies is characterized by thick, massive dark-grey muddy sandstones. Gravels are widely dispersed with small amounts of biogenic sediments such as plant debris, lignites, shell fragments and microfossils. Thin, laterally discontinuous sandstone laminae are commonly intercalated. This facies unit is interlayered with massive/crudely stratified or graded gravelstone (facies G1-b and G3). This facies occurs in sections of Ct-B-4 and -5 (Fig. 2), forming the upper part of the Chunbuk Formation.

Interpretation. The massive sedimentary structures and the co-occurrence of terrigenous materials with microfossils indicate deposition by hemipelagic settling. At the river mouth, sediment segregation is caused by the density difference between the inflowing water and the reservoir water (Bates, 1953; Colella *et al.*, 1987). If the inflowing water and reservoir water have the same density (homopycnal flow) or the inflowing water is less dense than the reservoir water (hypopycnal flow), the inflowing water will result in plane jet flow, transporting the sediment more basinward, while the coarse-grained bedload will be deposited at the river mouth to form Gilbert-type deltas (Bates, 1953). Continuous supply of terrigenous sediments from the fluvial system will cause rapid sedimentation and thick, homogeneous muddy sandstone can be formed. This facies is similar to the Facies III (homogeneous mudstone) of Choe (1986) and Choe and Chough (1988).

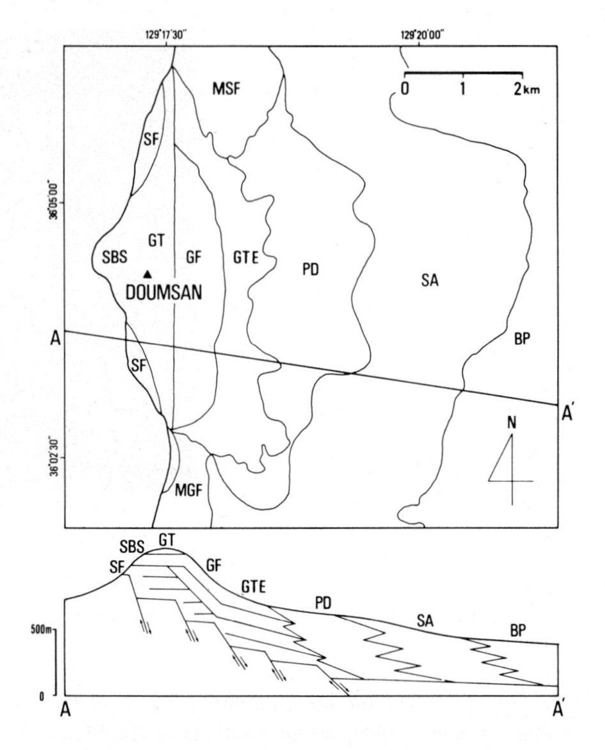

Fig. 9. Distribution and simplified cross-section of depositional environments in the Doumsan fan-delta system. The Chunbuk Formation comprises subaerial fan (SF), subaerial braided stream (SBS), Gilbert-type topset (GT), Gilbert-type foreset (GF) and Gilbert-type toeset (GTE). Prodelta (PD), slope apron (SA) and basin plain (BP) are also shown (after Chough *et al.*, 1990). Two other small fan-deltas, including Maesan fan-delta (MSF) and Malgol fan-delta (MGF), are outlined.

FACIES ASSOCIATIONS AND PALAEOENVIRONMENTS

The facies occurrence and associated bed geometry in the Chunbuk Formation show a systematic change depending on the stratigraphic sequence (Fig. 9). Bed geometries are based on either third or fourth orders of architectural elements (Allen, 1983; Miall, 1985). The facies associations (or facies assemblages) in these orders are characteristic of distinct environments. The Chunbuk Formation can be divided into five facies associations, indicating deposition on subaerial fan, braided stream, transitional zone between the subaerial and subaqueous parts (Gilbert-type topset), and submarine Gilbert-type fore- and toeset environments (Fig. 10). The characteristics of environments with regard to facies occurrence, distinct bed geometry, clast composition, spatial distribution and interactions with other environments are discussed in the following sections.

Facies association I: subaerial fan

Occurrence. This association is characterized by massive/crudely stratified breccias (facies B1-a and B1-b) with sheet-like geometry (Figs 10A, 11). It occurs in the lowermost part of the Chunbuk Formation and shows radial (radius less than 100 m) distributional pattern from the sediment origin (Fig. 9). Proximal to distal changes in sedimentary characteristics are evident: (1) maximum clast size decreases downfan; (2) in clast-supported breccia (facies B1-a), clay matrix decreases distally; (3) bed boundaries are generally diffuse in the proximal part, whereas they are rather sharp in the distal part; and (4) roundness of clasts slightly increases downcurrent.

Interpretation. Both the facies association and radial distribution pattern are most likely indicative of a subaerial fan (SF) environment where sheetfloods and debris flows form massive/crudely stratified breccias (facies B1-a and B1-b). Three small-scale, discrete subaerial fans are recognized along the faulted Cretaceous boundary where clast composition varies according to the immediately underlying basement rocks. According to Steel (1976) and Gloppen and Steel (1981), smaller alluvial fans with abundant debris-flow deposits commonly occur along

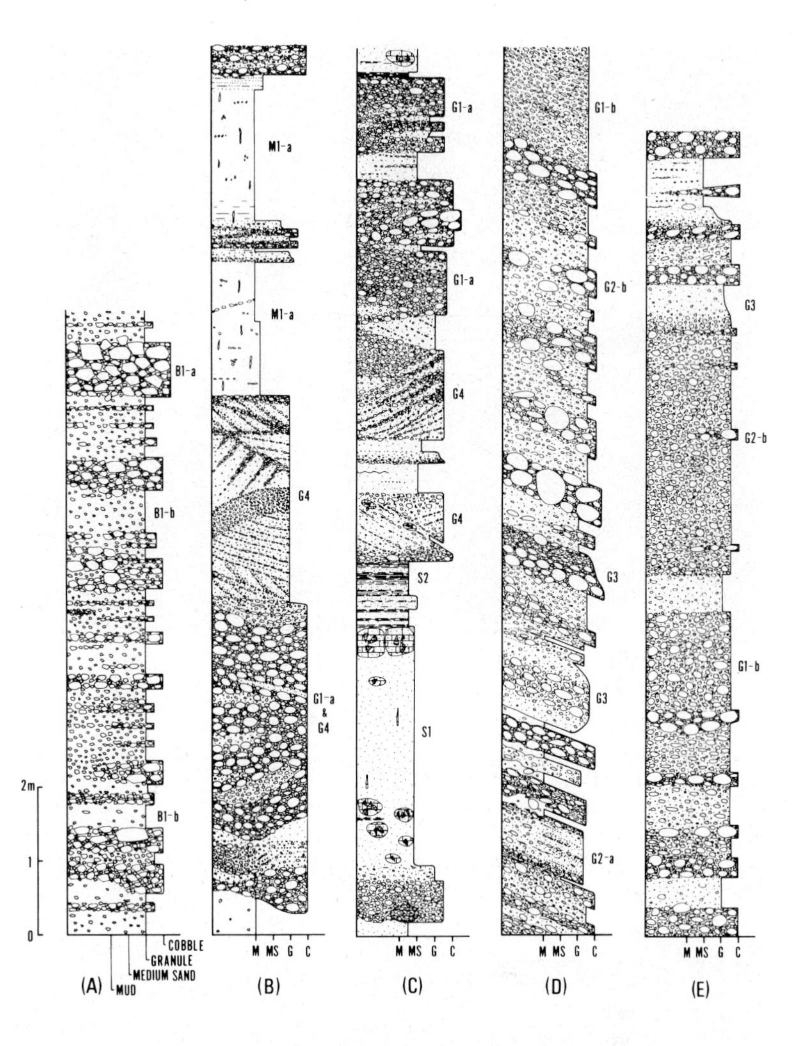

Fig. 10. Columnar sections of typical facies associations: (A) facies association I (subaerial fan; section Jr-1, Fig. 2); (B) facies association II (braided stream; section Jr-2, Fig. 2); (C) facies association III (Gilbert-type topset; section Jr-8, Fig. 2); (D) facies association IV (Gilbert-type foreset; section Ct-F-3, Fig. 2); and (E) facies association V (Gilbert-type toeset; section Ct-B-3, Fig. 2).

the basin margin with short drainage systems and steep slopes, whereas the larger fans are associated with less steep, large-scale drainage systems. Angular clasts and their clast composition also suggest short transport distance (Chough *et al.*, 1989, 1990; Hwang, 1989).

Facies association II: braided stream

Occurrence. This association is characterized by amalgamated units of massive/crudely stratified (facies G1-a) and cross-bedded (facies G4) gravelstones which commonly scour massive muddy sandstone (facies MS1-a) with channel geometry (Figs 10B, 12). The channel-fill sequence often shows fining-upward trends (Fig. 10B). This association mainly occurs in sections of Jr-2, -3 and -4 (Fig. 2), which form the lower part of the Chunbuk Formation.

Interpretation. Both the facies association and channel geometry indicate deposition in subaerial braided streams (SBS) in which migrating gravel bars and small-scale channel fills form massive/crudely stratified and cross-bedded gravelstones. On the other hand, massive muddy sandstones are interpreted as flood-plain deposits. This system is similar to the Facies Assemblage GII (Scott-type) transitional to the Facies Assemblage GIII (Donjek-type) of Miall (1978) and Rust (1978). The Scott-type is dominated by massive or crudely stratified gravelstone with minor amounts of cross-bedded gravelstone, sandstone and mudstone, commonly occurring in the proximal rivers (Miall, 1978; Rust,

Fig. 11. Photograph of typical subaerial fan deposits (section Jr-1, Fig. 2). Note sheet geometry of massive/crudely stratified breccias (facies B1-a and B1-b). Scale bar is 2 m long.

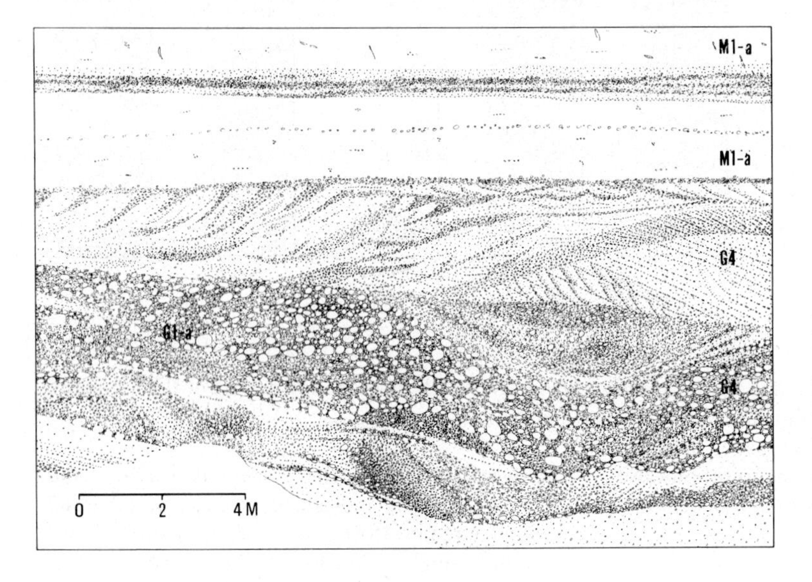

Fig. 12. Detailed sketch of part of the braided stream deposits (section Jr-2, Fig. 2). Gravelstones are massive/crudely stratified and cross-bedded (facies G1-a and G4), whereas muddy sandstones are generally massive and bioturbated (facies M1-a). Note fining-upward trend from gravelstone, gravelly sandstone to muddy sandstone.

1978). The Donjek-type is dominated by massive or crudely stratified gravelstone and cross-bedded gravelstone with minor amounts of sandstone and mudstone, which commonly occurs in the distal rivers. According to the architectural elements and facies associations, this association is similar to the Model 3 of Miall (1985) which occurs in large gravel-bed streams such as trunk rivers and in some large alluvial fans. According to Miall (1985), there are active channels filled with massive or crudely stratified gravelstones and cross-bedded gravelstones (Element Ch), and floodplains with sandy bedforms (Element Sb). This model commonly shows fining-upward sequences (Miall, 1985, 1988).

Although the facies association II was formed in braided streams, the sequence boundary with the Cretaceous basement shows a triangular geometry (Fig. 9), indicating that this association is part of an alluvial fan. According to Rust and Koster (1984), it is difficult to discriminate the fluvial system of wet-type alluvial fans from braided plains. However, alluvial fans can be distinguished from braided plain on the basis of depositional processes (i.e. sheetflood or debris-flow deposits), palaeocurrent pattern and configuration of a basin (Bull, 1972). The braided-stream deposits laterally grade into transitional zones between subaerial and subaqueous parts, forming the Gilbert-type topset (facies association III; Fig. 9).

Facies association III: Gilbert-type topset (transitional zone between subaerial and subaqueous parts)

Occurrence. This association is represented by massive/crudely stratified (facies G1-a) and cross-bedded (facies G4) gravelstones, and massive (facies S1) and laminated (facies S2) sandstones with sheet geometry (Figs 10C, 13). It occurs in sections of Jr-5, -6, -7 and -8 and Ct-T-1, -2, -3, -4 and -5 (Fig. 2), which form the lower and middle parts of the Chunbuk Formation. The sequence differs from the facies association II (braided-stream deposits) in that: (1) distinct large-scale channel geometry is absent except for small-scale secondary channels filled with trough cross-bedded gravelstone (facies G4); (2) mollusc fossil fragments of brackish to shallow-marine-type occur; (3) lignite-laminated sandstones (facies S2) are abundant; (4) wave ripples are present; and (5) this association shows coarsening-upward trends.

Interpretation. Bioturbated massive (facies S1) and lignite laminated (facies S2) sandstones are most likely indicative of deposition in transitional parts such as lagoons and salt marshes. Mollusc fossils of brackish to shallow-marine-type suggest mixed environments of fresh and salt waters. Lignite fragments may have originated from the adjacent salt marshes and settled from suspension. Abundant bioturbations and small-scale wave ripples are indicative of low-energy conditions with limited influence of waves and currents. Probable tidal deposits are rare and suggest deposition in microtidal settings. The massive/crudely stratified (facies G1-a) and cross-bedded (facies G4) gravelstones were deposited by migrating gravel bars and as small-scale channel fills. Sheet geometry of the gravelstone units indicates deposition in broad channels where the width/depth ratio is more than 100 (Friend, 1983).

The coarsening-upward sequence indicates prograding fluvial systems to shallow-marine environments, similar to shelf-type fan-delta systems or coastal alluvial fans (Flores, 1975; Galloway, 1976; Wescott & Ethridge, 1980; Hayes & Michel, 1982; Ethridge & Wescott, 1984; Kleinspehn *et al*., 1984; Killick, 1988; Orton, 1988; Rhine & Smith, 1988). Sedimentary facies varies significantly according to various factors acting on a delta: (1) intensity of

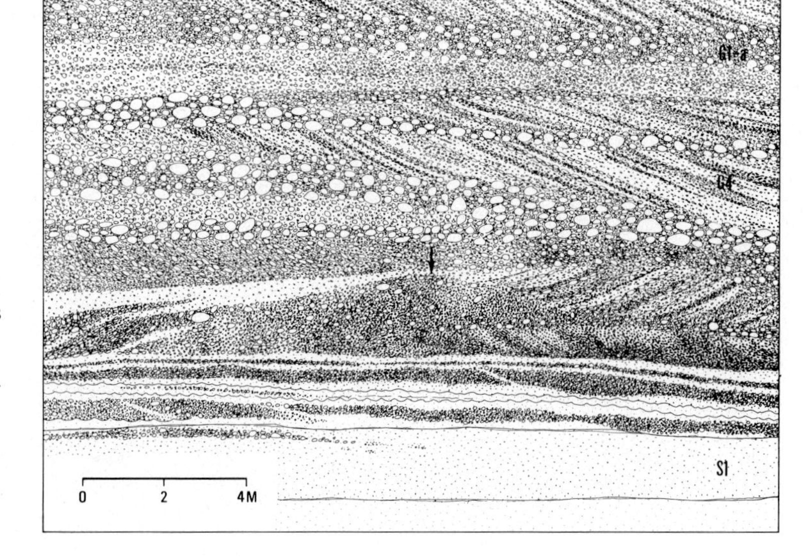

Fig. 13. Detailed sketch of Gilbert-type topset (section Ct-T-1, Fig. 2). Massive sandstone with wave ripples (facies S1) was formed in shallow-marine environments, whereas massive/crudely stratified and cross-bedded gravelstones (facies G1-a and G4) were deposited in braided streams. Note general coarsening-upward trend; remnants of gravel-bar geometry are shown in the middle part (arrow).

fluvial processes; (2) sediment dispersal pattern such as waves, tides and longshore currents; and (3) geometry of the basin controlled by tectonic activity (Wright & Coleman, 1974). The Doumsan system is partly enclosed by the volcanic ridges on the northeastern part (Choe, 1986), protected from extensive effects of waves and currents. Furthermore, the subaerial and transitional parts were dominated by fluvial processes and contain little indication of marine processes. Several episodes of delta progradation and basin subsidence resulted in various scale of coarsening-upward sequences (from 4 m up to 30 m in thickness).

Facies association IV: Gilbert-type foreset

Occurrence. This association is characterized by steeply inclined (> 20°) beds of massive/crudely stratified (facies G1-b), stratified (facies G2-a and G2-b) and graded (or inversely to normally graded) (facies G3) gravelstones with sheet geometry (Figs 10D, 14). Each flow unit is generally thin and laterally discontinuous. At the base of the foreset, outsized boulders (more than 1 m in long axis) are concentrated and thin muddy sandstone units are intercalated (Fig. 10D). This facies occurs in sections of Ct-T-3, -4, -5, Ct-F-1, -2, -3 and -4 (Fig. 2), which form the middle and upper parts of the Chunbuk Formation.

Interpretation. Coarse-grained sediments on the terminal edge of the transitional zone are resedimented and transported to the basin by various sediment gravity flows. The foreset beds were deposited mainly by small-scale slides (gravity slide or grain flow) with slight influence of density-modified grain flow, debris flow and gravelly high-density turbidity current (Postma, 1983, 1984, 1986; Postma & Roep, 1985; Colella *et al.*, 1987; Colella, 1988a). In the lower part, dominance of thick massive/crudely stratified (facies G1-b) and graded (facies G3) gravelstones implies that fluid turbulence played an important role. Outsized boulders were rolled and slid along the steep slope and accumulated at the base of the foreset. The foreset sequence lacks large-scale slope-failure structures that commonly occur in other delta-front units (Prior *et al.*, 1981; Postma, 1983, 1984; Postma & Roep, 1985).

The height of the foreset is more than 150 m. It is larger than many other examples (e.g. Postma, 1983; Postma & Roep, 1985; Colella *et al.*, 1987). The foreset prograded for about 1 km with a slope angle of about 20° and shows a radial distributional pattern from the point of sediment origin, west of the Doumsan mountain. Palaeocurrent pattern is similar to the local dip direction of the slope which also shows a radial pattern (Fig. 1). This association grades laterally into facies association V (toeset).

Interface between the topset and the foreset. The interface between the topset and the foreset is represented by the co-occurrence of facies associations III and IV with two distinct geometries (Figs 14, 15). The first type shows large truncation surfaces where the sequence of association III is steeply incised by that of association IV (Fig. 14). The interface extends

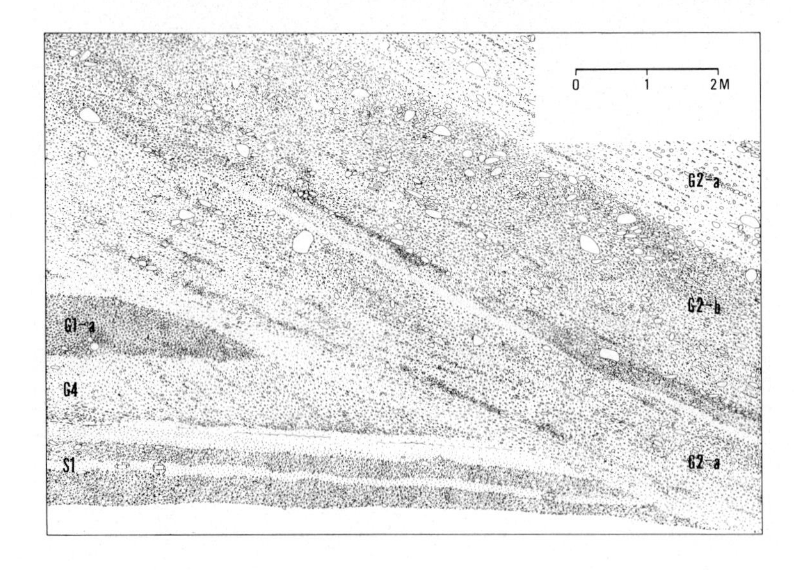

Fig. 14. Detailed sketch of part of a large-scale truncation surface in the foreset. The lower left is characterized by massive/crudely stratified and cross-bedded gravelstones (facies G1-a and G4) and massive sandstone (facies S1), similar to the Gilbert-type topset facies. The foreset is characterized by steeply inclined beds of stratified gravelstones (facies G2-a and G2-b).

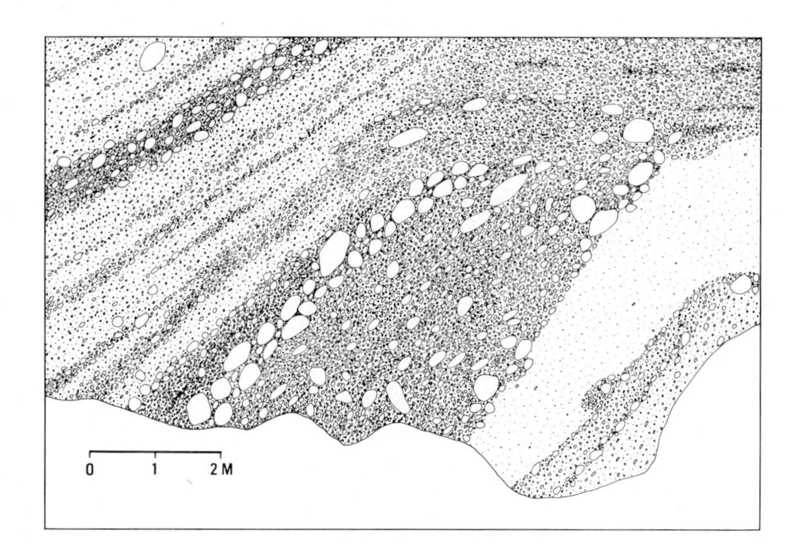

Fig. 15. Detailed sketch of the boundary between topset and foreset (section Ct-T-4, Fig. 2). Planar beds of the topset are transitional to steeply inclined foreset beds.

for several hundreds of metres downslope and several kilometres along the strike (Fig. 9). The second type is characterized by gradational change in bedding plane, in which the planar beds of association III are transitional to the steeply inclined beds of association IV (Fig. 15).

Geometry 1. The first type of interface differs from common Gilbert-type fan-deltas or microdeltas in which the topset lies on the foreset. It is similar to the back edge of the delta foreset units (BEFU) of Colella (1988a,b), forming the line along which the earliest foreset beds develop. According to Colella (1988a,b), the back edge of the delta foreset unit is marked by an abrupt change in slope of the underlying bedrock, and in the Crati Basin examples, it is coincident with the scarp of an extensional or strike-slip fault. However, in the Doumsan system, the underlying sequence is similar to that of the topset sequence, indicating that the early fan was formed in shallow-marine settings, showing characteristics of shelf-type fan-deltas or coastal alluvial fans. Small-scale sea-level changes and rapid sedimentation resulted in several coarsening-upward sequences in the transitional part. The basin was then subsided rapidly, and the unstable sediments at the topset edge were slumped, forming a steep tectonic slope. The foreset sequence was initially deposited on the scour surface and prograded further basinward where part of the association III overlies the foreset sequence (see Geometry 2). The large-scale of the truncation surfaces and their extensive lateral continuity, as well as the abnormally large-scale foreset units, also suggest tectonic movements, perhaps related to the right-lateral strike-slip fault systems in southeastern Korea (Reedman & Um, 1975).

Geometry 2. The second type of interface forms the boundary between the topset and the foreset resembling the sigmoidal progradational configuration of Mitchum *et al.* (1977) and Colella (1988a,b). According to Mitchum *et al.* (1977), this configuration implies relatively low sediment supply, rapid basin subsidence and/or rapid rise in sea level to allow deposition and preservation of the topset units. In the Doumsan system, large amounts of sediments were supplied from the adjacent highlands and there was a general fall of eustatic sea level during the foreset progradation (approximately 15 to 10 Ma). Thus, the sigmoidal geometry was mainly due to rapid basin subsidence.

Facies association V: Gilbert-type toeset

Occurrence. This association is characterized by deformed units of massive/crudely stratified gravelstones (facies G1-b) and massive muddy sandstone (facies MS1-b) (Figs 10E, 16). It occurs in sections of Ct-B-1, -2, -3 and -4, and Sh-1 (Fig. 2), which form the upper part of the Chunbuk Formation. Syndepositional deformations are represented by large-scale folds, faults and flame structures (Fig. 16). Small-scale chaotic units such as slide blocks and armoured mudstone balls also occur (Fig. 16). Clasts are less well-oriented than those of the foreset sediments. Each flow unit is generally thick and

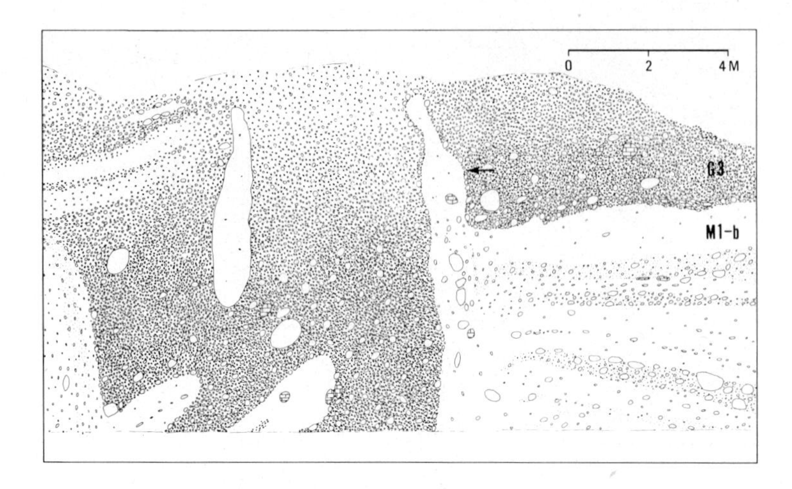

Fig. 16. Detailed sketch of deformed toeset bed (section Ct-B-4, Fig. 2). Note large-scale flame structures (arrow).

shows diffuse upper and lower boundaries (Fig. 10E).

Interpretation. This facies association most probably represents deposition in the toeset environment where the bedding plane is less steep (< 10°) than that of the foreset (> 20°). The gravelstones were deposited by gravelly high-density turbidity currents, whereas the muddy sandstones were formed by hemipelagic settling of terrigenous material separated at the river mouth. Although flow transformation from foreset to toeset is rather unclear, the gravelstones may have been deposited from the flow generated in the foreset. As the flow reaches to the toeset, flow velocity decreases abruptly with decreasing slope angle, resulting in 'hydraulic jump' (Porębski, 1984). An abrupt decrease in flow competence was probably responsible for the deposition of thick massive gravelstones.

Deformation at delta fronts is widely recognized in modern fjord deltas by Prior *et al.* (1981) and Prior and Bornhold (1988). Recently, Postma and Roep (1985) and Postma *et al.* (1988) observed deformed gravelstones at the base of Gilbert-type fan delta. According to Postma *et al.* (1988), the bottomset (or toeset in the Doumsan system) deformation is due to rotational sliding and resultant plastic flowage of the foreset and adjacent bottomset strata. The deformation propagates in the bottomset mainly as 'bulldozer' effect associated with the laminar flowage of the detached foreset strata (Postma *et al.*, 1988). However, abundant muds and rapid deposition of gravels as well as differential loading by the prograding foreset unit would also cause syndepositional deformation.

Slide blocks may have originated from calcite-cemented sandstone blocks which occur in the transitional part. The sediments that contain calcite-cemented sandstone blocks were avalanched at the terminal edge of the topset, and the blocks were rolled and slid down the steep foreset slope. Armoured mudstone balls indicate rounding and studding with gravels as they rolled. The slide blocks and armoured mudstone balls as well as outsized boulders were deposited in the toeset, prodelta and slope apron (Choe, 1986; Choe & Chough, 1988; Chough *et al.*, 1989, 1990).

SUMMARY

The Miocene Chunbuk Formation represents part of the Doumsan fan-delta system and displays marine, Gilbert-type fan-delta geometry and sedimentary facies (Fig. 17). It comprises tripartite depositional elements of subaerial, transitional and subaqueous parts. Near the Doumsan mountain, three small-scale subaerial fans (SF) occur along the faulted boundary (Fig. 9). The subaerial fan grades laterally into braided streams (SBS) that constitute part of a large-scale, wet-type alluvial fan. In the transitional part, various gravelstone facies were formed by migrating gravel bars and as small-scale channel fills, whereas sandstone facies were deposited in shallow-water environments (coastal lagoons or salt marshes). The coarsening-upward sequence from sandstone to gravelstone facies reflects prograding fluvial systems into a shallow-marine environment.

Initial development of the Gilbert-type foresets in the Doumsan system is most likely due to the effect

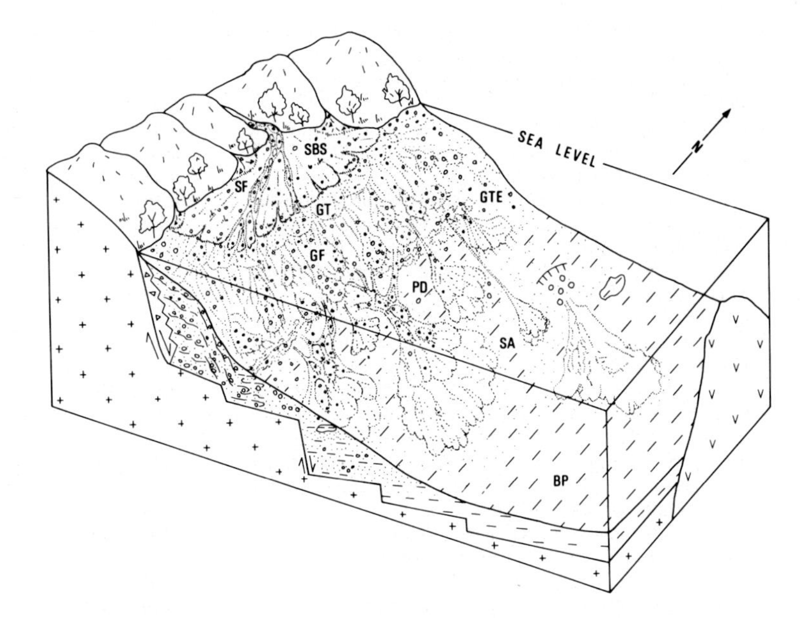

Fig. 17. Simplified model of depositional environments in the Doumsan fan-delta system. SF = subaerial fan, SBS = subaerial braided stream, GT = Gilbert-type topset, GF = Gilbert-type foreset, GTE = Gilbert-type toeset, PD = prodelta, SA = slope apron, BP = basin plain. After Chough *et al.* (1990).

of syndepositional tectonic subsidence. The large-scale truncation surface on which the foreset beds lie, is steeply incised in deposits of the transitional zone. This geometry indicates that an early shelf-type fan delta was formed in a shallow-marine setting. The basin then subsided rapidly and sediments on the topset edge were resedimented, forming a steeply inclined tectonic slope. Extensive lateral continuity of the truncation surface as well as the abnormally thick foreset units (more than 150 m in foreset height) are suggestive of rapid tectonic movement.

Foreset sediments were transported mainly by small-scale slides (gravity slides) or cohesionless debris flows forming stratified gravelstones (facies G2-a and G2-b). As the flow accelerates downslope, fluid turbulence gradually increases, depositing massive/crudely stratified (facies G1-b) and graded (facies G3) gravelstones at the base of the foreset. Outsized boulders, calcite-cemented sandstone blocks as well as mud blocks were rolled and slid down the steep slope. At the toeset, the flow decelerates with decreasing slope angle, causing a 'hydraulic jump'. Rapid sedimentation resulted in thick massive/crudely stratified gravelstones (facies G1-b). Fine fractions that are separated at the river mouth are settled from suspension forming thick, massive muddy sandstone (facies MS1-b). The bottomset sequence was extensively deformed showing large-scale folds, faults and flame structures.

ACKNOWLEDGEMENTS

This work was supported through grants to Chough by the Korea Science and Engineering Foundation and is based on Hwang's MSc thesis completed at the Seoul National University. M.Y. Choe, S.S. Chun, S.H. Yoon and Y.K. Sohn provided discussions in the field and laboratory work. The paper was improved by helpful criticism from M.R. Leeder and F. Massari, and in its final version, by A. Colella. All this help is gratefully acknowledged.

REFERENCES

Aalto, K.R. (1976) Sedimentology of a melange: Franciscan of Trinidad, California. *J. sedim. Petrol.* **46**, 913–929.

Allen, J.R.L. (1982) *Sedimentary Structures: Their Character and Physical Basis. Developments in Sedimentology*, 30A. Elsevier, Amsterdam, 593 pp.

Allen, J.R.L. (1983) Studies in fluviatile sedimentation: bars, bar-complexes and sandstone sheets (low-sinuousity braided streams) in the Brownstones (L. Devonian), Welsh Borders. *Sedim. Geol.* **33**, 237–293.

Ballance, P.F. (1984) Sheet-flow-dominated gravel fans of the non-marine middle Cenozoic Simmler Formation, central California. *Sedim. Geol.* **38**, 337–359.

Barg, E. (1986) Cenozoic geohistory of the southwestern margin of the Ulleung Basin, East Sea. Unpublished MSc thesis, Seoul National University, 173 pp.

Bates, C.C. (1953) Rational theory of delta formation. *Bull. Am. Ass. Petrol. Geol.* **37**, 2119–2162.

BONG, P.Y. (1982) Palynology and stratigraphy of Yeonil-Dongsanri area. *Report on Geoscience and Mineral Resources, Korea Institute of Energy and Resources* **13**, 19–24.

BONG, P.Y. (1985) Palynology of the Neogene strata in the Pohang sedimentary basin. Unpublished PhD thesis, Seoul National University, 239 pp.

BULL, W.B. (1972) Recognition of alluvial-fan deposits in the stratigraphic record. In: *Recognition of Ancient Sedimentary Environments*. (Ed. by J.K. Rigby and W.K. Hamblin). Spec. Publ. Soc. econ. Paleont. Mineral. Tulsa, 16, pp. 63–83.

CHOE, M.Y. (1986) The Hunghae Formation: coalescing slope aprons. Unpublished MSc thesis, Seoul National University, 207 pp.

CHOE, M.Y. & CHOUGH, S.K. (1988) The Hunghae Formation, SE Korea: Miocene debris aprons in a back-arc intraslope basin. *Sedimentology* **35**, 239–255.

CHOUGH, S.K. (1983) *Marine Geology of Korean Seas*. International Human Resources Development Corporation, Boston, 157 pp.

CHOUGH, S.K. & BARG, E (1987) Tectonic history of Ulleung Basin margin, East Sea (Sea of Japan). *Geology* **15**, 45–48.

CHOUGH, S.K., HWANG, I.G. & CHOE, M.Y. (1989) The Doumsan fan-fan delta system, Miocene Pohang Basin (SE Korea). *Field Excursion Guidebook*. Woosung Publ. Co., Seoul, 95 pp.

CHOUGH, S.K., HWANG, I.G. & CHOE, M.Y. (1990) The Miocene Doumsan fan-delta, SE Korea: a composite fan-delta system in back-arc margin. *J. sedim. Petrol.* **60** (in press).

CHUN, H.Y. (1982) Plant fossils from the Tertiary Pohang sedimentary basin, Korea. *Report on Geoscience and Mineral Resources, Korea Institute of Energy and Resources* **14**, 7–24.

COLELLA, A. (1988a) Pliocene–Holocene fan deltas and braid deltas in the Crati Basin, southern Italy: a consequence of tectonic conditions. In: *Fan Deltas: Sedimentology and Tectonic Settings* (Ed. by W. Nemec and R.J. Steel), pp. 50–74. Blackie and Son, London.

COLELLA, A. (1988b) Fault-controlled marine Gilbert-type fan deltas. *Geology* **16**, 1031–1034.

COLELLA, A., DE BOER, P.L. & NIO, S.D. (1987) Sedimentology of a marine intermontane Pleistocene Gilbert-type fan-delta complex in the Crati Basin, Calabria, southern Italy. *Sedimentology* **34**, 721–736.

COOK, H.E. (1979) Ancient continental slope sequences and their value in understanding modern slope development. In: *Geology of Continental Slopes* (Ed. by L.J. Doyle and O.H. Pilkey). Spec. Publ. Soc. econ. Paleont. Mineral, Tulsa, 27, pp. 287–306.

CURRY, R.R. (1966) Observation of Alpine mudflows in the Tenmile Range, Central Colorado. *Bull. geol. Soc. Am.* **77**, 771–776.

ETHRIDGE, F.G. & WESCOTT, W.A. (1984) Tectonic setting, recognition and hydrocarbon reservoir potential of fan-delta deposits. In: *Sedimentology of Gravels and Conglomerates* (Ed. by E.H. Koster and R.J. Steel). Mem. Can. Soc. Petrol. Geol. 10, pp. 217–235.

FLORES, R.M. (1975) Short-headed stream delta: model for Pennsylvanian Haymond Formation, West Texas. *Bull. Am. Assoc. Petrol. Geol.* **59**, 2288–2301.

FRIEND, P.F. (1983) Towards the field classification of alluvial fan architecture or sequence. In: *Modern and Ancient Fluvial Sediments* (Ed. by J.D. Collinson and J. Lewin). Spec. Publs. int. Assoc. Sedim. 6, pp. 345–354.

GALLOWAY, W.E. (1976) Sediment and stratigraphic framework of the Copper River fan delta, Alaska. *J. sedim. Petrol.* **46**, 726–737.

GILBERT, G.K. (1885) The topographic features of lake shores. *U.S. geol. Surv. 5th Annual Rept.*, 69–123.

GLOPPEN, T.G. & STEEL, R.J. (1981) The deposits, internal structure and geometry in six alluvial fan-fan delta bodies (Devonian-Norway) — a study in the significance of bedding sequences in conglomerates. In: *Recent and Ancient Nonmarine Depositional Environments: Models for Exploration* (Ed. by F.G. Ethridge and R.M. Flores). Spec. Publ Soc. econ. Paleont. Miner., Tulsa, 31, pp. 49–69.

GRAVENOR, C.P. (1986) Magnetic and pebble fabrics in subaquatic debris-flow deposits. *J. Geol.* **94**, 683–698.

GUSTAVSON, T.C. (1974) Sedimentation on gravel outwash fans, Malaspina Glacier Foreland, Alaska. *J. sedim. Petrol.* **44**, 374–389.

HARMS, J.C., SOUTHARD, J.B., SPEARING, D.R. & WALKER, R.G. (1975) Depositional environments as interpreted from primary sedimentary structures and stratification sequences. Soc. econ. Paleont. Miner., Tulsa, Short Course no. 2, 161 pp.

HARMS, J.C., SOUTHARD, J.B. & WALKER, R.G. (1982) Structures and sequences in clastic rocks. Soc. econ. Paleont. Miner., Tulsa, Short Course no. 9.

HARRISON, S. & FRITZ, W.J. (1982) Depositional features of March 1982 Mount St. Helens sediments flows. *Nature* **299**, 720–722.

HAYES, M.O. & MICHEL, J. (1982) Shoreline sedimentation within a forearc embayment, Lower Cook Inlet, Alaska, *J. sedim. Petrol.* **52**, 251–263.

HWANG, I.G. (1989) The Miocene Chunbuk Formation, Pohang Basin: Gilbert-type fan-delta system. Unpublished MSc thesis, Seoul National University, 244 pp.

KILLICK, M.F. (1988) Sedimentary and tectonic controls of fan-delta facies and development: an example from the Infracambrian of High Atlas, Morocco. In: *Fan Deltas: Sedimentology and Tectonic Settings* (Ed. by W. Nemec and R.J. Steel), pp. 212–225. Blackie and Son, London.

KIM, B.K. (1965) The stratigraphic and paleontologic studies on the Tertiary (Miocene) of the Pohang area, Korea. *J. Seoul National University, Sci. Tech. Ser.*, **15**, 32–121.

KIM, K.J. (1987) Organic-walled microfossils from the Neogene strata of H-well, Pohang, Korea. Unpublished MSc thesis, Seoul National University, 69 pp.

KLEINSPEHN, K.L., STEEL, R.J., JOHANNESSEN, E. & NETLAND, A. (1984) Conglomeratic fan-delta sequences, Late Carboniferous–Early Permian, western Spitsbergen. In: *Sedimentology of Gravels and Conglomerates* (Ed. by E.H. Koster and R.J. Steel). Mem. Can. Soc. Petrol. Geol. 10, 279–294.

KOSTER, E.H. (1978) Transverse ribs: their characteristics, origin and palaeohydraulic significance. In: *Fluvial Sedimentology* (Ed. by A.D. Miall). Mem. Can. Soc. Petrol. Geol. 5, pp. 161–186.

LEE, H.Y. (1982) Neogene foraminifera from southern

part of Euichang area, Korea. *Report on Geoscience and Mineral Resources, Korea Institute of Energy and Resources* **13**, 19–34.

Lee, Y.G. (1984) Micropaleontological (Diatom) study of the Neogene deposits in Korea. Unpublished PhD thesis, Seoul National University, 285 pp.

Lewis, D.W., Laird, M.G. & Powell. R.D. (1980) Debris-flow deposits of early Miocene age, Deadman stream, Marlborough, New Zealand. *Sedim. Geol.* **29**, 83–118.

Lindsay, J.F. (1968) The development of clast fabric in mud flows. *J. sedim. Petrol.* **38**, 1242–1253.

Lowe, D.R. (1979) Sediment gravity flows: their classification and some problems of application to natural flows and deposits. In: *Geology of Continental Slopes* (Ed. by L.J. Doyle and O.H. Pilkey). Spec. Publ. Soc. econ. Paleont. Miner., Tulsa, 27, pp. 75–82.

Lowe, D.R. (1982) Sediment gravity flows: II. Depositional models with special reference to the deposits of high-density turbidity currents. *J. sedim. Petrol.* **52**, 279–297.

Massari, F. (1984) Resedimented conglomerates of a Miocene fan-delta complex, southern Alps, Italy. In: *Sedimentology of Gravels and Conglomerates* (Ed. by E.H. Koster and R.J. Steel). Mem. Can. Soc. Petrol. Geol. 10, pp. 259–278.

Miall, A.D. (1977) A review of the braided-river depositional environment. *Earth Sci. Rev.* **13**, 1–62.

Miall, A.D. (1978) Lithofacies types and vertical profile models in braided river deposits: a summary. In: *Fluvial Sedimentology* (Ed. by A.D. Miall). Mem. Can. Soc. Petrol. Geol. 5, pp. 597–604.

Miall, A.D. (1985) Architectural-element analysis: a new method of facies analysis applied to fluvial deposits. In: *Recognition of Fluvial Depositional Systems and their Resource Potential* (Ed. by R.M. Flores *et al.*). Soc. econ. Paleont. Miner., Tulsa, Short Course no. 19, 33–81.

Miall, A.D. (1988) Architectural elements and bounding surfaces in fluvial deposits: anatomy of the Kayenta Formation (lower Jurassic), southwest Colorado. *Sedim. Geol.* **55**, 233–262.

Middleton, G.V. & Hampton, M.A. (1976) Subaqueous sediment transport and deposition by sediment gravity flows. In: *Marine Sediment Transport and Environmental Management* (Ed. by D.J. Stanley and D.J.P. Swift), pp. 197–218. Wiley and Sons, New York.

Mitchum, R.M., Vail, P.R., Jr. & Sangree, J.B. (1977) Seismic stratigraphy and global changes of sea level, Part 6: stratigraphic interpretation of seismic reflection patterns in depositional sequences. In: *Seismic Stratigraphy – Applications to Hydrocarbon Exploration* (Ed. by C.E. Payton). Mem. Am. Assoc. Petrol. Geol. 26, 117–133.

Nemec, W. & Steel, R.J. (1984) Alluvial and coastal conglomerates: their significant features and some comments on gravelly mass-flow deposits. In: *Sedimentology of Gravels and Conglomerates* (Ed. by E.H. Koster and R.J. Steel). Mem. Can. Soc. Petrol. Geol. 10, pp. 1–32.

Orton, G.J. (1988) A spectrum of middle Ordovician fan deltas and braidplain deltas, North Wales: a consequence of varying fluvial clastic input. In: *Fan Deltas: Sedimentology and Tectonic Settings* (Ed. by W. Nemec and R.J. Steel), pp. 23–49. Blackie and Son, London.

Pierson, T.C. (1981) Dominant particle support mechanisms in debris flows at Mt. Thomas, New Zealand, and implications for flow mobility. *Sedimentology* **28**, 49–60.

Porębski, S.J. (1984) Clast size and bed thickness trends in resedimented conglomerates: example from a Devonian fan-delta succession, southwest Poland. In: *Sedimentology of Gravels and Conglomerates* (Ed. by E.H. Koster and R.J. Steel). Mem. Can. Soc. Petrol. Geol. 10, pp. 399–411.

Postma, G. (1983) Water-escape structures in the context of a depositional model of a mass flow-dominated conglomeratic fan delta (Abrioja Formation, Pliocene, Almeria Basin, SE Spain). *Sedimentology* **30**, 91–103.

Postma, G. (1984) Mass-flow conglomerates in a submarine canyon: Abrioja fan delta, Pliocene, southeast Spain. In: *Sedimentology of Gravels and Conglomerates* (Ed. by E.H. Koster and R.J. Steel). Mem. Can. Soc. Petrol. Geol. 10, pp. 237–258.

Postma, G. (1986) A classification for sediment gravity flows based on flow characteristics during deposition. *Geology* **14**, 291–294.

Postma, G., Babić, L., Zupanič, J. & Røe, S.L. (1988) Delta-front failure and associated bottomset deformation in a marine gravelly Gilbert-type fan delta. In: *Fan Deltas: Sedimentology and Tectonic Settings* (Ed. by W. Nemec and R.J. Steel), pp. 91–102. Blackie and Son, London.

Postma, G. & Roep, T.B. (1985) Resedimented conglomerates in the bottomset of Gilbert-type gravel deltas. *J. sedim. Petrol.* **55**, 874–885.

Prior, D.B. & Bornhold, B.D. (1988) Submarine morphology and processes of fjord fan deltas and related high-gradient systems: modern examples from British Columbia. In: *Fan Deltas: Sedimentology and Tectonic Settings* (Ed. by W. Nemec and R.J. Steel), pp. 125–143. Blackie and Son, London.

Prior, D.B., Wiseman, W.J., Jr. & Bryant, W.R. (1981) Submarine chutes on the slope of fjord deltas. *Nature* **290**, 326–328.

Reedman, A.J. & Um, S.H. (1975). *The Geology of Korea*. Korea Institute of Energy and Resources, 139 pp.

Rhine, J.L. & Smith, D.G. (1988) The late Pleistocene Athabasca braid delta of northeastern Alberta, Canada: a paraglacial drainage system affected by aeolian sand supply. In: *Fan Deltas: Sedimentology and Tectonic Settings* (Ed. by W. Nemec and R.J. Steel), pp. 158–169. Blackie and Son, London.

Rust, B.R. (1972) Structure and processes in a braided river. *Sedimentology* **18**, 221–245.

Rust, B.R. (1978) Depositional models for braided alluvium. In: *Fluvial Sedimentology* (Ed. by A.D. Miall). Mem. Can. Soc. Petrol. Geol. 5, pp. 605–626.

Rust, B.R. & Koster, E.H. (1984) Coarse alluvial deposits. In: *Facies Model* (Ed. by R.G. Walker). *Geo. Sci. Can. Reprint Ser.* **1**, 53–70.

Schultz, A.W. (1984) Subaerial debris-flow deposition in the Upper Paleozoic Cutler Formation, western Colorado. *J. sedim. Petrol.* **54**, 759–772.

Smith, G.A. (1986) Coarse-grained nonmarine volcaniclastic sediment: terminology and depositional process. *Bull. geol. Soc. Am.* **97**, 1–10.

Smith, N.D. (1970) The braided stream depositional environment: Comparison of the Platte River with some

Silurian clastic rocks, north-central Appalachians. *Bull. geol. Soc. Am.* **81**, 2993–3014.

Smith, N.D. (1972) Some sedimentological aspects of planar cross-stratification in a sandy braided river. *J. sedim. Petrol.* **42**, 624–634.

Steel, R.J. (1976) Devonian basins of western Norway, sediment response to tectonism and varying tectonic context. *Tectonophysics,* **36**, 107–224.

Surlyk, F. (1984) Fan-delta to submarine fan conglomerates of the Volgian–Valanginian Wollastone Forland Group, East Greenland. In: *Sedimentology of Gravels and Conglomerates* (Ed. by E.H. Koster and R.J. Steel). Mem. Can. Soc. Petrol. Geol. 10, pp. 359–382.

Um, S.H., Lee, D.W. and Park, B.S. (1964) *Geological Map of Korea, Pohang Sheet* (1:50000). Geol. Surv. Korea, 21 pp.

Walker, R.G. (1978) Deep-water sandstone facies and ancient submarine fans: models for exploration for stratigraphic traps. *Bull. Am. Assoc. Petrol. Geol.* **62**, 932–966.

Walker, R.G. & Mutti, E. (1973) Turbidites facies and facies associations. In: *Turbidites and Deep-water Sedimentation* (Ed. by G.V, Middleton and A.H. Bouma). Soc. econ. Paleont. Miner. Pacific Section, Short Course pp. 119–157.

Wescott, W.A. & Ethridge, F.G. (1980) Fan-delta sedimentology and tectonic setting — Yallahs fan-delta, southeast Jamaica. *Bull. Am. Assoc. Petrol. Geol.* **64**, 374–399.

Winn, R.D., Jr. and Dott, R.H., Jr. (1977) Large-scale traction-produced structures in deep-water fan-channel conglomerates in southern Chile. *Geology* **5**, 41–44.

Wright, L.D. & Coleman, J.M. (1974) Mississippi river mouth processes: effluent dynamics and morphological development. *J. Geol.* **82**, 751–778.

Yoo, E.K. (1969) Tertiary Foraminifera from the PY-1 well, Pohang Basin, Korea. *J. geol. Soc. Korea*, **5**, 77–96.

Yoon, S. (1975) Geology and paleontology of the Tertiary Pohang Basin, Pohang district, Korea, Part 1. Geology *J. geol. Soc. Korea,* **11**, 187–214.

Yoon, S. (1976a) Geology and paleontology of the Tertiary Pohang Basin, Pohang district, Korea, Part 2. paleontology (mollusca), No. 1. systematic description of bivalvia. *J. geol. Soc. Korea,* **12**, 1–22.

Yoon, S. (1976b) Geology and paleontology of the Tertiary Pohang Basin, Pohang district, Korea, Part 2. paleontology (mollusca), No. 2. scaphopoda and gastropoda. *J. geol. Soc. Korea,* **12**, 63–72.

You, H.S (1983) The biostratigraphy of the Neogene Tertiary deposits, Korea. Unpublished PhD thesis, Seoul National University, 180 pp.

Yun, H.S. (1986) Amended stratigraphy of the Miocene formations in the Pohang Basin, Part 1. *J. Paleont. Soc. Korea,* **2**, 54–69.

Spec. Publs int. Ass. Sediment. (1990) **10**, 255–269

Sequence analysis of a marine Gilbert-type delta, La Miel, Albian Lunada Formation of northern Spain

J. GARCÍA-MONDÉJAR

Departamento de Estratigrafía, Geodinámica y Paleontología, Universidad del País Vasco, Apdo. 644, Bilbao 48080, Spain

ABSTRACT

The marine Gilbert-type delta of La Miel, at the base of the Albian Lunada Formation of north Spain, is about 2 km long, 1 km wide and 70 m thick, and consists of sandstones and small quartz-pebble conglomerates. From a section along the axis of the delta, six deltaic and two predeltaic sequences bounded by erosional unconformities have been distinguished. There are several genetic units of deltaic progradation, marine and fluvial aggradation and lateral marine accretion within the sequences, and each genetic unit is made up of one or more facies representing sedimentary environments that range in character from comparatively deep marine (tens of metres) to fluvial. The progradational Gilbert-type units have bottomsets and foresets but rarely topsets; the aggradational units are either shallow marine with corals or fluvial with braided channels; and the marine lateral accretion units are made up of pre-existing facies reworked by tides and waves.

Correlation of the La Miel sequences with others from different outcrops suggests an allocyclic origin for all of them. Short-term, relative sea-level variations, probably tectonically induced, are thought to have been responsible for both the sequences and the genetic units and facies within them. The sequences can be described by sequence stratigraphy (systems tracts) methodology.

The delta of La Miel is considered a part of a major system prograding towards the NW, the Valnera braidplain delta front, made up of several Gilbert-type contemporaneous lobes.

INTRODUCTION

High-angle deltas with the characteristic, tripartite internal structure (bottomset–foreset–topset) were first described from the shoreline deposits of Lake Bonneville (Gilbert, 1885). Subsequent descriptions of such Gilbert-type deltas were made in many other lacustrine deposits (e.g. Axelsson, 1967; Gustavson *et al.*, 1975; Stanley & Surdam, 1978; Farquharson, 1982; Dunne & Hempton, 1984; Kazancı, 1988). Only more recently have marine examples been reported (e.g. Hobday, 1974; García-Mondéjar, 1979; Walker & Harms, 1980; Prior *et al.*, 1981; Van der Meulen, 1983; Postma, 1984; Colella, 1984, 1988; Postma & Roep, 1985; Suter & Berryhill, 1985; Colella *et al.*, 1987; Ori & Roveri, 1987; Massari & Colella, 1988; Postma *et al.*, 1988; Postma & Cruickshank, 1988; Prior & Bornhold, 1988; Wood & Ethridge, 1988).

Several types of marine Gilbert-type deltas have been distinguished using the following characteristics: (1) the geometries of prograding foresets in fault-controlled situations (Colella, 1988); (2) the main sedimentary processes, fluvial or marine, forming both the topset and the transition zone (Colella, 1988); (3) the presence of gravitational failures on the delta front (Postma & Roep, 1985); (4) the lateral and vertical stacking of discrete delta bodies enveloped by marine transgressive facies (García-Mondéjar, 1979, 1988; Gawthorpe *et al.*, 1988; Zupanic *et al.*, 1988; (5) the geometry of the whole deltaic body when this is conditioned by specific relief in the substratum, such as large slump scars cut into shelf-edge deposits (Massari & Colella, 1988), Mediterranean rias (Clauzon & Rubino, 1988; Massari & Parea, 1988), other palaeovalleys (Muto, 1988), fiords (Corner *et al.*, 1988); and (6) the special volcanic nature, lava, of the material flowing into the sea (Porębski, 1988).

This study deals with lenticular coarse-clastic bodies in the Albian Lunada Formation of La Miel area, northern Spain (Fig. 1), which have been interpreted as a marine Gilbert-type delta belonging to a general braidplain delta front. The deltaic units in this model rest on erosional surfaces and are separated vertically and horizontally from each other by shallow-marine deposits, including coral limestones. The complete vertical section consists of several transgressive–regressive sequences, which have been related to sea-level fluctuations during an overall sea-level rise. Nevertheless, localized subsidence in an extensional regime is thought to have created the conditions for Gilbert-type delta construction. These include: high coastal-plain gradients, availability of coarse terrigenous sediments and confinement of the supplying river system in an elongate depression.

THE LUNADA FORMATION

The predominantly Mesozoic Basque-Cantabrian region of northern Spain is located between the Asturian Massif to the west and the Pyrenees to the east. It consists of three main domains from west to east: Peri-Asturian, Navarro-Cantabre and Basque (Fig. 1). The Peri-Asturian domain is composed mainly of Lower Cretaceous rocks, a significant proportion of these being made up of Aptian–Albian carbonates and siliciclastics (Urgonian Complex; Rat, 1959). In the central part of this domain (Valnera area), the Urgonian sedimentation is inferred to have filled an E–W trough with fluvial, shallow-marine and relatively deep-marine sediments. The early to middle Albian Lunada Formation (García-Mondéjar, 1979, 1982) is a representative unit of this basin-fill succession (Fig. 2). The formation consists of siliceous and calcareous

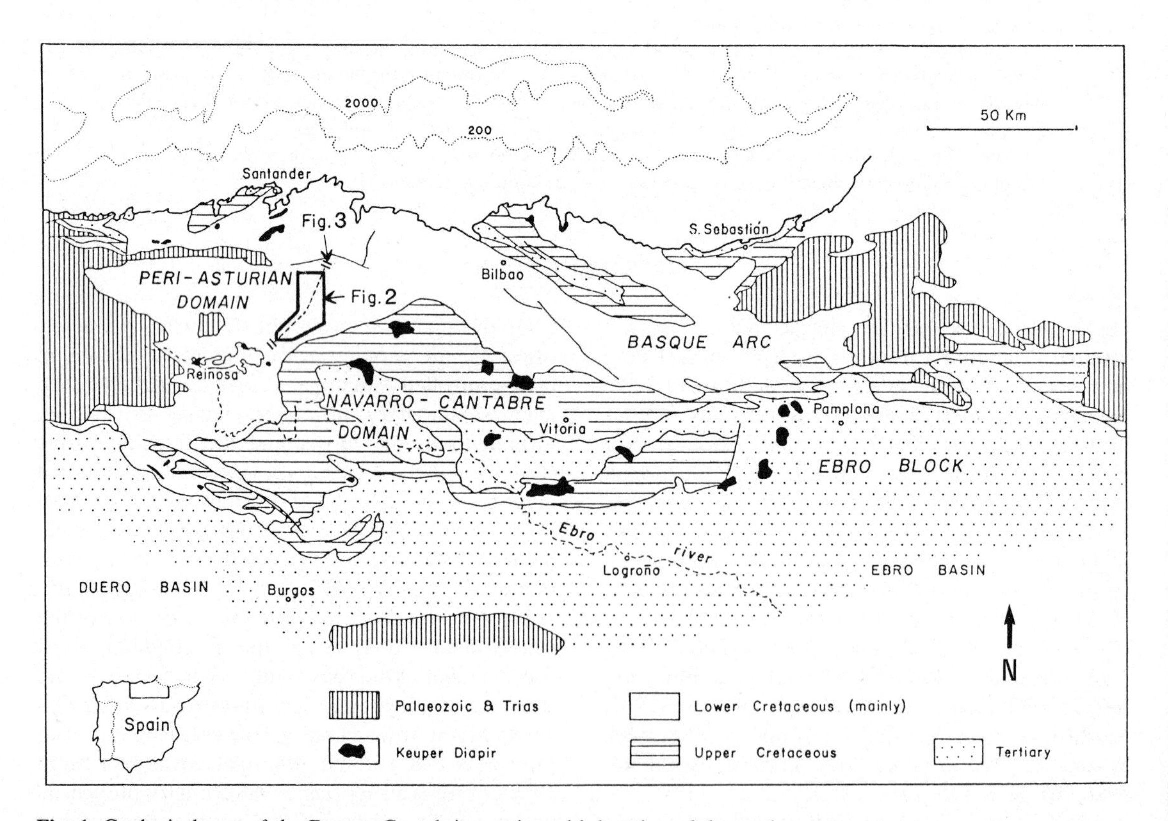

Fig. 1. Geological map of the Basque-Cantabrian region with location of the study area.

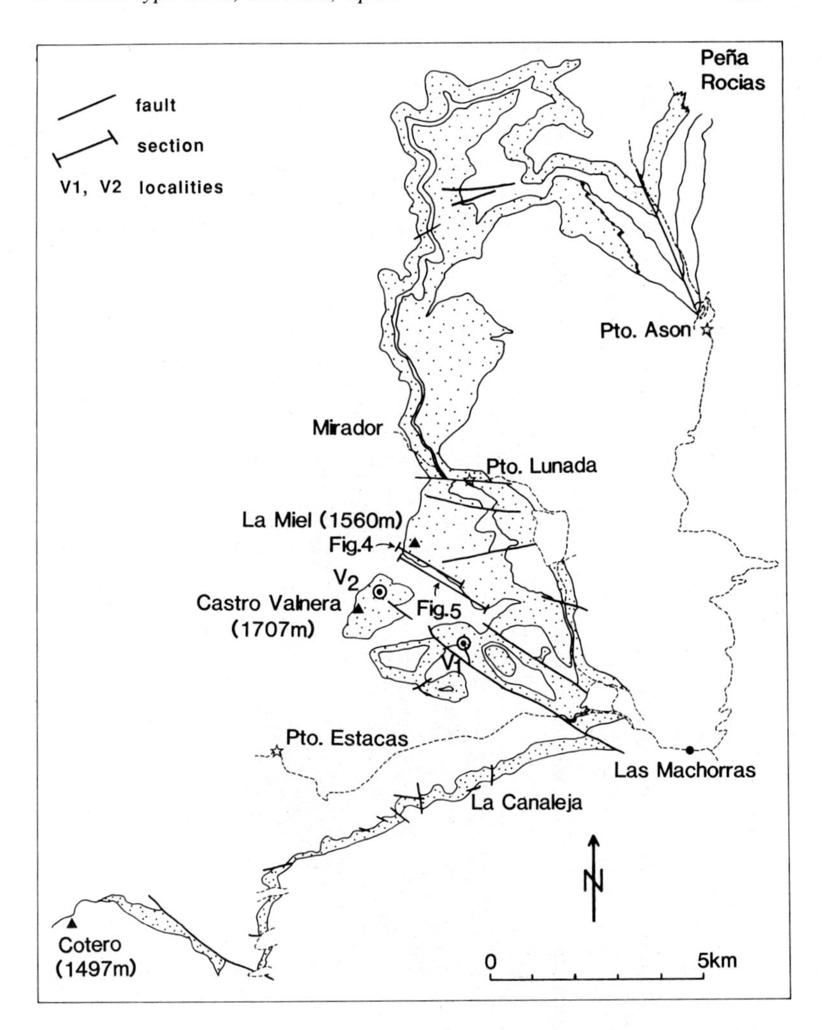

Fig. 2. Map of outcrops of the Lunada Formation (reference in Fig. 1). The locations of Figs 4 and 5 are shown.

sandstones (quartz arenites), quartz-pebble conglomerates and mudstones of 'transitional' (terrestrial/marine) environments, but includes also coral/rudist-rich (Urgonian) limestones of shallow-marine origin. It locally rests unconformably on eroded and slightly tilted limestones of the late Aptian–earliest Albian (Cantos Blancos Limestones Formation) in areas of diminutive subsidence, and conformably on the estuarine Rio Miera Sandstone Formation, in areas of strong subsidence (Fig. 3). The Lunada Formation grades both laterally towards the NE and vertically into Urgonian limestones (Figs 2, 3). It is made up of three members: the La Miel Member, with sandstones, mudstones, conglomerates and limestones; the Puerto de Lunada Member, with rudist-rich limestones; and the Rio Lunada Member, with sandstones, mudstones, limestones and conglomerates (Fig. 3; García-Mondéjar, 1982).

LA MIEL DELTA

Sandstones and conglomerates with Gilbert-type delta characteristics are common in the outcrops of the Lunada Formation between La Canaleja and Pto. de Lunada sections (Fig. 3). The delta construction, which is the subject of this paper, was developed in the La Miel area, particularly in the basal part of the La Miel Member (the major conglomeratic unit in Fig. 3). The southern face of La Miel mountain (1560 m high) shows a good outcrop of the

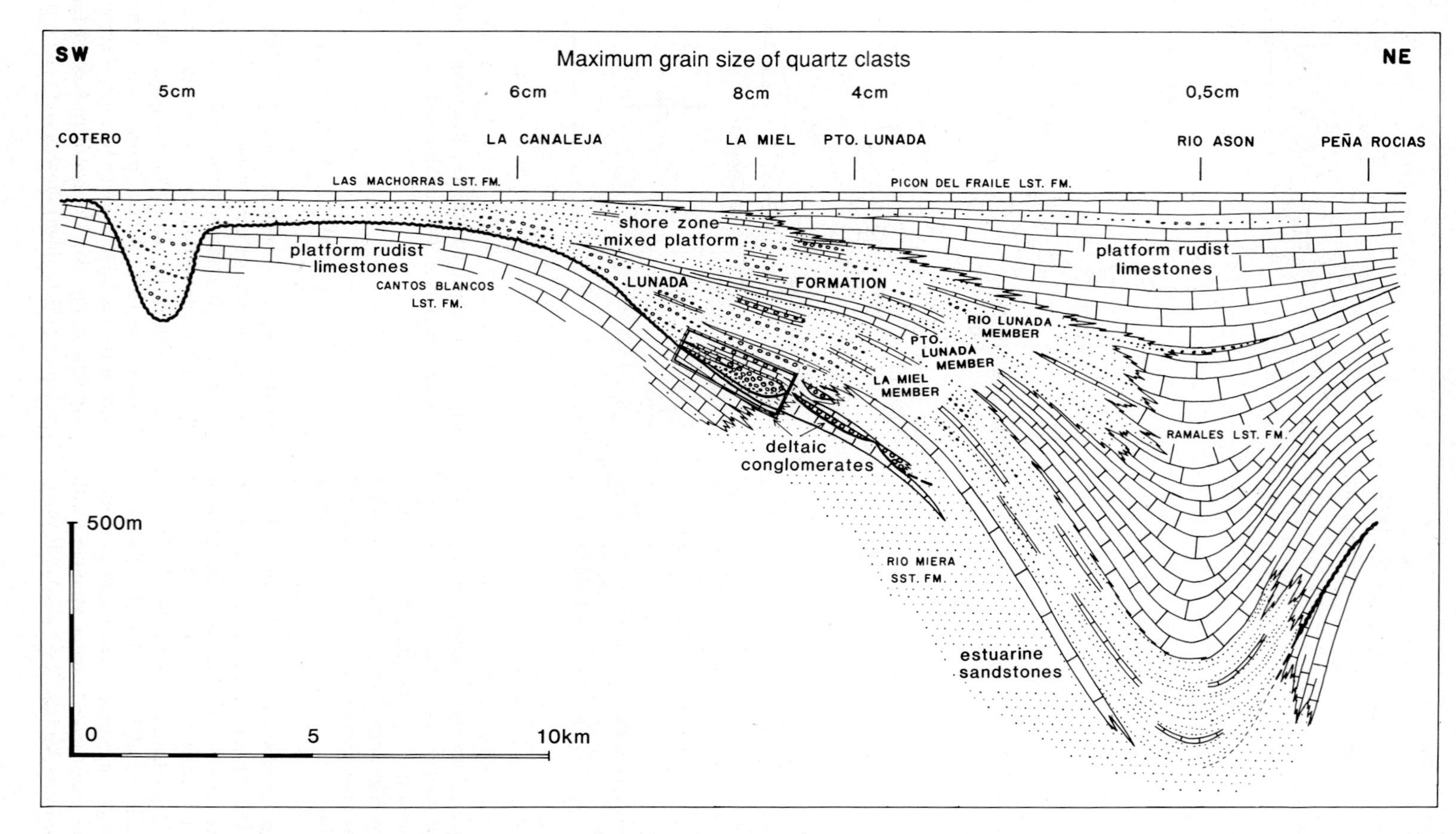

Fig. 3. SW–NE cross-section of the Lunada Formation showing stratigraphic relationships, members, sedimentary environments and the delta described in this paper (within rectangle). Maximum grain sizes of clasts in different parts of the Formation are shown.

lower part of the La Miel Member (Fig. 4). Several small normal faults disrupt the lateral continuity of the exposed facies, but a careful lithostratigraphic correlation has made it possible to establish the stratigraphic cross-section of Fig. 5a (see Fig. 2 for location). This section includes the southeasternmost outcrops of the La Miel escarpment, not represented in Fig. 4. Major erosional surfaces present in the section have been made use of to differentiate sequences, because of their importance in terms of sedimentary dynamics. Several predeltaic and deltaic sequences have been established (Fig. 5b), and these are considered the most important genetic units of the local stratigraphy. These sequences are used to describe the complete deltaic succession, together with other less important interspersed units. Two of the sequences have been labelled predeltaic (S_{00} and S_0) (Fig. 5b), as they do not contain any Gilbert-type units; they form a part of the underlying Rio Miera Sandstone Formation which was formed in an estuary with a western provenance for its feeder clastics, i.e. in diametric opposition to the SE provenance of the La Miel delta. Each of the remaining six sequences, S_1 to S_6, are considered constituent parts of the La Miel delta construction (Fig. 5b).

The units within the sequences have been termed progradational (P), aggradational (A) or of lateral accretion (L), according to Galloway and Hobday's (1983) classification of the major processes of basin-filling. The progradational units (P_1 to P_6) are conspicuous deltaic components, and each of them is thought to have caused a small, local regression of the shoreline when it was formed. The aggradational units (A_{00} to A_5) are all shallow marine except A_4, and were formed when water depth over the deltaic platform was sufficient to allow vertical growth. The lateral accretion units (L_1 to L_5) are also shallow marine, and result from the reworking of sediments eroded from the delta platform by tides and waves during periods of fluvial base-level lowering.

Apart from the main erosional surfaces of the section (1a–6a), which created the major gaps in the local stratigraphic record, several other surfaces of erosion or of non-deposition (00 to 6b) were formed intermittently. They separate units within the deltaic body, commonly corresponding to different sedimentary (sub-) environments. Most of these surfaces show intense bioturbation and are indicative of episodes of non-sedimentation.

The description that follows is based on observations from the stratigraphic section shown in Fig. 5a. Detailed lateral control of the facies in transversal sections has been impossible, except for the westernmost facies. Nevertheless, lateral equivalence in outcrops from the Valnera (V1 and V2 in Fig. 2) and the Pto. de Lunada and Mirador areas, has made it possible to establish the main three-dimensional characteristics of the major deltaic system of Valnera.

GENERAL CHARACTERISTICS OF THE SEQUENCES

Detailed environmental analysis of the deltaic materials considered here will be the subject of another paper, so that only a summary of the general characteristics of the sequences and their constituent genetic units, facies and sedimentary environments is given in this section (see Fig. 5 for location of all terms used).

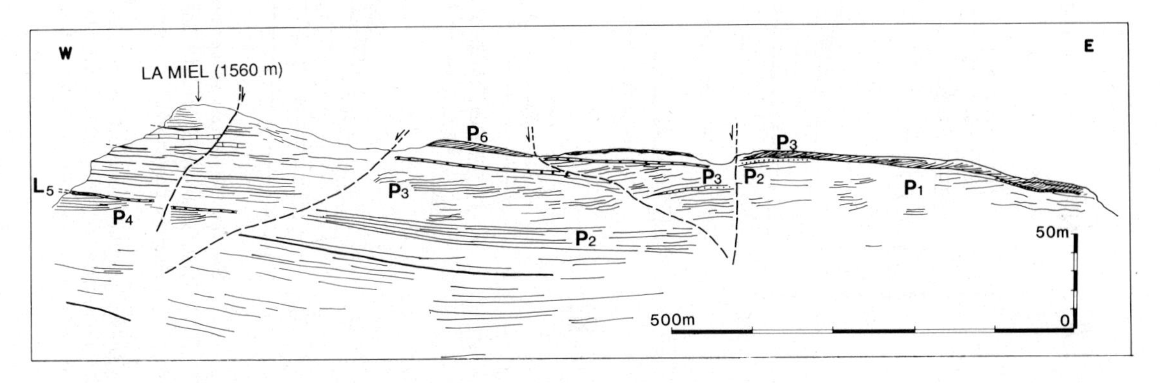

Fig. 4. Geological section of the southern face of La Miel mountain, with the lower part of the Lunada Formation (taken from photographs). Symbols refer to the units distinguished in Fig. 5.

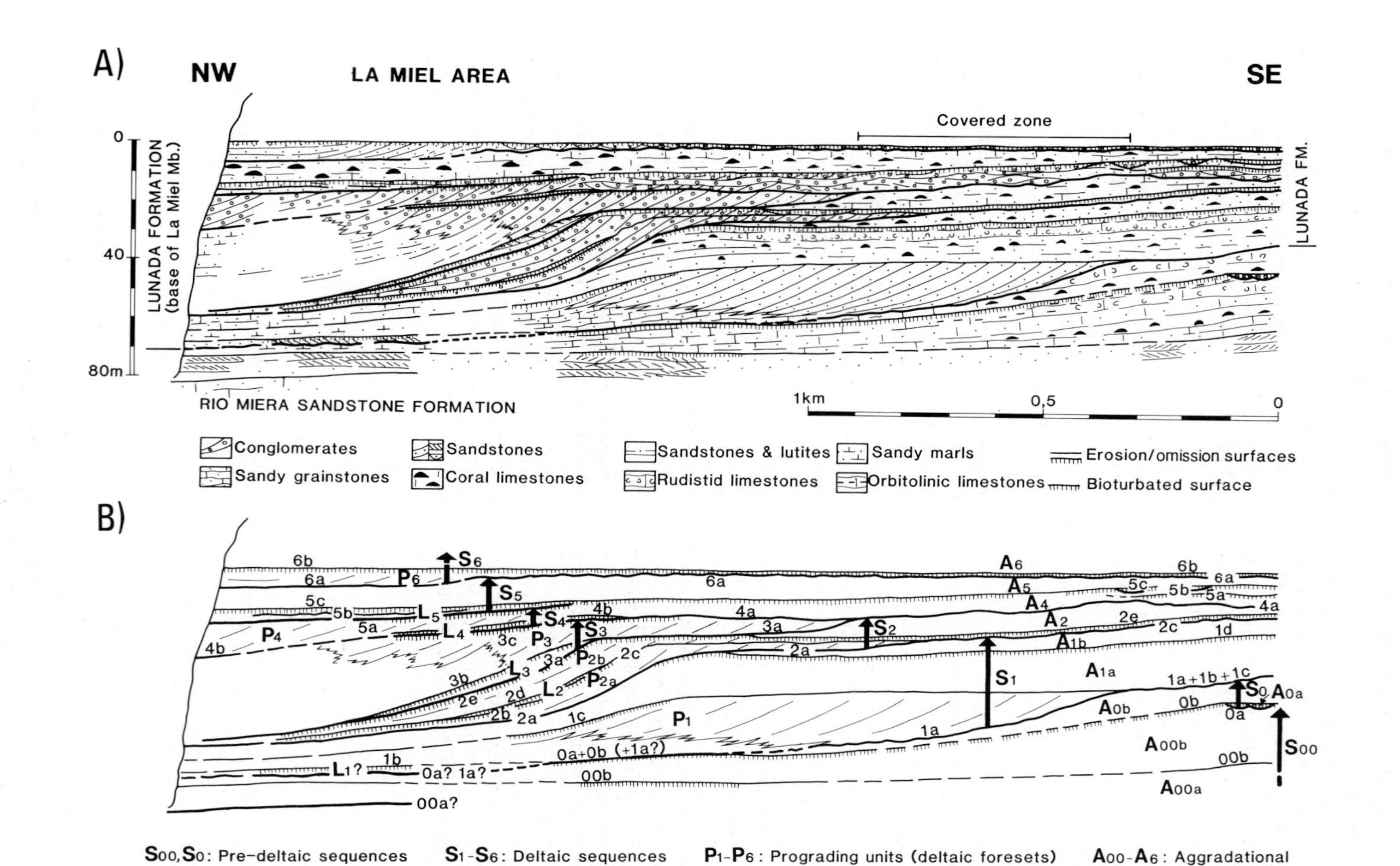

Fig. 5. (A) Stratigraphic cross-section representative of the lower part of the La Miel Member in the southwest escarpment of La Miel. The top of the last prograding unit with big foresets has been taken as a horizontal reference. The arrangement of facies make up an approximately longitudinal section of La Miel delta. (B) Scheme showing the main unconformities, sequences and genetic units distinguished in the La Miel delta.

Predeltaic sequences

Sequence S_{00}

Boundaries: 00a (non-erosional?) and 0a (erosional).
Genetic units: A_{00a} (aggradational) and A_{00b} (aggradational).
Internal discontinuity surfaces: 00b (omission).
Main facies in A_{00a}: grainstones and sandstones with trough cross-stratification and marine fauna; palaeocurrents towards the east.
Main facies in A_{00b}: grainstones and coral-rudist wackestones, fine-grained sandstones and sandy marls.
Sedimentary environments: subtidal estuarine with polarity east (A_{00a}); carbonate ramp deepening towards the west (A_{00b}).
Palaeogeographic implications: change from east to west in the direction of sloping of the local seafloor, probably due to synsedimentary tectonism.

Sequence S_0

Boundaries: 0a (erosional) and 1a (erosional).
Genetic units: A_{0a} (aggradational) and A_{0b} (aggradational).
Internal discontinuity surfaces: 0b (omission).
Main facies in A_{0a}: cross-stratified sandstones and conglomerates filling channels: pebbles up to 1.5 cm.
Main facies in A_{0b}; coral-rudistid *Chondrodonta* wackestones and fine grainstones.
Sedimentary environments: fluvial (A_{0a}); shallow marine: carbonate ramp deepening towards the east (A_{0b}).
Palaeogeographic implications: first arrival of coarse-grained siliciclastics of eastern provenance.

Deltaic sequences

Sequence S_1

Boundaries: 1a (erosional) and 2a (erosional).
Genetic units: L_1 (lateral accretion), P_1 (progradational), A_{1a} (aggradational) and A_{1b} (aggradational).
Internal discontinuity surfaces: 1b (omission), 1c (erosion) and 1d (omission).
Main facies in L_1: cross-stratified sandstones and skeletal grainstones with palaeocurrents towards the east.
Main facies in P_1: sandy foresets in a single set up to 18 m thick, and bottomsets made up of fine grainstones with marine fauna; palaeocurrents towards the west.
Main facies in A_{1a}: cross-laminated sandstones with coarsening-upwards sequences, coral-rudistid wackestones and skeletal grainstones.
Main facies in A_{1b}: cross-laminated sandstones with corals and coral-rudistid limestones.
Sedimentary environments: shallow marine (shoreline?) deepening with time (L_1); Gilbert-type delta lobe (P_1); and shallow, mixed siliciclastic and carbonate shelf (A_{1a} and A_{1b}).
Palaeogeographic implications: first occurrence of a Gilbert-type deltaic deposit from a source area to the SE.

Sequence S_2

Boundaries: 2a and 2c (erosional) and 3a and 4a (erosional).
Genetic units: P_{2a} (progradational), L_2 (lateral accretion), P_{2b} (progradational) and A2 (aggradational).
Internal discontinuity surfaces: 2b (omission), 2c (erosion) and 2d and 2e (omission).
Main facies in P_{2a}: sandy and conglomeratic megaforesets with pebbles up to 2 cm and relief up to 30 m; palaeocurrents towards the west.
Main facies in L_2: cross-stratified pebbly sandstones filling scoured depressions.
Main facies in P_{2b}: sandstones and conglomerates filling channels (east) and making up megaforesets up to 30 m high (west); palaeocurrents towards the west.
Main facies in A_2: marly sandstones and limestones very rich in corals, rudists and orbitolinas.
Sedimentary environments: Gilbert-type delta lobe (P_{2a} and P_{2b}); storm- and wave-influenced shoreline (L_2); and shallow siliciclastic and carbonate shelf (A_2).
Palaeogeographic implications: second episode of Gilbert-type delta construction with development of the highest bathymetric relief and turbidity-current underflow.

Sequence S_3

Boundaries: 3a (erosion) and 2e (omission); 4a (erosion) and 4b (omission).
Genetic units: L_3 (lateral accretion) and P_3 (progradational).
Internal discontinuity surfaces: 3b (omission).

Main facies in L_3: cross-stratified conglomerates and sandstones filling erosional depressions (pebbles up to 3 cm).
Main facies in P_3: conglomeratic and sandy foresets and bottomsets in a single mega-set up to 30 m thick; pebbles up to 5 cm across and palaeocurrents towards the west.
Sedimentary environments: storm- and wave-influenced shoreline deposition on a former delta slope (L_3); Gilbert-type delta lobe (P_3).
Palaeogeographic implications: third episode of Gilbert-type delta construction with 'descending' type progradation of foresets.

Sequence S_4

Boundaries: 4a (erosional) and 5a (erosional).
Genetic units: L_4 (lateral accretion), P_4 (progradational) and A_4 (aggradational).
Internal discontinuity surfaces: 4b (omission).
Main facies in L_4: coarse-grained sandstones with low-angle cross-stratification; foresets dipping to the east.
Main facies in P_4: conglomerates and sandstones forming a single set of cross-stratification up to 11 m thick; pebbles up to 4 cm across and palaeocurrents towards the west.
Main facies in A_4: cross-stratified conglomerates and sandstones filling channels; pebbles up to 3 cm.
Sedimentary environments: wave-influenced shoreline deposition (beach) (L_4); Gilbert-type delta lobe (P_4); fluvial (braided channels and bars) (A_4).
Palaeogeographic implications: fourth episode of Gilbert-type delta construction with less submarine relief to produce high foresets (lack of pronounced shelf break).

Sequence S_5

Boundaries: 5a (erosional) and 6a (erosional).
Genetic units: L_5 (lateral accretion) and A5 (aggradational).
Internal discontinuity surfaces: 5b (erosional) and 5c (omission).
Main facies in L_5: sandstones and conglomerates with a crude low-angle cross-stratification dipping to the east and filling wide channels; maximum pebble size 2.5 cm.
Main facies in A_5: sandy coral limestones with rudists and orbitolinas and calcareous sandstones.
Sedimentary environments: mouth-bar (shelf-type) delta lobe with marine reworking (L_5); shallow (subtidal) carbonate shelf (A_5).
Palaeogeographic implications: first episode of mouth-bar (shelf-type) delta construction on a gentle slope; levelling of the seafloor and subsequent transgression with carbonate deposition.

Sequence S_6 (only basal part)

Lower boundary: 6a (erosional).
Genetic unit: P_6 (progradational).
Internal discontinuity surface: 6b (omission).
Main facies in P_6: sandstones with granules in a single set of cross-stratification, 5.5 m thick, and sandstones and conglomerates filling channels; pebbles up to 2 cm and palaeocurrents towards the west.
Sedimentary environments: Gilbert-type delta lobe and feeder channels (P_6).
Palaeogeographic implications: fifth and last episode of Gilbert-type delta construction in La Miel section, with levelling of the seafloor.

AREAL EXTENT OF THE SEQUENCES

The sequences distinguished in La Miel section can be followed at least 2 km in a direction parallel to the delta migration (Fig. 5). Further northwest of this section there are no outcrops (present erosion), and further southeast of it the units are covered by younger successions of the Lunada Formation (Fig. 2).

There are some outcrops in a transversal (SE–NW) direction in which the lateral extent of the sequences can be tested. They are placed to the south of the type section (Castro Valnera, localities V_1 and V_2), and to the north of it (Pto. Lunada and Mirador sections) (Fig. 2). In V_1, situated 1 km to the south of the eastern end of La Miel section, excellent exposures of the basal part of La Miel Member exist and the sequences 1, 2, (3 + 4 + 5) and 6 have been identified. No Gilbert-type delta lobes characterize any of the sequences; instead, sandstones and conglomerates on top of several lithologies with marine fauna are seen filling channels 2–5 m deep and more than 100 m wide, particularly in the interval correspondent with the sequence S_2. These channels replace each other laterally, suggesting that they conform to an original braided pattern. Therefore, the section in V_1 demonstrates that 'shelf-type' sedimentary conditions with mouth-bar sequences and channels occurred laterally to the Gilbert-type delta area present in La Miel.

The locality V_2 is placed between the Castro Valnera and La Miel summits, 1 km to the south-west of the western end of La Miel section (Fig. 2). The stratigraphic section in it is very similar to the one in the locality V1, i.e. reduced in thickness and of shelf-type, whereas in the section of La Miel Gilbert-type delta lobes and relatively great thicknesses predominate. The succession in V_2 is also similar to others in the Valnera area, the only difference with them being the presence of a small Gilbert-type deltaic unit in the first sequence S_1 (a 6 m thick set of cross-stratification with palaeocurrents towards the west). The importance of this succession is that it represents an intermediate stage between the pure shelf-type and Gilbert-type end members of the deltaic sedimentation of the area.

In the zones between La Miel and Pto. Lunada and between Pto. Lunada and Mirador (Fig. 2), there are three conglomeratic units at the base of La Miel Member which are considered time-equivalent to the delta of La Miel (Fig. 3). They are discrete, single sets of cross-stratified conglomerates (maximum pebble size 3 cm), up to 8 m thick and 1 km long, with palaeocurrents roughly towards the west. They correspond to the sequence S_4 in the type section, and a lower sequence S_3 is also represented in one of the outcrops. These conglomeratic units reveal that several time-equivalent Gilbert-type deltaic lobes appear separatedly in a N–S direction, oblique to the NW–SE axial direction of the La Miel prograding delta.

Well-correlated data from outcrops to the north and south of La Miel section demonstrate that some of the sequences distinguished in the type section can be followed a minimum distance of 5 km in a N–S direction. Nevertheless, the units within the sequences are of more local variability, especially the deltaic prograding lobes.

With all data of foreset progradation taken from the different sequences in the studied outcrops, a regional chart of palaeocurrents has been elaborated (Fig. 6). Local models of distribution of measures are quite unipolar, with reasonably little dispersion, and there is also relatively good agreement between palaeocurrents from different outcrops on a regional scale, all measures suggesting a general palaeoslope towards the northwest.

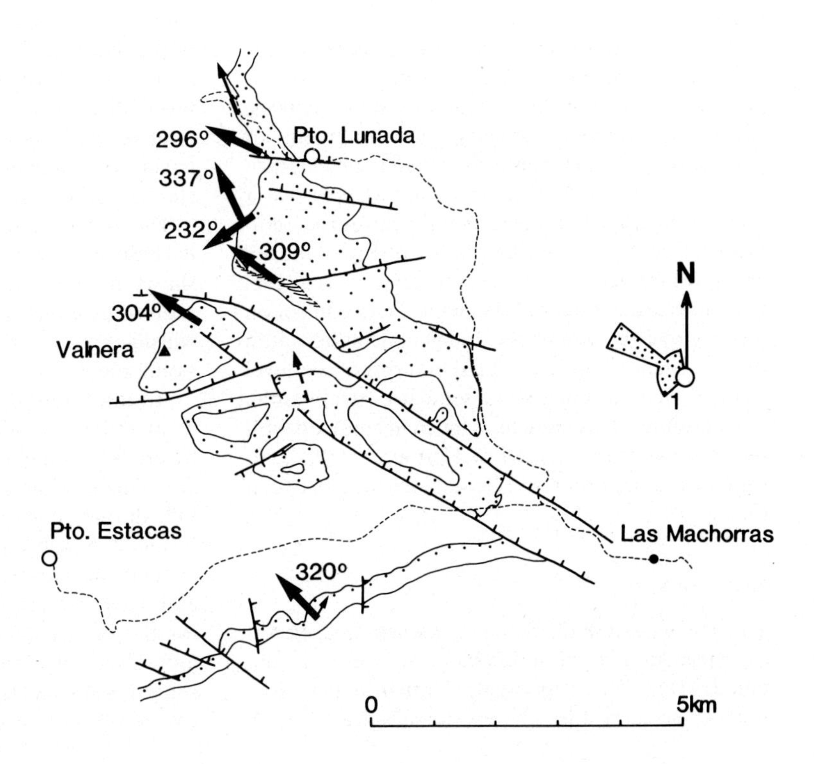

Fig. 6. Regional distribution of palaeocurrents in the Valnera braid-plain delta. Vectors represent average values of foreset orientations in Gilbert-type units, and the rose diagram gives the preferred orientation of these vectors.

ORIGIN OF THE SEQUENCES

Several kinds of sequences have been reported in marine, Gilbert-type deltas, which have been attributed to one or more of the following causes: active synsedimentary tectonism, sediment loading and concomitant subsidence, variations in the rate of deposition, lateral shifts in the location of the stream mouths and relative, short-term sea-level fluctuations (e.g. Postma, 1984; Suter & Berryhill, 1985; Ori & Roveri, 1987; Colella, 1988; Muto, 1988; Postma & Cruickshank, 1988; Zupanic *et al.*, 1988).

Most likely the appearance and development of the La Miel delta was primarily controlled by tectonism, although no palaeofaults directly related to it have been found. Sediment loading does not account for the sudden interruptions in the La Miel delta progradation or aggradation, and changes in the rate of sediment supply would not explain, on their own, the abrupt oscillations in the fluvial base level implied by the sequence boundaries. Correlation of the sequences from low-subsidence to high-subsidence areas suggests that an allocyclic mechanism would explain their origin better than an autocyclic one such as lateral shifts of stream mouths.

Relative, short-term sea-level fluctuations, perhaps tectonically induced, seem the best hypothesis to account for the origin of the sequences. During episodes of lowstand, the emergent or shallow-water areas would have been subjected to erosion or no sedimentation, and deeper-water areas would have received mainly reworked sediments. Conversely, during phases of highstand all areas would have aggraded. If relative sea-level variations are considered the main direct control on the origin of the sequences, and genetic units within these sequences are interpreted in terms of sea-level change, then a small-scale sequence stratigraphy model results. This model has much in common with the 'systems tracts' model of Haq *et al.* (1987), in spite of the big difference in scale existing between them.

Sequence S_1

The first deposit of the deltaic sequences interpreted as representative of a lowstand sea level is the unit L_1 (Fig. 5b), supposedly formed at the same time as or soon after the erosional surface 1a. A subsequent, relative, sea-level rise would have produced the omission surface 1b, along with accommodation space enough to allow the formation of the first Gilbert-type deltaic body, P_1. Progradation of the foresets of P_1 converted the surface 1b into a downlap surface and buried it.

A new, rapid pulse of relative sea-level rise brought about slope-and probably shelf-starved conditions, and the omission surface 1c appeared as a consequence. Two aggradational units A_{1a} and A_{1b} and the omission surface 1d were then formed, probably because of two successive phases of decrease in the rate of relative sea-level rise, punctuated by a pulse of increase.

Making comparisons with the model of Haq *et al.* (1987), the lateral accretion deposit L_1 would represent a lowstand systems tract (somehow equivalent to a lowstand fan), and the prograding unit P_1 would be comparable to a lowstand, wedge systems tract (prograding complex). The surface 1c would be a transgressive surface, the unit A_{1a} a transgressive systems tract and the unit A_{1b} another transgressive or, more probably, a highstand systems tract.

Sequence S_2

The erosional surface 2a separates marine deposits of the sequence S_1 from fluvio-deltaic deposits of the sequence S_2 (Figs 5a and 5b), so that a relative sea-level fall is invoked to account for its origin. A quick relative sea-level rise must have followed, creating again accommodation space in the outer shelf and starvation in both outer shelf and slope.

The deltaic unit P_{2a} has a downlap surface at its base and an erosional truncation at its top. It shows a steady-state type of progradation which represents either very fast lateral sedimentation or stillstand sea-level conditions, or both causes acting simultaneously.

An interruption of the arrival of clastic materials created the omission surface 2b, which was followed by erosion in the upper slope and platform (surface 2c). This erosion is attributed to a relative sea-level fall, during which the lateral accretion unit L_2 was formed. A subsequent relative sea-level rise can explain the appearance of the omission surface 2d and accommodation space in the platform enough for the steady-state progradation of a new deltaic unit, P_{2b}. This unit has some channel facies which represent fluvial topset deposits. A new, rapid, relative sea-level rise created the omission surface 2e and enabled the deposition of the aggradational unit A_2.

Both units P_{2a} and P_{2b} could be considered representative of a lowstand, wedge systems tract in Haq *et al.* (1987) terminology, although the eastern part of P_{2b} would better correspond to a transgressive systems tract. The unit L_2 was formed during a short-lived phase of relative sea-level fall, which occurred under general conditions of lowstand sea level. The surface 2e seems to be transgressive and the overlying unit A_2 has characteristics of a transgressive systems tract.

Sequence S_3

Its lower boundary is the erosional surface 3a (Fig. 5b), attributed to a relative sea-level fall as it separates marine (below) from deltaic facies (above). The lateral accretion unit L_3 is a shoreline deposit left by a subsequent sea-level rise. Once the platform was flooded, this rise created additional accommodation space along with the omission surface 3b.

Relative stillstand sea-level conditions explain the steady state or even descending type of delta progradation that characterizes the (more than 1 km long) unit P_3. As the delta remained inactive the omission surface 3c was formed, suggesting another pulse of marine flooding.

The lateral accretion unit L_3 represents a lowstand systems tract (lowstand wedge). As the surface 3c is transgressive, the unit P_3 can be attributed to a transgressive systems tract; nevertheless, a highstand systems tract would be a better option if the equally transgressive surface 3b was considered more penetrative than the 3c.

Sequence S_4

Fluvial erosion on deltaic and marine sediments of the platform created the surface 4a (Fig. 5b), and platform-derived clastics subsequently formed the wedge-shaped deposit L_4 in the slope area. This represents a relative sea-level fall succeeded by a relative sea-level rise. The omission surface 4b was the result of a new pulse in relative sea-level rise, and when relative stillstand sea-level conditions were reached, a steady-state delta progradation formed the unit P_4.

Unit L_4 is another example of lowstand wedge systems tract. Unit A_4 would represent both transgressive and highstand systems tract. And the unit P_4, finally, is considered a part of a highstand systems tract.

Sequence S_5

The lateral accretion unit L_5 was formed subsequently to the relative sea-level falls which respectively created the surfaces 5a and 5b by means of fluvio-deltaic erosion. The surface 5c, at the top of L_5 (Fig. 5b), is interpreted as an abandoned wave-cut platform converted into an omission surface with increase in the rate of relative sea-level rise. The aggradational unit A_5, finally, resulted from the continued relative sea-level rise.

Marls and limestones onlapping the youngest foreset of the unit P_4 (to the north of its NW end in Fig. 5b), are attributed to a lowstand systems tract. The unit L_5 represents the first deposit of the transgressive systems tract, which ended with the transgressive surface 5c. And the unit A_5 shows characteristics of both transgressive and highstand systems tracts.

Sequence S_6

The erosional surface 6a likely represents a relative sea-level fall as fluvio-deltaic materials overlay marine deposits (Fig. 5). A subsequent relative sea-level rise created room enough in the outer platform to favour progradation of the unit P_6, which probably took place during a relative stillstand in sea level. Later on, a wave-cut platform replaced the former delta plain, and it was soon converted into an omission surface, 6b, when the relative sea level rose again very fast. Then a new aggradational unit, A_6, began forming as the rate of sea-level rise decreased.

The unit P_6 represents a lowstand wedge systems tract in its western part and a transgressive systems tract in its eastern part. The omission surface 6b is a subsequent transgressive surface, and the unit A_6 corresponds to both a transgressive and a highstand systems tract.

Concluding remarks

Three processes common to all sequences are worth emphasizing. First, available marine space for deltaic progradation in the outer platform was created in a pulsatory way after previous sea-level falls, and so quickly in each case that no transgressive deposits were able to form. Second, the repetitive relative sea-level falls seem to have been very quick processes as well. And third, progradation of the deltaic units was accomplished in conditions of relative sea-level stillstand or nearly so (angular upper boundaries of the foresets, approximately horizontal tops of the sets). The three processes appear as anomalous when compared with the relatively normal sedimentation deduced from the aggradational units; this is because they suggest activity during short periods of time, and because they involve either sudden arrivals of coarse terrigenous sediments or sudden stops. Probably local tectonism within the general extensional regime of the area, acting in a pulsatory way, was responsible for the sudden creation of tilted areas or gradient variations, differently oriented slopes and other morphotectonic features, and all this had an immediate effect on the relative behaviour of the sea level. Examples of this influence in footwall-sourced fans from neotectonic extensional settings have been provided by Leeder *et al.* (1988). These authors also suggested that bathymetric gradients in coastal fans will be primarily controlled by tectonic subsidence, and that footwall-derived fan deltas commonly prograde basinwards as Gilbert-type deltas or as 'unsteady-state' deltas lacking a subaerial prograding platform.

GENERAL SEDIMENTARY MODEL

Constructing a three-dimensional model with the more relevant features of the example of La Miel requires data on the transversal evolution of at least some units. These data have been found only in the unit P_4, from the sequence S_4 (Fig. 5b). In the longitudinal section of La Miel, P_4 has progradational characteristics along approximately 800 m, without any important change but the constant increment in thickness from 5.11 m; nevertheless, it shows a sudden sedimentary interruption in the northwestern end of the section. Another outcrop of this unit in a parallel section, located 200 m to the NE of the type section and 500 m from its NW end, shows a thickness even less than 2 m. This means that the unit has a quick disappearance towards the north and that, at least in its final part, it was completely unconfined (not channelled). The lateral control in the locality V_2 (Fig. 2), indicates that the unit P_4 also disappears as such in a maximum distance of about 1 km towards the southwest. Besides, the elongation of the set forming the unit towards the NW coincides approximately with the palaeocurrent direction calculated from its axial zone (N310°E). All this defines a lobate form for this unit in the area of La Miel.

Two time-equivalent, similar Gilbert-type lobes were formed at 800 m to the north of La Miel summit and in the area between Pto. Lunada and Mirador, respectively (Figs 2 and 3). They appear separated in the approximately S–N available section, and only the northernmost one is seen confined in its more proximal part (filling an erosional depression on top of coral limestones).

From all these data it appears that individual Gilbert-type units were developed as lobes of a larger delta system (Valnera), and that the sequential superposition of lobes in the area of La Miel was the result of the peculiar sedimentary and morphotectonic conditions which prevailed there for some time. A three-dimensional reconstruction of this major system of Valnera, interpreted as a braidplain delta front, is shown in Fig. 7. The Gilbert-type lobes attained maximum radii of about 1.3 km, and prograded actively during rather short periods of time because of the abundant sediment supply. High gradients in the fluvial area must be implied to account for this. The reconstructed shape of the lobes remains the theoretical model of an inertia-dominated effluent (Wright, 1977), in which the resultant lunate bar has longitudinally a Gilbert-type profile.

The close lateral spacing of the lobes suggests an association with a multi-channel, bedload (braided) fluvial system. The channels in this model, from a few tens to a few hundred of metres in width, would have transported sand and gravel as bedload; longitudinal and transverse bars would have formed horizontal and cross-stratified beds, respectively. These channels are thought to have been cut into sands and sandy carbonate sediments, as a consequence of relative falls in sea level. Modern examples of large-scale, braided distributary channels incised directly into a platform as a result of a eustatic sea-level drop, have been described from the modern Gulf Coast setting (Suter & Berryhill, 1985). Ancient cases of deep distributaries have been attributed to either normal autocyclic processes, such as lateral migration or avulsion (McCabe, 1977; Mossop & Flach, 1983; Hopkins, 1985), or tectonically related, allocyclic processes (alluvial-fan pulses, Dutton, 1982). Both auto- and allocyclic processes were probably involved in the development of the Valnera braidplain delta system; for example, the migration of individual channels probably controlled

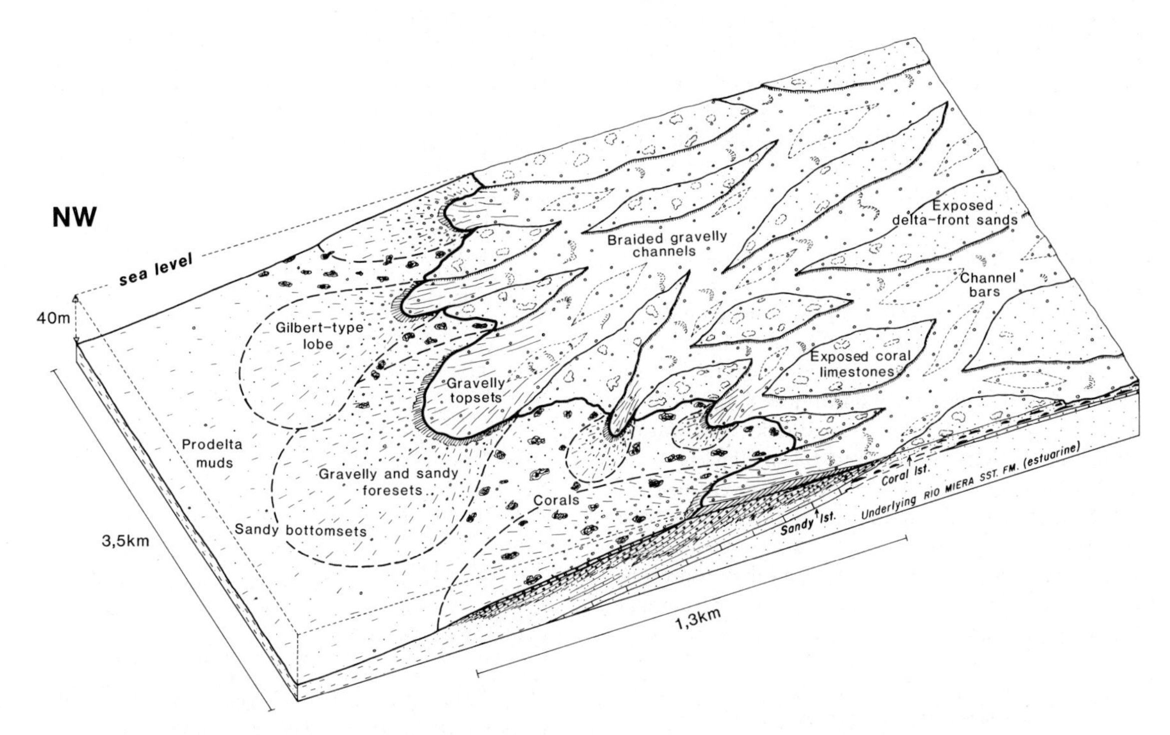

Fig. 7. The Valnera braidplain delta model, corresponding to the lowermost part of La Miel Member in the Valnera zone. The lobes in this model can be considered individual Gilbert-type deltaic systems, sometimes made up of several laterally and vertically stacked, smaller, Gilbert-type deltaic units.

the delta lobes (mouth bars), whereas relative sea-level drops caused general progradation of the composite multilobe delta front (braidplain delta complex).

The progradation of the distributary mouths created coarsening-upwards sequences of both low-angle and foreset (Gilbert-type) lobes, mainly dependent on the local depth of the sea bottom. Low-angle mouth-bar sections seen in the area of La Miel and in other outcrops, are attributed to a rather slow progradation into shallow subhorizontal sea bottoms; in this process even corals were able to colonize, intermittently, the sandy sea bed.

Palaeocurrent analysis suggests a general migrating trend of the Gilbert-type foresets towards the NW, so that a zone of more subsidence must have existed in that direction. Although the distal sediments of the system are now lost by erosion, regional considerations suggest that they never were too deep and that they were formed in an intraplatform trough.

The appearance of thick Gilbert-type units just to the northeast of Castro Valnera (Fig. 2), seems to have been controlled by the development in that area of higher rates of subsidence than in surrounding areas. Two alpine, normal faults, SE–NW oriented, lie now in what once was the boundary between two areas with different subsidence, demonstrated by thickness inequalities and rapid changes of facies. One of these palaeogeographic features was the lateral change of the underlying Cantos Blancos Limestone Formation to the Rio Miera Sandstone Formation (Fig. 3). Still another, although a bit older, was the important growth fault that affected the Rio Miera Formation in the area of Valnera, prior to the deposition of the Cantos Blancos Formation (García-Mondéjar, 1979); it was a SE–NW oriented fault and had its hanging wall in the northeast.

According to all this, it is proposed that a probable tectonic line or zone of weakness in the basement could have created a SE–NW oriented, depressed area to the northeast of Castro Valnera in the early Albian (Fig. 2). Extensional movements during the deposition of the rest of the Lunada Formation, probably maintained both the sedimentary trough and its polarity to the NW. Small incremental accommodations in synsedimentary faults of higher tectonic levels, should have induced the relative sea-level oscillations which are thought to have controlled the sequential arrangement of the La Miel deltaic complex.

Palaeogeographical considerations establish the provenance of the deltaic terrigenous materials to be from the SE or SSE of the region (García-Mondéjar, 1979), so that the source area must have been located in the Palaeozoic of the Sierra de La Demanda, near Burgos, and/or the Palaeozoic of the southern part of the Ebro Block, now buried under the Tertiary of the Ebro Basin (Fig. 1).

The deltas of the Lunada Formation were formed immediately after the first action of the early Albian tectonic movements of the Basque-Cantabrian region (García-Mondéjar, 1979), and they represent the beginnings of an important phase of regional transgression (Fig. 3).

ACKNOWLEDGEMENTS

This paper has been written under research project 121.310–26/87 from the Universidad del País Vasco and PGV 8806 (1988) from the Gobierno Vasco. Thanks are given to Begoña Bernedo and María Jesús Sevilla for typing the text, and to Luis Miguel Agirrezabala for helping with the preparation of the figures. The manuscript has benefited from the critical review of Ian Reid.

REFERENCES

Axelsson, V. (1967) The Laitaure delta — a study of deltaic morphology and processes. *Geogr. Ann.* **49**, 1–127.

Clauzon, G. & Rubino, J.L. (1988) Why proximal areas of the Mediterranean Pliocene rias are filled by Gilbertian fan-deltas. *Abstr. Int. Works. Fan Deltas, Calabria*, 13–14.

Colella, A. (1984) Marine Gilbert-type deltas in the Lower (?) Pleistocene deposits of the Crati Valley (Calabria, southern Italy). A preliminary report. *Abstr. 5th European Regional Meeting of Sedimentology, IAS, Marseille*, 112–113.

Colella, A. (1988) Pliocene–Holocene fan deltas and braid deltas in the Crati Basin, southern Italy: a sequence of varying tectonic conditions. In: *Fan Deltas: Sedimentology and Tectonic Settings* (Ed. by W. Nemec and R.J. Steel), pp. 50–74. Blackie and Son, London.

Colella, A., De Boer, P.L. & Nio, S.D. (1987) Sedimentology of a marine intermontane Pleistocene Gilbert-type fan-delta complex in the Crati Basin, Calabria, Southern Italy. *Sedimentology* **34**, 721–736.

Corner, G.D., Nordahl, E., Munch-Elligsen, K. & Robertsen, K.R. (1988) Sedimentology of the Holocene Alta delta, Northern Norway. *Abstr. Int. Works. Fan Deltas, Calabria*, 19–20.

Dunne, L.A. & Hempton, M.R. (1984) Deltaic sedimentation in the Lake Hazar pull-apart basin, southeastern Turkey. *Sedimentology* **31**, 401–412.

Dutton, S.P. (1982) Pennsylvanian fan-delta and carbonate deposition, Mobeetie Field, Texas Panhandle. *Bull. Am. Assoc. Petrol. Geol.* **66**, 389–407.

Farquharson, G.W. (1982) Lacustrine deltas in a Mesozoic alluvial sequence from Camp Hill, Antarctica. *Sedimentology* **29**, 717–725.

Galloway, W.E. & Hobday, D.K. (1983) *Terrigenous Clastic Depositional Systems*. Springer Verlag, New York, 423 pp.

García-Mondéjar, J. (1979) *El Complejo Urgoniano del sur de Santander*. Tesis Doctoral, Universidad de Bilbao, 673 pp.

García-Mondéjar, J. (1982) Aptiense-Albiense. In: *El Cretácico de España*, pp. 63–84. Universidad Complutense, Madrid.

García-Mondéjar, J. (1988) The marine Gilbert-type delta of La Miel (Albian Lunada Formation, Northern Spain). *Abstr. Int. Works. Fan Deltas, Calabria*, 24–25.

Gawthorpe, R.L., Hurst, J.M. & Sladen, C.P. (1988) Geometry and evolution of Miocene, footwall-derived fan deltas, Ras Budran area, Gulf of Suez (East). *Abstr. Int. Works. Fan Deltas, Calabria*, 26–27.

Gilbert, G.K. (1885) The topographic features of lake shores. *U.S. geol. Surv., Fifth Annual Report.*, 69–123.

Gustavson, T.C., Ashley, G.M. & Boothroyd, J.C. (1975) Deformational sequences in glaciolacustrine deltas. In: *Glaciofluvial and Glaciolacustrine Sedimentation* (Ed. by A.V. Jopling and B.C. Mac Donald). Spec. Publ. Soc. econ. Paleont. Miner., Tulsa, 23, 264–280.

Haq, B.U., Hardenbol, J. & Vail, P.R. (1987) Chronology of fluctuating sea levels since the Triassic. *Science* **235**, 1156–1167.

Hobday, D.K. (1974) Interaction between fluvial and marine processes in the lower part of the Late Precambrian Vadsø Group, Finnmark. *Norges geol. Unders.* **303**, 39–56.

Hopkins, J.C. (1985) Channel-fill deposits formed by aggradation in deeply scoured, superimposed distributaries of the Lower Kootenai Formation (Cretaceous). *J sedim. Petrol.* **55**, 42–52.

Kazanci, N. (1988) A bouldery fan-delta succession in the Pleistocene–Holocene Burdur basin, Turkey: the role of basin-margin configuration in sediment entrapment and differential facies development. *Abstr. Int. Works. Fan Deltas, Calabria*, 30–31.

Leeder, M.R., Ord, D.M. & Collier, R. (1988) Development of alluvial fans and fan deltas in neotectonic extensional settings: implications for the interpretation of basin fills. In: *Fan Deltas: Sedimentology and Tectonic Settings* (Ed. by W. Nemec and R.J. Steel), pp. 173–185. Blackie and Son, London.

MASSARI, F. & COLELLA, A. (1988) Evolution and types of fan-delta systems in some major tectonic settings. In: *Fan Deltas: Sedimentology and Tectonic Settings* (Ed. by W. Nemec and R.J. Steel), pp. 103–122. Blackie and Son, London.

MASSARI, F. & PAREA, G.C. (1988) Wave-influenced Gilbert-type bodies prograding into drowned valley and gravelly beach sequences in the hinterland of the Gulf of Taranto: the role of sea-level changes. *Abstr. Int. Works. Fan Deltas, Calabria*, 38.

MCCABE, P.J. (1977) Deep distributary channels and giant bedforms in the Upper Carboniferous of the Central Pennines, northern England. *Sedimentology* **24**, 271–290.

MOSSOP, G.D. & FLACH, P.D. (1983) Deep channel sedimentation in the Lower Cretaceous McMurray Formation, Athabasca Oil Sands, Alberta. *Sedimentology* **30**, 493–509.

MUTO, T. (1988) Gilbert-type deltas prograding within fan valleys: a spectrum of facies associations in coastal fans controlled by relative sea-level changes. *Abstr. Int. Works. Fan Deltas, Calabria*, 43–44.

ORI, G.G. & ROVERI, M. (1987) Geometries of Gilbert-type deltas and large channels in the Meteora Conglomerate, Meso-Hellenic basin (Oligo-Miocene), Central Greece. *Sedimentology* **34**, 845–859.

PORĘBSKI, S.J. (1988) Generation of Gilbertian deltas by lava flowing into the sea: an Oligocene example from King George Island (South Shetland Islands). *Abstr. Int. Works. Fan Deltas, Calabria*, 49–50.

POSTMA, G. (1984) Mass-flow conglomerates in a submarine canyon: Abrioja fan-delta, Pliocene, southeast Spain. In: *Sedimentology of Gravels and Conglomerates* (Ed. by E.H. Koster and R.J. Steel). Can. Soc. Petrol. Geol. Mem. 10, 237–258.

POSTMA, G., BABIĆ, L., ZUPANIČ, J. & RØE, S.L. (1988) Delta-front failure and associated bottomset deformation in a marine, gravelly Gilbert-type fan delta. In: *Fan Deltas: Sedimentology and Tectonic Settings* (Ed. by W. Nemec and R.J. Steel), pp. 91–102. Blackie and Son, London.

POSTMA, G. & CRUICKSHANK, C. (1988) Sedimentology of a late Weichselian to Holocene, terraced fan delta, Varangerfjord, northern Norway. In: *Fan Deltas: Sedimentology and Tectonic Settings* (Ed. by W. Nemec and R.J. Steel), pp. 144–157. Blackie and Son, London.

POSTMA, G. & ROEP, T.B. (1985) Resedimented conglomerates in the bottomsets of Gilbert-type gravel deltas. *J. sedim. Petrol.* **55**, 874–885.

PRIOR, D.B. & BORNHOLD, B.D. (1988) Delta-front slope failures. *Abstr. Int. Works. Fan Deltas, Calabria*, 53–54.

PRIOR, D.B., WISEMAN, W.J. & GILBERT, R. (1981) Submarine slope processes of a fan delta, Howe Sound, British Columbia. *Geo-Mar. Lett.* **1**, 85–90.

RAT, P. (1959) Les pays crétacés basco-cantabriques (Espagne). Thèse. *Publ. Univ. Dijon* **XVIII**, 525 pp.

STANLEY, K.O. & SURDAM, R.C. (1978) Sedimentation on the front of Eocene Gilbert-type deltas, Whashakie Basin, Wyoming. *J. sedim. Petrol.* **48**, 557–573.

SUTER, J.R. & BERRYHILL, H.L. JR. (1985) Late Quaternary shelf-margin delta, northwest Gulf of Mexico. *Bull. Am. Assoc. Petrol. Geol.* **69**, 77–91.

VAN DER MEULEN, S. (1983) Internal structure and environmental reconstruction of Eocene transitional fan-delta deposits, Monllobat-Castigaleu Formations, southern Pyrenees, Spain. *Sedim. Geol.* **37**, 85–112.

WALKER, R.G. & HARMS, J.C. (1980) Fan-delta deposition, Minturn Formation, McCoy area. *Proc. Rocky Mt. Sect. Fall Field Conf., Soc. econ. Paleontol. Miner., Tulsa.*

WOOD, M.L. & ETHRIDGE, F.G. (1988) Sedimentology and architecture of Gilbert- and mouth bar-type fan deltas. Paradox Basin, Colorado. In: *Fan Deltas: Sedimentology and Tectonic Settings* (Ed. by W. Nemec and R.J. Steel), pp. 251–263. Blackie and Son, London.

WRIGHT, L.D. (1977) Sediment transport and deposition at river mouths: a synthesis. *Bull. geol. Soc. Am.* **88**, 857–868.

ZUPANIČ, J., BABIĆ, L., RØE, S.L. & POSTMA, G. (1988) Construction and destruction of Gilbert-type deltas during transgression. An example from the Eocene Promina Formation, Yugoslavia. *Abstr. Int. Works. Fan Deltas, Calabria*, 58–59.

Ancient Alluvial Deltas — Effects of Varying Climate and Water Level

Spec. Publs int. Ass. Sediment. (1990) **10**, 273–280

Climatically triggered Gilbert-type lacustrine fan deltas, the Dead Sea area, Israel

D. BOWMAN

Ben-Gurion University of the Negev, Beer Sheva 84105, Israel

ABSTRACT

The marginal facies of the Upper Pleistocene, lacustrine Lisan Formation in the Dead Sea area consists of a moderately dipping, slightly cemented and crudely bedded boulder conglomerate, composing a sequence of fan deltas and alluvial fans at canyons exits. Overlying these are Gilbert-type sequences which demarcate the very top marginal facies of the Lisan Formation. Various sources, such as pollen, archaeology, paleosols and lacustrine sediments provide evidence of the wetter character at the end of the Lisan stage 18 000–14 000 yr BP.

Gilbert-type sequences were observed in small catchment areas (Wadi Mor, W. Boqeq) and in medium ones (W. Rahaf). However, these sequences were not observed in the largest catchment areas as those of W. Zeelim and W. Arugot. It is suggested that the small drainage basins were most sensitive to the increased humidity, i.e. through their denser vegetative cover, erosion was retarded and run-off decreased, resulting in deposition of Gilbert-type units. The large catchment areas, notwithstanding the vegetative cover, retained their flashflood regime, thus filtering out any evidence of hydrological change.

The sharp textural and structural differences between basal fan deltas and the superimposed Gilbert-type sequences reflect the change in the hydrological regime, i.e. confinement of sediment traction and decrease in the dispersal energy. The unbridging wedge-shaped geometry of the Gilbert-type units further precludes a tectonic origin and emphasizes the climatic trigger for depositing Gilbert-type sequences.

INTRODUCTION

Classic lacustrine Gilbert-type fan deltas, such as found at Lake Bonneville, Utah (Gilbert, 1885) and in fluvioglacial settings, are characterized by the tripartite cross-sets: subhorizontal topsets, steep, coarse bedload foresets and fine-grained, suspension-load bottomset units. These facies groups are distinguished on the basis of their internal stratification and textural characteristics. The steeply inclined foresets are a crucial recognizable feature, indicating coarse bedload and average low-current velocity with a possible low depth ratio, i.e. deep basin (Jopling, 1963). Gilbert-type fan deltas prograde into a low-energy water body by the influx of bedload, within a steep-gradient river-dominated environment. Turbulent jet diffusion (Wright, 1977) is responsible for the sediment mixing and for the progressive basinward outflow deceleration, as well as for the confined lateral dispersion of the sediment.

Except for Frostick and Reid (1989), Gilbert-type fan deltas have not been mentioned along the Dead Sea margins. Gilbert-type structures have usually been ignored as climatic indicators and were regarded, as most alluvial fans, to be sedimentary responses to fault movements (Miall, 1984). It is in fact extremely difficult to identify the climatic control on basin sedimentation (Steel, 1974, 1976), or on the depositional architecture, unrelated to tectonics. This issue has been discussed previously by Leeder *et al.* (1988) and Frostick and Reid (1989). Multistage fan-delta development triggered alternately by climatic changes and by tectonics has been described by Kazancı (1988).

The aim of this study is to document Gilbert-type lacustrine fan deltas in the Dead Sea area and to examine their origin, while focusing on the climatic trigger.

STUDY AREA

The Dead Sea area occupies a section of the Dead Sea rift which is part of the large Syrian African rift-valley system (Fig. 1). The area is extremely arid: mean annual precipitation is 50 mm. The watersheds of some of the drainage systems reach an altitude of 800–1000 m, and have a mean annual precipitation exceeding 600 mm. The rift valley attained its present form during late Pliocene to early Pleistocene (Bentor & Vroman, 1960; Neev & Emery, 1967; Begin *et al.*, 1974). The fault scarp bordering it to the west is marked by steep cliffs composed of dolomite and limestone of Cenomanian and Turonian age. Its lower part (Fig. 2) exhibits clear evidence of abandoned shorelines of Lake Lisan (Bowman, 1971), the last of a series of water bodies which intermittently occupied extensive areas of the rift valley. Lake Lisan, a low-energy, tideless lake, was the precursor of the Dead Sea, dating 50 000–14 000 yr BP (Kaufman, 1971; Neev & Hall, 1977; Begin *et al.*, 1985). Its highest level reached approximately 230 m above the present level of the Dead Sea (405 m below MSL), and its salinity was 7–17%. The canyons along the margins of the rift were flooded by Lake Lisan. Fan deltas gradually formed at their outlets, composed mainly of the redeposited lacustrine Lisan sediments, originally deposited in the canyons.

Sneh (1979) suggested a four-facies model for characterizing the textural and structural spectra of the lacustrine Lisan sediments. His proximal facies, adjacent to the border-fault scarp, was described as a slightly cemented, poorly sorted and crudely bedded boulder conglomerate, composed of subangular to subrounded clasts, without cut-and-fill structures. Manspeizer (1985) defined this marginal coarse unit as proximal alluvial-fan facies. In front of this marginal coarse-boulder conglomerate and basinwards the deep lacustrine facies is composed of 40 m of fine-grained clasts and chemical precipitates divided into the lower, mainly clay laminated unit and the upper aragonite-rich 'White Cliff' member (Begin *et al.*, 1974).

Lake Lisan, which was about 200 km long, contracted at 14 000 yr BP, either due to lake-floor subsidence, or desiccation following a warmer and more arid period. Thereafter the sublacustrine fans became exposed and post-Lisan alluvial fan complexes radiated eastwards (Bowman, 1978). The margins of the Dead Sea Rift are ideal for morpho-sedimentary study, since they are well preserved and accurately dated.

METHODS

The internal structure of the fan deltas was examined from Ein Gedi to the Zin Valley (Fig. 1) along spectacular exposures (Fig. 2), provided through post-Lisan to recent fan dissection. As physical examination of the fan exposures was not possible due to the vertical exposures, field sketches of the cliffs and analysis of photomosaics were carried out. Dying out of the fans caused gradually poorer outcrop conditions at the basinward distal areas.

RESULTS

Basal fan delta

The proximal facies of Sneh (1979) is poorly sorted and texturally immature. It shows matrix-to-clast-supported frameworks and is overwhelmingly of low porosity and devoid of clast orientation or imbrication. Its coarse clast mode is cobble, with local outsized boulders (0.5–1.5 m). Although barely graded and massive, a gentle (3–6°) eastward dip is irregularly manifested by some crude layering. The lack of grading and imbrication and the poor stratification suggest a disorganized, partly debris-flow-dominated unit (Walker, 1978), making impossible any distinction between top-, fore- and bottomset units. Brief pauses in the coarse alluvial supply are marked by transgressive beach facies, shown by sandy to pebbly landward-dipping backshore units and by planar-laminated lacustrine clays and aragonite up to a few metres thick. These beach intercalations suggest that the basal unit is a fan delta. This basal fan delta has been thereafter truncated by an impinging lake level indicated by a major 3–8° eastward-dipping erosional unconformity (Fig. 3), draped with an up to 1-m thick fine-laminated lake sediment, well-sorted coarse beach sand and granules. A scoop-shaped scar, 27° steep, filled with these transgressional laminated sediments at the front of the basal fan delta, indicates a detachment surface, following gravitational instability.

Overlying Gilbert-type fan delta

Steeply-inclined (24–34°) eastward-dipping planar beds crop out at Wadi Mor (Fig. 3) and at Wadi Rahaf for a minimum thickness of 17 m, indicating lakeward propagation of foresets without interference of recurrent, lacustrine sedimentation. The foreset beds consist of planar couplets composed

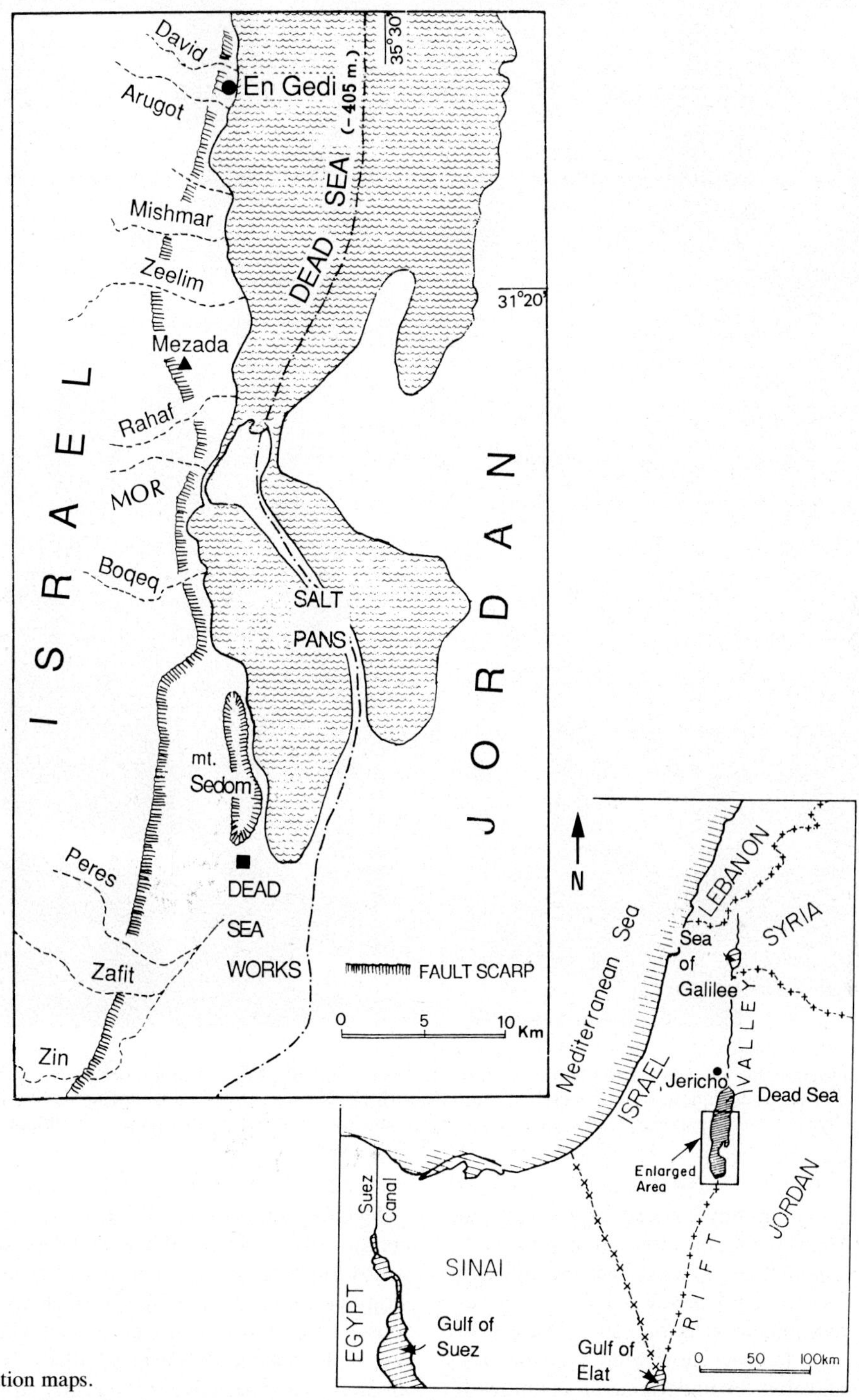

Fig. 1. Location maps.

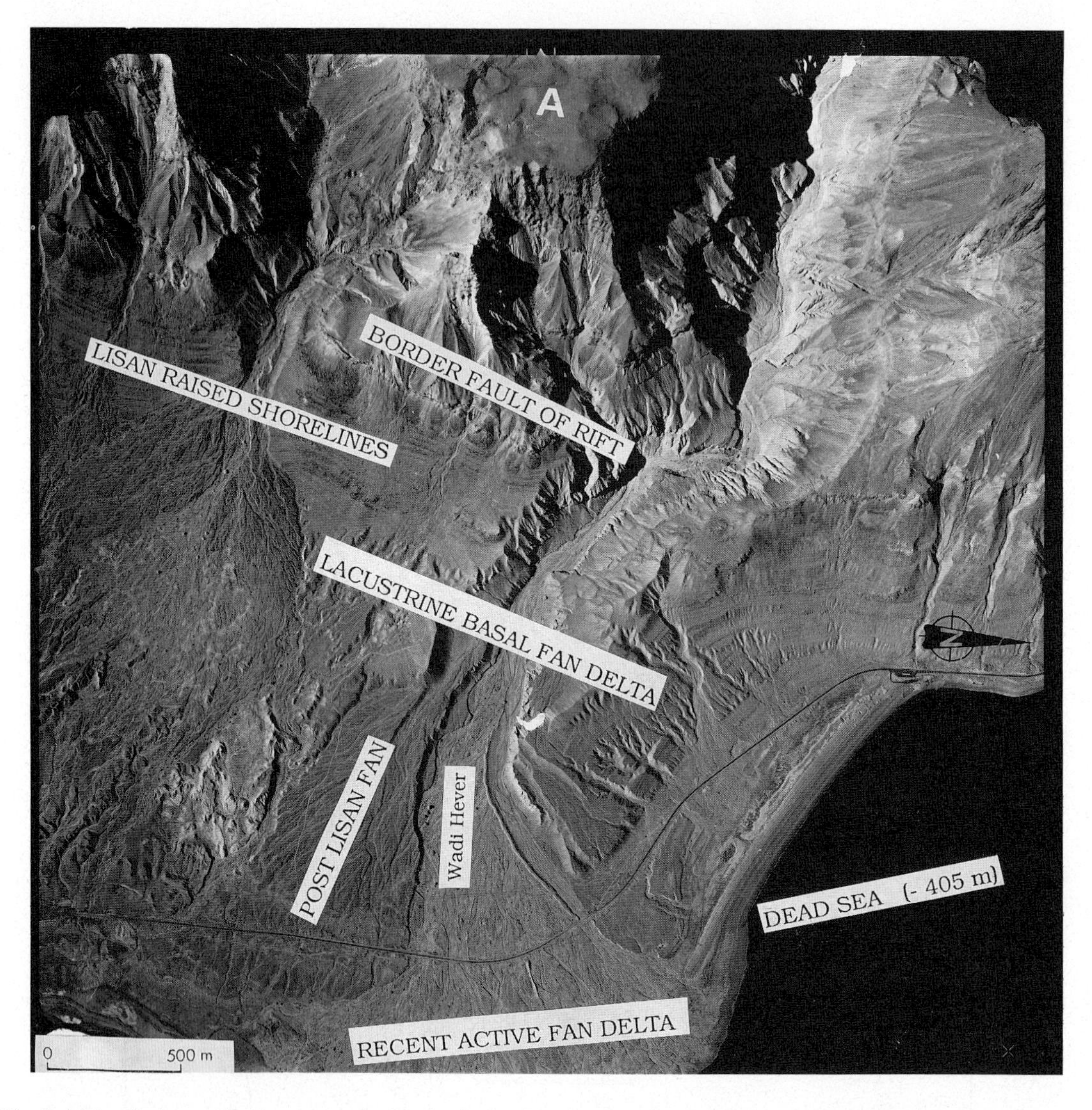

Fig. 2. Main physiographic elements at the border fault of the Dead Sea Rift. Altitude difference between the Dead Sea and the plateau of the Judean desert (A) is nearly 500 m. Wadi Hever is entrenched in a lacustrine basal fan delta, providing thereby spectacular exposures. Post-Lisan alluvial fans are free of beach terraces and show braided texture.

of well-segregated, tightly packed clast-supported cobbles and small boulders, alternating with well-sorted granule-pebbly units, which suggest effective sorting and winnowing processes in a shallow wave-dominated environment (Clifton, 1973; Nemec & Steel, 1984). The foresets show some planar cross-stratification and abut the underlying basal unit with a high angular contact. No clear bottomsets have been observed. Missing bottomsets have likewise been reported by Colella (1988).

The foreset beds display a sharp erosional unconformity, which predated the deposition of the gently dipping ($< 1°$) 9 m-thick, well-sorted and stratified capping unit (Fig. 5) which consists of the topset beds. On the surface, raised-beach terraces are well preserved (Bowman, 1978). The foresets are thus bounded between the gently dipping topsets and the moderately dipping basal fan delta, resulting in a planar, cross-stratified megastructure (Fig. 4). The rhythmic coarsening of both foresets and top-

Fig. 3. The basal fan-delta deposits (1) of Wadi Mor and the superimposed, Gilbert-type foresets (2). Note the sharp subhorizontal unconformity in between.

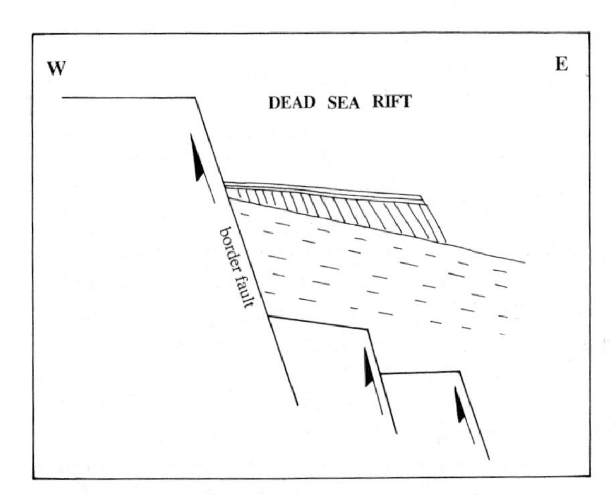

Fig. 4. Field relations between the gently dipping basal fan deposits (dashed lines) and overlying Gilbert-type fan-delta sediments and the border fault (schematic, not to scale).

sets, displayed by cyclic alternations of bouldery–cobbly units with granule–pebbly layers, indicates short-term pulsational depositional events.

The small dimensions of the lacustrine fan deltas of Wadi Mor (0.25 km^2) and of Wadi Rahaf (1.25 km^2) do not leave room for any significant downstream textural change other than some decrease in the boulder and cobble size. Based on lack of downfan grain-size grading and the overall small fan dimensions, the energy gradient dissipation was likely to be low.

The overall wedge-shaped geometry and foreset–topset organization are consistent with the classical model of a coarse-grained Gilbert-type fan delta. The fan deltas prograded from high locations in the canyons at the border fault towards the low-energy Lake Lisan.

The Gilbert-type fan delta sharply contrasts, texturally and structurally, with the underlying sheet-like proximal Lisan facies described by Sneh (1979). The topsets indicate the fluvial underwater extension; the steep foresets, almost at angle of repose, suggest bedload transportation and dumping in a basin of weak littoral and fluviatile dispersal energy. The contrast in density between the fluviatile fresh water and the saline Lake Lisan favoured flow separation which contributed towards formation of the mud-poor Gilbert-type fan deltas.

PALAEOCLIMATIC IMPLICATIONS

The last glacial optimum, $\approx$ 18 000 yr BP, was a rather dry period (Gat & Magaritz, 1980), evident even in dune-forming processes (Sarnthein, 1978). This stage was followed by a wetter, but still semi-arid period, indicated by the uppermost calcic horizon in the loess of southern Israel, dated 13 000 yr BP (Magaritz, 1986; Goodfriend & Magaritz, 1988). The wetter period 17 000–13 000 yr BP is further denoted by the alluvial–colluvial complex of Fazael, by the high water table (gleying) at Qadesh Barnea, eastern Sinai, as well as by the paleosol at Gebel Maghara, Northern Sinai (Goldberg, 1981). A fresh-

Fig. 5. The topsets of the Gilbert-structure at Wadi Mor. Note cyclic gravelly couplets (1/2).

water diagenetic zone of calcite cementation adjacent to the Lisan shorelines also indicates a humid climate between 14 000 and 13 000 yr BP (Druckman *et al.*, 1987). Furthermore, widespread Epipalaeolithic sites from Lebanon to Suez and to Wadi Feiran (Goldberg & Bar-Yosef, 1982; Goldberg, 1986) indicate definitely wetter conditions during 15 000–12 000 yr BP. A late Würmian pluvial phase, 18 000–11 000 yr BP, with mild winter rains and some summer rains was also corroborated by the presence of pollen (Horowitz, 1979). This suggests Mediterranean maquis and steppe vegetation south to Beer Sheva valley and on the Central Negev Highlands, respectively. During this semi-pluvial, but still relatively dry period, the 'White Cliff' member (Begin *et al.*, 1974) was deposited. Being the top unit of the marginal facies of the Lisan Formation, the Gilbert-type fan deltas are generally consistent with this 18 000–14 000 yr BP palaeoclimatic history.

The lake stage must have been high during deposition of the high Gilbert-type fan deltas and could of course be lower during the formation of the basal fan deltas. However, depth ratio is only one factor determining the different fan-delta structures. The Gilbert-type fan deltas along with the underlying basal fan deltas compose a facies sequence which might also be indicative of changes in the river flow regime, i.e. towards a lower-concentrated and more fluviatile type of flow, with weaker dispersal energy. Such a fingerprint at the rift margin correlates with the deep-water facies sequence, i.e. the 'Hamarmar' clay-enriched formation, capped by the 'White Cliff' aragonite unit. The fall in stream energy, indicated by both the marginal and the deep-lake facies sequences, reflects lowering of the tractive force and of the dispersa' capability, which are essential for forming steep Gilbert-type foresets (Jopling, 1963, 1965). Weakening of the stream power is further indicated by the improved sorting of the foreset and topset units as compared to the poor sorting of the basal fan delta. The angular character of the foresets points to the same hydrological weakness. A decrease in dispersal force does not, however, contradict the simultaneous high lakestand, as explained by Begin (1988), who suggested a reduction of some 10°C in the water temperature and a substantial reduction in lake evaporation, as causes for the highest Lisan level.

SUMMARY AND CONCLUSIONS

The overall geometry of the Gilbert-type fan deltas is wedge-shaped or of a veneer form, i.e. starting thin by the bounding fault and gradually increasing in thickness basinwards (Fig. 6). Thus, the fan delta is not thickest at the bounding fault, i.e. no bridging was attempted to re-establish equilibrium following tectonic activity (Steel & Gloppen, 1981). The thinning-out architecture towards the boundary fault rules out a tectonic control. A non-tectonic approach is further corroborated by the geochemical model of Katz *et al.* (1977), who concluded that the composition of the Lisan water needed 400–600 m of water

Fig. 6. Wedging out of the Gilbert-type fan-delta deposits (2) westwards towards the border fault (A — top-sets, B — foresets, 1 — basal fan delta).

depth, suggesting that the tectonics of the deep northern Dead Sea basin predated the deposition of the Lisan Formation.

The best-developed Gilbert-type fan deltas occur in the small catchment area of Wadi Mor (10 km^2), Wadi Boqeq (21 km^2) and at the basin of Wadi Rahaf (66 km^2). Decrease of the stream discharge in small basins during wetter semi-arid conditions must have stemmed from the effect of increased precipitation which significantly improved vegetational cover, retarded erosion and decreased run-off, flood peaks and sediment yield (Schumm, 1968). However, not all the small drainage basins (Wadi Zafit, 33 km^2; Wadi Peres, 23 km^2; Wadi Mishmar, 18 km^2) reveal Gilbert-type cappings. The inconsistent occurrences of Gilbert-type fan deltas at small drainage basins corresponds to the inconsistency in the number of fan terraces at adjacent drainage basins of similar lithology and catchment area (Bowman, 1988).

In large catchment areas, the denser vegetation cover was ineffective in limiting run-off and stream power under flash-flood conditions. The largest catchment areas, Wadi Zeelim (236 km^2) and Wadi Arugot (229 km^2), show therefore, no continuous capping by Gilbert-type fan deltas. Here only embryonic local foreset units are observed. These large catchment areas possess a hydrologic filtering capability (Steel, 1976), i.e. the size of the drainage basin acts as a hydrological regulator of the fan structure. Graf (1987), who studied temporal sediment storage changes in the Colorado plateau region, with 150–750 mm mean annual precipitation, draws similar conclusions related to different scales of basins: the smallest basins (< 1 km^2) stored little sediment and appeared insensitive to environmental changes. It is the mid-drainage area of 1–1000 km^2 where the stored sediment increased rapidly and radical changes occurred in stream power in response to land management or climatic changes. In reaches of 1000–10 000 km^2 stored sediment declined rapidly and commonly approached zero. Thus the mid-basins were the areas of major fluctuations in erosion and deposition.

Textural, morphological and structural characteristics, aided by the dated framework and supplemented with abundant palaeoclimatic data, suggest that climate might have triggered and controlled the marginal facies sequence of the uppermost Lisan. Gilbert-type fan deltas are thus also fingerprints of palaeoclimatic–hydrologic fluctuations. Further study is recommended to define the threshold conditions for filtering out climatic fluctuations.

ACKNOWLEDGEMENTS

I am grateful to C.J. Dabrio, S. Porębski and I. Reid for their constructive comments and suggestions.

REFERENCES

Begin, Z.B. (1988) The rise and retreat of Lake Lisan: A review. *The Israel Geological Society Annual Meeting*, En-Boqeq, 10–11.

Begin, Z.B., Broecker, W., Buchbinder, B., Druckman, Y., Kaufman, A., Magaritz, M. & Neev, D. (1985) Dead Sea and Lake Lisan levels in the last 30 000 years. *Preliminary Report Israel Geol. Surv., GS1/29/85*, 18 pp.

Begin, Z.B., Ehrlich, A. & Nathan, Y. (1974) Lake Lisan, the Pleistocene precursor of the Dead Sea. *Israel Geol. Surv. Bull.* **63**, 30 pp.

Bentor, Y.K. & Vroman, A. (1960) *The Geological Map*

of Israel on a 100000 Scale, Ser. A, sheet 16, Mount Sodom, explanatory text. 117 pp.

Bowman, D. (1971) Geomorphology of the shore terraces of the Late Pleistocene Lisan Lake (Israel). *Palaeogeogr. Palaeoclimatol. Palaeoecol.* **9**, 183–209.

Bowman, D. (1978) Determination of intersection points within a telescopic alluvial fan complex. *Earth Surf. Proc.* **3**, 265–276.

Bowman, D. (1988) The declining but non-rejuvenating base level — The Lisan Lake, the Dead Sea area, Israel. *Earth Surf. Proc.* **13**, 239–249.

Clifton, H.E. (1973) Pebble segregation and bed lenticularity in wave-worked versus alluvial gravel. *Sedimentology* **20**, 173–187.

Colella, A. (1988) Gilbert-type fan deltas in the Crati Basin (Pleistocene, southern Italy): sedimentology and tectonic setting. In: *Fan Deltas: Sedimentology and Tectonic Settings* (Ed. by W. Nemec and R.J. Steel), pp. 50–74, Blackie and Son, London.

Druckman, Y., Magaritz, M. & Sneh, A. (1987) The shrinking of Lake Lisan as reflected by the diagenesis of its marginal oolitic deposits. *Israel J. Earth Sci.* **36**, 101–106.

Frostick, L.E. & Reid, I. (1989) Climatic versus tectonic controls of fan sequences: Lessons from the Dead Sea, Israel. *J. Geol. Soc. London* **146**, 527–538.

Gat, J.R. & Magaritz, M. (1980) Climatic variations in the eastern Mediterranean Sea area. *Naturwissenschaften* **67**, 80–87.

Gilbert, G.K. (1885) The topographic features of lake shores. *Fifth Ann. Rept., US geol. Surv.*, 69–123.

Goldberg, P. (1981) Late Quaternary stratigraphy of Israel: An eclectic view. In: *Prehistoire du Levant*, Colloques Internationaux CNRS **598**, 55–56.

Goldberg, P. (1986) Late Quaternary environmental history of the southern Levant. *Geoarchaeology* **1**, 225–244.

Goldberg, P. & Bar-Yosef, O. (1982) Environmental and archaeological evidence for climatic changes in the southern Levant. *British Archaeol. Rep Serv.* **133**, 399–414.

Goodfriend, G.A. & Magaritz, M. (1988) Paleosols and late Pleistocene rainfall fluctuations in the Negev desert. *Nature* **332**, 144–146.

Graf, W.L. (1987) Late Holocene sediment storage in canyons of the Colorado Plateau. *Bull. geol. Soc. Amer.* **99**, 261–277.

Horowitz, A. (1979) *The Quaternary of Israel*. Academic Press, London, 394 pp.

Jopling, A.V. (1963) Hydraulic studies on the origin of bedding. *Sedimentology* **2**, 115–121.

Jopling, A.V. (1965) Hydraulic factors controlling the shape of lamina in laboratory deltas. *J. sedim. Petrol.* **35**, 777–791.

Katz, A., Kolodny, Y. & Nissenbaum, A. (1977) The geochemical evolution of the Pleistocene Lake Lisan — Dead Sea system. *Geochim. Cosmochim. Acta* **41**, 1609–1629.

Kaufman, A. (1971) U-series dating of Dead Sea Basin carbonates. *Geochim. Cosmochim. Acta* **35**, 1269–1281.

Kazanci, N. (1988) Repetitive deposition of alluvial fan and fan-delta wedges at a fault-controlled margin of the Pleistocene–Holocene Burdur Lake Graben, southwestern Anatolia, Turkey. In: *Fan Deltas: Sedimentology and Tectonic Settings* (Ed. by W. Nemec and R.J. Steel), pp. 186–196, Blackie and Son, London.

Leeder, M.R., Ord, D.M. & Collier, R. (1988) Development of alluvial fans and fan deltas in neotectonic extensional settings: Implication for the interpretation of basin-fills. In: *Fan Deltas: Sedimentology and Tectonic Settings* (Ed. by W. Nemec and R.J. Steel), pp. 173–185. Blackie and Son, London.

Magaritz, M. (1986) Environmental changes recorded in the Upper Pleistocene along the desert boundary, southern Israel. *Palaeogeogr. Palaeoclimatol. Palaeoecol.* **53**, 213–229.

Manspeizer, W. (1985) The Dead Sea Rift: Impact of climate and tectonism on Pleistocene and Holocene sedimentation. In: *Strike-Slip Deformation, Basin Formation and Sedimentation* (Ed. by K.T. Biddle and N. Christie-Blick). Spec. Publ. Soc. econ. Paleont. Miner., Tulsa, 37, 143–158.

Miall, A.D. (1984) *Principles of Sedimentary Basin Analysis*. Springer, 490 pp.

Neev, D. & Emery, K.O. (1967) The Dead Sea, depositional processes and environments of evaporites, Israel. *Israel Geol. Surv. Bull.* **41**, 147 pp.

Neev, D. & Hall, J.K. (1977) Climatic fluctuations during the Holocene as reflected by the Dead Sea levels. In: *Terminal Lakes*. (Ed. by D.C. Greer). *Proceedings from International Conference on Desertic Terminal Lakes, Ogden, Utah*, 53–60.

Nemec, W. & Steel, R.J. (1984) Alluvial and coastal conglomerates: Their significant features and some comments on gravelly mass-flow deposits. In: *Sedimentology of Gravels and Conglomerates*. (Ed. by E.H. Koster and R.J. Steel). Mem. Can. Soc. Petrol. Geol. 10, pp. 1–31.

Sarnthein, M. (1978) Sand deserts during glacial maximum and climatic optimum. *Nature* **272**, 43–46.

Schumm, S.A. (1968) Speculations concerning paleohydrologic controls of terrestrial sedimentation. *Bull. geol. Soc. Amer.* **79**, 1573–1588.

Sneh, A. (1979) Late-Pleistocene fan deltas along the Dead Sea Rift. *J. Sedim. Petrol.* **49**, 541–552.

Steel, R.J. (1974) New red sandstone floodplain and piedmont sedimentation in the Hebridean Province, Scotland. *J. sedim. Petrol.* **44**, 336–357.

Steel, R.J. (1976) Devonian basins of western Norway — sedimentary response to tectonism and to varying tectonic context. *Tectonophysics* **36**, 207–224.

Steel, R.J. & Gloppen, T.G. (1981) The deposits, internal structure and geometry in six alluvial fan, fan-delta bodies (Devonian-Norway) — a study in the significance of bedding sequence in conglomerates. In: *Recent and Ancient Nonmarine Depositional Environments: Models for Exploration*. (Ed. by F.G. Ethridge and R.M. Flores). Spec. Publ. Soc. econ. Paleont. Mineral, Tulsa, 31, pp. 49–69.

Walker, R.G. (1978) Deep-water sandstone facies and ancient submarine fans: Models for exploration for stratigraphic traps. *Am. Assoc. Petrol. Geol.* **62**, 932–966.

Wright, L.D. (1977) Sediment transport and deposition at river mouths: a synthesis. *Bull. geol. Soc. Amer.* **88**, 857–868.

Spec. Publs int. Ass. Sediment. (1990) **10**, 281–295

Pleistocene glacial fan deltas in southern Ontario, Canada

I. P. MARTINI

Department of Land Resource Science, University of Guelph, Guelph, Ontario, Canada

ABSTRACT

Stratified ice contact and outwash deposits were commonly formed at the margins of the Pleistocene ice-sheets which covered North America. Local advance and retreat of the glacial terminus mimicked reactivation of faults and erosion in tectonically active areas. Examples are presented in this paper of foresetted fan deltas which formed in lakes and seas near glacier terminus, fed by eskers or other englacial streams, and at the end of outwash and 'valley trains' of various lengths. The fan deltas treated here have similar foresets, characterized by massive to parallel-bedded gravelly layers alternating with openwork gravel and coarse sand lenses, mostly emplaced by mass flow. They vary, however, in types of topsets (bouldery channel deposits or washed beach gravels), and in the complexity of lateral and vertical facies transitions which is in part due to rapid change in recurring strong fluvial floods and to rapid water-level changes of lakes or sea.

INTRODUCTION

As the Pleistocene glaciers retreated for the last time from Ontario (about 13 000 yr BP), the Great Lakes and the glacial Champlain Sea in eastern Ontario and Quebec underwent rapid changes in water level and emersion (Fig. 1; Hough, 1958; Dreimanis, 1969; Lewis, 1969; Prest, 1970; Chapman & Putnam, 1984; Karrow & Calkin, 1985). Deposits of such lakes and seas are well-exposed in sand and gravel pits and some outcrops.

The objective of this paper is to show various settings of fan deltas which developed under the influence of Pleistocene continental glaciers in Ontario. It is of interest to determine whether significantly different sedimentary architectures formed under the various conditions and to establish the importance of various sediment gravity flows in forming certain facies associations. Previously described (by others or by me, as indicated) exposures have been re-examined and in part reinterpreted focusing on their Gilbert-type delta characteristics (Gilbert, 1885; Nemec & Steel, 1988).

GLACIO-LACUSTRINE AND MARINE FAN DELTAS

Examples are reported here of lacustrine and marine fan deltas developed either in direct contact with the glacier, fed by glacier streams (Allan Park, Westmeath-Osceola and Fonthill), or at the end of narrow bedrock valleys which funnelled glacial meltwater (Campbellville and Joe Lake).

The following first three examples are of fan deltas developed at or very near the glacier terminus.

Allan Park Station (Esker-fed, lacustrine fan delta)

Description

A sand and gravel pit exposes up to 12 m of a highly variable sand and gravel deposit (0.5 km long, 2 km wide and 35 m thick) (Fig. 2; Cowan *et al.*, 1979; Sharpe, 1982). The deposit shows three major facies associations (after Sharpe, 1982):

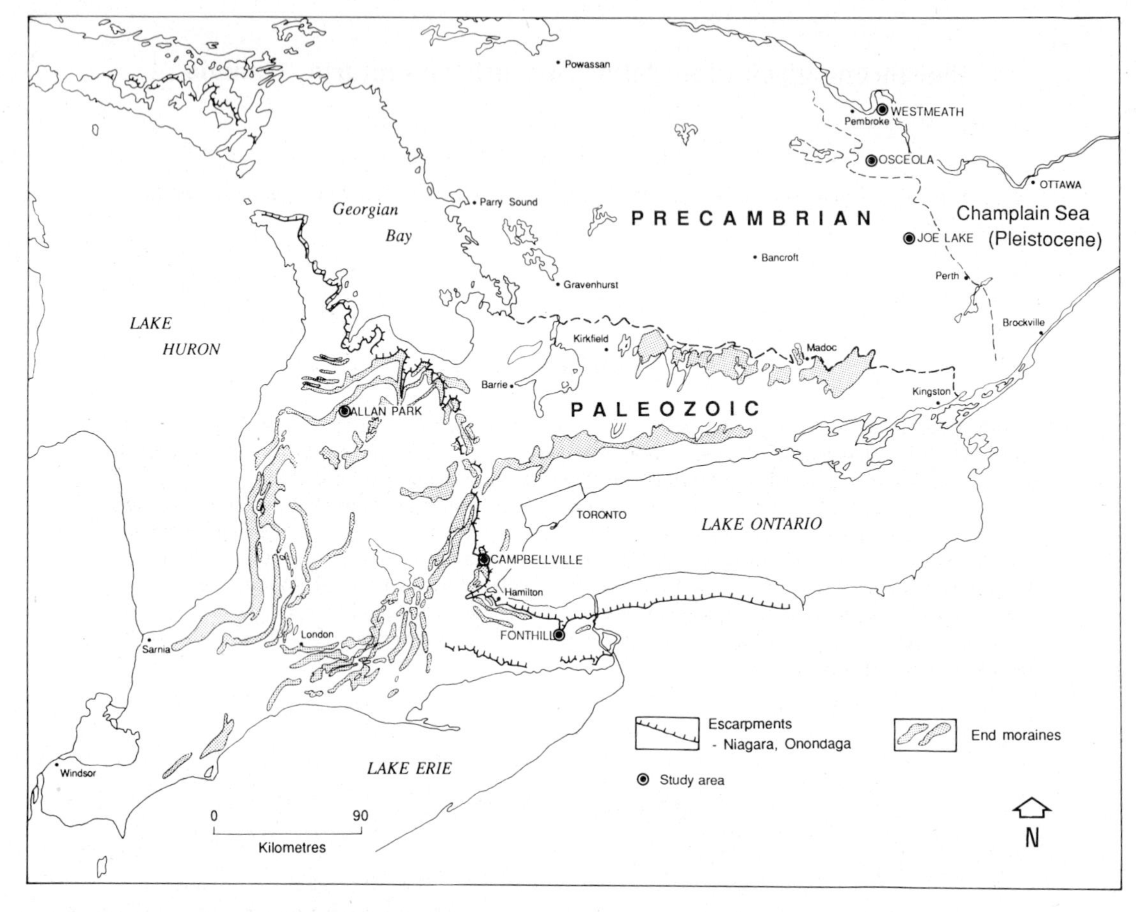

Fig. 1. Pleistocene map of southern Ontario and localities studied.

1 A gravelly facies association characterizes coarse foresets, and consists of an irregular interlayering of massive, poorly sorted, coarse gravels (Gm facies) with local, open framework (openwork) gravel lenses (up to 30 cm thick), massive to faintly laminated, poorly sorted, pebbly coarse sand to medium sand (Sm facies), trough cross-bedded sand (St facies), and few layers of ripple-drift cross-laminated sand (Sr facies) (Figs 2C, units 1, 2, and 2D-3). The layers show sharp bases, and fining-upward sequences, although local inverse to normal grading occurs. The beds vary in thickness from 10 cm to 3 m (Gm), and are steeply dipping, up to 30° (Fig. 2C). Cuts and fills are absent. The composition of the pebbles is primarily Silurian dolostone (local bedrock) and metamorphic rocks transported by glaciers from the Precambrian Shield. The pebbles are subrounded and subspherical, and do not show well-developed imbrication. Sharpe (1982) reported pebble a-axis orientation subparallel to the dip of the beds, and trending to the SSW (Fig. 2D-3).

2 A sandy facies association forms finer foresets and toeset beds, and consists of a regular alternation of thin beds of medium-grained, massive sand layers (Sm facies), ripple-drift cross-laminated medium to fine sand (Sr facies) and minor plane-bedded medium sand (Sh facies), and sandy gravel (fine pebbles) layers (Gh facies) (Figs 2C, units 2, 3, and 2D-2). Thin silty drapes occur. Most layers have sharp bases. Flame structures and other load deformations occur. Local small cuts and fills are present. Layers of this facies association have steep dips (up to 30°) when they are in contact with the gravelly facies association, and become more gently dipping (10–15°) to subhorizontal in more distal areas (Fig. 2C, unit 3).

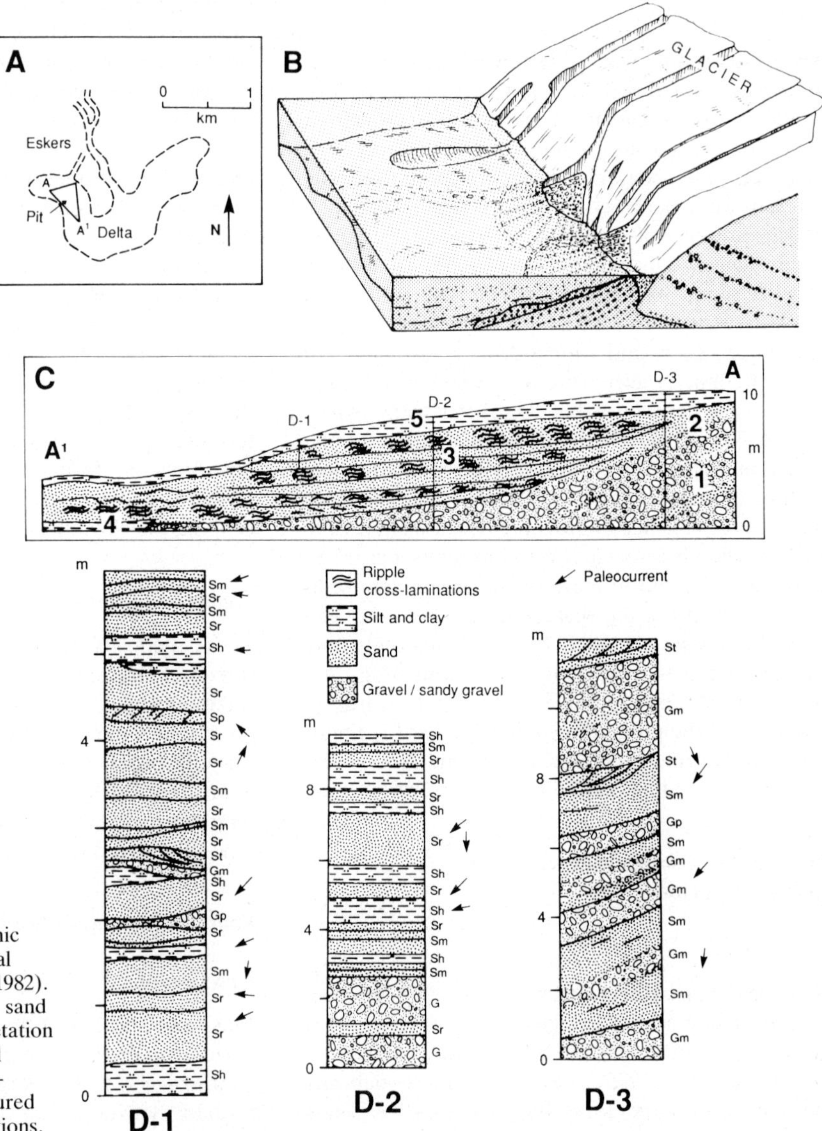

Fig. 2. Allan Park: stratigraphic information and environmental interpretation (after Sharpe, 1982). A. Map of the margins of the sand and gravel deposit; B. Interpretation of ice marginal environmental setting; C. Longitudinal cross-section with location of measured sections; D. Stratigraphic sections.

3 Sharpe (1982) reported that during a phase of excavation of the pit 'two broad channels of about 10 m width and 7·5 m depth were exposed trending to the southeast' (Figs 2C, unit 5, and 2D-1). The facies association of these channels consisted of medium to thin beds of alternating ripple-drift, cross-laminated sand (Sr facies), few massive sands (Sm facies), cross-stratified sandy gravels (St, Sp and Gp facies) and rare, thin, massive, pebble to small-cobble gravels (Gm facies) (Fig. 2D-1). The gravel facies were usually contained in secondary (50–80 cm deep) cuts and fills. Flame structures and pseudonodules (balls and pillows) were present in some sandy layers.

4 Fine-grained, silty and clayish bottomsets are poorly exposed in the pit (Fig. 2C, unit 4).

Palaeoenvironment

The geomorphology and Pleistocene geology of the area show that this deltaic system was fed by two confluent eskers and it consists of coalescing fans which prograded into a temporary shallow glacial lake (Figs 2A, 2B). The geometry of the gravelly foresets, the lack of well-developed apposition fabric and the presence of some inverse- to normally graded layers suggest subaqueous deposition by mass flow (Fig. 2C, unit 1). The foresets (Fig. 2C, units 1, 2) grade laterally into sandy layers which become increasingly finer and ripple-drift cross-laminated (Fig. 2C, unit 3) away from the sediment injection points and were probably deposited from hyperpycnal flows. The rhythmic sedimentation of the sandy units (Fig. 2D-2) suggest floods related to ice-melting events. However, it is not possible to recognize daily or seasonal cycles, both because of possible irregular floods associated with unusually warm and rainy periods, and because of the incomplete sedimentary record in the shallow lake. The sandy and gravelly facies association of the channel-fills may be considered to represent topset units (Fig. 2C, unit 5), although it is not clear whether it records an emersion of the prograding deltaic body or cut-and-fill events during rapid draining of the lake as the glacier retreated.

Westmeath–Osceola area (esker-fed, marine fan delta)

Description

Sand and gravel deposits are present in Eastern Ontario rising as small ridges on the otherwise flat terrain previously occupied by the glacial Champlain Sea (Chapman, 1975). Several of these deposits do not show large foresets, have massive sands and sandy gravels interstratified with trough cross-bedded sandy and gravelly units, and are interpreted as subaqueous 'outwash fans' (Rust & Romanelli, 1975; Rust, 1977; Naldrett, 1988; Sharpe, 1988; Gorrell, 1988). Others, such as those of the Westmeath–Osceola area (Figs 1, 3), near the edge of the Pleistocene sea, have well-developed large foresets and are considered to be fan deltas (Barnett & Kennedy, 1988).

The deposits exposed at Westmeath and Osceola have foresets up to 8–10 m thick, steeply inclined (up to 30° in direction 140° at Westmeath; 22° towards 220° at Osceola). The foresets are composed of interlayers of (1) crudely bedded, poorly sorted sandy gravel with clasts primarily of crystalline Precambrian rocks, not well-defined apposition fabric except in a few pebble clusters (Fig. 4); (2) coarse to very coarse sand with disseminated fine pebbles primarily placed parallel to the foreset surfaces; and (3) open-framework gravel lenses (10–30 cm thick) showing local high-upstream imbrication of pebbles.

The foresets are overlain by a subhorizontal deposit (topsets) characterized by a bouldery bed overlain by sandy gravel locally showing low dips up to 7°, and thin openwork gravel lenses. Stratified, lensing, washed gravel layers are locally present at the top and on the flanks of the deposit on cut terraces. Barnett and Kennedy (1988) reported a few *Macoma balthica* shells from some of these terraces, and an occurrence of diamicton partially capping the Westmeath sequence.

Palaeoenvironment

These deposits were formed in coalescing ice-proximal foresetted fans. The topsets show fluviatile sediments which were in part reworked by waves of the glacial Champlain Sea, and sparsely fossiliferous gravelly beaches on wave-cut terraces. The sandy diamicton which caps the sequence at Westmeath is interpreted by Barnett and Kennedy (1988) to be flow till derived from a proximal glacier terminus.

Fonthill (ice-contact fan delta)

Description

This sand and gravel deposit forms a large isolated hill on top of the Niagara Escarpment (Figs 1, 5A; Feenstra, 1981; Feenstra & Frazer, 1988). It was formed in front of an 'ice stream' (more actively moving portion of the glacier) which was funnelled through a narrow re-entrance in the 50–70 m-high bedrock Niagara Escarpment. A high and steep ice-contact slope marks the northern side of the deposit (Figs 5D, 6A). Gentler and terraced slopes are present to the south. An elongated ridge trends to the northwest, and stands at an elevation lower than the top of the main deposit.

The deposit shows several facies associations (Figs 5A, 5D):

1 A gravelly ridge (about 3 m high) tops the sequence (Fig. 5A-1), and is composed primarily of round carbonate pebbles.

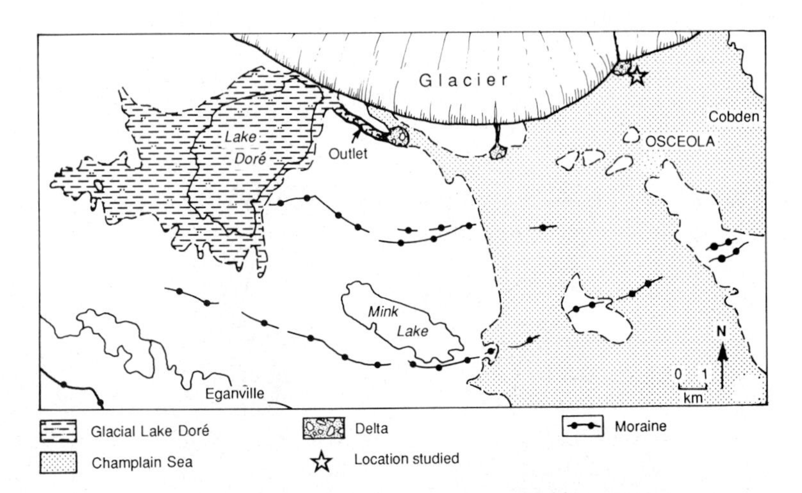

Fig. 3. Interpreted position of the ice-contact fan deltas at Osceola (after Barnett & Kennedy, 1988).

Fig. 4. Characteristic poorly-sorted sandy gravel foresets at Osceola.

2 An upper terrace contains the main deposit which is composed of steep foresets of sandy gravels with thin lensing interlayers of coarse sand and openwork gravel lenses (Fig. 5A-2). Two sets of foresets are present. One set dips to the south (22° in direction 190°) and shows features much similar to, but finer than previously described foresets (Fig. 6C); the other dips to the northeast (26° with azimuth 60°), in the direction of the glacial source. The northeasterly dipping foresets differ from all others described in this chapter as they are composed of irregular lenses of sandy gravel and openwork gravel, draped by coarse sand (Fig. 6B). Thin, basal inverse grading occurs in some gravelly beds (Fig. 6D).

3 A lower terraced sand body shows both massive sandy layers and cross-bedded (several metres thick) sandy gravel units (channel-fills) (Figs 5A-3, 5C). The sand is coarse grained, fairly poorly sorted with small pebbles disseminated throughout.

4 The poorly exposed northwest-trending ridge is most likely composed of sandy gravel as indicated by geophysical prospecting (Fraser, pers. comm.).

5 Poorly exposed fine sand rims the southern quadrants of the main deposit, and this grades into the silt and clay of the antecedent lacustrine plain.

Palaeoenvironment

This deposit has been referred to in the past as a kame delta (Feenstra, 1981; Chapman & Putnam,

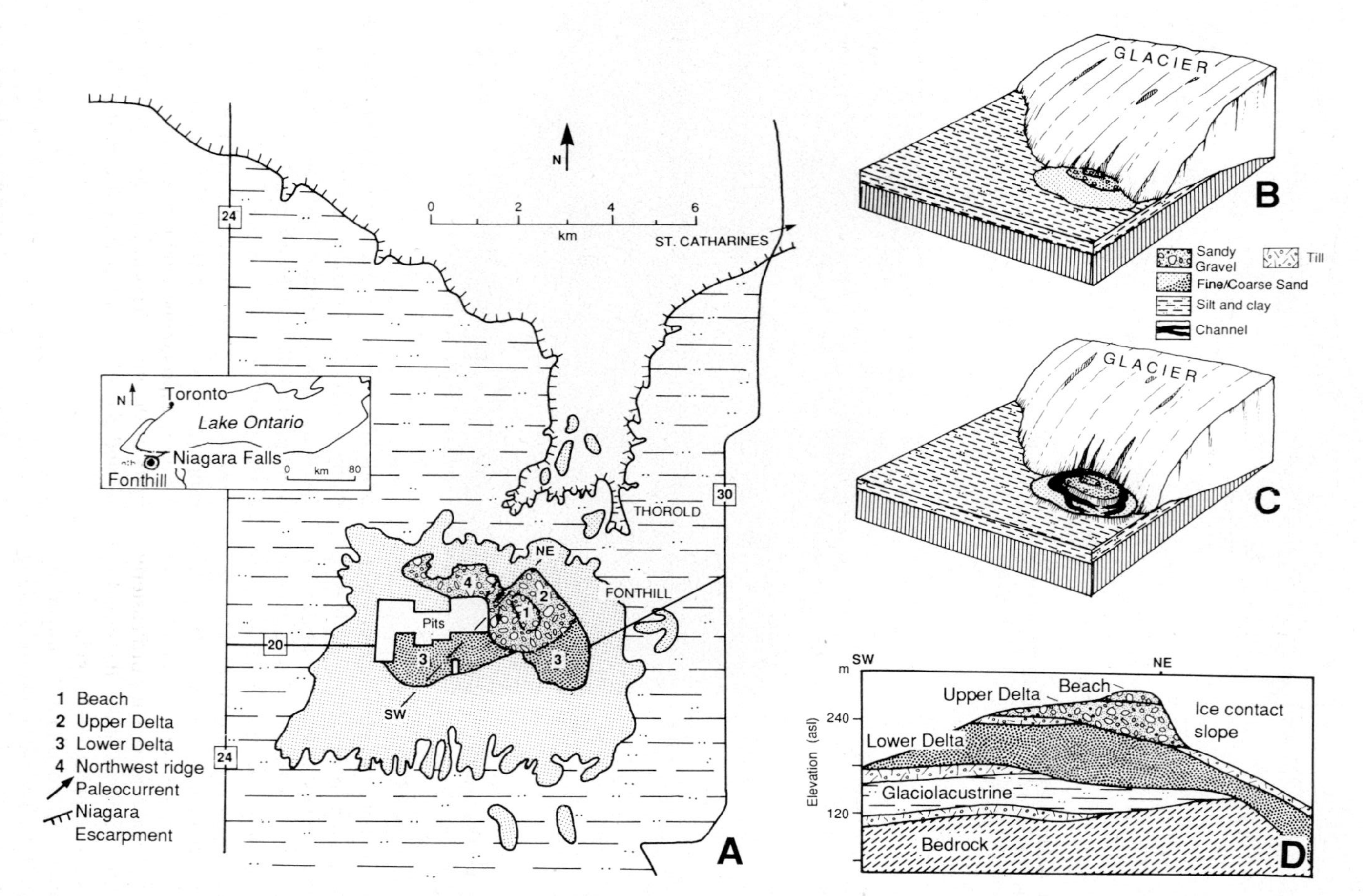

Fig. 5. Fonthill: stratigraphic information and environmental interpretation. A. Location map of deposit; B. Formation of the ice-contact fan delta; C. Formation of braided-stream deposits during slight retreat of glacier and lowering of lake level (after Frazer, pers. comm.); D. Longitudinal cross-section showing interpreted architecture of deposit (after Feenstra & Frazer, 1988).

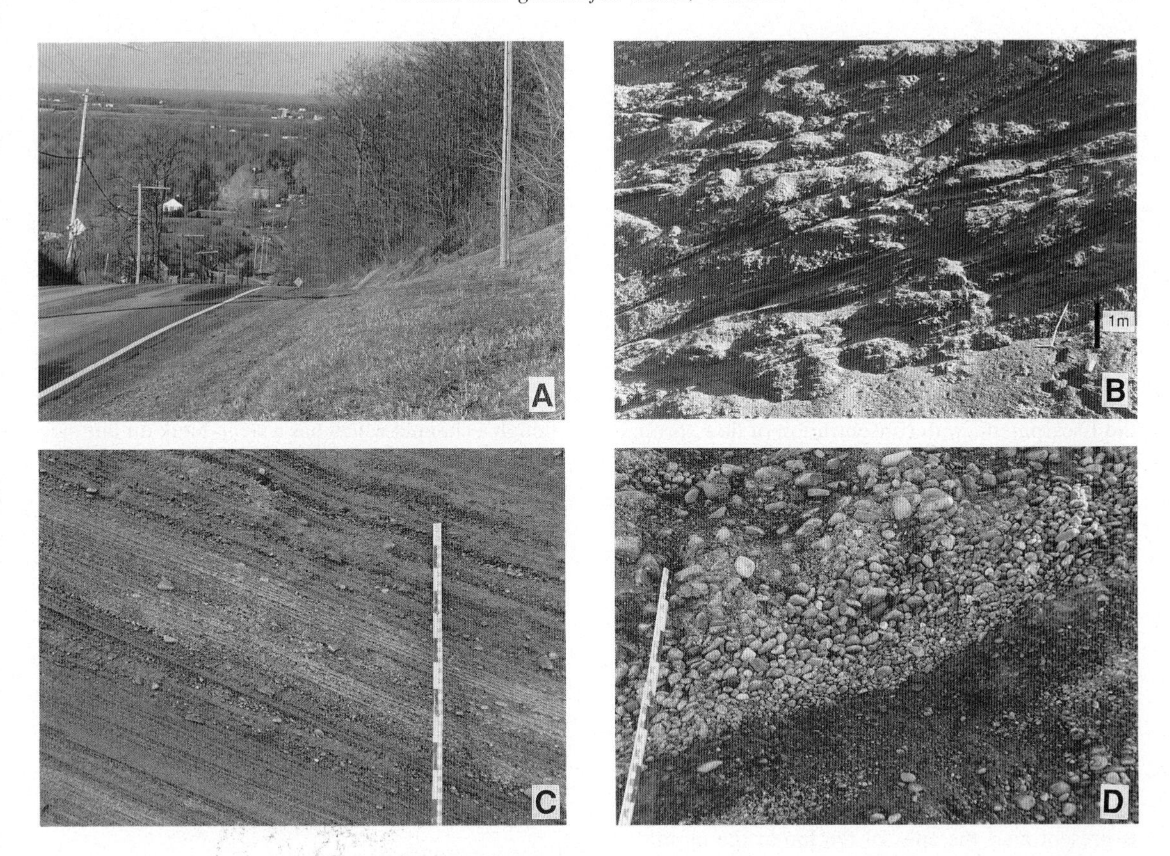

Fig. 6. Characteristic features of the Fonthill deposit: (A) Northern steep ice-contact slope; Lake Ontario can be seen in the background; (B) Northeastward (sourceward) dipping foresets (backsets ?) showing lensing layers of coarse pebbly sand and sandy gravels; (C) Regularly laminated southward-dipping foresets of pebbly sands to sandy gravels; (D) Inversely to normally graded openwork gravel lenses in the northeastward dipping foresets.

1984; Feenstra & Fraser, 1988). The divergent dips of the exposed foresets (association 2) suggest that the northwestward dipping ones may be backsets associated with the northwest trending ridge (association 3) which could be interpreted as a feeding subglacial conduit (Feenstra, pers. comm.; Jopling & Richardson, 1966). This interpretation is supported by the lensing bedding characteristics and local well-developed imbrication, although the deposit is somewhat better organized than one would expect in such an environment (Jopling & Richardson, 1966). The whole deposit is most likely the product of coalescing fan deltas fed by supraglacial or englacial streams discharging at a very complex ice margin (Figs 5B, 5C).

Fraser (pers. comm.) has suggested that after a slight retreat of the glacier and drop in lake level, glacial braided streams flowed between the receding glacier margin and the main body of the pre-existing gravelly fan which stood as an island (Fig. 5C). These streams formed the gravelly channel-fills and part of the lower sands which surround the main deposit. Fluctuating lake levels have permitted also wave reworking of part of the materials at the top of the sequence and on wave-cut terraces on its flanks.

The following two examples are of fans fed by glacial meltwater and developed away from the glacier terminus, at the downstream end of valleys cut into bedrock. They have many features of alluvial fans, but they are associated with lacustrine deposits and contain well-developed foresets, hence they have fan-delta characteristics.

Campbellville (outwash fan delta)

Description

The Niagara Escarpment is an erosional positive bedrock feature (up to 50–80 m high) on the otherwise flat landscape of southern Ontario. It rims the western and southern side of the Lake Ontario basin. Locally the Escarpment has outliers separated from the main trend by narrow dry valleys. As the glacier retreated towards the centre of the Lake Ontario basin (Lake Ontario glacial lobe), funnelling of meltwater occurred between the glacier itself and the Escarpment, with concentration of flow through the narrow outlier valleys (Figs 7A, 7C). Relatively large fans developed at the downstream end of the outlier valleys (now dry valleys) and prograded into temporary shallow lakes bounded at one side by the glacier. The deposits of one such a fan are exposed in large sand and gravel pits near Campbellville (Figs 1, 7A; Martini, 1972, 1982).

A variety of facies associations occur in the fan (Figs 7D, 7E):

1 A bouldery, massive to crudely bedded, subhorizontal thick (3–5 m) unit with dolomitic (locally derived) and crystalline (derived from the distant Precambrian shield) clasts (up to 60 cm in diameter) is present near the apex of the fan at the mouth of the dry valley.

2 A variable sandy–gravelly facies association forms two distinct types of deposits. (a) One is characterized by subhorizontal to slightly inclined massive sandy gravels (2 m thick units) with some disseminated small boulders. (b) The second forms thick (8–10 m), steeply inclined (18–20°) foresets where poorly sorted sandy gravels alternate with pebbly sand and openwork gravel lenses (Figs 7D, 7E, 8A). The openwork lenses are coarse grained (pebbles to small boulders), and show coarsening towards the downdip nose, pebble imbrication (up to 70°) behind the coarse nose and irregularly imbricated pebbles in the main body (Fig. 8A).

3 A pebbly sand facies association is found in two main unit types: (a) One characterizes steeply dipping (18°) foresets composed of poorly sorted, coarse to very coarse sand with disseminated pebbles. Local intraset, downdip-oriented, low-angle inclined laminations occur (Fig. 8C). In places, many shallow (in the order of a few centimetres) and short (in the order of tens of centimetres) lenses of openwork granules characterize the foresets. In some exposures, the base of the foresets (toesets) is marked by trough cross-beds which show palaeoflow trending at high angle from the direction of dip of the foresets. (b) The second type is subhorizontal, has pervasive trough cross-bedding and occurs as stacked units of about 4 m each recognizable one from the other because of slight differences in particle size (Figs 7D, 7E, 8D).

4 A sorted fine to medium sand facies with good to fair sorting and some rare fine pebbles and granules is found in thin to medium, subhorizontal beds which show plane beds, cross-beds and ripple cross-laminations and minor scours and fills. In one instance, a 2 m high cut and fill shows a cross-section though a channel which has a steep bank on one side and a grading bank in the other, as expected to form at a bend of a stream. This facies is found only in the distal portions of the deposit.

5 A silt–clay facies occurs interstratified with the pebbly sand and sorted sand facies associations. It occurs in two types of units: (a) One shows massive, lensing, subhorizontal, locally deformed medium beds with disseminated fine and occasional large pebbles (lonestones). (b) The second consists of thin horizontal units (order of 40–60 cm) of well-developed varve-like rhythmites with sparse, pebble-size lonestones. Some units are intensely deformed.

Palaeoenvironment

This deposit is a complex, fluvial-dominated fan affected by strongly variable flood conditions and lacustrine water levels. Feeding torrential streams cut and filled channels with poorly-sorted bouldery deposits (facies 1) near the apex (Figs 7B, D). These grade downslope and laterally into massive (facies 2a) and cross-bedded (facies 3b) pebbly sand of large braided streams, which in turn grade into more distal, sorted sand (facies 4) of streams and deltas, and finally into silt–clay lacustrine material (facies 5) (Figs 7B, 7D, 7E). These facies are cut longitudinally by large channels (small valleys) which are in part filled with large-scale gravelly (facies 2b) and pebbly sand (facies 3a) foresets resembling those of Gilbert-type deltas.

The features of the gravelly foresets (facies 2b) and of the openwork gravelly lenses they contain, such as inverse to regular vertical grading and coarsening and better imbrication of clasts towards the noses of the lenses, record modified debris flows and grain flows (Lowe, 1976, 1982; Nemec & Steel, 1984; Colella, 1988).

The features of the more gently dipping sandy

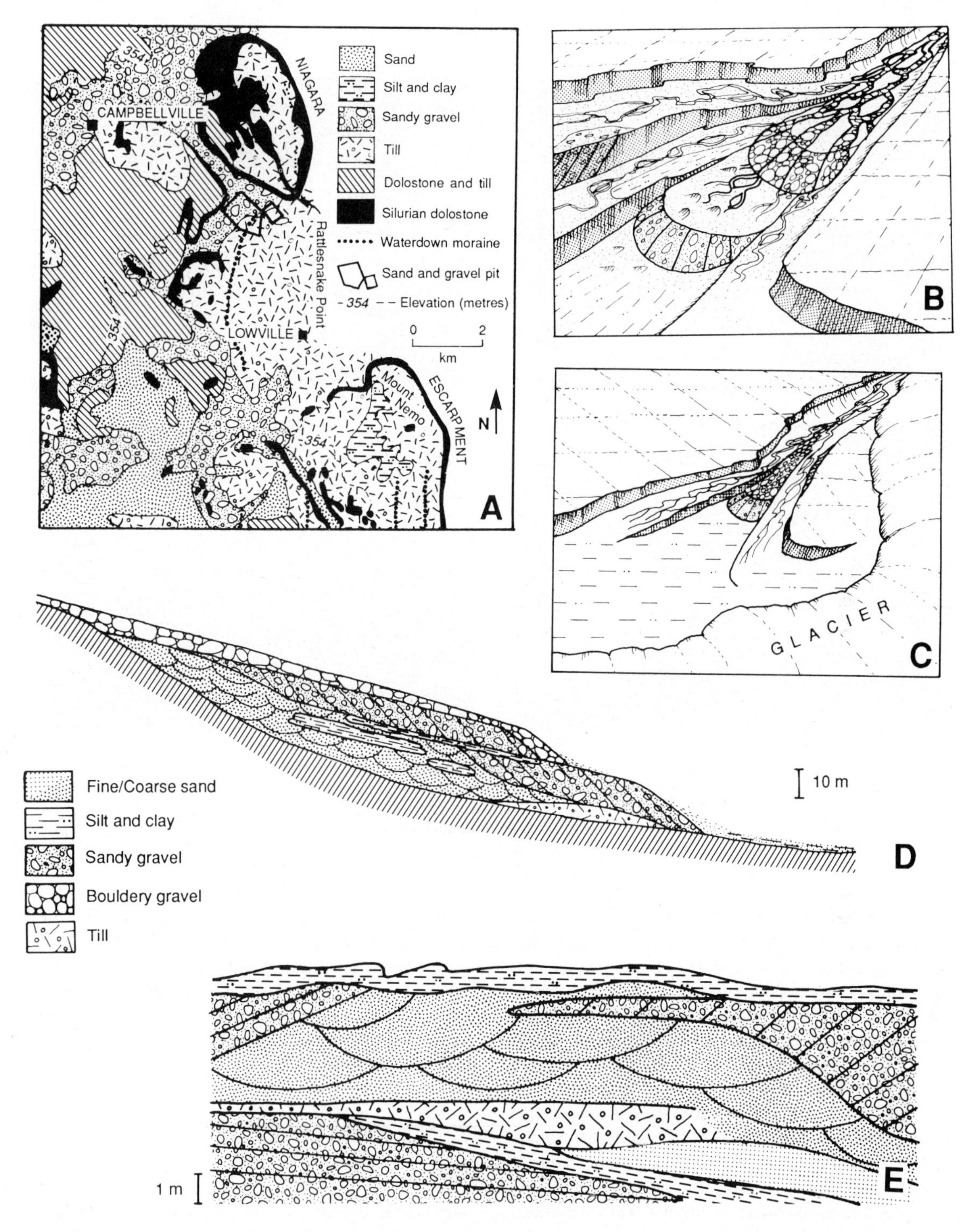

Fig. 7. Campbellville: stratigraphic information and environmental interpretation. A. Pleistocene map (after Karrow, 1986); note the narrow dry valley between Rattlesnake Point outlier and the main Silurian dolostone escarpment; B. Interpreted schematic distribution of deep channels and facies associations; C. Interpreted schematic reconstruction of the glacial position during a stage of development of the fan; D. Idealized facies relationships and schematic longitudinal cross-section (approximate horizontal distance 1.5 km, thickness about 40 m); E. Schematic transverse cross-section in the middle-lower part of the fan, some of the sandy gravel to pebbly sand foresets may be fills of large fluvial channels (approximate distance 200 m, thickness, 25 m).

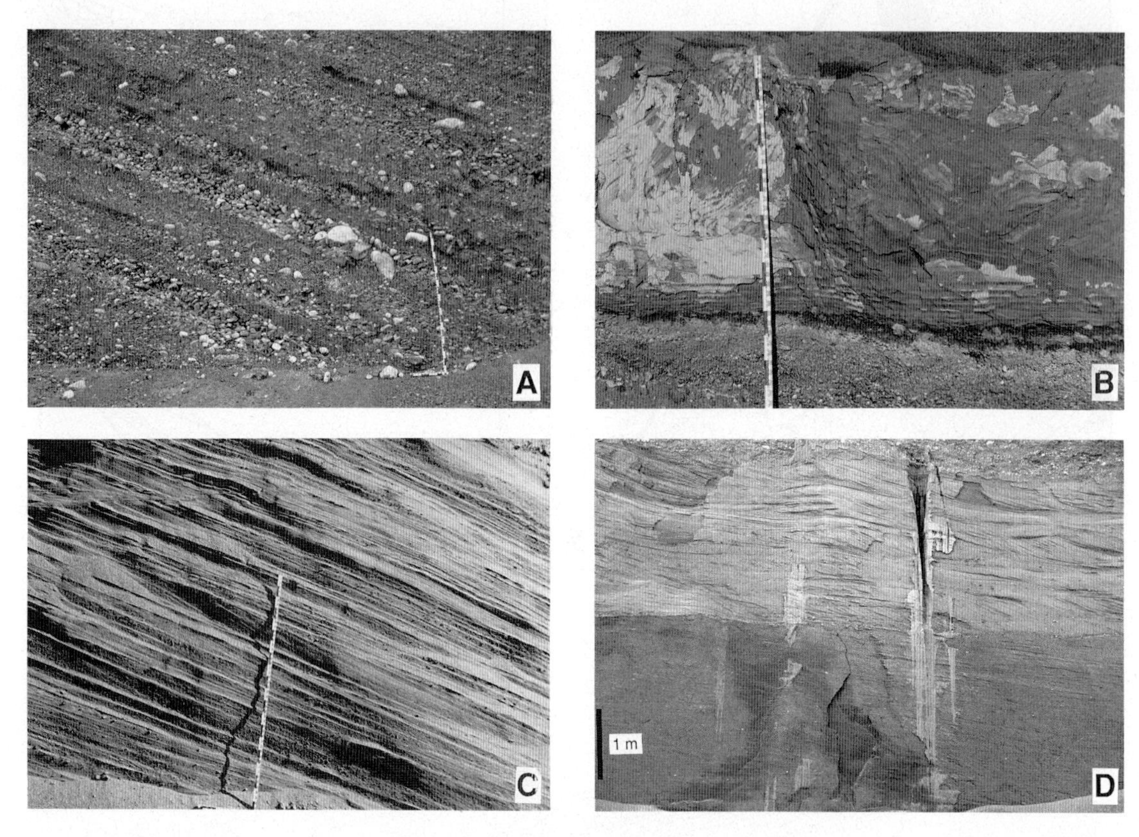

Fig. 8. Characteristic feature of the Campbellville deposit: (A) Steeply inclined gravelly foresets with openwork gravel lenses; note the coarse clasts at the nose of the lenses and the steep imbrication of some pebbles; (B) Lacustrine deposit showing varve-like rhythmites at the base and deformed lacustrine silt; (C) Sandy foresets with intraset low-angle cross-laminations; (D) Trough cross-bedded coarse sand to pebbly sand deposited in braided streams (thickness of photographed outcrop is about 8 m).

foresets (facies 3a), such as granule openwork lenses developed in few centimetres deep cuts and fills, down-dip low inclination of intraset laminations, and the trough cross-bedded character of their toesets, suggest instead relatively high-energy traction currents. Some of these foreset beds represent fluvial fills of large channels (Ori & Roveri, 1987), others which have associated 'varved' interlayers may be the result of attached hyperpycnal flows of stream-modified Gilbert-like deltas (Postma & Roep, 1985).

The foresetted units are capped by subhorizontal, gravelly and sandy, massive and cross-bedded units of braided streams (facies 1, 2a, 3b). The lacustrine, very fine sand and silt deposits (facies 5a, 5b), locally varved, occur at different levels indicating recurrence of glaciolacustrine conditions. The numerous erosional events that this deposit has experienced make it difficult to establish whether the lacustrine phases were continuous across the fan body or they developed in separated pools on it. The fairly well-sorted fine to medium sandy facies (facies 4) with internal fining-upward sequences and some asymmetric cut and fill, is associated with the lacustrine phases, and represents alluvial deposits prograding into shallow lakes.

Joe Lake (valley-train fan delta)

Description

The sand and gravel deposit of Joe Lake is composed of metamorphic clasts and formed at the end of a relatively narrow, long fault-bounded valley cut into Precambrian rocks (Figs 1,9). The deposit contains large foresets, up to 20 m thick, with dips up to 25–30° in direction 300°. These poorly-sorted, sandy

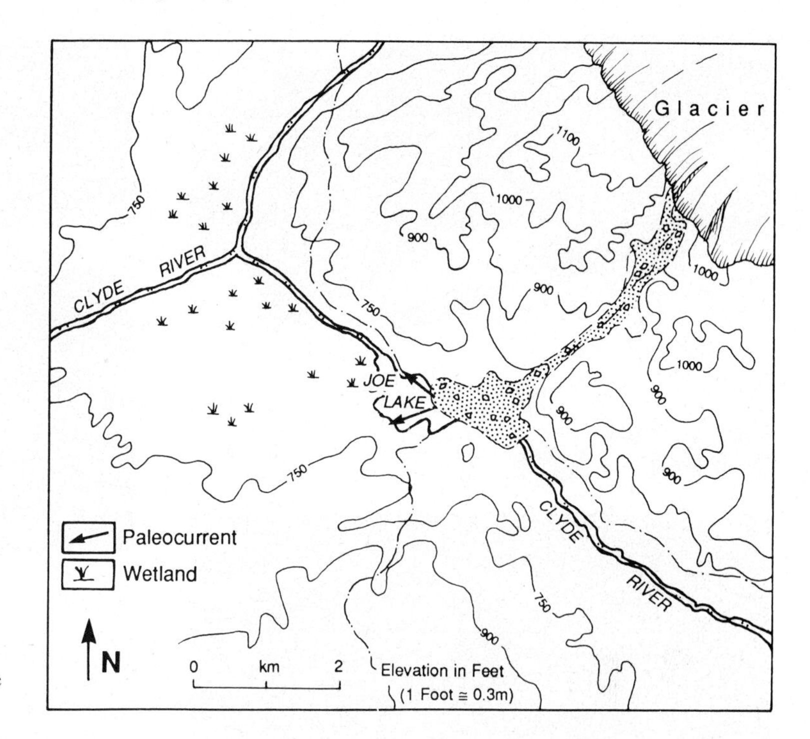

Fig. 9. Interpreted position of the valley train and fan delta at Joe Lake (after Barnett & Kennedy, 1988).

gravel foresets differ from those previously described primarily because they contain some large boulders occurring in isolation or in clusters. Some boulders have a quasi-vertical orientation, some load through thin sandier units which mark the reactivation surfaces between gravelly foresets (Figs 10A, 10B).

The transition between the coarse-grained foresets and the antecedent lacustrine deposits is characterized by interbedding of poorly sorted sandy gravels, coarse sand locally cross-bedded, and lenticular well-sorted fine sand and silt showing plane and ripple lamination and some small-scale load structures (Fig. 10C). This toesets facies association is capped by a bouldery gravel layer.

Palaeoenvironment

This is a fluvial-dominated fan delta which formed at the end of a torrential valley train. Coarse debris flows and possible downward rolling of boulders formed the foresets. The toeset beds show interlayering of foresets and more distal lacustrine materials. It is possible that parts of the toeset deposits formed later during the shoaling-out phase of the lake and were covered by bouldery gravels partially washed by waves.

DISCUSSION

Independently of the setting, foresets of the fan deltas described in this paper are dominated by poorly sorted sandy gravels and pebbly sands emplaced by sediment gravity flows, such as high-concentration debris flows and grain flows (Postma & Roep, 1985; Postma & Cruickshank, 1988). Slumping and turbulent resuspension of foreset sediments do not appear to have been active in these small-scale Pleistocene deposits. Variations between layers of each foresetted unit may be related to variation in floods of the feeding meltwater stream, and to the amount and type of material reaching, at any one time, the brink of the prograding slope. Such a process resembles that forming rhythmic foresets of large subaqueous dunes influenced by the arrival of the crests and troughs of ripples, thus of coarser and finer sediment, at the brink of the precipitation slope (Dalrymple, pers. comm.), except that in glacial streams the bedforms and sediment load are much more variable.

Grain-flow processes are clearly recorded in the openwork gravel lenses which are present in all coarse-grained foresets (Lowe, 1976, 1982; Postma & Roep, 1985). It is not clear, however, how the

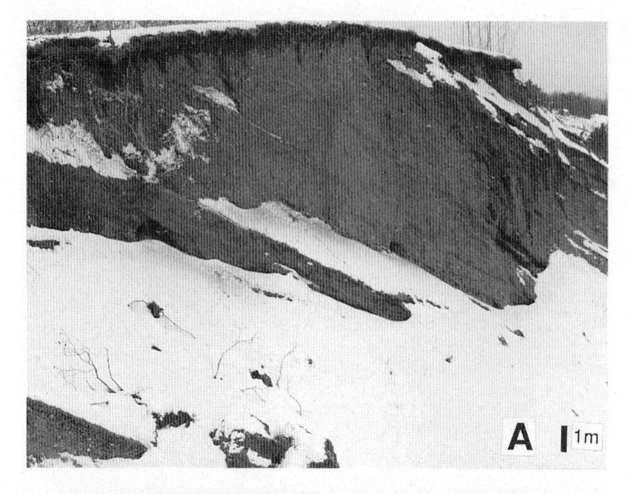

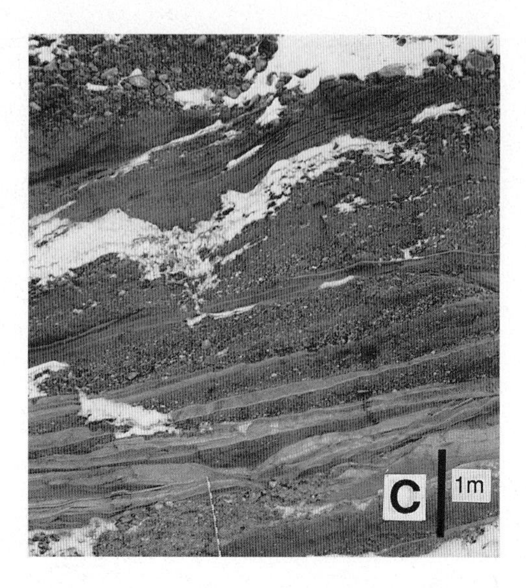

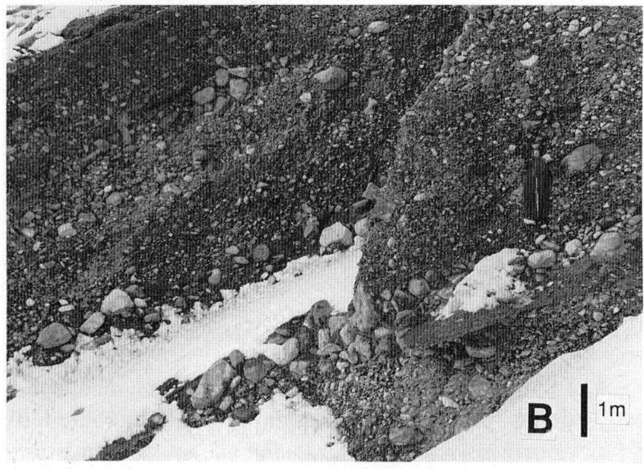

Fig. 10. Joe Lake deposit: (A) Steeply inclined foresets of poorly-sorted sandy gravel; (B) Coarse-grained foresets, note quasi-vertical position of boulders; (C) Toeset deposits showing interlayering of coarse and fine sediments in a section quasi-perpendicular to direction of progradation of delta.

open framework character, that is, the removal of the fines, can develop on these slopes. Removal of fines from mounds of pebbles is readily observable during intermediate flood stages in rivers (Martini & Ostler, 1973), and during receding storm stages in gravelly beaches, but it is unlikely that it may proceed to any significant extent during the short grain-flow transport along foresets. Therefore, either the lenses form by other processes, for instance, downslope movements of discrete particles and local trapping of more pivotable clasts, or, more likely, they reflect recurring injection of already-washed gravels from the feeding fluvial conduits. If this interpretation were correct, should they be more frequent in the middle and upper part of the foresets? This is the case in the Pleistocene exposures described here, but in other larger foresets similar openwork lenses have been reported throughout the delta face, including its lower part, and have been interpreted as 'flow slide deposits' (Colella *et al.*, 1987).

Furthermore, interlayering of gravelly sand or sandy gravels typical of the foresets occur with better-sorted, finer, more distal lacustrine deposits in the toeset zone, but the change from one to the other is abrupt. Few granules and pebbles escape the prograding coarse foresets to be deposited as isolated clasts in the finer lacustrine layers. This plus the fact that mud, scars and other deformation structures are absent or seldom observed on the granular foresets, suggest that the well-sorted fine materials are not distal facies of turbidites initiated as slumps along the relatively short foresets. Rather, the removal of the fines from the coarse fraction had occured in highly turbulent feeding channels, and whereas the proximal part of the foresets are generated by mass redistribution of coarse bedload carried to the oversteepened edge of the subaqueous slope during large floods, the fines of the toesets and bottomsets derive from submerging turbid (hyperpycnal) flows formed by suspended loads (Postma & Roep, 1985; Colella *et al.*, 1987). This process does not inhibit recognition of superimposed seasonal cyclicity within the same lacustrine sediments. The characteristic winter layer of varve-like rhythmites is recognizable because of its well-sorted fine-clay composition and uniform thickness which is not significantly influenced by variations in the 'summer' sandy and silty interlayers. A few fine dropstones are also present in most bottomset units.

In the case of the sandy foresets at Campbellville and at Allan Park, the separation of the two types of flows (mass and hyperpycnal flows) is not very clear. The characters of the sandy foresets suggest that persistent attachment of swiftly moving hyperpycnal flows were responsible for cutting shallow scours, deposition of plane and cross-laminated units and may have accelerated the movement of traction carpets on the inclined depositional surfaces. These foresetted sandy deposits may represent a stream-modified Gilbert-type delta, although at Campbellville some of them may represent fills of large channels of the subaerial part of the fan.

Topset facies rest unconformably over the foresets. They are fluvial dominated, with some wave reworking on those deposits formed at the margins of large lakes (Fonthill) or seas (Westmeath and Osceola). In the case of the ice-contact fans, the glacier represented ephemeral upstream portions, in some cases the only true subaerial part of the fan delta. Terminal and ground moraines, eskers and other similar glacial deposits may represent signatures of such fan-head setting (Brodzikwski & van Loon, 1987). Not only was the glacier an integral part of the fan, but by surging and melting back it mimicked, in a sort of short-duration geological experiment, unstable landforms of active tectonic areas.

The analysis of these small foresetted fan deltas provide better understanding and diagnostic characters for interpreting ancient sequences, but can their glacial nature be recognized as well? Many Pleistocene examples are reported in the literature, although the deposits may or may not have been named fan deltas (e.g. Clemmensen & Houmark-Nielsen, 1981; Elhers, 1983; Eyles & Eyles, 1989). However, to my knowledge no specific report exists of pre-Pleistocene Gilbert-type fan deltas developed in cold-climate environments. The glacial settings of the Pleistocene cases can be inferred from surficial geology and geomorphology, for instance the presence of ice-contact slopes, isolated mound-like distribution of the deposits, moraines, drumlins or other glacial features. However, major internal sedimentological features of the fans diagnostic of glacial environments are limited to polymictic lithology and highly variable size of the clasts, many being rafted from distant areas, and the presence of varve-like rhythmites and lonestones in lacustrine deposits. Only at Westmeath has flow till been found in the fan itself. Indeed, if it were not for the stratigraphic context, glacial legacy on Gilbert-type fan deltas could not be readily recognizable in ancient deposits.

CONCLUSIONS

In Southern Ontario the ice-sheets advanced and retreated as lobes funnelled along valleys which are now the Great Lakes and the Ottawa River Valley formerly occupied by the glacial Champlain Sea. During the last retreat, ice lobes melted back into the basins, generating optimum conditions for discharge of meltwater and sediments into surrounding temporary glacial lakes and marine embayments. Fan deltas developed in association with eskers and other sediment-laden englacial or periglacial meltwater streams. They vary in shape, area (from a few km^2 to tens of km^2), and thickness (10–80 m). The examples examined in this paper illustrate foresetted fan deltas formed both in direct contact with glaciers in marine and lacustrine settings, and developed at the mouth of narrow valleys which funnelled meltwater floods.

Foresets of these fan deltas are characterized by: (1) uniform dips, approximating angle of repose of various granular materials; (2) presence of massive sandy gravels alternating with openwork gravel lenses and massive to laminated coarse sands, all emplaced by mass-flow processes; (3) presence of shallow (order of centimetres) cut and fills with granules and fine pebbles, down-dipping, low-angle intraset laminae (at Campbellville), and ripple-drift-laminae (at Allen Park) in sandy foresets which indicate attachment and sedimentation from hyperpycnal turbulent to supercritical flows in a modified Gilbert-type delta. Some of the pebbly sand foresets of Campbellville may be part of fills of large subaerial channels.

There is an abrupt transition between the poorly sorted gravelly and coarse-sand units of the foresets and the more distal, well-sorted, lacustrine fine sands and silts, although interlayering occurs between them (Allan Park, Campbellville and Joe Lake). Most likely the separation of the various grain size populations occurred in the turbulent fluvial environment of feeding conduits, and whereas the bedload population was responsible for construction of the foresets through slope processes, the suspended load population continued as a high-density underflow (hyperpycnal flow) and formed parts of the toeset and the lacustrine bottomset deposits.

ACKNOWLEDGEMENTS

I would like to thank P.J. Barnett for suggesting locations of well-exposed foresetted sedimentary bodies in eastern Ontario, J.Z. Fraser for introducing me to the Fonthill deposit, D.R. Sharpe for information and discussions on Allan Park, G. Gorrell for information on the Pleistocene geology of eastern Ontario, and S. Sadura for helping in making the manuscript a bit more readable. Suggestions by L.B. Clemmensen and an anonymous reviewer helped in focusing and improving the final version of the paper. Financial support was provided by the National Science and Engineering Research Council of Canada (Grant A7371).

REFERENCES

Barnett, P.J. & Kennedy, C.C. (1988) Deglaciation, marine inundation and archaeology of the Renfrew-Pembroke Area. *Guidebook Excursion D. XII INQUA Congress*, pp. 46–56.

Brodzikwski, K. & Van Loon, A.J. (1987) A systematic classification of glacial and preglacial environments, facies and deposits. *Earth Sci. Rev.* **24**, 297–381.

Chapman, L.J. (1975) *The Physiography of the Georgian Bay–Ottawa Valley Area*. Geoscience Report 128, Ontario Division of Mines, 33 pp.

Chapman, L.J. & Putnam, D.F. (1984) *The Physiography of Southern Ontario, 3rd Edn*. Ontario Geological Survey, special volume 2. Ontario Ministry of Natural Resources, 270 pp.

Clemmensen, L.B. & Houmark-Nielsen, M. (1981) Sedimentary features of a Weichselian glaciolacustrine delta. *Boreas* **10**, 229–245.

Colella, A. (1988) Gilbert-type fan deltas in the Crati Basin (Pliocene–Holocene, Southern Italy), In: *Int. Works. Fan Deltas: Excursion Guidebook* (Ed. by A. Colella), pp. 19–77. *Università della, Calabria, Consenza.*

Colella, A., De Boer, P.L. & Nio, S.D. (1987) Sedimentology of a marine intermontane Pleistocene Gilbert-type fan-delta complex in the Crati Basin, Calabria, Italy. *Sedimentology* **34**, 721–736.

Cowan, W.R., Sharpe, D.R., Feenstra, B.H., & Gwyn, Q.H.J. (1978) Glacial geology of the Toronto–Owen Sound area. In: *Toronto 78 Field Trips Guidebook* (Ed. by A.L. Currie and W.O. Mackasey), Geological Society of America, pp. 1–16.

Dreimanis, A. (1969) Late-Pleistocene Lakes in the Ontario and Erie Basins. *Proc. 12th Conference on Great Lakes Research, International Associations for Great Lakes Research*, 170–180.

Ehlers, J. (Ed.) (1983) *Glacial deposits in North-west Europe*. Balkema, Rotterdam, 470 pp.

Eyles, C.H. & Eyles, N. (1989) The upper Cenozoic White River 'tillite' of southern Alaska: subaerial slope and fan-delta deposits in a strike-slip setting. *Bull. Geol. Soc. Am.* **101**, 1091–1102.

Feenstra, B.H. (1981) *Quaternary Geology and Industrial Minerals of the Niagara–Welland Area, Southern Ontario*. Ontario Geological Survey, Open File Report 5361.

Feenstra, B.H. & Frazer, J.Z. (1988) Fonthill Kame-Delta. In: *Quaternary History of Southern Ontario, Guidebook for Field Excursion A-11*. (Ed. by P.J. Barnett and R.I. Kelly), *XII INQUA Congress*, pp. 34.

Gilbert, G.K. (1885) The topographic features of lake shores. *Ann. Rept. U.S. geol. Surv.* **5**, 75–123.

Gorrell, G. (1988) Lanark Esker Environment. In: *Quaternary History of Southern Ontario, Guidebook for Field Excursion A-11*. (Ed. by P.J. Barnett and R.I. Kelly), *XII INQUA Congress*, pp. 68–69.

Hough, J.L. (1958) *Geology of the Great Lakes*. University of Illinois Press, Urbana, 313 pp.

Jopling, A.V. & Richardson, E.V. (1966) Backset bedding developed in shooting flow in laboratory experiments. *J. sedim. Petrol.* **36**, 821–825.

Karrow, P.F. (1986) *Quaternary Geology of the Hamilton Area, Southern Ontario*. Ontario Geological Survey, Map 25098. Quaternary Geology Series, scale 1:50 000.

Karrow, P.F. & Calkin, P.E. (eds.), 1985. *Quaternary Evolution of the Great Lakes*. Spec. Pap. Geol. Assoc. Can. **30**, 257 pp.

Lewis, C.F.M. (1969) Late Quaternary history of lake levels in the Huron and Erie basins. *Proc. 12th Conference on Great Lakes Research, International Association for Great Lakes Research*, pp. 250–270.

Lowe, D.R. (1976) Grain flow and grain flow deposits. *J. sedim. Petrol.* **46**, 188–199.

Lowe, D.R. (1982) Sediment gravity flows: II. Depositional models with special reference to the deposits of high-density turbidity currents. *J. sedim. Petrol.* **52**, 279–297.

Martini, I.P. (1972) Campbellville outwash-alluvial fan. In: *Niagaran Stratigraphy: Hamilton, Ontario* (Ed. by J.T. Sanford, I.P. Martini and R.E. Mosher), Michigan Basin Geol. Soc. Field Guidebook, pp. 24–25.

Martini, I.P. (1982) Campbellville gravel pit. In: *IAS 11th International Congress on Sedimentology – Excursion 11A: Late Quaternary Sedimentary Environments of a Glaciated Area: Southern Ontario* (Ed. by P.F. Karrow, A.V. Jopling and I.P. Martini) pp. 88–92. McMaster University, Hamilton.

Martini, I.P. & Ostler, J. (1973) Ostler lenses: possible environmental indicators in fluvial gravels and conglomerates. *J. sedim. Petrol.* **43**, 418–422; Errata: *J. sedim. Petrol.* **44**, 274.

NALDRETT, D.L. (1988) The late glacial–early glaciomarine transition in the Ottawa Valley: evidence for a glacial lake? *Geographie physique et Quaternaire* **42**, 171–179.

NEMEC, W. & STEEL, R.J. (1984) Alluvial and coastal conglomerates: their significant features and some comments on gravel mass-flow deposits. In: *Sedimentology of Gravels and Conglomerates* (Ed. by E.H. Koster and R.J. Steel) Mem. Can. Soc. Petrol. Geol. **30**, 91–103.

NEMEC, W. & STEEL, R.J. (Eds.) (1988) *Fan Deltas: Sedimentology and Tectonic Settings*. Blackie and Son, London, 444 pp.

ORI, G.G. & ROVERI, M. (1987) Geometries of Gilbert-type deltas and large channels in the Meteora Conglomerate, Meso-Hellenic basin (Oligo-Miocene), central Greece. *Sedimentology* **34**, 845–859.

POSTMA, G. & CRUICKSHANK, C. (1988) Sedimentology of a late Weichselian to Holocene terraced fan delta, Varangefjord, northern Norway. In: *Fan Deltas: Sedimentology and Tectonic Settings* (Ed. by W. Nemec and R.J. Steel), pp. 145–157. Blackie and Son, London.

POSTMA, G. & ROEP, T.B. (1985) Resedimented conglomerates in the bottomsets of Gilbert-type gravel deltas. *J. sedim. Petrol.* **55**, 874–885.

PREST, V.K. (1970) Quaternary Geology of Canada. In: *Geology and Economic Minerals of Canada* (Ed. by R.J.W. Douglas), Geol. Surv. Can. Econ. Geol. Rpt. No. 1, 676–764.

RUST, B.R. (1977) Mass-flow deposits in a Quaternary succession near Ottawa, Canada: diagnostic criteria for subaqueous outwash. *Can. J. Earth Sci.* **14**, 175–184.

RUST, B.R. & ROMANELLI, R. (1975) Late Quaternary subaqueous deposits near Ottawa, Canada. In: *Glaciofluvial and Glaciolacustrine Sedimentation*. (Ed. by A.V. Jopling and B.C. McDonald), Spec. Publ. Soc. econ. Paleont. Miner., Tulsa, **23**, pp. 177–192.

SHARPE, D.R. (1982) Allan Park Kame-Delta. In: *IAS 11th International Congress on Sedimentogy – Excursion 11A: Late Quaternary Sedimentary Environments of a Glaciated Area: Southern Ontario*. (Ed. by P.F. Karrow, A.V. Jopling and I.P. Martini) pp. 53–60. McMaster University, Hamilton.

SHARPE, D.R. (1988) Brazeau Glaciomarine Fan. In: *XII INQUA Congress – Quaternary History of Southern Ontario. Guidebook for Field Excursion A-11* (Ed. by P.J. Barnett and R.I. Kelly), pp. 69–71.

Spec. Publs int. Ass. Sediment. (1990) **10**, 297–309

Diurnally and seasonally controlled sedimentation on a glaciolacustrine foreset slope: an example from the Pleistocene of eastern Poland

K. MASTALERZ

Institute of Geological Sciences, University of Wrocław, Cybulskiego 30, 50–205 Wrocław, Poland

ABSTRACT

Large-scale glaciolacustrine foresets from a Pleistocene terminal-moraine complex near Neple, eastern Poland, were examined to determine foreset-forming processes. The foresets studied reach 5 m in thickness and consist of inclined beds with dip angles markedly decreasing from 30–40° in the proximal part to 5–15° distally. The foreset beds can be classified into the following lithofacies: gravel, graded sand, massives and, massive to laminated sand, laminated sand, silt, scour-and-fill facies, and contorted deposits. The main foreset-forming processes were: cohesionless debris flow, turbulent density flow, tractional deposition and suspension fallout. The bed-scale rhythmicity in the foresets is believed to represent diurnal fluctuations of depositional conditions controlled by air temperature, meltwater discharge and sediment supply (diurnal rhythms). On a larger scale, foreset segments show decreasing bed thickness and dip angle as well as grain size passing from their older to younger portions, and they are interpreted as the record of seasonal variations (annual cycles).

INTRODUCTION

The foreset unit is the element of a Gilbert-type delta which records most of the progradational style of its development. There have been numerous sedimentological studies of slipface sedimentation on smaller-scale foreset forms of aeolian and subaqueous dunes and microdeltas (Allen, 1963, 1965, 1968, 1984; Jopling, 1964, 1965, 1966; Collinson, 1970; Smith, 1972; Hunter, 1977, 1981, 1985; Cant & Walker, 1976; Buck, 1985). The majority of these forms constitute very simple foresets displaying either one or two facies only. Such a foreset architecture reflects a uniform sedimentary system, with slipfaces dominated by grain-flow processes.

On large, high-angle slopes as those of Gilbert-type deltas, sedimentation is often more complex, which results in a more heterogeneous foreset structure. Downslope avalanching and suspension sedimentation on such slopes may alternate with mass flows (Wescott & Ethridge, 1983; Postma, 1984; Postma & Roep, 1985; Postma *et al.*, 1988; Colella, 1988a, b; Colella *et al.*, 1987). In some cases high-density turbidity currents dominate the frontal faces of Gilbert-type deltas (Postma & Roep, 1985; Postma & Cruickshank, 1988). Although fan-delta and Gilbert-type delta facies analysis has received much attention recently (e.g. Nemec & Steel, 1988; Adriani *et al.*, 1988), there are few detailed studies on large-scale foreset bedding (e.g. Stanley & Surdam, 1978; Postma & Roep, 1985; Postma & Cruickshank, 1988; Postma *et al.*, 1988; Colella *et al.*, 1987), and some processes acting in these depositional environments remain poorly understood.

The purpose of this paper is to describe and to interpret the architecture and facies of a foreset unit formed in a late Pleistocene periglacial zone (Podlasie, eastern Poland, Fig. 1). From studies of modern periglacial environments, it is known that these areas are characterized by strong seasonal and diurnal climatic oscillations (Small, 1978). Sedimentation in such areas can be strongly influenced by the ablation of ice as controlled by a change of thermal conditions (Axelsson, 1967; Boothroyd & Ashley, 1975; Church & Gilbert, 1975; Gustavson *et al.*, 1975; Ostrem, 1975; Smith, 1978; Hammer & Smith, 1983; Fenn *et al.*, 1985; Ashworth & Fergusson, 1986). In winter, when temperature lies well below 0°C, weathering, erosion and transportation processes become inactive. In contrast, summer periods are characterized by thaw and increasingly higher discharges in glacier-fed streams. In addition, diurnal freeze–thaw cycles influence weathering and hydrological processes (Gustavson, 1974; Church & Gilbert, 1975; Small, 1978; Smith, 1978; Hammer & Smith, 1983; Fenn *et al.*, 1985). The foreset sequence studied near Neple provides an example of diurnally and seasonally controlled sedimentation as inferred from small-scale and large-scale rhythmicity.

GEOLOGICAL SETTING

The Podlasie region, eastern Poland (Fig. 1), was affected by two large glacial episodes, i.e. the Elsterian and Saalian Glaciation during the Pleistocene. The Warthian Stadial of the Saalian Stage was the period of the greatest importance with respect to the volume and variability of sediments which cover the Podlasie area (Nowak, 1974). The front of the main Warthian ice-sheet was situated 30–40 km NNW of the Krzna River valley (Fig. 1), and it is marked by a prominent ridge of terminal moraines (Rühle & Mojski, 1965; Nowak, 1974). A widespread system of coalescent glacial outwash was built in front of the glacier and southwards it fed a proglacial fluvial-valley network.

The deposits of the Warthian Stadial in Podlasie can be subdivided into three main groups: glacial, fluvioglacial and lacustrine (Fig. 2). The glacial lodgement till forms a relatively uniform layer over the whole area, usually reaching several metres in thickness. The fluvioglacial deposits consist predominantly of trough and planar cross-bedded sands and gravels which were deposited in terminal moraines, on outwash fans, and within a braided

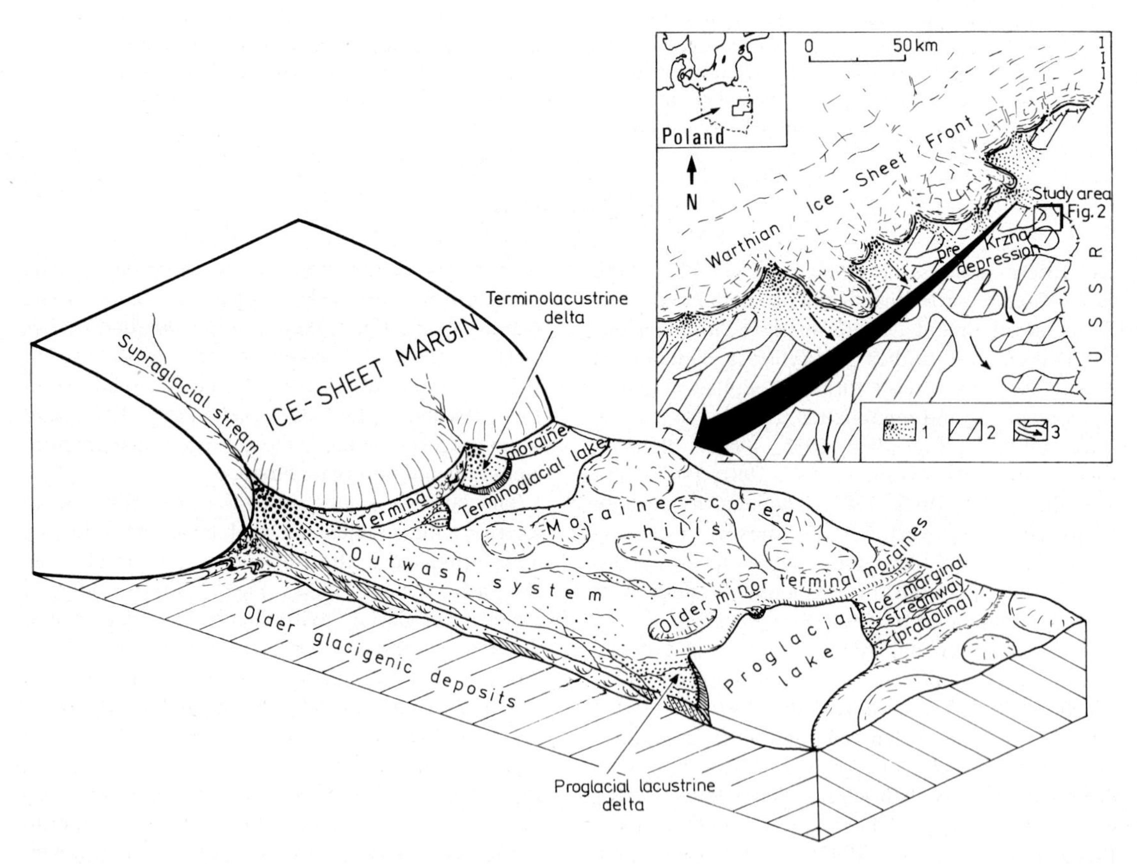

Fig. 1. Palaeogeographic sketch map of east-central Poland during the Warthian Stadial, late Pleistocene (modified from Rühle & Mojski, 1965): 1 — outwash fans; 2 — elevated areas built of boulder tills; 3 — directions of water run-off in main valleys. Block diagram shows a reconstruction of main depositional settings of the Podlasie region, inferred from the deposits of the Warthian age.

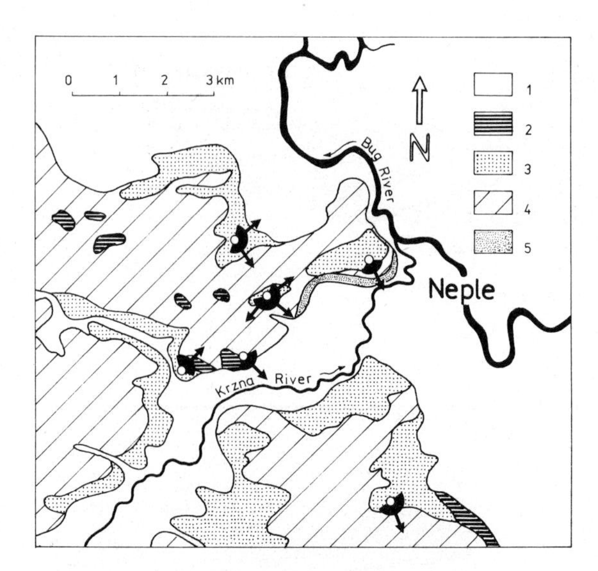

Fig. 2. Geological sketch map of the vicinity of Neple, eastern Podlasie (modified from Nowak, 1974) with vector means and ranges of palaeoflow indicators in fluvioglacial deposits: 1 — Holocene deposits; 2–4 — Warthian Stadial: 2 — lacustrine deposits, 3 — fluvioglacial deposits, 4 — boulder tills; 5 — older fluvial deposits.

river system of a proglacial valley network. The lacustrine deposits consist predominantly of heterolithic facies comprising laminated silts and minor clays and sands, the latter being rarely burrowed.

Deposits of the Warthian age are exposed in several sand pits located on elevations along the northern margin of the Krzna valley (Fig. 2). In a sand pit near Neple village, a completely preserved large-scale foreset unit can be seen. It is included in the complex of a minor terminal moraine, being connected with the 'temporary' ice-front stand of the Warthian ice-sheet (Nowak, 1974). This complex comprises boulder till and overlying coarse-grained fluvioglacial deposits showing numerous cryogenic structures. Southeastward palaeotransport directions (Fig. 2) are indicative of a glacier-driven sedimentary system in this region (Rühle & Mojski, 1965).

STRUCTURE AND DEVELOPMENT OF THE NEPLE FORESET SEQUENCE

Foreset facies

The Neple foreset sequence reaches 5 m in thickness, 35 m in radius, and at least 25 m in width (Fig. 3). It is composed of thin, SSE-dipping beds. The inclination of the dominantly sandy beds diminishes gradually from 30 to 40° in the proximal part of the foreset to 5–15° in its distal portion (Fig. 3). The foreset unit is truncated and overlain by younger, poorly sorted clayey sands and gravels of glacially influenced drift. The foreset beds pass laterally and distally into trough cross-bedded pebbly sands. In the proximal to medial part, the foreset strata directly overlie the thick layer of an older glacial till. Bottomset beds, which might belong to a tripartite Gilbert-type delta system, are not exposed.

The formation of particular foreset beds was controlled by textural features of the sediment carried to the frontal slope as well as flow-type and sedimentation mode. A combination of these factors resulted in different sedimentary structures, bed boundaries and geometries, and these allow the foreset beds to be subdivided into several facies (Table 1). Sandy beds (Sg) dominate the foresets studied; however, facies-type varies down the foresets. Gravel beds (G) are the most common in the proximal foreset unit, whereas the silty ones become more abundant and thicker distally (Figs 4, 5).

The beds of Sg and the majority of G facies are inferred to be the result of downslope avalanching of loose clastic material. The transport mechanism may be similar to that of laminar, high-concentration cohesionless flows (Postma, 1986), well known from slipfaces of natural aeolian and subaqueous dunes (Smith, 1972; Lowe, 1976; Hunter, 1977, 1985; Buck, 1985) as well as from laboratory flume studies (Bagnold, 1954; Allen, 1965; Jopling, 1965). The mechanism of grain support in avalanches usually promotes clast and/or density segregation, which results in graded beds (cf. Bagnold, 1954; Allen, 1965, 1984; Buck, 1985; Hunter, 1985). Some gravel beds (G) showing crude stratification were deposited from strong tractive currents.

The origin of massive sand (Sm) cannot be clearly defined. Steeply inclined, structureless beds are reported from aeolian and subaqueous dunes, and interpreted as the result of sand flows of relatively well-sorted material (Hunter, 1985; Buck, 1985). Similar effects may also be connected with a viscous granular shear regime as suggested by Bagnold (1954) (see also Allen, 1984), regardless of initial sediment sorting. Lowe (1982, 1988) discussed high-rate, suspended-load fallout from highly-concentrated flows as a possible mechanism leading to the formation of massive sand beds. However, sand deposited in this way is loosely packed and it

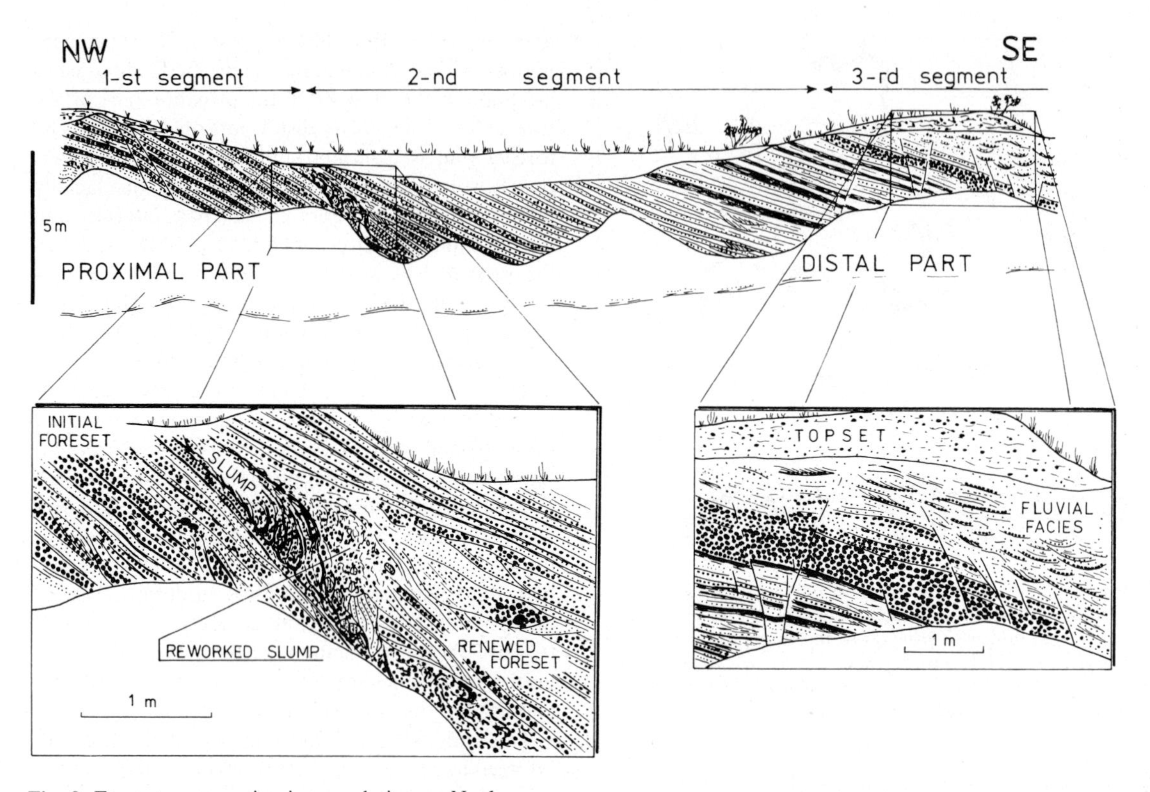

Fig. 3. Foreset cross-section in a sand pit near Neple.

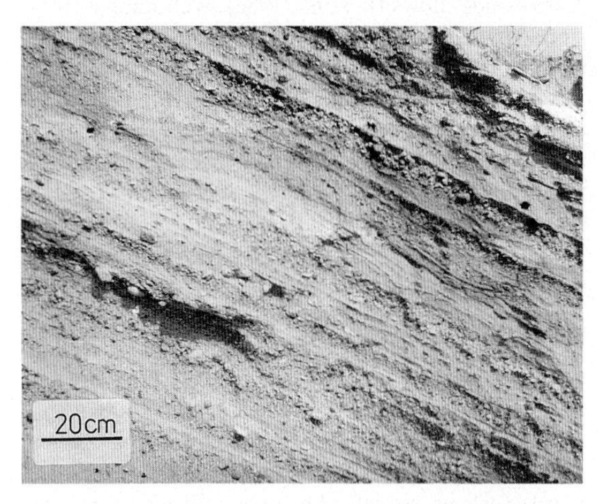

Fig. 4. Sequence of foreset beds of G, Sg and Sr facies; proximal part of the foreset unit. The true dip of the foreset beds is 30° toward SE.

can be easily liquefied (see discussion by Lowe, 1982, p. 286); thus preservation of such a sediment on steep slopes is rather doubtful.

Massive to laminated sand (Sml), the lower member of which is normally graded, represents deposition from turbulent density-current underflows which result in a simple T_{abc} succession. Sml beds showing massive lower members seem to be a result of rapid suspension sedimentation (cf. Bouma, 1962; Lowe, 1982, 1988) from a highly concentrated sand flow, followed by tractive sedimentation supported by suspension fallout from waning density currents.

Laminated sands (Sr) are inferred to represent sedimentation from tractive currents flowing down the frontal slope. Backflow ripples were generated by upslope eddies connected with the formation of flow-separation cells. The orientation of ripple cross-lamination (Fig. 6) suggests that the currents usually flowed down the slope and that the separation cells were of a simple two-dimensional nature (cf. Jones & McCabe, 1980; Allen, 1984).

Silty beds (ST), the most common (30%) in the foreset unit, represent periods of 'normal' conditions, typical of the prevailing time-span of the foreset development, with slow sedimentation of fines from dilute suspensions. However, the quiet

Table 1. Lithofacies of the Neple foreset

Facies	Description	Interpretation
Gravel (G)	Pebble gravel, clast supported Graded, less commonly lacking internal organization or crudely stratified Locally upslope a (p) a (i) imbricate pebbles Beds 2–20 cm thick, tabular to wedge-shaped Sometimes one-pebble thick stringers of gravel (Fig. 4, 5A)	Cohesionless debris flow Strong tractive currents
Graded sand (Sg)	Very coarse- to medium-grained sand, sometimes scattered granules Graded, with inverse and inverse to normal grading prevailing Beds 2–12 cm thick, tabular, with sharp or gradational, flat boundaries (Fig. 5B, D)	Cohesionless debris flow
Massive sand (Sm)	Medium- to coarse-grained sand Ungraded, structureless Beds 2–10 cm thick, with gradational, flat to undulating boundaries (Fig. 5C, D)	Cohesionless debris flow High-rate suspension fallout from highly concentrated sand flow?
Massive to laminated sand (Sml)	Medium- to coarse-grained, less commonly fine-grained sand Bipartite beds with massive to normally graded lower member passing upwards into finer grained member with diffuse, parallel or ripple cross-lamination Beds 1–8 cm thick, usually gradational boundaries and wavy tops (Fig. 5C)	Turbulent density flow High-rate suspension fallout + traction/suspension deposition?
Laminated sand (Sr)	Coarse- to fine-grained sand Ripple cross- to parallel lamination Beds 1–6 cm thick, with flat bases and ripple-shaped tops, often gradational boundaries (Fig. 5C, D)	Traction currents
Silt (ST)	Silt, silty sand to silty clay Ungraded to indistinctly normally graded, with diffuse subparallel lamination Beds from a few mm to 6 cm thick, with gradational bases and sharp tops, subparallel boundaries (Fig. 5C, D, E)	Slow fallout from suspension
Scour-and-fill (SC)	Pebble gravel to coarse-grained sand Cross-bedded Lens-shaped infills, with sharp, concave-up bases and flat tops, 15–25 cm in depth, up to 100 cm in lenght (Fig. 5F)	Scouring and infilling by localized, strong traction currents
Contorted packets (D)	Heterolithic packets gravel to silt beds Intensely contorted, asymmetrical fold-like deformations, chaotic brecciated structure Thick 30–40 cm, irregular in shape packets of beds (Fig. 3)	Slumping and/or creeping

fallout was sometimes accompanied by weak tractive currents, indicated by the diffuse, thin laminae within some silty beds.

Scour-and-fill structures (SC) resulted from strongly erosive and localized eddies of powerful currents, scouring asymmetrical depressions which were filled with coarse-grained, cross-bedded sediment when the current waned.

Contorted packets (D) are related to slumping and/or creeping of the foreslope sediment. These

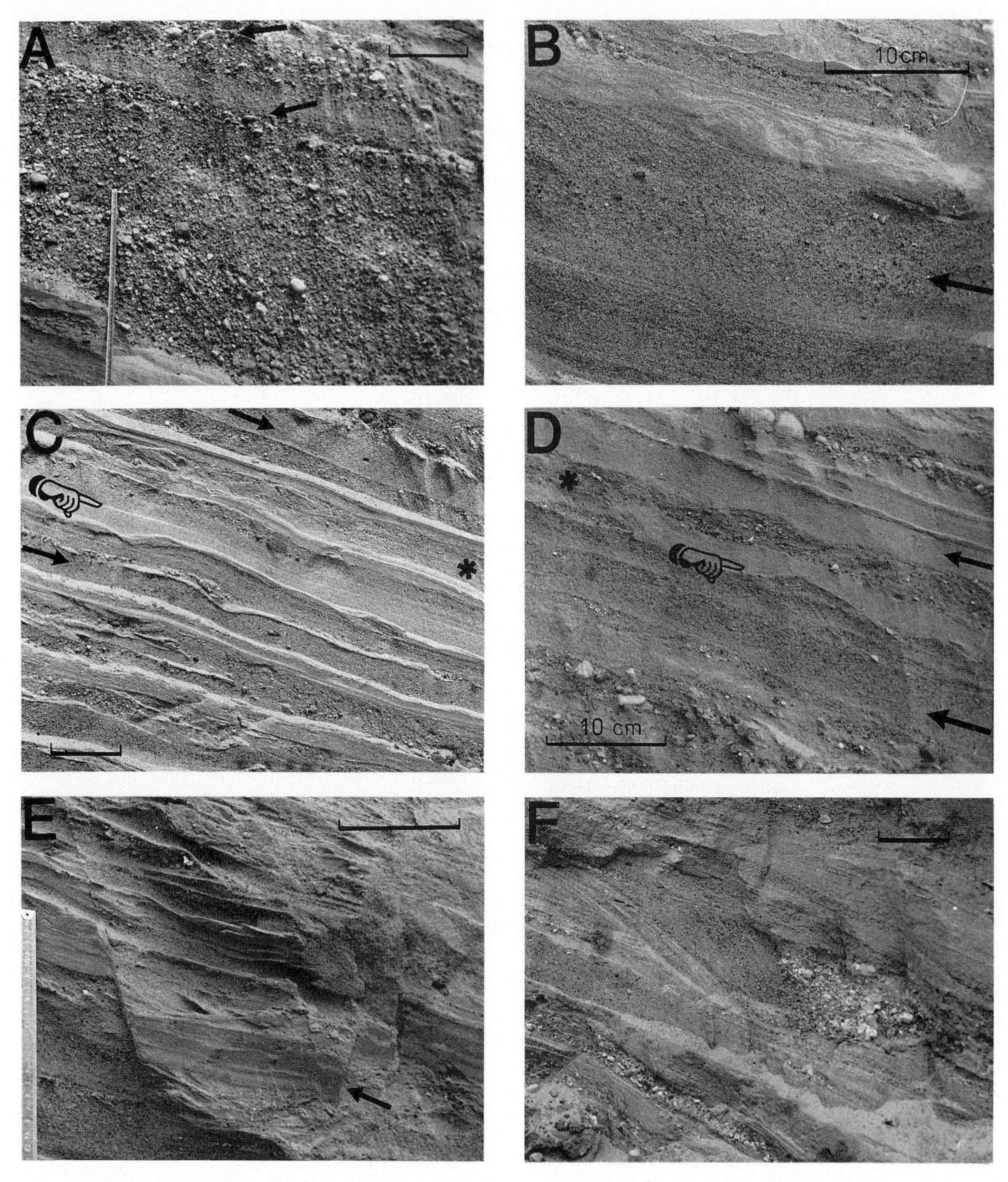

Fig. 5. Details of the Neple foreset lithofacies (bar scale in each photo — 10 cm):
(A) Amalgamated gravel beds (G facies) covering the distal foreset unit. Note upslope, upcurrent imbricate pebbles at the tops of inversely graded units (arrows). (B) Sequence composed of laminated sand bed passing gradationally into inverse to normally graded bed (arrow) and topped with rippled sandy-silt bed. Medial part of the foreset unit. (C) Sequence composed of massive sands (arrows) and ripple cross-laminated beds (asterisk) with thin silty veneers. Note the massive to laminated bed in the centre of the photo. Medial part of the foreset unit. (D) Composite sequence including graded (thick arrow), massive (thin arrow) and ripple cross-laminated (asterisk) beds. Note well-defined backflow ripples related to formation of flow separation. Medial part of the foreset unit. (E) Complex of interlayered sandy laminated and silty beds. Note the multistorey silty bed near the bottom of synsedimentary trough. Distal foreset unit. (F) Scour-and-fill structure. Distal foreset unit.

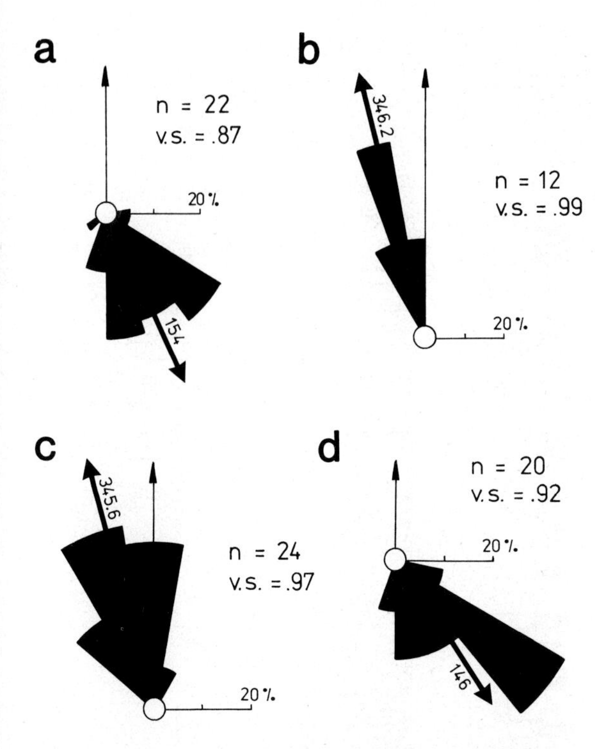

Fig. 6. Palaeotransport directions from ripple cross-lamination (a) backflow ripples, (b) imbricate pebbles, (c) within the foreset and trough cross-bedding, (d) from in-channel facies at the front of foreset unit.

processes did not play a constructional role in the formation of the foreset but they modified it. Some parts of the slumped masses, behaving coherently during the downslope displacement, display gentle folding, whereas other slumped beds disintegrated into brecciated and chaotic sediment (Fig. 3).

Palaeocurrents

The foreset strata dip generally toward the SSE showing the direction of foreset progradation. However, the dip azimuth changes gradually from 140° in the proximal to 170° in the distal foreset.

Palaeoflow indicators (ripple cross-lamination, scour-and-fill structures, imbricate pebbles and trough cross-bedding) show SSE palaeotransport directions coincident with the general palaeoslope azimuth (Fig. 6). The resultant vector means range from 146° to 165.6° depending on the structure. Backflow ripples display a vector mean of 346.2°. Individual measurements are relatively weakly dispersed; vector strengths of the computed vector means are high and range from 0.87 to 0.99. This suggests a relatively simple transport system on the foreslope with downslope flowing currents. It is noteworthy that the trough cross-bedded deposits in front of the foreset also show SSE palaeotransport direction (Fig. 6).

Sedimentary rhythms

Four sections were selected to study the succession of the foreset beds by means of Markov chain analysis (Fig. 7). The measured sections range from 113 to 247 cm in stratigraphic thickness and contain a total of 186 facies transitions. The method used in this paper takes into account transitions between discrete states (i.e. different foreset beds), applied in the embedded chain method. However, there are also possible transitions between the beds of the same facies. Thus, the structure of the resulting transition matrix is similar to that of regular Markov chains (Ethier, 1975; Orford, 1978).

The transition matrices constructed on the basis of measured sections display a distinct asymmetry (Table 2). The positive entries of the difference matrix indicate transitions which appear in the succession more frequently than randomly expected. The simple facies couplets connected by modal transitions (GST, SgST, SmST, GSr, SmSr) dominate in the sequence studied. The transition diagram (Fig. 8) displays a structure composed of three clusters of facies. The central cluster is closely related to ST facies. The remaining elements of this group (G, Sg, Sm) show a strong tendency to pass into ST facies. The second cluster concentrates around Sr facies. The beds of Sr facies show affinity to G and Sm ones and readily form GSr and SmSr depositional rhythms. In addition, Sr facies display a tendency to form repetitions composed of Sr beds only. Sml facies is the only component of the third cluster. This facies tends to form repetitive SmlSmlSml... transitions (Fig. 8).

The detailed study of the cyclic structure of the foreset facies succession confirms these results. The sections measured contain 107 elementary, fining-up sequences (Fig. 7). Bipartite depositional rhythms (e.g. GST, SgST, SmSr) are the most common and constitute 66 (63%) elementary sequences. The repetitions of individual facies (e.g. SmlSml) resulted in 33 elementary cycles. The remaining cycles are three- (7) and four-member (1) sequences.

The Markov property of the succession studied is not stationary. Transition diagrams obtained for

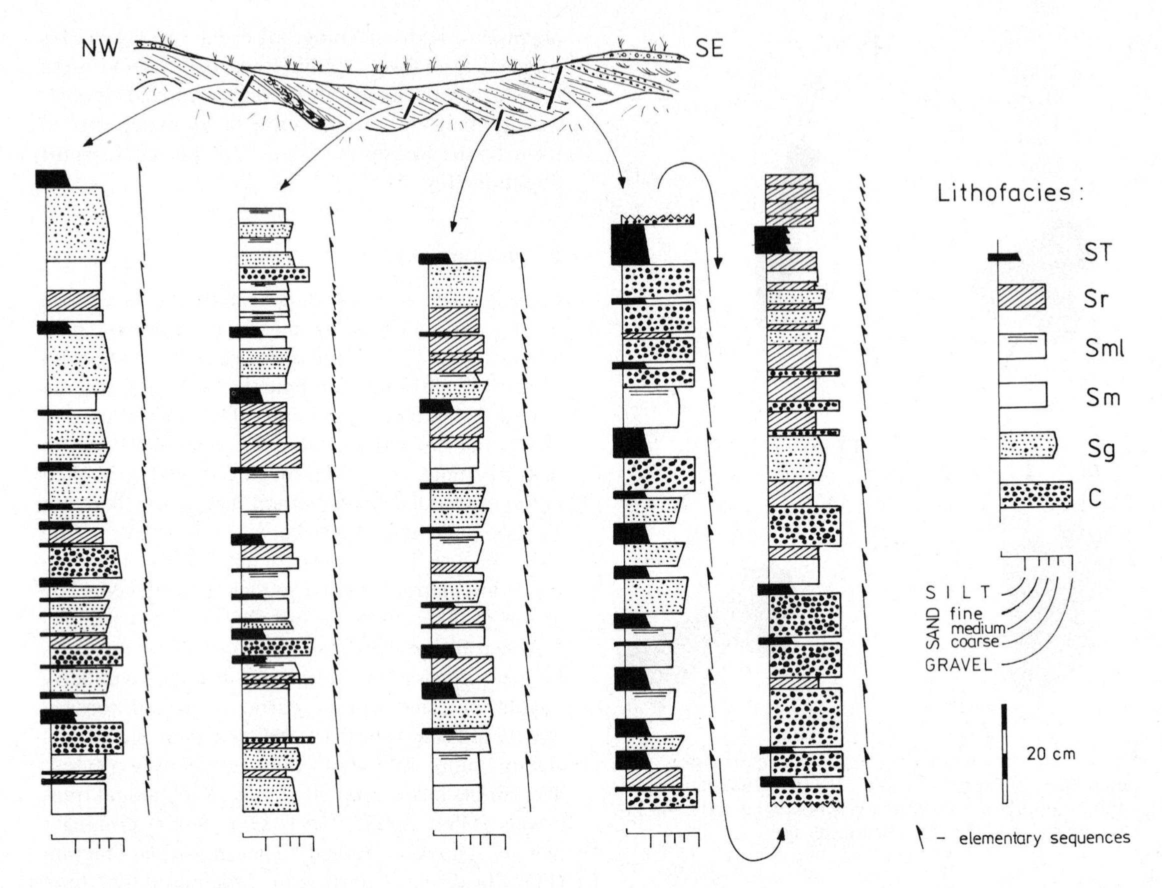

Fig. 7. Measured sections of the Neple foreset unit (for locality see Figs 2, 3).

Table 2. One-step transition frequency and difference matrices (difference probabilities in brackets) for combined section data.

Facies	G	Sg	Sm	Sml	Sr	ST	Total
	Facies						
Gravels (G)	— (−0.11)	1 (−0.11)	1 (−0.06)	— (−0.10)	8 (0.16)	12 (0.22)	22
Graded sands (Sg)	1 (−0.08)	2 (−0.09)	2 (−0.04)	3 (0.00)	5 (−0.03)	17 (0.24)	30
Massive sands (Sm)	1 (−0.06)	3 (−0.02)	1 (−0.06)	1 (−0.05)	7 (0.13)	8 (0.06)	21
Massive to laminated sands (Sml)	2 (0.01)	1 (−0.10)	1 (−0.05)	6 (0.26)	1 (−0.14)	6 (0.03)	17
Laminated sands (Sr)	7 (0.08)	6 (0.00)	2 (−0.06)	2 (−0.05)	9 (0.05)	11 (−0.02)	37
Silts (ST)	10 (0.06)	16 (0.10)	12 (0.09)	6 (0.00)	8 (−0.06)	7 (−0.19)	59

$\chi^2 = 67.6$

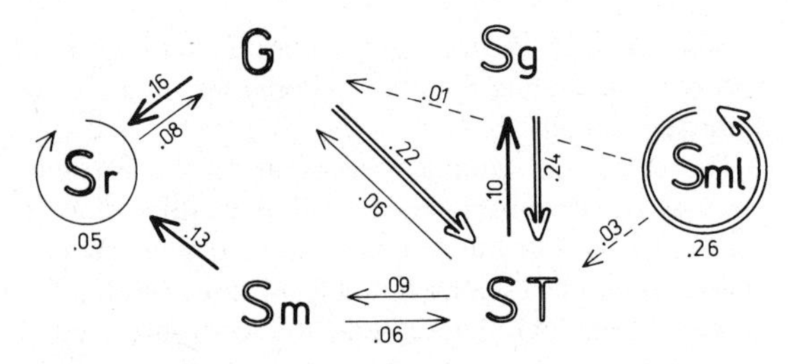

Fig. 8. Transition pattern for data from all sections measured along the foreset unit. Arrows indicate directions of preferable upward/distal transitions. Data from the difference matrix, Table 2.

particular sections differ from each other as well as from the general diagram. However, the tendency to form simple sedimentary rhythms is common for all parts of the foreset. The consistency of the two-member Markov chain structure may suggest the simple, oscillating pattern of changeability of the foreslope sedimentary conditions. Each sedimentary rhythm comprises beds which contrast with each other with respect to their textural/structural features. Each rhythm usually starts with sediment gravity-flow deposits (G, Sg, Sm, Sml beds) and finishes with suspension fallout (ST) or traction deposits (Sr).

The geological setting and stratigraphic position of the foreset near Neple allow us to suggest that its sedimentation took place in a periglacial environment (Rühle & Mojski, 1965). The position of the foreset sequence within the complex of a small terminal moraine (Nowak, 1974) indicates the proximity of ice front during its formation. Diurnal oscillations of hydrological conditions, and consequently sediment transport and sedimentatiön rate are believed to have resulted in the rhythmic structure of the foreset. Elevated daytime temperatures may have promoted ablation, resulting in increased meltwater discharge (cf. Church & Gilbert, 1975; Hammer & Smith, 1983; Østrem, 1975; Smith, 1978; Fenn *et al.*, 1985, Fig. 9). Sediment-laden fluvioglacial streams lost their load as they entered the standing body of water. This resulted in high sedimentation rates and the formation of the steep depositional slope dominated by sediment gravity-flow processes. At night, the temperature decrease lowered the stream discharge and sediment supply to the slipface. Consequently, the textural composition of sediment load also changed, with a drastic

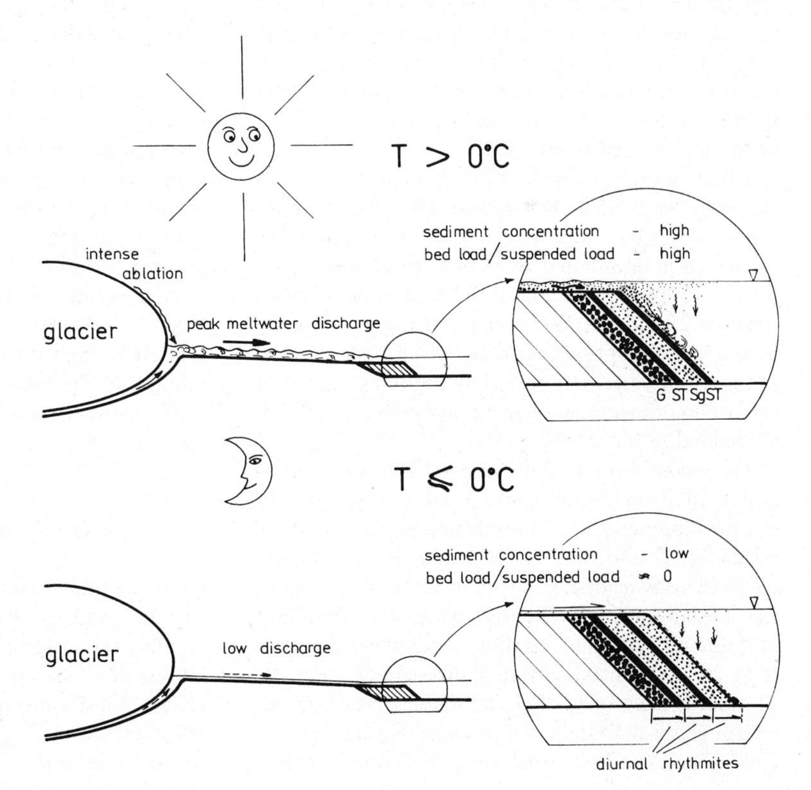

Fig. 9. Model of diurnally controlled sedimentation of the foreset unit near Neple.

decrease in volume and grain size of the bedload. These conditions resulted in the dominance of fallout from dilute suspensions and the formation of ST beds on the lee slope (Fig. 9).

The differences in structure between the individual sedimentary rhythms in the foreset succession (Fig. 7) are believed to be effects of longer time scales — daily or weekly fluctuations in weather conditions. Relatively warm periods resulted in an overall increase in sedimentation rate and the development of well-defined foreset rhythmites (e.g. GST, SgST, SmST). However, warmer nights might have promoted tractive transport and sedimentation instead of suspension fallout, which resulted in GSr, SgSr or SmSr diurnal couplets. Colder periods, in turn, resulted in truncated rhythms composed on Sml or Sr repetitions. On the other hand, longer precipitation periods could effectively mask diurnal cycles (cf. Gustavson, 1972, 1974).

Large-scale trends

The cross-section of the Neple foreset displays some well-defined large-scale trends (Fig. 3). The grain size decreases towards the distal part of the foreset. The frequency and average thickness of Sg, Sm, and G beds declines distally. The reverse trend is displayed by the beds of Sr and ST facies. Additionally, this is accompanied by the consistent increase of the silt/sand ratio in ST beds, and the general decrease of the foreset bed inclination.

Three large segments can be distinguished within the succession of foreset strata. The first segment encompasses the oldest part of the foreset unit (Fig. 3), which is dominated by pebble to granule sized material (Fig. 4). The dip of the foreset beds varies between 25 and 40°. The sediment becomes gradually finer grained approaching more distal beds. Such a sequence of beds is truncated by the thick packet of intensely contorted, gravelly to silty beds and chaotic brecciated sediments (Fig. 3).

The second foreset segment begins with contorted strata (deformational packet) succeeded by the steeply dipping (25–40°) sandy and gravelly beds of facies Sg, G and Sm. The dip of the beds slightly decreases downslope, so the beds tend to be tangential. Distally, the gravel beds become less abundant and thinner, and the bed dips successively diminish to 5–20° in the distal portion of this segment (Fig. 3). The occurrence of graded and massive sands is decreasing and the laminated ones increasing successively. Scour-and-fill structures, backflow ripples, as well as thick compound packets of fine-grained sediment (including ST and Sr beds) become more abundant distally.

The third and youngest segment of the foreset sequence is composed of a few relatively thick gravel beds (Fig. 3). They are massive to inversely graded and in some places contain a-axis upslope imbricate pebbles (Fig. 5A). These beds are covered directly by trough cross-bedded pebble sands at the front of the foreset sequence.

The origin of the Neple foreset segments is inferred to have been controlled by seasonal changes in depositional conditions. The intense summer sedimentation of coarse-grained material resulted in the formation of the oldest foreset segment (Figs 3, 4, 10). The ensuing cooling in the late summer and autumn limited the discharge and bedload transport so that mainly silts were deposited. Depositional processes completely came to a halt during the winter, when the surface of the foreset delta became frozen (Fig. 10).

The thick packet of contorted deposits, which begins the sequence of the second foreset segment, is inferred to be related to spring thaw which induced the downslope creeping of the surface sediment layer, and slump development (Figs 3, 10). Intensified water discharges in summer caused increased sediment transport and further foreset progradation. The depositional slope was dominated by the avalanching of cohesionless sediment. Tractive currents accompanied by continuous suspension fallout become increasingly important during the decline of the melting season.

In the third thaw season a vigorous input of coarse-grained sediment resulted in a set of thick gravel beds covering the surface of the distal foreset unit (Figs 3, 5A, 10). This depositional episode was related to high discharge leading to a considerable infilling of the lake. The feeding streams expanded southeastwards and brought the foreset formation to the end.

CONCLUDING REMARKS

Annual sedimentary rhythms are well known from various modern and ancient periglacial environments. The magnitude of observed diurnal oscillations of discharge and sediment concentration in glacier-fed streams (Church & Gilbert, 1975; Smith, 1978; Hammer & Smith, 1983; Fenn *et al.*, 1985) suggest that diurnally controlled sedimentation can

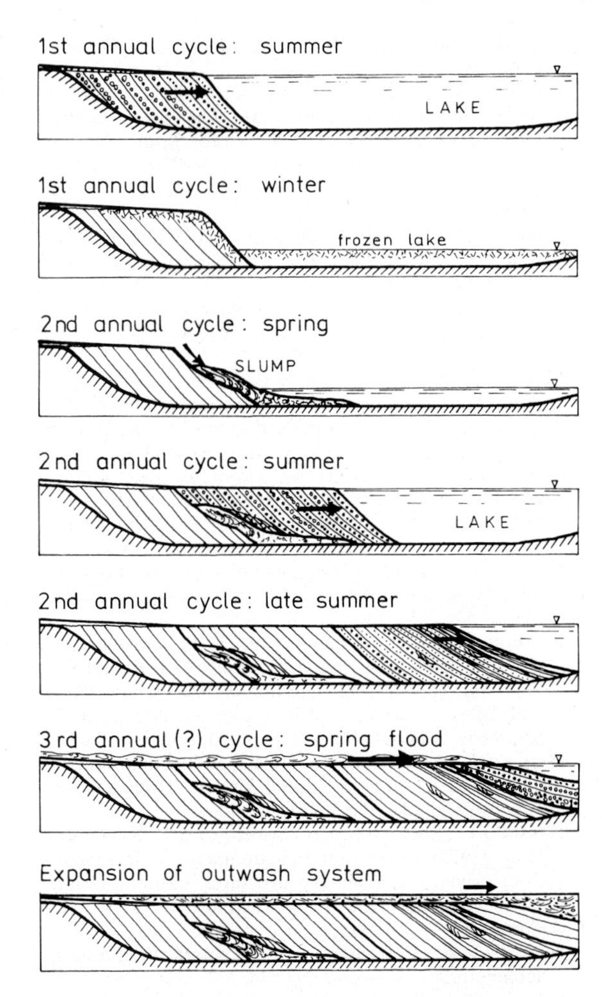

Fig. 10. Model of seasonally controlled development of the Neple foreset unit.

take place in proglacial settings. This is supported by examples of diurnally controlled sedimentation in glaciomarine settings (e.g. Domack, 1984). In addition, fluctuations of comparable time spans do have considerable impact on slipface sedimentation on sand waves in tidal environments (Allen, 1980, 1984; Allen & Homewood, 1984) or large bed forms in fluvial channels (Smith, 1972; Harms *et al.*, 1975).

The large-scale foreset near Neple originated in a periglacial environment, being included in the minor terminal moraine related to the 'temporary' stand of the Warthian ice sheet. It was formed at the mouth of a glacier-fed stream which entered a small lake (ice-marginal?). The lake depth reached at least 5 m, as indicated by the thickness of the foreset. Principal sedimentary processes operating on the lee slope of this depositional form can be classified into four groups: cohesionless debris flow, turbulent density flow, tractional deposition and suspension fallout. Sediment gravity flows dominated during the formation of the proximal foreset sequence. The role of tractive sedimentation and slow fallout from suspension increased considerably in the later stage of the foreset progradation. The lee slope was occasionally modified by sediment creeping and slumping.

Sedimentary processes on the Neple foreset sequence are inferred to be related to a combination of factors, with diurnal and seasonal hydrological oscillations controlled by freeze–thaw rhythms being the most important. They are recorded in the structure of the foreset sequence consisting of small-scale rhythmites which are organized into a few large-scale cycles. High variability in internal structure of the rhythmites, relatively large and highly variable thickness of ST beds with prevailing ungraded, massive or compound structure (untypical of simple, waning flow deposits) and terminoglacial conditions during the foreset formation suggest the diurnal nature of these deposits. Daily peaks of discharge and sediment concentration induced 'highly-energetic' pulses of sedimentation on the lee slope, resulting in layers of G, Sg, Sm or Sml lithologies. 'Normal' conditions were characterized by slow sedimentation from dilute suspension and resulted in ST beds. The complete foreset sequence developed during three melting seasons when the sediment supply was intensified. During winter seasons the hydrological and sedimentary processes became inactive due to general freezing.

The model of diurnally/seasonally controlled foreset development has allowed the estimation of the lobe progradation rate as well as the duration of melting seasons. The average thickness of 0.06 m for a diurnal rhythmite couplet suggests that the daily rate of the lobe-crest progradation was in the order of 0.12–0.24 m, assuming 30–15° for a slipface inclination. The second segment of the Neple foreset sequence represents a complete annual cycle; its sedimentation lasted during the whole melting season. This segment attains 9 m in stratigraphic thickness, which equals at least 150 diurnal rhythms. Thus, it is believed that the duration of this melting season was at least 5 months.

ACKNOWLEDGEMENTS

I would like to express my thanks to George Postma and two anonymous reviewers for their helpful suggestions and constructive criticism that greatly improved earlier versions of the manuscript.

REFERENCES

ADRIANI, P., CAPPADONA, P. & COLELLA, A. (1988) *Abstr. Int. Works. Fan Deltas, Calabria,* 59 pp.

ALLEN, J.R.L. (1963) The classification of cross-stratified units with notes on their origin. *Sedimentology*, **2**, 93–114.

ALLEN, J.R.L. (1965) Sedimentation to the lee of small underwater sand waves — an experimental study. *J. Geol.* **73**, 95–116.

ALLEN, J.R.L. (1968) *Current Ripples*. North Holland Pub. Co., Amsterdam, 433 pp.

ALLEN, J.R.L. (1980) Sand waves: a model of origin and internal structure. *Sedim. Geol.* **26**, *281–328.*

ALLEN, J.R.L. (1984) *Sedimentary Structures — Their Character and Physical Basis.* Elsevier, Amsterdam, 663 pp.

ALLEN, P.A. & HOMEWOOD, P. (1984) Evolution and mechanics of a Miocene tidal sandwave. *Sedimentology* **31**, 63–81.

ASHWORTH, P.J. & FERGUSON, R.I. (1986) Interrelationships of channel processes, changes and sediments in a proglacial braided river. *Geogr. Ann.* **68A**, 361–371.

AXELSSON, V. (1967) The Laiture Delta. *Geogr. Ann.* **49A**, 1–127.

BAGNOLD, R.A. (1954) Experiments on a gravity-free dispersion of large solid spheres in a Newtonian fluid under shear. *Roy. Soc. London Proc.* **A225**, 49–63.

BOOTHROYD, J.C. & ASHLEY, G.M. (1975) Processes, bar morphology, and sedimentary structures on braided outwash fans, northeastern Gulf of Alaska. In: *Glaciofluvial and Glaciolacustrine Sedimentation* (Ed. by A.V. Jopling and B.C. McDonald). Spec. Publ. Soc. econ. Paleont. Mineral., Tulsa 23, pp. 193–222.

BOUMA, A.H. (1962) *Sedimentology of Some Flysch Deposits*. Elsevier, Amsterdam, 168 pp.

BUCK, S.G. (1985) Sand-flow cross strata in tidal sands of the Lower Greensand (early Cretaceous), southern England. *J. sedim. Petrol.* **55**, 895–906.

CANT, D.J. & WALKER, R.G. (1976) Development of a braided-fluvial facies model for the Devonian Battery Point Sandstone, Quebec. *Can. J. Earth Sci.* **13**, 102–119.

CHURCH, M. & GILBERT, R. (1975) Proglacial fluvial and lacustrine environments. In: *Glaciofluvial and Glaciolacustrine Sedimentation* (Ed. by A.V. Jopling & B.C. McDonald). Spec. Publ. Soc. econ. Paleont. Mineral., Tulsa 23, pp. 22–100.

COLELLA, A. (1988a) Pliocene–Holocene fan deltas and braid deltas in the Crati Basin, southern Italy: a consequence of varying tectonic conditions. In: *Fan Deltas: Sedimentology and Tectonic Settings* (Ed. by W. Nemec and R.J. Steel), pp. 61–85. Blackie and son, Glasgow, 444 pp.

COLELLA, A. (1988b) Gilbert-type fan deltas in the Crati Basin (Pliocene–Holocene, southern Italy). In: *Int Works Fan Deltas, Calabria Excursion Guidebook* (Ed. by A. Colella), pp. 19–77. Università delle Calabria, Cosenza.

COLELLA, A., DE BOER, P.L. & NIO, S.D. (1987) Sedimentology of a marine intermontane Pleistocene Gilbert-type fan-delta complex in the Crati Basin, Calabria, southern Italy. *Sedimentology* **34**, 721–736.

COLLINSON, J.D. (1970) Bedforms of the Tana River, Norway. *Geogr. Ann.* **52A**, 31–56.

DOMACK, E.W. (1984) Rhythmically bedded glaciomarine sediments on Whidbey Island, Washington. *J. sedim. Petrol.* **54**, 589–602.

ETHIER, V.G. (1975) Application of Markov analysis to the Banff Formation (Mississippian), Alberta. *Math. Geol.* **7**, 47–61.

FENN, C.R., GURNELL, A.M. & BEECROFT, I.R. (1985) An evaluation of the used suspected sediment rating curves for the prediction of suspended sediment concentration in a proglacial stream. *Geogr. Ann.* **67A**, 71–82.

GUSTAVSON, T.C. (1972) *Sedimentation and Physical Limnology in Proglacial Malaspina Lake, Alaska*. Dept. Geol. Univ. Mass. Amherst, Tech. Rept. no 5-CRC, 48 pp.

GUSTAVSON, T.C. (1974) Sedimentation on gravel outwash fans, Malaspina Glacier foreland, Alaska. *J. sedim. Petrol.* **44**, 374–389.

GUSTAVSON, T.C., ASHLEY, G.M. & BOOTHROYD, J.C. (1975) Depositional sequences in glaciolacustrine deltas. In: *Glaciofluvial and Glaciolacustrine Sedimentation* (Ed. by A.V. Jopling and B.C. McDonald) Spec. Publ. Soc. econ. Paleont. Miner., Tulsa **23**, pp. 264–280.

HAMMER, K.M. & SMITH, N.D. (1983) Sediment production and transport in a proglacial stream: Hilda Glacier, Alberta, Canada. *Boreas* **12**, 91–106.

HARMS, J.C., SOUTHARD, J.B., SPEARING, D.R. & WALKER, R.G. (1975) Depositional environments as interpreted from primary sedimentary structures and stratification sequences. Soc. econ. Paleont. Mineral., Tulsa, Short Course no. 2, 161 pp.

HUNTER, R.E. (1977) Basic types of stratification in small aeolian dunes. *Sedimentology* **24**, 361–387.

HUNTER, R.E. (1981) Stratification styles in aeolian sandstones: some Pennsylvanian to Jurassic examples from the western interior, USA. In: *Nonmarine Depositional Environments: Models for Exploration* (Ed. by F.G. Ethridge and R.M. Flores). Spec. Publ. Soc. econ. Paleont. Miner., Tulsa 31, pp. 315–329.

HUNTER, R.E. (1985) Subaqueous sand-flow cross strata. *J. sedim. Petrol.*, **55**, 886–894.

JONES, C.M. & MCCABE, P.J. (1980) Erosion surfaces within giant fluvial cross-beds of the Carboniferous in northern England. *J. sedim. Petrol.* **50**, 613–620.

JOPLING, A.V. (1964) Laboratory studies of sorting processes related to flow separation. *J. geophys. Res.* **69**, 3403–3418.

JOPLING, A.V. (1965) Hydraulic factors controlling the shape of laminae in laboratory deltas. *J. sedim. Petrol.* **35**, 777–791.

JOPLING, A.V. (1966) Origin of cross-laminae in a laboratory experiment. *J. geophys. Res.* **71**, 1123–1133.

LOWE, D.R. (1976) Grain flow and grain flow deposits. *J. sedim. Petrol.*, **46**, 188–199.

LOWE, D.R. (1982) Sediment gravity flows: II. Depositional models with special reference to the deposits of high-density turbidity currents. *J. sedim. Petrol.* **52**, 279–297.

LOWE, D.R. (1988) Suspended-load fallout rate as independent variable in the analysis of current structures. *Sedimentology*, **35**, 765–776.

NEMEC, W. & STEEL, R.J. (Eds.) (1988) *Fan Deltas: Sedimentology and Tectonic Settings*. Blackie and Son, London, 444 pp.

NOWAK, J. (1974) *Explanations to the Geologic Map of Poland 1:25 000, Biała Podlaska Quadrangle*. Geological Publ. Co., Warsaw, 35 pp.

ORFORD, J.D. (1978) A comment on the derivation of conditional vector entropy from lithologic transition tally matrices. *Math. Geol.* **10**, 97–102.

ØSTREM, G. (1975) Sediment transport in glacial meltwater streams. In: *Glaciofluvial and Glaciolacustrine Sedimentation* (Ed. by A.V. Jopling and B.C. McDonald). Spec. Publs Soc. econ. Paleont. Miner., Tulsa 23, pp. 101–122.

POSTMA, G. (1984) Mass-flow conglomerates in a submarine canyon: Abrioja fan-delta, Pliocene, southeast Spain. In: *Sedimentology of Gravels and Conglomerates* (Ed. by E.H. Koster and R.J. Steel). Mem. Can. Soc. Petrol. Geol. 10, pp. 237–258.

POSTMA, G. (1986) Classification for sediment gravity-flow deposits based on flow conditions during sedimentation. *Geology* **14**, 291–294.

POSTMA, G, BABIĆ, L., ZUPANIČ, J. & RØE, S.-L. (1988) Delta-front failure and associated bottomset deformation in a marine, gravelly Gilbert-type fan delta. In: *Fan Deltas: Sedimentology and Tectonic Settings* (Ed. by W. Nemec and R.J. Steel), pp. 91–102. Blackie, and Son, London, 444 pp.

POSTMA, G. & CRUICKSHANK, C. (1988) Sedimentology of a late Weichselian to Holocene, terraced fan delta, Varangerfjord, northern Norway. In: *Fan Deltas: Sedimentology and Tectonic Settings* (Ed. by W. Nemec and R.J. Steel) pp. 144–157. Blackie and Son, London.

POSTMA, G. & ROEP, T.B. (1985) Bottomset-modified Gilbert-type deltas (Espiritu Santo Formation, Pliocene, Vera Basin, SE Spain). *J. sedim. Petrol.* **55**, 874–885.

RÜHLE, E. & MOJSKI, J.E. (Eds.) (1965) *Geological Atlas of Poland. Stratigraphic and Facial Problems. Fasc. 12 – Quaternary*. Geological Institute Publ. Co., Warsaw.

SMALL, R.J. (1978) *The Study of Landforms*. Cambridge University Press, Cambridge, 502 pp.

SMITH, N.D. (1972) Some sedimentological aspects of planar cross-stratification in a sandy braided river. *J. sedim. Petrol.* **42**, 624–634.

SMITH, N.D. (1978) Sedimentation processes and patterns in a glacier-fed lake with low sediment input. *Can. J. Earth Sci.* **15**, 741–756.

STANLEY, K.O. & SURDAM, R.C. (1978) Sedimentation on the front of Eocene Gilbert-type deltas, Washakie Basin, Wyoming, *J. sedim. Petrol.* **48**, 557–574.

WESCOTT, W.A. & ETHRIDGE, F.G. (1983) Eocene fan delta-submarine fan deposition in the Wagwater Trough, east-central Jamaica. *Sedimentology* **30**, 235–247.

Spec. Publs int. Ass. Sediment. (1990) **10**, 311–331

Wave-dominated Gilbert-type gravel deltas in the hinterland of the Gulf of Taranto (Pleistocene, southern Italy)

F. MASSARI* *and* G. C. PAREA†

* *Dipart. di Geologia, Università di Padova, via Giotto 1, 35100 Padova, Italia*
† *Istit. di Geologia, Università di Modena, C.so Vittorio Emanuele II, 59, 41100 Modena, Italia*

ABSTRACT

A number of Gilbert-type gravel and sand–gravel deltas developed in the hinterland of the Gulf of Taranto during the Pleistocene. Progradation took place in broad erosional depressions interpreted as drowned valleys cut during a previous relative lowstand of base level. The resulting confined bodies may represent constrained lowstand deltas.

Topsets mostly consist of beach deposits, suggesting that deltas experienced high nearshore wave power. Erosional surfaces commonly interrupt the continuity between topsets and foresets and result in alternating sigmoid and oblique offlap; this is regarded as the expression of constructional and destructional phases in delta growth, related to small fluctuations of relative base level or to the shifting of stream mouth(s).

Purely gravitative processes initiated by slumping or sliding apparently played only a minor role on the delta front. Instead, the commonly observed merging of wave-worked topsets into foreset beds and the textural characters of the latter suggest that sediment dispersal on to the foreset slope was frequently initiated in the nearshore area by storm-induced, high-density surging flows which were subsequently driven by gravity down the foreset slope. Some segments of the foreset are characterized by diffusely stratified gravel and sands with antidune-like bedforms and scour-filling backset beds, the latter thought to reflect development of chutes and pools, with sedimentation related to upstream-migrating hydraulic jumps. A line source rather than a single or multiple point source(s) is thought to have been active, leading to the formation of a debris apron on a channel-deficient foreset slope. Ubiquitous scour-filling backset beds at the toe are thought to record hydraulic jumps which accompanied the flow deceleration at the slope break.

Changes in the geometry and internal organization of the deltaic bodies, as recognized in sections parallel and transverse to the direction of progradation, show that a continuum exists from gravel-beach sequences of increasing ramp height to Gilbert-type bodies.

INTRODUCTION

Most examples of marine Gilbert-type deltas or fan deltas described in the literature concern fluvial-dominated types prograding into a standing body of water of relatively low energy, such as fjords or sea embayments (Prior *et al.*, 1981; Postma, 1984a,b; Postma & Roep, 1985; Prior & Bornhold, 1986; Colella *et al.*, 1987; Corselli *et al.*, 1985; Prior & Bornhold, 1988). In these systems, reworking by waves is minimal and the presence of a point source is reflected by a system of chutes and splays radiating from the major stream mouth and forming an extensive subaqueous sediment cone.

Wave-influenced Gilbert-type bodies have recently been documented in the Crati basin by Colella (1987, 1988). In addition, Colella (1984) pointed out that an alternation of sigmoid and oblique offlap in Gilbert-type bodies may be regarded as the expression of constructional and destructional phases in delta growth, related to fluctuations of base level or shifting of the feeding stream mouth.

In this paper some examples of wave-dominated Gilbert-type deltas are described where highly concentrated storm-induced flows are thought to change into gravity-driven flows, commonly in supercritical regime, as they spread down the foreset slope. These systems show a chute-deficient delta front and dominant feeding from a line source rather than a point source.

GEOLOGIC AND GEOMORPHOLOGIC SETTING

Coarse-grained marine progradational sequences, including Gilbert-type deltas and beach sequences, developed during the Pleistocene along the microtidal Ionian coast extending between the front of the uplifting Apennine chain and the Apulian foreland. Their substrate is the Plio–Pleistocene fill sequence of the Bradanic foredeep (Fig. 1). During the Quaternary, the Bradanic area was subject to progressive uplift due to isostatic adjustment after thrust emplacement (Ciaranfi *et al.*, 1983). Interference between uplift and glacioeustatic fluctuations of sea level led to the development of a series of marine terraces.

The wave climate of the present-day Ionian coast of Italy is characterized by periodic storm events, mostly concentrated in the winter season. Winds blowing from the SE display maximum fetch (about 1500 km) and tend to predominate during storms; an active northward longshore drift is mostly promoted by winds blowing from the SSE which have a fetch of about 800 km.

The Apenninic rivers (from S to N: Sinni, Agri, Cavone, Basento and Bradano) debouching on the southern segment of the coast of the Gulf of Taranto, form a series of small bedload deltas. Although these fluvial systems carry a large amount of debris to the sea, fluvial-dominated deltaic deposits represent a very small portion of the present-day coastal sediments. This is probably a consequence of the lack of important radiating distributary systems, the flash-flood regime of the rivers, and active and continuous reworking by waves. It may reason-

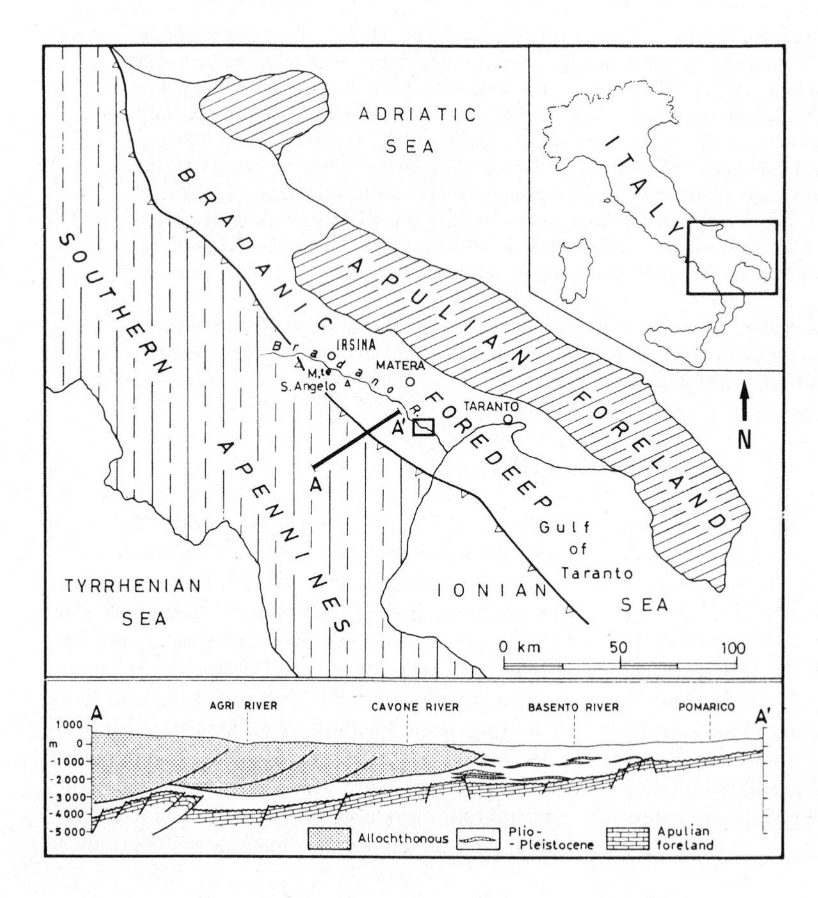

Fig. 1. Index map showing schematic geology of southern Italy (section taken from Sella *et al.*, 1988). Studied Gilbert-type bodies are distributed along Bradano river. Irsina body (details in Fig. 2) and M.S. Angelo body are located in upper reaches of Bradano valley; Cugnola Volta and Cozzo del Presepio bodies in lower reaches (squared area, details in Fig. 3).

ably be supposed that a similar setting existed during the progradation of the studied systems; these should not therefore be considered fan deltas, as they were probably fed by braided fluvial channels.

GILBERT-TYPE DELTAIC BODIES

A number of Gilbert-type deltaic bodies have been identified in the basin of the Bradano river, showing distinct thickening in the direction of progradation and flat-topped lenticular geometry with broadly concave-up base in cross-section. They appear as gravel bodies mainly confined in segments of elongated depressions cutting down in most cases into Calabrian sediments (Figs 2, 3). These depressions are interpreted as drowned palaeo-Bradano valleys incised during relative lowstand(s) of sea level. The geometry of the bodies is thus inherited from a previous morphologic cycle. In the case of the Irsina body (Fig. 2), the dip directions of foreset beds distinctly converge towards the axis of the trough, suggesting the confining influence of the valley flanks.

Evidence of wave working in the topsets of the bodies is striking and implies a high-energy environment, like that existing on open or nearly open coasts. Due to the relative low relief and great erodibility of the dissected landscape invaded by the sea, and the gentle seaward tilting of the Calabrian substrate, valley drowning is not believed to have resulted in the formation of deep rias, which would have led to a significant lowering of wave energy. Instead, smoothing of the ridges between valleys by shoreface erosion during base-level rise, and deposition of large volumes of sediments near river mouths during the subsequent progradational stage, coupled with very efficient longshore drift, rapidly led to significant straightening of the coastline. Wave-worked gravels were thus also the dominant deposits at the river mouths. Deltaic progradation into the valleys led to progressive thickening of gravel bodies, whose lower part, developing below the storm wave base, was built out as gravity-dominated foresets. In this situation, a continuum existed from gravel-beach sequences of increasing ramp height to Gilbert-type bodies, the former grading into the latter both along the direction of progradation and from lateral to axial parts of the bodies.

Some samples from the bottomsets of the Cugnola Volta body yielded a nannofossil assemblage

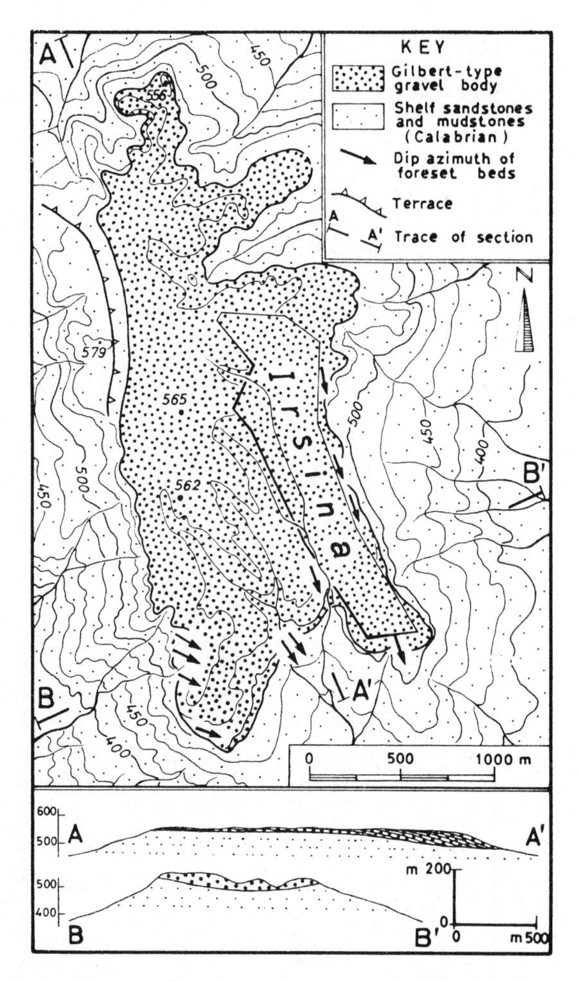

Fig. 2. Geologic map and sections of Irsina body.

not older than the *Calcidiscus macintyrei* zone and correlatable to the *Pseudoemiliania lacunosa* zone of Rio *et al.* (1990), due to the presence of *Gephirocapsa* sp. 3. Although the studied bodies are erosional remnants, the Irsina body may be tentatively correlated to the M.S. Angelo body (location in Fig. 1) whose top surface corresponds to Brückner's (1980) terrace T11, tentatively dated at 775–690 Ka. The other Gilbert-type bodies show a sedimentary cover of variable thickness, whose terraced top surfaces span from 500–400 Ka BP to 250 Ka BP, according to Brückner (1980). Consequently, many uncertainties exist regarding their ages.

Field data suggest that the Irsina (Fig. 2) and M.S. Angelo bodies represent the fill of quite straight, elongated valley segments, whereas the Cugnola

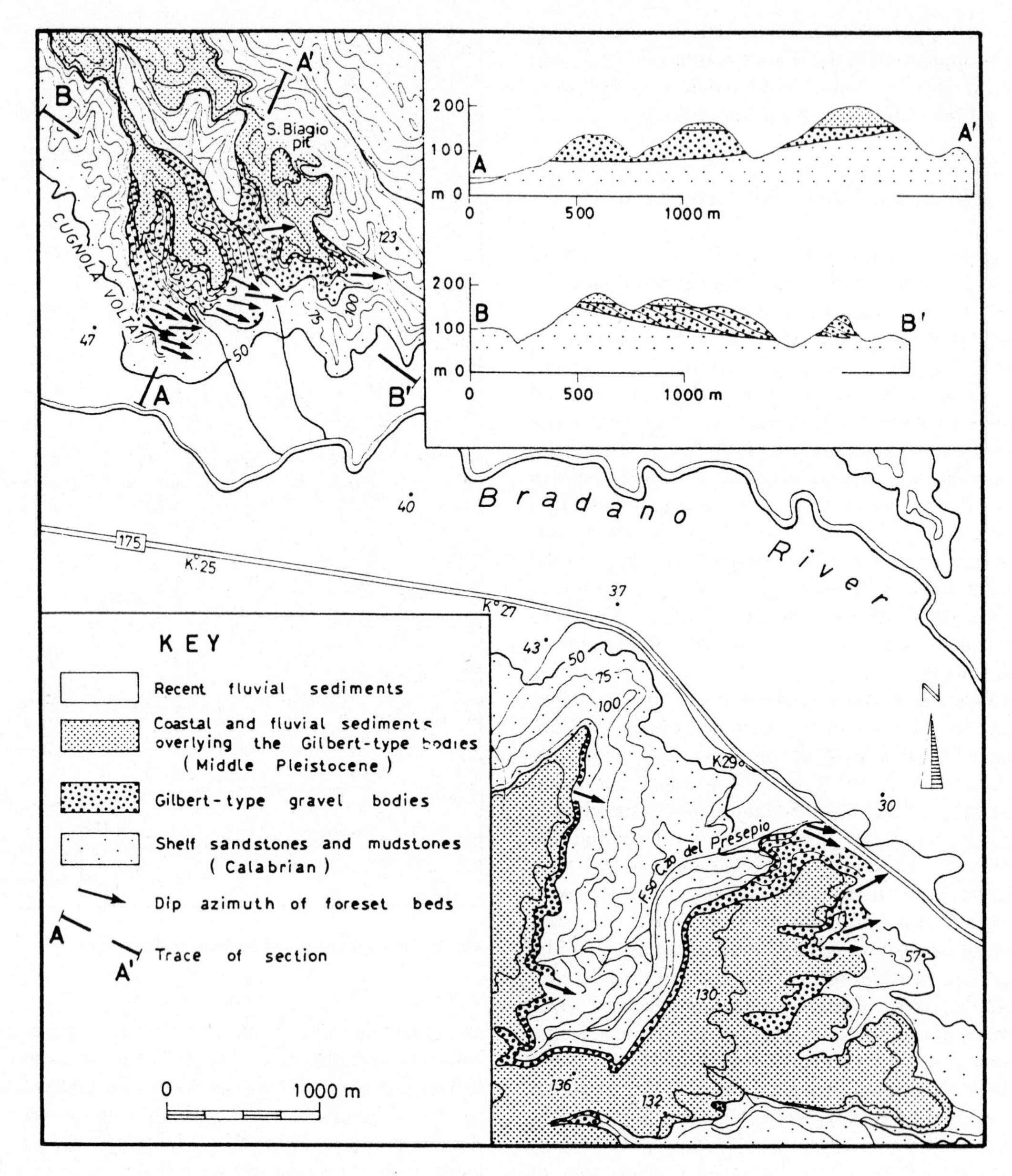

Fig. 3. Geologic map and sections of Cugnola Volta (top left) and Cozzo del Presepio (bottom right) bodies (location of area shown in Fig. 1). S. Biagio body is part of sedimentary cover of Cugnola Volta body.

Volta and Cozzo del Presepio bodies (Fig. 3) are moulded on broad bends of a sinuous valley. Consequently, it was considered meaningless to define the width and length of the bodies in the second case (Table 1). Due to the deep later dissection of the bodies, the measured lengths obviously provide only a minimum figure of the real distances achieved by progradation.

Unlike the S. Biagio body, which is characterized by mixed lithology, the other Gilbert-type bodies (Irsina, M.S. Angelo, Cugnola Volta and Cozzo del Presepio) are essentially composed of gravel and comparably larger. Gravels are usually in the pebble–cobble size range, rarely coarser; clasts are

Table 1. Summary of Gilbert-type bodies in the basin of the Bradano river

Bodies	Maximum thickness (m)	Width (m)	Length in the direction of progradation (m)
Irsina	90	960	1000
M.S. Angelo	80	1100	5570
Cugnola Volta	90	—	—
Cozzo del Presepio	50	—	—
S. Biagio	13	—	—

well rounded to subrounded; the lithologies reflect the Apennine source rocks and include limestones, cherts, sandstones, rare ophiolites, some granite and vein quartz.

Gravel bodies

Wave-worked topset beds

Wave-worked topsets are far more common than fluvial topset beds. They form ramps gently dipping towards the basin and either merge into foreset beds with progressive increase in dip angle (sigmoid geometry; Fig. 7), or form a distinct sequence downlapping onto subhorizontal erosional surfaces (see below) marking the topset–foreset boundary (Fig. 6).

Topmost layers are subhorizontal to gently inclined in the landward direction, and consist of dominantly seaward imbricated, large discs, commonly floating in a sandy matrix. They grade to and interfinger with thinner and more regularly bedded and well-sorted layers, rich in disc-shaped pebbles (Fig. 4), locally grouped in erosively bounded sets with low-angle mutual relationships. These layers dip at very low angles in the seaward direction and show dominant seaward dip of clast imbrication. They grade into and interfinger downdip with progressively thicker-bedded, coarser-grained, clast-supported gravel layers which form the lower part of the ramp. In the case of downlapping geometry, the beds show a definite downdip increase in the amount of spheres and local openwork texture, especially in coarser-grained beds (Fig. 5). The fabric is commonly chaotic in the coarser layers, with clast imbrication dipping in both landward and seaward directions, but is more regular in fine-grained gravels, with imbrication dipping consistently upslope, commonly at high angles.

In addition to minor erosional surfaces occurring in the topset, more prominent truncation surfaces

Fig. 4. Section oriented SSE–NNW, normal to palaeoshoreline (land to right) showing upper beachface gravels in topset of Irsina body. Note planarity of stratification, high degree of pebble segregation, flatness of pebbles and well-developed seaward-dipping imbrication. Rod shows scale in decimetres. Village of Irsina.

Fig. 5. Openwork gravel layers with predominance of equant pebbles in inferred lower beachface of Cozzo del Presepio body. Note normal grading in bed below centre. Rod shows scale in decimetres. Abandoned quarry.

commonly exist near the topset–foreset boundary and to some extent affect the foreset of the bodies (Figs 6, 7). Arcuate scars 1–3.5 m high, affecting the uppermost foreset (Fig. 7, arrow) and locally also the lowermost topset, are quite rare. They show steep head slopes, are locally draped by thin mud laminae or by plane-laminated sands and are generally filled by subsequent progradation of foreset beds.

Another group of surfaces displays a shape broadly convex upwards and in the direction of progradation; they are subhorizontal in the onshore area, where they angularly truncate foreset beds and locally also topset beds, and tend to become conformable downdip with foreset beds. The portions of the body separated by these surfaces may show differences in stratification attitude; changes in dip direction of foreset beds range from a few degrees up to 35°. The surfaces may be draped by a coarse gravel lag or a layer of thoroughly burrowed sand with sparse pebbles. Topset and foreset beds overlying these surfaces either downlap on to them (Fig. 6) or lie on them with conformable contacts (Fig. 7). Within this group of erosional surfaces two types may be distinguished, according to whether they are accompanied by a vertical shift in the position of the topset–foreset boundary, or not.

It should be stressed that the first stage of progradation of Gilbert-type bodies is characterized by a sequence similar to the topset of fully developed bodies. It displays progressive increase in ramp height in the direction of progradation, and only in a later stage, when the body encroaches on deeper water, gravity-dominated foresets develop. The same geometrical relationships exist between the thicker, fully developed Gilbert-type sequence in the axial part of the body and the thinner coeval sequence in the lateral position.

Interpretation

The previously described topset beds display a number of features which are quite comparable to those usually found in the beachface gravels of progradational beach sequences, particularly of high-ramp type, as described in the marine terraces of the same area (Massari & Parea, 1988; the term beachface is intended by the authors to include both the emergent portion and the submerged, seaward-sloping ramp of the beach). Prominent, wave-induced shape-sorting effects are typical of high beachface ramps with relatively reflective behaviour.

Topmost layers with large discs and sandy matrix are regarded as washover deposits (Massari & Parea, 1988). The thinner- and regularly bedded layers with good sorting are believed to represent the swash zone, whereas the thicker-bedded division displays features similar to those of the inferred submerged ramp (lower beachface) deposits of progradational gravel-beach sequences.

Minor erosional surfaces affecting the topset of the body may record the impact of storms on the beachface, leading to erosional truncation of the upper beachface profile and offshore redistribution of removed material. Arcuate scars are thought to reflect limited slope failures, concentrated in the uppermost part of the foreset and possibly induced by the impact of storm waves or oversteepening due to localized accumulation.

Erosional surfaces separating bundles of beds with different stratification attitude may result from the shifting of stream mouth(s), so that a portion of the body is subject to erosion and sediment is supplied to a new site of deltaic construction. Convex-up, regular, erosional surfaces implying a vertical shift of the topset–foreset boundary are thought to represent ravinement surfaces cut by landward shifting shoreface erosion as a response to small relative rises in base level, which followed equally small relative drops. The latter two types of erosional surfaces record destructional episodes leading to the formation of a shoreface platform in the upper part of the body. Subsequent seaward accretion of the beachface, as a result of resumed progradation, rapidly eliminated the platform and finally restored a sigmoid geometry of the topset/foreset transition.

The observed changes in geometry and internal organization of the body in sections parallel and transverse to the direction of progradation suggest that the evolution of the progradational sequence is strongly influenced by the gradient of the substrate and that a continuum exists from gravel-beach sequences of increasing ramp height to Gilbert-type bodies.

Fluvial topset beds

In a single locality (northwestern edge of the M.S. Angelo body — see location in Fig. 1), a section transverse to the direction of progradation shows a topset sequence consisting of the aggradational fill of a broad depression cut in the foreset beds, with a relief of about 10 m and a concave-up erosional base. The sequence, mostly consisting of clast-

Fig. 6. Photograph (encircled shovel 107 cm high for scale) and sketch of upper part of Cozzo del Presepio body. Section is subparallel to direction of progradation. Beachface gravel layers downlap on prominent erosional surfaces located at leftward-increasing altitudes. Abandoned quarry.

supported gravels, is characterized by a trend fining and thinning upwards from large-scale trough and locally planar cross-bedded pebble–cobble gravels (sets up to 3 m thick) associated with minor trough cross-bedded sands, to predominantly horizontally bedded tabular pebble–gravel units accompanied by some medium-scale planar cross-beds.

Trough cross-bedded fluvial gravel topsets merging with sigmoid geometry into foreset beds are occasionally observed in some segments of sequences 8–10 m thick, intermediate in character between high-ramp beach sequences and Gilbert-type bodies. In this case, trough cross-bedded gravels persist to the uppermost foreset slope and are replaced downdip by massive, sand-rich pebble–cobble foreset layers.

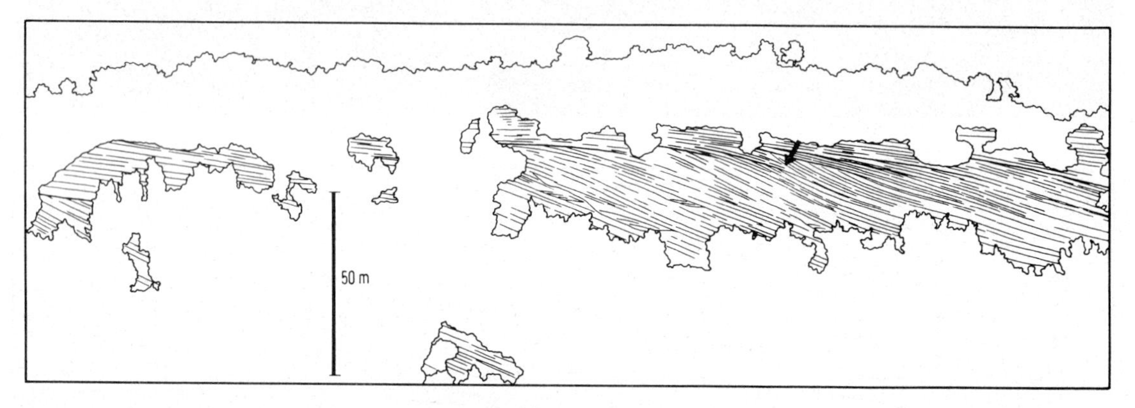

Fig. 7. Photograph and sketch of Cugnola Volta body. Section is oriented NNW–SSE and exposed on eastern flank of main valley cutting through body (see Fig. 3). Topsets consist of beach deposits. Note that topset–foreset boundary is remarkably lower in altitude to left, and is alternately erosional (oblique offlap) and transitional (sigmoid offlap). Note also slump scar (arrow) and some scour-filling backset beds interbedded with foreset beds.

Interpretation

The M.S. Angelo section enables the recognition of the effects of the relative fluctuations of base level on the feeding fluvial system. The upward-fining topset sequence is thought to represent the fluvial aggradational fill of a small valley. This was presumably incised into the foreset beds during a phase of lowered relative base level, and the fill may represent the early effect of a small, subsequent relative rise, leading to backfilling of the valley. The trough and planar cross-beds may result from migration and accretion of gravel bars with sinuous to linear fronts down the stream bed during major floods. The large scale of these structures and the coarseness of the deposits in the lower part of the fill sequence suggest the activity of high-gradient streams and high flow depths due to confinement of flow by valley walls. The observed upward fining and thinning suggest progressive lowering of stream gradients and decreased flow depth, with transition to a shallow braided system dominated by longitudinal bars. Relationships between fluvial topsets and foresets will be discussed later.

Foreset beds

Foreset beds are mostly planar in geometry, with high lateral persistence and very low-angle wedging-out. Dip angles range from 15° to 32°, averaging 22°.

Textural features of gravels may change from one segment of the foreset to another. Some segments show evidence of wave-induced segregation of different morphometric populations before the resedimentation on the foreset. In this case beds appear dominated by disc-shaped or spherical pebbles, or characterized by a random mixing of the two populations, generally with large spheres floating in a 'matrix' of disc-shaped small pebbles (Fig. 8). Openwork gravel layers consisting of remarkably equant large pebbles and cobbles occur locally. Pebbles with encrusting bivalves are rarely found. Other segments of the foreset do not show evidence of shape segregation. Foresets unconformably overlain by fluvial topsets at the northwestern edge of the M.S. Angelo body locally show a number of specific textural features, including layers consisting of an ill-sorted mixture of sand and gravel, without any trace of shape sorting, lower degree of pebble rounding and presence of outsize clasts.

Generally speaking, two different types of foreset beds have been recognized: those lacking internal stratification, including ungraded, normally graded, inversely graded and inverse to normally graded layers, and those displaying diffuse stratification, including beds characterized by thick planar lamination or low-angle cross-lamination, sometimes with broadly undular upper profiles, and backset beds. Observations on 146 layers within the first type give the following results: layers are mostly clast-supported and show a thickness range of 20–150 cm (average 50 cm); 27% are ungraded, 41% inversely graded, 23% normally graded, 5% inverse-to-normally graded, 4% ungraded with basal inversely graded band. Massive beds with pebbles floating in a sandy matrix are generally rare, but may be common in some segments of the bodies. Lateral changes from one type of grading to another, or from ungraded to graded units, occur within a few metres of each other and with no clear systematic trend. However, on a larger scale, normal grading seems to predominate in the uppermost low-gradient part of the foreset, whereas inverse grading is more common in the more steeply inclined portion.

Inversely graded clast-supported gravel layers, sometimes with inverse grading confined to a basal band and the bulk of the bed ungraded, commonly have strong and consistent preferred fabric; they show a-axis imbrication dipping upslope, commonly at high angles (measured from the bedding plane) or with angles steepening upwards (Fig. 9) and locally flattening at the top.

Foresets of the second type are characterized by diffuse stratification and consist of bands of different grain size, in most cases displaying crude planar to broadly curved lamination and locally bounded by low-angle erosional surfaces. Sand content is usually high. Finer-grained bands consist of sand or granules, whereas coarser ones may range from thin gravel bands to pebble–cobble rows in a sand matrix (Fig. 10). Thin, repeated, inversely graded layers occur locally. Outsize clasts and clay chips are relatively common. Upslope-dipping clast imbrication is usually well developed, with generally low dip angles. The a-axis fabric is either unimodal [a(p), a(i)], or bimodal (Fig. 11(a)).

Careful tracing of individual stratification band boundaries shows that they may deviate from planarity; sand-dominated strata especially may

Fig. 8. Vertical view (section slightly oblique to direction of progradation) of bimodal foreset beds of M.S. Angelo body. Note high sphericity of pebbles and cobbles embedded in a 'matrix' of small pebbles, suggesting mixing of beach-derived clast populations. Rod shows scale in centimetres.

Fig. 9. Vertical view (section slightly oblique to direction of progradation) of foreset beds of Cozzo del Presepio body. Finer-grained layer below rod shows remarkable upward steepening of imbrication angles. Rod shows scale in decimetres. Abandoned quarry.

Fig. 10. Vertical view (section slightly oblique to direction of progradation) of diffusely stratified facies. Foreset of Cozzo del Presepio body.

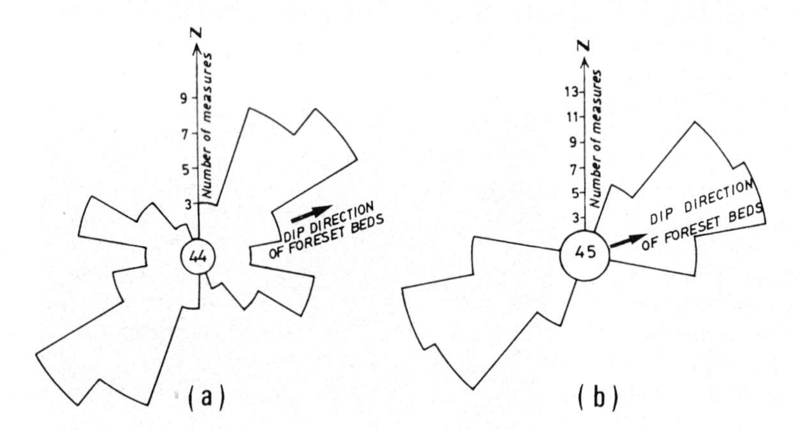

Fig. 11. Azimuthal roses of clast long axes in (a) diffusely stratified facies and (b) individual backset bed within set of backset beds. Backset beds form an angle of 22° relative to foreset bedding planes. Foreset of Cozzo del Presepio body.

pinch and swell and locally show trochoidal profiles with wavelength ranging from about 1.5 m to 7 m. Internal bands are either in phase with the wave profiles or are horizontal to low-angle cross-bedded; in the latter case they usually dip upslope at angles of less than 10° (relative to major bedding planes).

Solitary sets of backset beds 0.65–2.7 m thick are locally intercalated with foreset beds (Fig. 7). They are confined in spoon-shaped, more or less irregular scours up to 17 m wide, with sharp erosional bases and long axes oriented downslope and measuring up to 23 m. Backset beds consist of upslope-dipping well-defined cross-beds of geometry ranging from planar to gently concave- or convex-up, to occasionally sigmoid; their angles relative to the foreset bedding planes tend to increase with grain size and average 25–30°. The backset beds occur in a wide range of grain sizes, from coarse sands to pebble gravels with sparse cobbles. Within individual sets the average grain size tends to decrease remarkably in the upslope direction. Individual backset beds are usually ungraded, but occasionally show inverse grading; in addition, clasts predominantly show a-axis imbrication (Fig. 11(b)), dipping upcurrent commonly at high angles relative to the bedding planes of backset beds, or with upward steepening angles.

As noted by Postma (1984b) in a similar context, rapid changes in facies not only occur between beds, but also within individual beds. Lateral changes within a few metres not only occur from one type of grading to another, as outlined above, but also from internally stratified planar to undular, to backset bedded units.

Layers of fine sand or sandy mud, sometimes showing bioturbation, occur locally between gravelly foreset beds, more commonly in the lower part of the bodies. In most cases they consist of a number of sand–mud couplets; sand bands display planar lamination and/or ripple cross-lamination of unidirectional or bidirectional type (the latter only in the upper part of the foreset) with mud drapes commonly preserving the ripple profiles, or locally medium-scale high-angle cross-lamination (sets up to 15 cm thick). The dip of unidirectional cross-lamination is most commonly downslope, rarely along-slope. Convolute lamination and dewatering structures are other local features.

Pelitic layers, commonly with parallel lamination, are in most cases no more than 10 cm thick, but packets up to 1 m thick, locally with thin sandy interbeds, are sometimes found in the lower part of the foreset slope. They are conformable with foreset beds or drape small-scale arcuate slump scars. They are themselves often truncated by multiple, small slump scars.

Segments of folded mudstone layers occasionally occur embedded within foreset beds, especially in the lower part of the body; some ends of truncated mud layers also appear folded into the overlying gravel bed (see also Postma, 1984a).

A large-scale scar cutting into prodelta muds occurs near the toe of the Irsina body. The scar, the lower part of which does not outcrop, is filled with a set more than 3 m thick of gravel foresets which display an outward decreasing inclination.

U-shaped troughs up to 1.5 m deep, with steep to vertical walls, locally cut into foreset beds and are plugged by massive pebble–cobble gravels, usually coarser than the surrounding foreset gravels. The fabric is generally chaotic in the central part of the fill (plug), but a distinct imbrication may occur near the sides, suggesting that clasts could only be distinctly oriented in the boundary shear zone.

Interpretation

The a-axis imbrication in inversely graded beds attests to the importance of interparticle impact during granular shear. Upward steepening of imbrication angles is possibly produced in the last phase of motion of highly concentrated dispersions, when the sediment is strongly sheared immediately prior to frictional freezing. Flattening of imbrication dip at the top may be produced by the shear of an overriding flow.

Dominant normal grading in the upper low-gradient part of the foreset may reflect greater average flow turbulence, whereas dominant inverse grading in the steeper lower part may reflect the influence of gravity, leading to Bagnoldian dispersions.

The diffusely stratified facies probably represents deposition from traction carpets at the base of highly concentrated, supercritical flows, probably of surging type, occasionally implying migration and aggradation of long-wavelength antidunes. As suggested by Hiscott and Middleton (1979), the local occurrence of thin, repeated, inversely graded bands in this facies may imply intense shearing of a basal layer of grains with progressive freezing of successively generated traction carpets.

A general interpretation concerning the nature of flows depositing foreset beds and the genesis of backset beds will be given in later sections.

Individual sand–mud couplets forming fine-grained beds are thought to be deposited in most cases by waning-energy diluted flows directed downslope; the local occurrence of wave ripples suggests occasional reworking by oscillatory flows at depths above the wave base.

Thick pelitic packets may record small base-level rises or phases of switching of the river mouth, which may render part of the delta face temporarily inactive, allowing the settling of fine-grained sediments. Thinner layers on the other hand may simply represent fair-weather settling of fines following a low-frequency high-energy storm or a major flood of the debouching river.

The Irsina scar is thought to represent a shallow rotational slide due to loading of the prograding delta front on prodelta muds (see also Prior & Bornhold, 1988). The U-shaped troughs may represent chutes filled with debris flows originating from small slope failures, as in the cases described by Postma (1984b) and Postma *et al.* (1983). Small retrogressive slumps probably evolved downslope into debris flows which effectively plugged the contemporaneously formed troughs, thus preserving the steep sides. In modern examples (Prior *et al.*, 1981; Prior & Bornhold, 1986, 1988), chutes usually originate in arcuate re-entrants with steep head slopes cut into mouth-bar fronts, and exhibit an elongated gully section and an extensive downslope hummocky zone.

Toeset and bottomset beds

The foreset beds in the lower part of the bodies show a progressive asymptotic decrease in dip angle, gradually merging downslope into toeset beds, whose inclination is in the range of 4–6°. Toeset beds show significant differences in stratification pattern with respect to foreset beds, being characterized by extreme lenticularity. Gravels fill a pattern of ubiquitous, mutually intersecting, spoon-shaped scours, elongated in the inferred flow direction, 0.7–7 m wide and up to 1.6 m deep, with width/depth ratio averaging 15 (Figs 12, 13). The scours cut in fine-grained lithologies locally show steep vertical and sometimes undercut walls, especially at the upcurrent side, with mudstone layers occasionally sticking out from the walls, as noted by Postma (1984a), and rip-up mudstone clasts, sometimes of large size. In the proximal toeset, the troughs are deeper and filled with single sets of gravelly or sandy–gravelly backset beds (Fig. 13), whereas in the distal part they are flatter and may show lenticular sandy fills with distinct lamination draped over the sides of the trough (Fig. 14).

The characteristics of backset beds are comparable to those already described. Some scours are draped by mud and later filled with gravel, suggesting that a time gap characterized by little sedimentation separated scouring from coarse-grained filling. The clast fabric in gravelly backset beds is characterized by long-axis imbrication and shows high variability in orientation in different parts of the scour fill. The direction of pebble imbrication in the central part of the fill is generally parallel to the scour axis, but near the trough walls the dip of imbrication may show gradual to abrupt changes, and the ab planes of clasts generally tend to arrange themselves parallel to the trough walls (a feature also noted by Postma, 1984a). The different orientation is probably due to boundary shear effects. The range of variation of the dip azimuths of imbrication within individual scour fills may be as much as 60°. In addition, the dip angles of imbrication are commonly very high, either throughout the backset beds, or increasing in their upper part with an upward steepening pattern, and vertically oriented a-axes are not uncommon.

Thick units of clast-supported structureless pebble–cobble gravels may have been emplaced by cohesionless debris flows. Planar layers of diffusely laminated sands with granule or pebble rows are a subordinate facies of the toeset: they locally display low-angle backset lamination and sometimes broadly

Fig. 12. Sketch from photograph of toeset beds of Irsina body (section oriented WNW–ESE). Flow is on average towards observer slightly to left. Note ubiquitous, mutually intersecting shallow scours and broadly convex-up surfaces in upper part. Abandoned quarry southwest of village of Irsina.

Fig. 13. Sketch from photograph of toeset beds (right) and proximal bottomset beds (left) of Cugnola Volta body (section oriented NNE–SSW). Flow inferred from pebble imbrication is to left. Toeset beds consist of ubiquitous scour-filling backset beds. Abandoned quarry.

Fig. 14. Distinctly laminated sandy lenticular infill of a broad scour in distal toeset of Cugnola Volta body. Bar shows scale in decimetres. Abandoned quarry.

undular upper profiles. Displaced, complexly folded gravel masses up to 5 m thick locally occur in the toe area.

Gravelly toeset beds die out very rapidly in the direction of flow. They are replaced by interbedded to interlaminated fine sand and mud, with a distinct basinward decrease in the sand/mud ratio and thickness of sand layers (bottomset beds). Sand layers show sharp bases and tops and variable lateral persistence; wedging-out may reflect the filling of shallow scours or the presence of a gently convex-up top. Structures include planar to low-angle to gently convex-up lamination, and unidirectional ripple cross-lamination, sometimes of the climbing type. No evidence of oscillatory flows was found. Mud veneers may occur interlaminated with the sand and commonly preserve the ripple profiles at the bed tops (Fig. 15). Bioturbation is rare and represented by U-shaped or very thin branching burrows. Intraformational conglomerates consisting of rounded pebble-sized mud balls in a matrix of coarse sand, abundant shell fragments of nearshore molluscs and sparse exotic pebbles rarely occur in association.

Evidence of mass displacement is widespread in the bottomset and includes: (1) rotated slices showing upslope-dipping stratification; (2) slump scars filled with festoon-like beds showing distinct wedging-out towards the scar sides; (3) small-scale arcuate growth faults. These different processes are interpreted as effects of depositional loading of the prograding delta front (Prior & Bornhold, 1988).

Interpretation of the structures will be given in a later section.

Fig. 15. Detail of layers showing sand/mud interlaminations in bottomset of Cugnola Volta body. Note pelitic (lighter) lamina inserted within climbing ripple lamination and sharp contact between rippled top of sandy band and pelitic drape. Bar is 12 cm long. Abandoned quarry.

Sand-gravel body

A body characterized by mixed lithology, with sand-dominated clinoforms, forms a younger cycle in the sedimentary cover of the Cugnola Volta body. It outcrops in a pit (Fig. 3: S. Biagio pit) where only the topset and part of the foreset unit are exposed. The base of the body has been reached by borings at about 5 m below the bottom of the pit.

As in gravel bodies, wave-worked topset beds overlie foreset beds, either with erosional contact or gradual transition (Fig. 16). In the former case, the boundary is overlain by variable thicknesses of a heterolithic facies consisting of subhorizontal sand and gravel beds with planar lamination, and bimodal trough cross-bedding with directions of 40° and 220°. Truncation surfaces at the topset–foreset boundary occur at different levels (two of which appear in Fig. 16), marking different steps in the growth of the body. Both the gently inclined topmost gravels and foreset beds consistently dip towards 120°.

The foreset beds are mostly made up of sand and secondarily of pebbly sand and gravel. Scour-filling sets of backset beds, ranging in thickness from less than 10 cm to 114 cm, are an important component of the foreset beds (Fig. 16). The scours show a wide range of variation in geometry from short deep features (Fig. 17), commonly with irregular base and deeply indented and locally undercut sides, to more regular features, which may develop downslope over distances of up to 15 m. The typical characteristics are those already described in gravelly bodies, with some additional ones.

1 The dip angles of backset beds may reach 50° with respect to the foreset bedding planes at the downslope end of the scours.

2 The sands in the backset beds are very well sorted and the gravels commonly include typical beach-derived morphometric populations.

3 Backset beds are usually thickest and coarsest near the lower erosional surface and become progressively finer-grained and thinner towards the upper boundary of the set (Fig. 17; see also Nielsen *et al.*, 1988).

4 Successive backset beds may show remarkably different grain sizes, sometimes with an alternating pattern (Fig. 18). However, the average grain size tends to decrease remarkably upslope within the set, with earlier bands usually dominated by clast-supported gravels and younger bands commonly consisting of crudely laminated sand with sparse granules or small pebbles. Concurrently, the dip angles of backset beds commonly tend to decrease gradually upslope in a fan-like pattern, so that the latest-formed topmost layers of the structure become conformable with foreset beds, and may lie with a toplap truncating relationship on the earlier formed, steeply inclined bands (Figs 16, 17). These late-stage units display the typical features of diffusely stratified facies. In addition, the upper surface of some sets, in sections both parallel and normal to the foreset slope, shows a profile broadly convex-upwards and

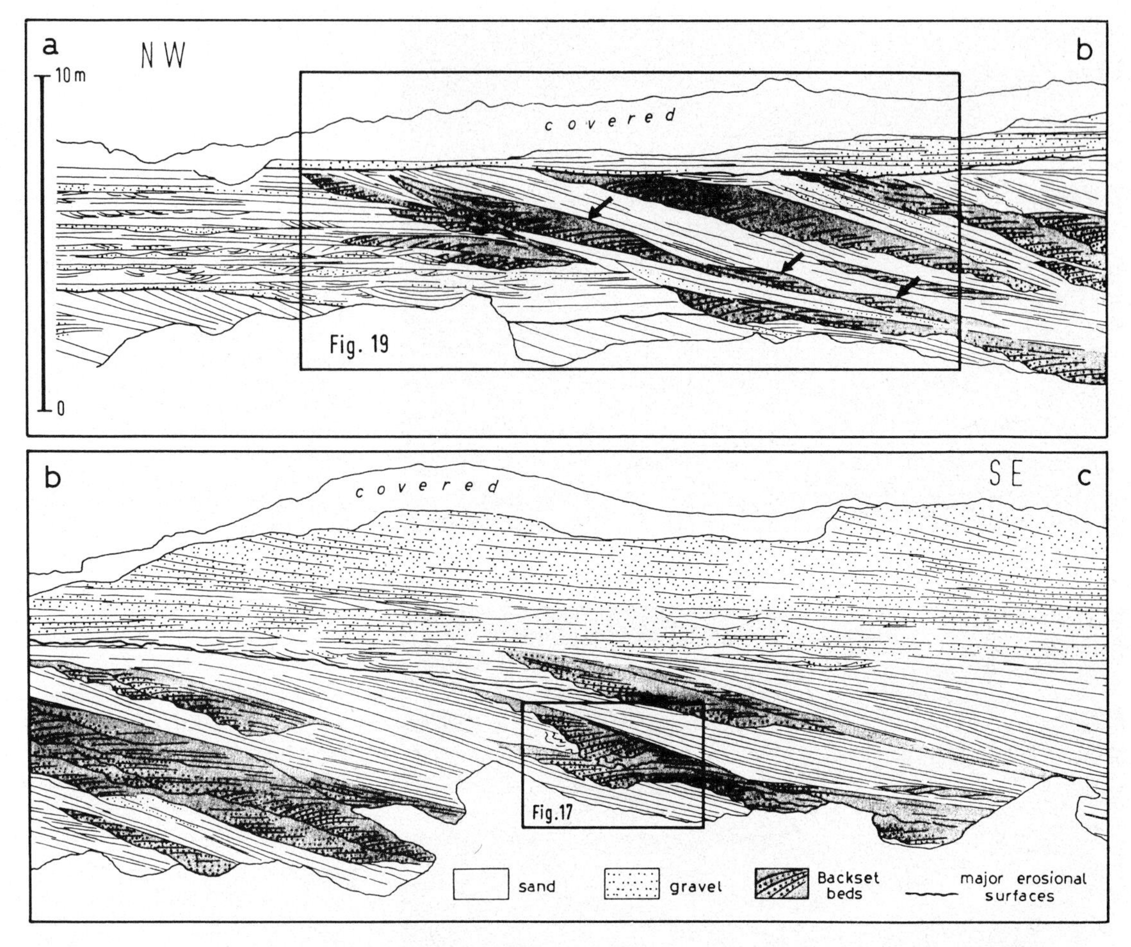

Fig. 16. Sketch from photograph of front of S. Biagio pit (location shown in Fig. 3) showing complexity of relations between topsets and foresets and abundance of scour-filling backset beds, locally showing broadly undular upper boundary (arrows). At northwestern end of section sets of backset beds 'climb' to top of foreset slope and can be locally traced to some extent into topset platform, where they interfinger with subhorizontal shoreface layers. Shoreface aggradation in this zone was not accompanied by significant progradation of foreset slope. Two major surfaces at different levels mark topset/foreset boundary; they are erosional except in southeastern end of section where topset beds merge into foreset beds.

in places undular, with wavelengths of up to 10 m (arrows in Fig. 16).

5 Outsize mudstone clasts, commonly imbricate, occur locally.

6 Local deformational structures at the boundaries of the scours include convolute lamination, distorted bedding and flame structures (Fig. 17), and suggest that the trough walls were subject to liquefaction and plastic deformation.

It is important to note that scour-filling backset beds locally 'climb' to the top of the foreset slope: near the northwestern end of the section, they may be traced to some extent into the topset platform, where they interfinger with subhorizontal heterolithic layers (Figs 16, 19).

Towards the southeastern end of the section the backset beds decrease greatly in abundance and a general fining of foreset beds may be noted in the direction of progradation (Fig. 16). On this side of the pit a rhythmic pattern is given by an alternation of sandy foreset beds and pelitic layers; individual foreset beds commonly consist of a fining-upward sequence (Fig. 20), which may include, as a lowermost division, a thin set of backset beds consisting of pebbly sandstone; this grades upwards, through flattening of backsets and decreasing grain size, into

Fig. 17. Two stacked sets of scour-filling backset beds in foreset of S. Biagio body. Note regularity of bedding, concentration of pebbly fraction in lower part of infill, and upslope progressive decrease in dip angle of backset bedding. Note also truncation of backset beds by planar-laminated layers at downslope end of upper set. Bar shows scale in decimetres. See Fig. 16 for location.

Fig. 18. Alternation of sandy and gravelly bands in scour-filling set of backset beds. Foreset of S. Biagio body (section slightly oblique to progradation). Bar 12 cm long lies on base of set.

planar-laminated sand, rippled fine sand and a pelitic band which may preserve the ripple profiles. Dewatering structures are common in the fine sands. Ripples are either of the climbing type, migrating downslope, or with internal structures suggesting combined or oscillatory flow. These waning-flow sequences may be repetitive, and the local occurrence of burrows at their top, as well as the presence of a pelitic band, suggest quiet periods separating major pulses of sediment emplacement on the foreset slope.

Interpretation

The gently inclined gravel topsets are interpreted as beachface deposits, whereas the subhorizontal heterolithic facies is regarded as shoreface gravel and sand, with a bimodal pattern of trough cross-bedding indicating longshore drift. As in the case of the gravel bodies, the geometrical relationships between topsets and foresets suggest alternating phases of active progradation of the system during which the beachface ramp merged into the foreset

Fig. 19. Detail of section shown in Fig. 16. Shovel 107 cm high for scale. See Fig. 16 for location and comments.

Fig. 20. FU sequence showing upward transition from low-angle backset lamination (in pebbly sandstone) to plane lamination, to small-scale cross-lamination. Note preserved ripple profiles with downslope-dipping cross-lamination at top of sequence. Section is oriented NNW–SSE and true dip of foresets is 18° towards 120°. Bar shows scale in decimetres. S. Biagio pit.

slope, and destructional phases during which the progradation stopped and a shoreface platform developed, due to removal of sediments from the upper part of the body. Major erosional surfaces are thought to be generated by shoreface retreat during small relative rises in base level following equally small relative drops.

The relationships between topset and foreset deposits near the northwestern end of the section are particularly complex (Figs 16, 19). Shoreface platform deposits are especially thick in this zone and show a fining-upward trend; in addition, they interfinger with the backset beds of the adjacent foreset unit. These relations suggest that the shoreface deposits were subject to aggradation with slightly deepening trend, and concomitant occasional scouring at the platform edge due to the emplacement of backset beds. During this stage, progradation of the foreset slope was clearly insignificant, and a shoreface platform is thought to have been developing, due to shoreface retreat. At the resumption of progradation, the topset–foreset boundary was initially at a level somewhat higher than that of the upper major erosional surface, and subsequently shifted to progressively lower levels. This surface actually removes an increasingly large portion of the foreset beds landwards (Fig. 16). The erosional event was then followed by a new phase of progradation. An overall transgressive trend is indicated by the stepped rise of the topset/foreset contact (Fig. 16).

The organization of the body closely resembles that of the spit sequence recently described by Nielsen *et al.* (1988). However, the geometry and structures indicate that the progradation of the system was distinctly in a seaward direction; in addition, the persistence of progradation, although in an overall transgressive context, implies high sediment supply feeding an actively growing system. The heterolithic composition of the S. Biagio body may be interpreted either as the result of the overall transgressive trend leading to significant lowering of river gradients, or as the expression of the fact that the sedimentation site was located laterally with respect to a river mouth.

NATURE OF FLOWS DEPOSITING FORESET BEDS

The commonly observed sigmoid geometry of the transition between wave-worked topset beds and foreset beds suggests that the generation of flows active on the foreset slope is often closely related to the storm processes moulding the topset of the system. Storm-driven currents are among the highest-energy flows in nearshore dynamics and are able to carry even coarse sediments into deeper water. In addition, the steep offshore gradient, especially in the case of sigmoid offlap, substantially reduces the width of the zone of shoaling wave effects, so that the impact of incoming storm-generated waves and flows takes place with higher energy, and offshore mass-transport may be considerably more common than in the case of prograding coastal plains (Clifton, 1988). Thus, it may reasonably be thought that periodic storm-driven flows can contribute sediment surges onto foreset slopes. This is confirmed by the common presence of beach-derived morphometric populations of pebbles and good sorting of sands in foreset beds. The envisaged process was probably favoured during constructional phases by the absence of a shoreface platform.

As already outlined, evidence of slope failure and of chute-like features on the delta front is scarce. It may therefore be assumed that, in most cases, a line source for sediment emplacement on the delta front was involved, and that slope failures were only a subordinate mechanism promoting gravity flows.

In the case of foresets overlain by fluvial topsets (northwestern edge of the M.S. Angelo body), the unconformable relation between topsets and foresets prevents direct analysis of the possible flow mechanisms involved. However, as already stated, a continuity between fluvial topsets and foresets (sigmoid geometry) is occasionally observed in some segments of gravel sequences intermediate in character between high-ramp beach sequences and Gilbert-type bodies. Such continuity most probably also existed in that part of the typical Gilbert-type bodies directly influenced by the river mouth(s). In the observed cases, the progressive change in the dip angle of beds at the transition topset–foreset, lack of slump scars, persistence of trough-cross-bedding to the uppermost foreset slope and its change downdip into the massive structure of foreset beds, suggest that, during major floods, a continuity may have existed at stream mouth(s) from heavily-laden traction currents into gravity-driven underflows.

ORIGIN OF BACKSET BEDS

Scour-filling gravel and sand showing backset bedding are described on the foreset and toeset of Gilbert-type bodies by Postma (1984a), Postma *et al.* (1983), Postma and Roep (1985); Colella *et al.* (1987). It is here believed that these structures developed when the transported sediment concentration was very high, and that the genesis of scour-filling backset beds on the foreset slope of Gilbert-type systems may reflect the upstream migration of chutes and pools. Flows generating these bedforms have been observed in flume experiments under high slopes and very violent flow conditions (Simons *et al.*, 1965). Steeply dipping backset beds have been experimentally produced by Jopling and Richardson (1966). Angles of up to 30° have been observed with fine sand, but steepness may depend on grain size (Barwis & Hayes, 1985). In some of Hand's (1974) experiments on density currents, chutes and pools developed at densiometric Froude numbers greater than 1.0, sedimentation being related to upstream-migrating hydraulic jumps. Strong erosion of the bed takes place: this may be explained by highly increased turbulence near the bed, when the front of a breaking wave becomes a submerged hydraulic jump capable of migrating upstream, while sediment accretion occurs in the subcritical region downcurrent of the jump (Hand, 1974).

The observed upward transition of backset beds to layers displaying planar stratification and/or broadly undular bedforms with long trochoidal profiles may involve the conversion of chutes and pools to antidunes and possibly plane beds. This evolution suggests progressive waning of flow energy in time. The presence of sandy bands intercalated with gravelly units in backset beds (Fig. 18) suggests the action of surging flows characterized by a series of pulses in succession, and the dimensions of the structures imply flows of long duration, as well as exceptionally high rates of sediment supply.

The possibility should be considered that these structures were sometimes generated by surging flows related to retrogressive flow slides. The Cabezo unit, described by Postma and Roep (1985) as a thick, trough-filling backset-bedded unit deposited at the toe of the Espiritu Santo Gilbert-type delta, may represent a large-scale example of this type of origin. However, the fact that backset beds locally climb to the edge of the shoreface platform (S. Biagio system, Fig. 16) suggests that the flows generating backset beds may originate in the nearshore area. It may therefore be assumed that backset beds in the foreset of the bodies are preferentially originated by either storm-induced or flood-induced underflows: in both cases, a gravity-driven flow of relatively long duration and probably similar physical behaviour is involved. However, it should be noted that the sets are particularly concentrated in some segments of the foreset and are rare or completely missing in other segments. They seem to be preferentially associated with segments built out during phases of relative highstands of base level, probably characterized by more intense reworking by storm-driven flows. A preferential link of backset beds with these flows is also suggested by the textural characters of sediments and by their local close association with wave-worked shoreface sands and gravels. In addition, very similar structures were described in a spit sequence by Nielsen *et al.* (1988). The scarcity or absence of a mud fraction in the flow is probably an important factor, as inertial stresses must dominate over viscous stresses.

Storm- or (?)flood-induced, offshore-directed highly concentrated flows may undergo acceleration down the steep gradient of the foreset slope and transformation into density underflows in supercritical regime. Chute-and-pool bedforms are thought to be generated, as the flow is subject to hydraulic jumps after strong acceleration under the influence of gravity. The scours produced by high turbulence at hydraulic jumps are progressively extended upslope following upcurrent shifting of hydraulic jumps and are backfilled by progressive upslope migration of sediment deposition (backset beds). Scour excavation and filling must have been almost simultaneous to preserve the steep-sided and locally indented walls. The scour-filling setting of backset beds prevents their erosion by subsequent flows of decreasing energy.

Amalgamation of successive units may remove the upper part of the sequence, but under favourable conditions a complete fining-upward sequence may be preserved, backset beds being followed by structures recording the progressive waning of flow energy, and transition from chute-and-pool flow to antidunes and/or plane bed and then lower flow regime.

It is generally stated that lamination related to supercritical flows are characterized by a faintness due to the absence of grain sorting. However, as stated by Barwis and Hayes (1985), avalanching is unnecessary for grain sorting, as well-defined laminae are commonly deposited by supercritical flows, probably as a result of shear sorting.

The toeset area of the studied systems is dominated by mutually intersecting scours which appear filled in most cases with gravelly or sandy–gravelly backsets. We believe that these ubiquitous scours are the expression of hydraulic jumps, through which flows achieving supercritical conditions down the foreset slope had to pass when they reached the base of the steep slope (Komar, 1971). The concurrent enlargement and dilution of the flow and strong reduction in velocity led to a sudden decrease in competence and the coarsest bed material was dumped as high-angle backset beds immediately downstream from the jump (Jopling & Richardson, 1966). As noted by Mutti and Normark (1987), in the case of highly concentrated flows carrying coarse grain sizes, settling of the coarsest fraction through the flow takes place within the hydraulic-jump zone as flow competence decreases. On the other hand, flows containing a muddy component sufficient to maintain a driving force for the current beyond the hydraulic jump, produce less intense scouring, and in this case most of the sediment generally bypasses the hydraulic-jump zone and is carried farther downcurrent in a flow surge which keeps most of the sediment in motion. The scour-filling coarse-grained backsets in the proximal toeset and the scour-draping sands and muds in the distal toeset and proximal bottomset may respectively represent the expression of the former and latter conditions.

Scours related to hydraulic jumps must be very common at the toe of Gilbert-type systems. Mutually intersecting shallow scours in the toe area of Gilbert-type or slope-apron systems have been described by Stanley and Surdam (1978), Orombelli and Gnaccolini (1978), Postma *et al.* (1983), Surlyk (1984), Postma (1984a), Postma and Roep (1985), Prior and Bornhold (1986), Postma and Cruickshank

(1988). The hypothesis of the occurrence of hydraulic jumps and of backward fill of scours in the slope-break area and just beyond it was clearly outlined by Postma and Roep (1985). The Cabezo unit described by the authors may represent an exceptionally large-scale example.

CONCLUDING REMARKS

The studied systems appear significantly different in behaviour from the fluvial-dominated Gilbert-type and slope-apron systems such as the Arctic examples described by Prior and Bornhold (1988). With respect to them, the studied bodies show: comparative rarity of (1) steep-walled chutes cutting foreset slopes, (2) large-scale soft-sediment deformations, and (3) arcuate slump scars; absence of mud-supported debris flows; importance of wave-working and storm activity in shaping the delta shoreline and promoting gravity flows down foreset slopes. It should be noted in addition that the dispersal of coarse-grained materials is confined to within a relatively short distance from the toe of the system, as indicated by abrupt wedging-out of toeset gravels and rapid transition to sandy/muddy layers. The system was clearly fed by poorly efficient flows.

The above features reflect the activity of a line source rather than a single or multiple point source(s), and the fact that mud-poor sand–gravel mixtures are involved. As a result, gravel flows are characterized by strong frictional forces between the clasts; this leads in turn to decreased mobility of flows, and increased stability of the slope (Colella *et al.*, 1987). In this context, slope failures are relatively rare and involve only small volumes of sediments, which move over short distances and come to rest upon the steep foreset slope or at its toe. The importance of wave reworking in the topset and of storm-induced flows in the sediment dispersal on to the foreset contrasts with the widespread conviction that Gilbert-type progradation generally takes place in low-energy basins. The depositional setting at river mouths in the studied systems may actually have been similar to that of the present-day mouths of the Apennine rivers debouching into the Gulf of Taranto, which are flanked by gravel beaches, and it strongly contrasts with that of most fan deltas, where littoral deposits, if they exist, probably have very little preservation potential due to continuous erosion and reworking by shifting distributaries.

The studied Gilbert-type bodies fill broadly trough-shaped depressions oriented at high angles to the palaeoshoreline and interpreted as drowned palaeovalleys. Entrenchment of these valleys is thought to have occurred as a result of a fall in sea level; the lack of aggrading alluvial bodies suggests a very rapid and major subsequent rise, so that the valleys were quickly drowned and the relatively steep nearshore slopes inherited from the previous cycle favoured the Gilbert-type progradation. These confined bodies may represent constrained lowstand deltas. In the absence of strong tectonic control, rapid rises in sea level leading to drowning of previously incised valleys followed by a stillstand, may be more important than generally believed in the development of Gilbert-type deltas. The Pliocene systems described by Clauzon and Rubino (1988) are other clear examples of the importance of this context in favouring Gilbert-type progradation.

Changes in the geometry and internal organization of the gravel bodies, as recognized in sections parallel and transverse to the direction of progradation, show that a continuum exists from gravel-beach sequences of increasing ramp height to Gilbert-type bodies.

The alternation of sigmoid and oblique offlap indicates that the delta growth was punctuated by progradational, destructional and aggradational phases, related to small fluctuations of relative base level or shifting of the feeding-stream mouth(s). The former cause is favoured in the case of abrupt vertical shifts of the topset–foreset boundary. Although the present data do not allow a choice between tectonic or high-frequency eustatic control, the latter seems to be more probably involved, as short-lived up-and-down movements are highly unlikely in a region which underwent an isostatically-driven uniform uplift during the Pleistocene.

ACKNOWLEDGEMENTS

This work was financially supported by a 40% grant from the Italian Ministry of Education. D. Rio and C. Fioroni are gratefully thanked for the examination of nannofossil assemblages, and G.G. Ori for the critical comments. We are indebted to L.H. Nielsen and G. Ghibaudo, who read the earlier versions of the manuscript and provided useful suggestions which have improved it considerably. In addition, we are grateful to F. Todesco for the careful drawings and to C. Brogiato for precise execution of photographs.

REFERENCES

Barwis, J.N. & Hayes, M.O. (1985) Antidunes on modern and ancient washover fans. *J. sedim. Petrol.* **55**, 907–916.

Brückner, H. (1980) Marine Terrassen in Süditalien. Eine quartärmorphologische Studie über das Kustentiefland von Metapont. *Düsseldorfer Geogr. Schriften* **14**, 222 pp.

Ciaranfi, N., Ghisetti, F., Guida, M., Iaccarino, G., Lambiase, S., Pieri, P., Rapisardi, L., Ricchetti, G., Torre, M., Tortorici, L. & Vezzani, L. (1983) Carta neotettonica dell'Italia meridionale. *Publ. 515 P.F. Geodinamica*, 62 pp.

Clauzon, G. & Rubino, J.L. (1988) Why proximal areas of Meditterranean Pliocene rias are filled by Gilbert deltas? *Abstr., Int. Works. Fan Deltas, Calabria*, 13–14.

Clifton, H.E. (1988) Foreshore and nearshore facies in prograding shoreline deposits. Theme and variations with special consideration of fan-delta systems. *Abstr., Int. Works. Fan Deltas, Calabria*, 15–16.

Colella, A. (1984) Marine Gilbert-type deltas in Lower(?) Pleistocene deposits of Crati valley (Calabria, southern Italy); a preliminary note. *Abstr., Int. Ass. Sediment., 5th Eur. Reg. Meeting, Marseille*, 112–113.

Colella, A. (1987) 'Fluvial-dominated' and 'wave-influenced' Gilbert-type fan deltas in the extensional Crati Basin (Pleistocene, southern Calabria). *Abstr., Int. Assoc. Sediment., 8th Reg. Meeting, Tunis*, 524–525.

Colella, A. (1988) Pliocene–Holocene fan deltas and braid deltas in the Crati Basin, southern Italy: a consequence of varying tectonic conditions. In: *Fan Deltas: Sedimentology and Tectonic Settings* (Ed. by W. Nemec and R.J. Steel), pp. 50–74, Blackie and Son, London.

Colella, A., De Boer, P.L. & Nio, S.D. (1987) Sedimentology of a marine intermontane Pleistocene Gilbert-type fan-delta complex in the Crati Basin, Calabria, southern Italy. *Sedimentology* **34**, 721–736.

Corselli, C., Gnaccolini, M. & Orombelli, G. (1985) Depositi deltizi pliocenici allo sbocco della Val Brembana (Prealpi Bergamasche). *Riv. Ital. Paleont. Stratigr.* **91**, 117–132.

Hand, B.H. (1974) Supercritical flow in density currents. *J. sedim. Petrol.* **44**, 637–648.

Hiscott, R.N. & Middleton, G.V. (1979) Depositional mechanics of thick-bedded sandstones at the base of a submarine slope, Tourelle Formation (Lower Ordoviciian), Quebec, Canada. Spec. Publ. Soc. econ. Paleont. Mineral., Tulsa, **27**, 307–326.

Jopling, A.V. & Richardson, E.V. (1966) Backset bedding developed in shooting flow in laboratory experiments. *J. sedim. Petrol.* **36**, 821–825.

Komar, P.D. (1971) Hydraulic jumps in turbidity currents. *Bull. geol. Soc. Amer.* **82**, 1477–1488.

Massari, F. & Parea, G.C. (1988) Progradational gravel beach sequences in a moderate- to high-energy, microtidal marine environment. *Sedimentology* **35**, 881–913.

Mutti, E. & Normark, W.R. (1987) Comparing examples of modern and ancient turbidite systems: problems and concepts. In: *Marine clastic sedimentology* (Ed. by J.K. Leggett and G.G. Zuffa), pp. 1–38, Graham and Trotman, Oxford.

Nielsen, L.H., Johannessen, P.N. & Surlyk, F. (1988) A late Pleistocene coarse-grained spit-platform sequence in northern Jylland, Denmark. *Sedimentology* **35**, 915–937.

Orombelli, G. & Gnaccolini, M. (1978) Sedimentation in proglacial lakes: a Würmian example from the Italian Alps. *Z. Geomorph. N.F.* **22**, 417–425.

Postma, G. (1984a) Mass-flow conglomerates in a submarine canyon: Abrioja fan-delta, Pliocene, southeast Spain. In: *Sedimentology of Gravels and Conglomerates* (Ed. by E.H. Koster and R.J. Steel). Mem. Can. Soc. Petrol. Geol. 10, pp. 237–258.

Postma, G. (1984b) Slumps and their deposits in fan-delta front and slope. *Geology* **12**, 27–30.

Postma, G. & Cruickshank, C. (1988) Sedimentology of a Late Weichselian to Holocene terraced fan delta, Varangerfjord, northern Norway. In: *Fan Deltas: Sedimentology and Tectonic Settings* (Ed. by W. Nemec and R.J. Steel), pp. 144–157, Blackie and Son, London.

Postma, G. & Roep, T.B. (1985) Resedimented conglomerates in the bottomsets of Gilbert-type gravel deltas. *J. sedim. Petrol.* **55**, 874–885.

Postma, G., Roep, T.B. & Ruegg, G.H.J. (1983) Sandy–gravelly mass-flow deposits in an ice-marginal lake (Saalian, Leuvenumsche Beek Valley, Veluwe, The Netherlands) with emphasis on plug-flow deposits. *Sedim. Geol.* **34**, 59–82.

Prior, D.B. & Bornhold, B.D. (1986) Sediment transport on subaqueous fan-delta slopes, Brittania beach, British Columbia. *Geo-Mar. Lett.* **5**, 217–224.

Prior, D.B. & Bornhold, B.D. (1988) Submarine morphology and processes of fjord fan deltas and related high-gradient systems: modern examples from British Columbia. In: *Fan Deltas: Sedimentology and Tectonic Settings* (Ed. by W. Nemec and R.J. Steel), pp. 125–143, Blackie and Son, London.

Prior, D.B., Wiseman, W.J. Jr. & Bryant, W.R. (1981) Submarine chutes on the slopes of fjord deltas. *Nature* **290**, 326–328.

Rio, D., Raffi, I. & Villa, G. (1990) Calcareous nannofossil quantitative distribution patterns during Pliocene and Pleistocene in the western Mediterranean (ODP site 107-653-Tyrrhenian Sea). *Initial Reports Deep Sea Drilling Project*, Washington (U.S. Government Printing Office).

Sella, M., Turci, C. & Riva, A. (1988) Petroleum geology of the 'Fossa Bradanica' (foredeep of the southern Apennine thrust belt). *Atti 74° Congr. Soc. Geol. Ital., Sorrento, Prestampe*, 49–57.

Simons, D.B., Richardson, E.V. & Nordin, C.F. Jr. (1965) Sedimentary structures generated by flow in alluvial channels. In: *Primary Sedimentary Structures and their Hydrodynamic Interpretation* (Ed. by G.V. Middleton). Spec. Publ. Soc. econ. Paleont. Mineral., Tulsa, **12**, 34–52.

Stanley, K.O. & Surdam, R.C. (1978) Sedimentation on the front of Eocene Gilbert-type deltas, Washakie Basin, Wyoming. *J. sedim. Petrol.* **48**, 557–573.

Surlyk, F. (1984) Fan-delta to submarine fan conglomerates of the Volgian–Valanginian Wollaston Forland Group, east Greenland. In: *Sedimentology of Gravels and Conglomerates* (Ed. by E.H. Koster and R.J. Steel). Mem. Can. Soc. Petrol. Geol., 10, pp. 359–382.

Non-alluvial Deltas

Spec. Publs int. Ass. Sediment. (1990) **10**, 335–351

Lava-fed Gilbert-type delta in the Polonez Cove Formation (Lower Oligocene), King George Island, West Antarctica

S. J. PORĘBSKI *and* R. GRADZIŃSKI

Instytut Nauk Geologicznych, Polska Akademia Nauk, Senacka 3, 31-002 Kràkow, Poland

ABSTRACT

The Oberek hydroclastic delta of the Polonez Cove Formation (Lower Oligocene, King George Island, South Shetland Islands) originated through coalescence of two, Gilbert-type bodies, more than 600 m in radius each, that prograded on to a *c*. 20 m deep marine storm-dominated shoreface. A deltaic body begins with a basal unit (3.5 m thick) of flat-bedded, petromict volcaniclastic breccia overlain by a foreset-bedded unit (16–23 m) of basaltic breccia and hyaloclastite (angle of dip up to 35°), in turn, overlain by a topset (min. 5 m) of sheet-like lava-breccia. This stratigraphy is interpreted to reflect initiation of delta growth by laharic debris flows, followed immediately by incursions of basaltic lava, that on passing through the shoreline, produced a subaqueous hydroclastic prism and a subaerial lava platform on top of the prism.

The foreset unit consists of six lithofacies: (A) ill-bedded cobble breccia; (B) bedded pebble breccia; (C) lag breccia; (D) interbedded breccia and sandy hyaloclastite; (E) ill-stratified, bimodal hyaloclastite; and (F) graded-laminated hyaloclastite. The lithofacies are arranged to form three major segments of the foreset unit: inner, indistinct planar (facies A); medial, planar (facies B, C); and outer, tangential having upper planar ends (facies E) and wide, low-angle toes (B, C) passing downwards into well-developed bottomset beds (B, D, F). This architecture records accretion of the delta slope through mass-gravity processes, changing in importance with continued delta growth from grain free-falling and avalanching (inner, indistinct planar segment) to slump-generated fluidal sediment gravity flows (outer tangential segment). Such change is considered to reflect primarily the increasing grain-size range of hydroclastic detritus made available for remobilization, in response to changing feeding-lava behaviour towards lower viscosities and smaller, possibly surging discharges.

INTRODUCTION

Since work by Fuller (1931), many authors have inferred that flows of basaltic lava from land into a body of standing water become fragmented and may produce deltaic bodies showing an internal stratigraphy analogous to that of alluvial Gilbert-type deltas (Waters, 1960; Nelson, 1966; Swanson, 1967; Jones & Nelson, 1970; Snavely *et al.*, 1973; see also examples described by Holmes, 1978). This view has been substantiated by J.G. Moore and his diving team who observed the growth of such a lava-fed delta generated during the 1971 eruption of Kilauea (Moore *et al.*, 1973).

The available descriptions of Gilbert-type hydroclastic deltas provide, however, little information as to their actual growth styles, range of processes operating on a delta face and the resulting depositional products, as well as to an overall facies context of such deltas. These issues are addressed throughout in this paper, in which we discuss the structure and evolution of a marine hydroclastic delta from the lower Oligocene Polonez Cove Formation on King George Island, South Shetland Islands. This study demonstrates some major similarities between the depositional styles of hydroclastic deltas and coarse-grained alluvial deltas, and emphasizes the dominant role of mass-gravity transport in the growth of such Gilbert-type systems, whether fed by lava streams or fluvial discharges.

HYDROCLASTIC DELTAS – BACKGROUND

A land-derived lava flow crossing the shoreline undergoes quenching and hydroclastic fragmentation that ranges from non-explosive granulation to explosive comminution (Fisher & Schmincke, 1984; see also Kokelaar, 1986), depending mainly on the viscosity and flow rate of the lava and the wave/tidal energy of the receiving basin. Consequently, the hydroclastic components may vary greatly in proportions, encompassing shattered lava masses, pillowed and angular litho- and hyaloclasts and glassy sand. These components, admixed with epiclastic detritus, tend to accumulate as a steep-flanked talus that progrades offshore and forms a foreset-bedded body capped by a platform of subaerially frozen lava (Moore *et al.*, 1973).

As in alluvial Gilbert-type systems, the contact between the breccia foreset and sheet-lava topset marks the position of water level in the host basin. This contact may record subtle fluctuations in the water level, such as tidal range (Furnes & Fridleifsson, 1974), whereas in vertically stacked hydroclastic delta bodies, it may reflect major changes of base level, due to tectonic or eustatic causes (Jones & Nelson, 1970).

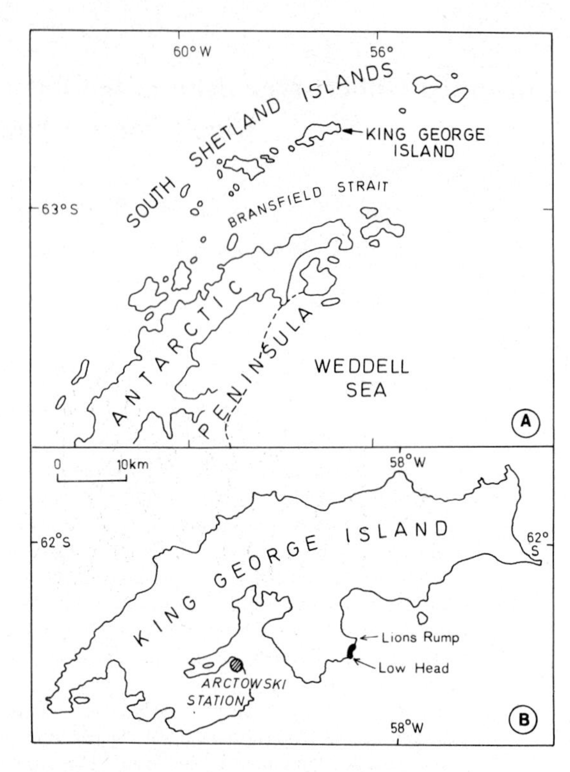

Fig. 1. Index map showing the location of study area (shaded in Map B).

THE POLONEZ COVE FORMATION

The Polonez Cove Formation (Birkenmajer, 1982) is one of the few, chiefly sedimentary intercalations within the Upper Cretaceous–Tertiary volcanic–volcaniclastic succession of King George Island – the largest segment in the South Shetland Island arc (Fig. 1A). The arc records mainly calc-alkaline volcanism related to the oceanic subduction along the Pacific margin of the Antarctic Peninsula (Weaver *et al.*, 1982).

In its type area (Fig. 1B), the Polonez Cove Formation is more than 90 m thick and crops out from beneath the insular ice cap in a steep coastal cliff, *c.* 3 km long and over 200 m high (Fig. 2). The formation shows erosive contacts with underlying, upper Cretaceous basalts (Mazurek Point Formation) and with overlying acidic volcanic and volcaniclastic rocks (Boy Point Formation), dated at more than 24 Ma (Birkenmajer *et al.*, 1986).

The Polonez Cove Formation commences with a sheared diamictite unit (Krakowiak Glacier Member in Fig. 2), rich in exotic clasts, interpreted as the subglacial deposits of a grounded (marine-based) continental ice sheet (Birkenmajer, 1982; Porębski & Gradziński, 1987). The diamictites are overlain by a shallow-marine, fossiliferous, volcaniclastic assemblage of mainly conglomerates and sandstones, which is inserted at two stratigraphic levels by Gilbert-type deltaic prisms. The latter are composed of basaltic breccia and hyaloclastite and show foreset heights up to 25 m. The better-exposed upper prism, referred to as the Oberek delta (ODS in Fig. 2), will be the main objective of this study.

The formation is faulted in places (offsets up to 25 m) and cut by plugs and dykes of which the largest (Chopin Dyke) yielded a K–Ar age of emplacement at *c.* 22 Ma (Fig. 2). Coccolith assemblages (Gazdzicka & Gazdzicki, 1985) and radiometric K–Ar dates set the upper age-limit of the Polonez Cove Formation at about 30 Ma (Birkenmajer, 1988). Details of lithostratigraphy, palaeontology and sedimentology are given by Birkenmajer (1982), Gazdzicki and Pugaczewska (1984) and Porębski and Gradziński (1987).

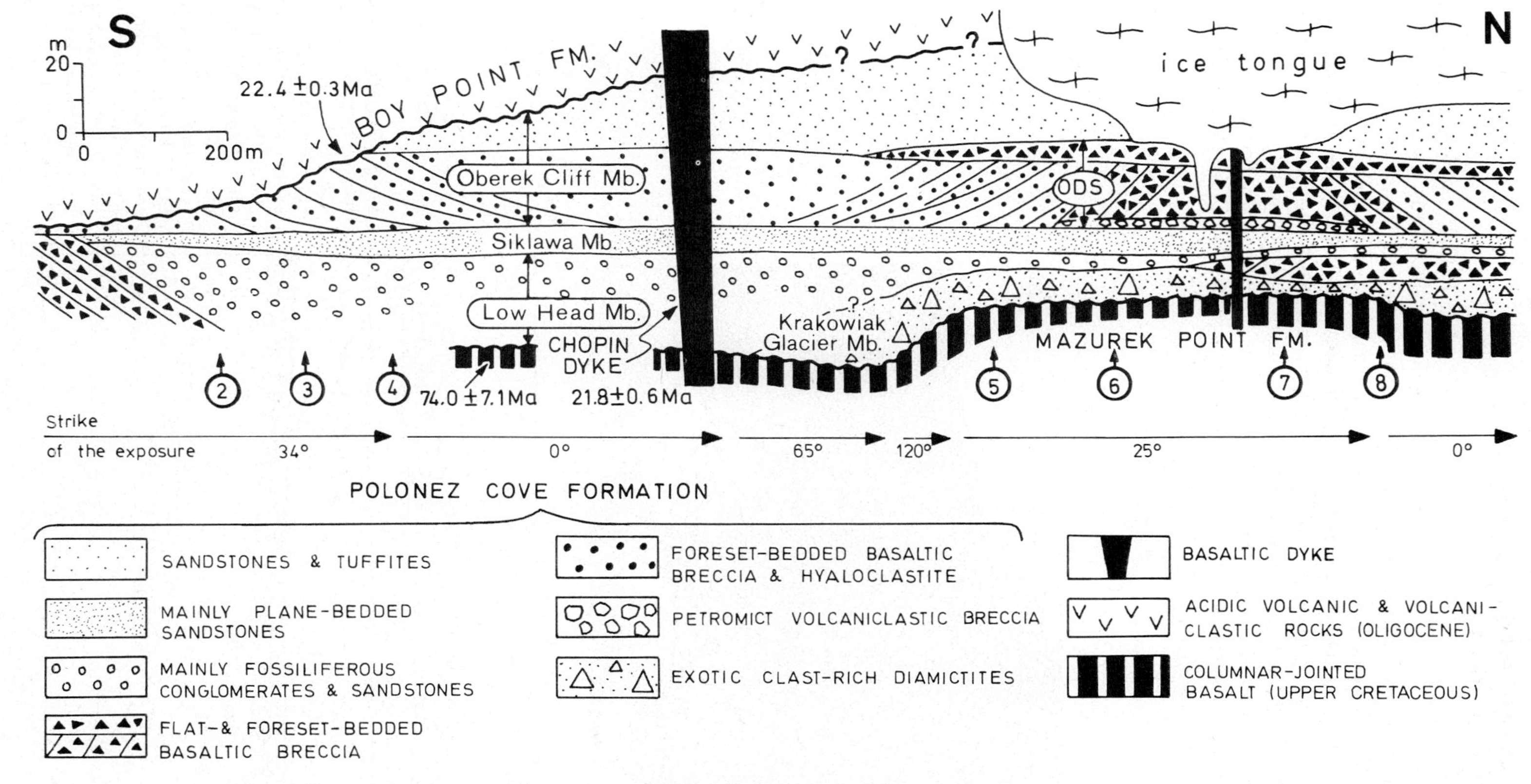

Fig. 2. Cross-section through coastal cliff at the type locality of the Polonez Cove Formation, showing the stratigraphic framework and geometry of main rock units, relevant K–Ar ages of volcanic rocks (after Birkenmajer *et al.*, 1986), and location of logged sections (encircled numbers). Lithostratigraphic names after Birkenmajer (1982). ODS–Oberek delta system.

THE HOST ENVIRONMENT OF THE OBEREK DELTA

The Siklawa Member, which forms the immediate substratum to the Oberek delta, is a tabular well-bedded unit (Figs 2, 3), 4–10 m thick, composed of plane-bedded sandstone (50% of the aggregate thickness of all sections measured) intercalated with sharp-based pebbly sandstones (25%), granule conglomerates (20%) and mudstones (5%).

Lithofacies

The *plane-bedded sandstones* occur in fine- to medium-grained beds, 1–70 cm thick, which may amalgamate locally into bedsets, up to 2 m thick, intercalated with thin (below 1 cm) and discontinuous mud drapes. The beds are flat, or rarely ripple cross-laminated. Two bedding planes carrying symmetrical ripples were noted. In places, the beds reveal scattered outsized clasts, up to cobble size, of both local and exotic provenance. The local volcanic

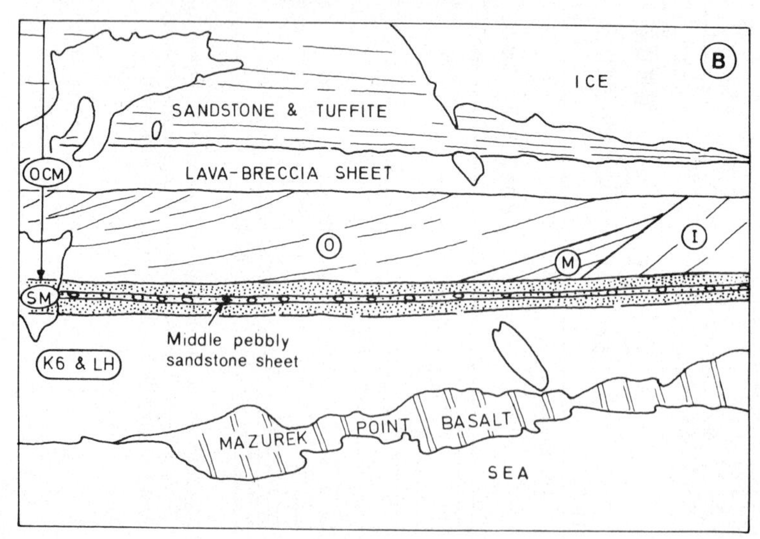

Fig. 3. (A) Oblique air-photo of cliff face at Mazurek Point (stations 5 to 6 in Fig. 2), showing the northern body of the Oberek delta and its surroundings. Width of photo is *c.* 300 m. (B) Redrawn from (A) to show stratigraphic units referred in text: KG and LH — diamictites, conglomerates and sandstones of the Krakowiak Glacier and Low Head members (undifferentiated); SM — Siklawa Member; OCM — Oberek Cliff Member; I, M and O inner, medial and outer segments of the foreset unit.

basement is represented by abundant basaltic debris; exotic detritus includes mainly crystalline and sedimentary-clast types, some of which can be matched with distant sources on the Antarctic Peninsula and the inner areas of the Weddell Sea (Birkenmajer & Wieser, 1985).

Sharp-based pebbly sandstone beds tend to be amalgamated into persistent sheets (Figs 3, 9), up to 150 cm thick and 600 m (possibly up to 1.5 km) in lateral extent. The beds show disrupted textures with pebble- to cobble-size, volcanic and exotic clasts, angular mudstone rafts, diamict clots and articulated bivalve shells, all dispersed throughout an unsorted sandy groundmass, which also bears recognizable admixtures of 'fresh' pyroclasts. Inverse grading and a crude flat lamination are present locally. Some of the beds reveal tops moulded into large, near-symmetrical bedforms, up to 0.3 m high and 4–7 m in wavelength. These are either internally structureless, or show rare traces of low-angle lamination dipping in various directions.

The *granule conglomerates* tend to occur in thin beds and convex-up lenses, up to 60 cm high, which appear to be sections through large ripples. Some lenses show cross-stratification indicating palaeoflow towards the NW. The *mudstone* forms impersistent beds, up to 56 cm thick, which commonly show erosive tops and bear densely clustered burrows, each having inside the pholadid bivalve *Penitella* (Gazdzicki *et al.*, 1982). Rare benthic foraminifers were noted in some mudstone beds.

Interpretation

The lithofacies assemblage of the Siklawa Member provides evidence of dominantly sheet-flow conditions (plane-bedded sandstones), rapid deposition from catastrophic flows driven by a combined density-hydraulic mechanism (sharp-based pebbly sandstones) and megaripple-phase reworking of the seabed (granule conglomerates). These high-energy events alternated with periods of slow suspension settling of fines, possibly dropstone fallout from floating ice, and the colonization of firm bottom muds by opportunistic bivalve burrowers (*Penitella*-bearing mudstones).

This facies assemblage is therefore interpreted to record mainly storm sedimentation on a lower shoreface. The altitude of the topset/foreset contact in the overlying Gilbert-type bodies indicates water depth in the range of 16–23 m. Regional data discussed elsewhere (Porębski & Gradziński, 1987) suggest that this environment abutted eastwards against a rugged, rocky volcanic coast dominated by high-energy wave processes.

SEDIMENTOLOGY OF THE OBEREK DELTA

The Siklawa Member is abruptly overlain by the rudaceous Oberek Cliff Member, whose lower part comprises the studied Gilbert-type delta (Figs 2, 3).

As seen in the N-trending cliff wall, the Oberek delta system consists of two Gilbert-type bodies that face westwards and coalesce north of the Chopin Dyke (Fig. 4; zone of coalescence either inaccessible or covered by scree). The *northern body* is more than 600 m in radius, has foresets up to 23 m in height, and reveals a tripartite stratigraphy. It commences with a basal, flat-bedded lense of petromict volcaniclastic breccia, overlain by a foreset-bedded unit of basaltic breccia and hyaloclastite, in turn overlain by a topset of sheet-like lava-breccia. The *southern body* is more than 800 m in radius, up to 22 m in height, and shows only a foreset unit preserved. This body is truncated at the top by a subhorizontal erosion surface which, in turn, is followed upwards by a thick, probably marine, sandstone–tuffite sequence (Figs 2, 3), which crops out in the least accessible cliff wall and thus was not studied in detail.

Basal petromict-breccia unit

The basal part of the northern deltaic body is a flat-bedded lenticular unit, up to 3.5 m thick and about 500 m wide, whose beds are composed of petromict volcaniclastic breccia and intercalated with thin coarse sandstones (Fig. 5). The breccia beds show a highly disrupted, ill-sorted framework of pebbles and cobbles and an abundant tuffaceous matrix of coarse to granule sandstone (Fig. 6). Clasts, up to 70 cm in diameter, are mostly angular to subrounded fragments of basalt, andesite and red tuff. Bed thicknesses range between 50 cm and 130 cm and show a significant positive correlation with maximum particle size (mean of ten largest clasts locally per bed; Fig. 7). The beds are either disorganized, or exhibit a coarse inverse to normal grading and a crude flat stratification at top (Figs 5, 6).

Towards the south, over a distance of *c.* 250 m, the basal breccia unit shows: (1) a 1.5 times decrease in the bed thicknesses; (2) a 2.5-times decrease in

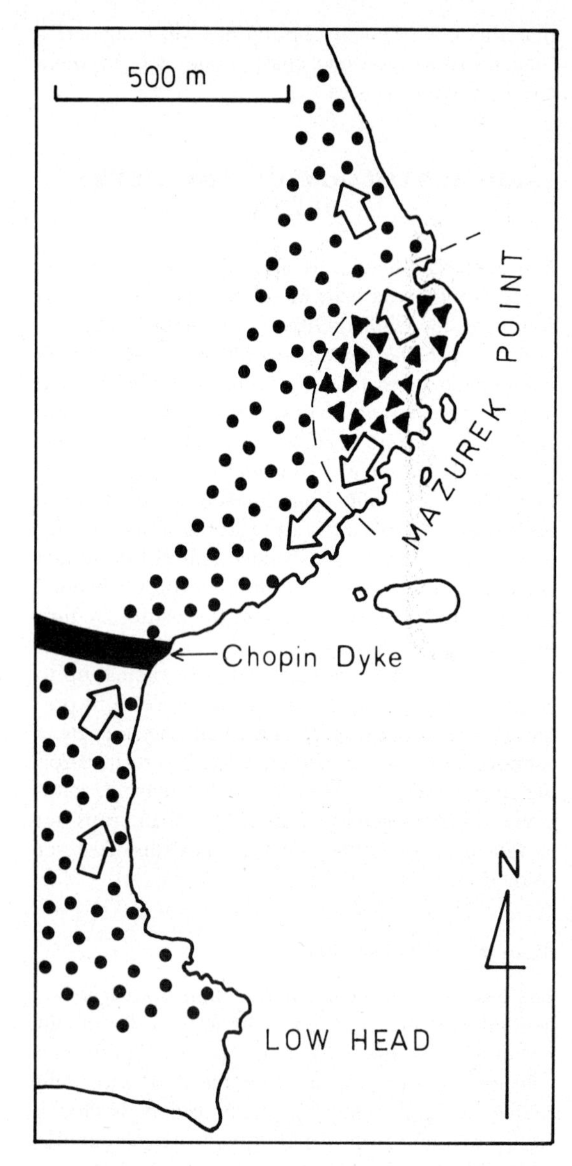

Fig. 4. Map of the studied cliff section, showing distribution and foreset-dip azimuths of the foreset-bedded basaltic breccia (triangles) and breccia/hyaloclastite (dots).

the maximum particle sizes; (3) a 2.5-times decrease in the thickness of the sandstone interbeds; and also (4) a tendency for better development of clast-size grading in the breccia beds (Fig. 5).

Foreset unit

Composition

The bulk of the foreset-bedded deposits comprises granule-to-cobble size, litho- and hyaloclasts of dark-grey, aphanitic olivine basalt, and a glassy sand now strongly altered to palagonite. Most of the clasts are subequant in shape, angular to subrounded and reveal glassy rims; rare half-rounds appear to be fragments of broken pillows. Basaltic hornblende and plagioclase crystaloclasts and sideromelane shards, are locally recognizable in fine-grained matrix. However, they rarely escaped an extreme halmyrolitic alteration, whose products (including palagonite, smectite–illite, zeolites, chalcedon and calcite), impregnate both the matrix and the larger litho- and hyaloclasts. This heavily blurs the original grain-to-grain and framework-to-matrix relationships, gives a false impression of originally high lutite content in the rock and causes severe difficulties in recognition of primary textures and structures.

Non-volcanic constituents occur in trace amounts and include strained quartz grains, mudstone and sandstone intraclasts, as well as single rounded clasts of crystalline and sedimentary rocks whose range corresponds to the debris of the basal, glacial diamictites.

Lithofacies

Six lithofacies were distinguished within the foreset and bottomset deposits of the Oberek delta (Fig. 8).

(A) Ill-bedded cobble breccia. This facies forms the core part of the northern foreset-bedded body. Beds are up to several metres thick and are internally unstratified. They display tightly fitted frameworks of massive to slightly vesicular basaltic clasts, 6–8 cm in diameter, and a minor sand-sized matrix (Fig. 6). This clast-supported breccia grades locally into large, irregular masses of crackle or 'jigsaw-puzzle' breccia and contains scarcely dispersed lumps, up to 1 m in size, of a heavily fractured basalt.

(B) Bedded pebble breccia. This facies comprises wedge-shaped beds, up to 200 cm thick, which are either amalgamated and barely demarcated by discontinuous, planar to convex-up surfaces, or separated by gravel lags and thin, sandy hyaloclastite

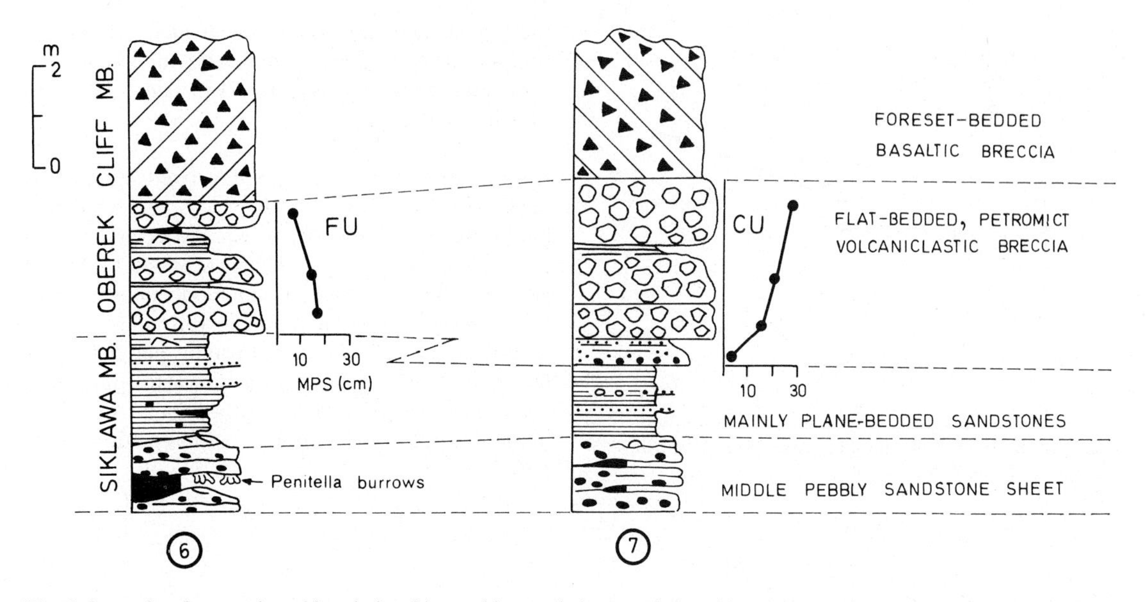

Fig. 5. Logs showing stratigraphic relationships and internal characteristics of basal, petromict breccia unit. See Fig. 2 for location of the sections measured.

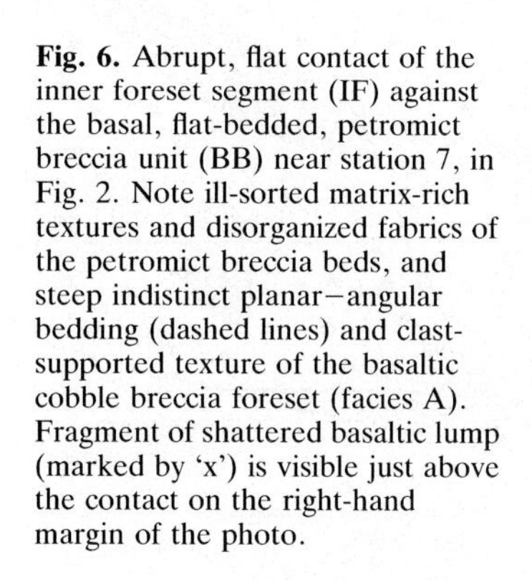

Fig. 6. Abrupt, flat contact of the inner foreset segment (IF) against the basal, flat-bedded, petromict breccia unit (BB) near station 7, in Fig. 2. Note ill-sorted matrix-rich textures and disorganized fabrics of the petromict breccia beds, and steep indistinct planar–angular bedding (dashed lines) and clast-supported texture of the basaltic cobble breccia foreset (facies A). Fragment of shattered basaltic lump (marked by 'x') is visible just above the contact on the right-hand margin of the photo.

lenses. The beds display a clast-supported, ill-sorted framework of pebbles, 1–3 cm in diameter, with rare scattered cobbles (60 cm in one bed) and a minor, lithic to glassy sand matrix (Fig. 9). Larger clasts reveal a strong, internal fracturing and commonly have indistinct outlines rimmed by a crushed aggregate. The beds tend to be internally massive, and ungraded. Some exhibit a thin basal zone of inverse grading and slightly sand-enriched tops.

(C) Lag breccia. This facies is a lag-type deposit composed of clasts up to cobble size, that occur in one-clast-thick layers and matrix-free lenticular concentrates. Despite common lateral pinch-out, these layers tend to occur along persistent levels which mark best the foreset bedding (Fig. 10). Some of

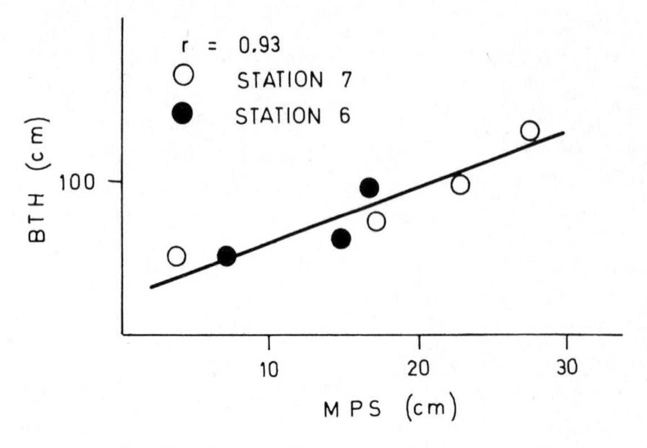

Fig. 7. Plot showing positive correlation between bed thickness (BTH) and maximum particle size (MPS) in the basal petromict breccia unit.

the lenticular concentrates appear as nearly *in situ* brecciated lava tongues, in which the individual rock fragments were only slightly displaced relative to one another (Fig. 9).

(D) Interbedded breccia and sandy hyaloclastite. This facies consists of interbedded clast-supported fine pebble-breccia (beds 10–40 cm thick) and coarse to medium sand-sized hyaloclastite (beds 1–20 cm) (Fig. 11). The breccia beds commonly show coarse-tail normal grading in upper levels and a crude stratification, delineated by bedding parallel alignments of larger clasts. The hyaloclastite interbeds are either structureless or faintly parallel stratified; neither ripple forms nor cross-lamination were noted.

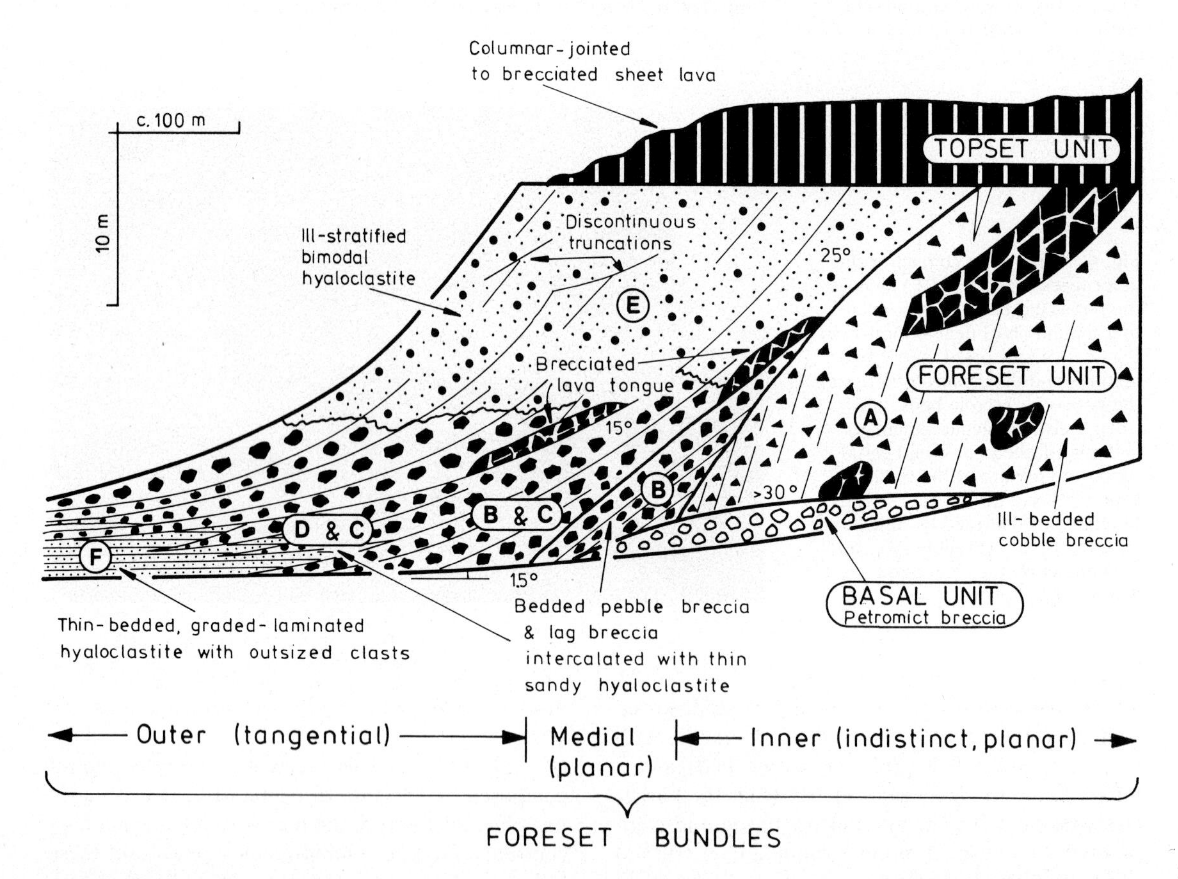

Fig. 8. Composite cross-section through Oberek hydroclastic delta body, showing general stratigraphy, foreset-bed geometry and lithofacies distribution.

Fig. 9. Close-up view of the bedded pebble breccia (facies B), showing clast-supported ill-sorted texture and pillowy fractured nature of some larger clasts (marked by 'x'). Thin matrix-free breccia intercalation (facies C) is slightly fragmented lava tongue. B and C — lithofacies codes as in text.

Fig. 10. View of a cliff wall made up of tangential foreset beds (outer foreset segment north of station 5 in Fig. 2). Width of photo is *c*. 40 m. Note alternation of bedded pebble breccia with lag breccia (facies B and C) in the lower portions of the foreset and the crudely bedded nature of its upper parts which consist of ill-stratified pebble hyaloclastite (facies E). Dashed line marks the base of the foreset which is underlain by the marine sediments of the Siklawa Member — prominent bench (arrowed) is the middle amalgamated pebbly sandstone marker indicated in Figs 3, 5. B, C and E — lithofacies codes.

(E) Ill-stratified pebble hyaloclastite. This facies comprises poorly sorted, highly bimodal beds composed of an abundant, strongly palagonitized granule–sand groundmass which bears variable admixtures of scattered coarser clasts (less than 5%). Beds are up to 150 cm thick and show common signs of amalgamation. Crude parallel stratification is locally discernible.

(F) Graded-laminated hyaloclastite. The facies comprises beds, 1–15 cm thick, of well-sorted medium- to fine-grained hyaloclastite (Fig. 12). The beds commonly have granule-enriched bases, overlain by faintly parallel-laminated, rarely cross-laminated sandy parts which, in turn, may be capped by millimetre thick mud drapes. Outsized clasts, including rare exotic components, are scatterd locally along the bedding planes.

Types of foreset bundles

The lithofacies described are assembled into three

Fig. 11. Proximal bottomset beds composed of thick pebble breccia bed (facies B) followed upwards by interbedded breccia and sandy hyaloclastite (facies D), all resting on the plane-bedded sandstones of the Siklawa Member (SM) along a sharp slightly wavy contact (dashed line) (near Station 3 in Fig. 2). Note thinning/fining-upward trends in the bottomset. B and D — lithofacies code.

Fig. 12. Thin-bedded graded-laminated hyaloclastite (facies F), followed upwards by amalgamated pebble-breccia beds (facies B) in a toeset/bottomset transition (station 4 in Fig. 2) B and F — lithofacies codes.

types of foreset bundles, differing in geometry and facies suite (Fig. 8): 1. indistinct planar (facies A); 2. planar (B, C); and 3. tangential (E, B, C), passing gradually into a well-developed bottomset (B, D, F). All three types can be recognized in the northern deltaic body, whereas the southern body is made up of the tangential type only.

Northern deltaic body

In the northern deltaic body the three bundle types occur in an offlapping succession from type (1) to (3), representing three successive increments of the delta growth, each different from the preceding one (Figs 3, 8). Accordingly, they are referred to below as the inner, medial and outer segments of the foreset unit.

Inner foreset segment. The preserved interior of the northern deltaic body appears as a fan-shaped west-facing prism composed of the ill-bedded cobble breccia (facies A). The prism is 16–20 m high and at least 220 m in radius, has a flat base and sharply defined flanks (Figs 3, 4, 8). The south flank dips at 25° towards 250° and truncates the underlying foreset beds inclined at 35–30° towards the WSW. The north flank exhibits an apparent angle of dip 12° towards the NW. The core of the prism appears nearly unbedded (Fig. 6) and the foreset bedding of a discontinuous-planar geometry becomes better discernible and more densely spaced towards the prism's flanks.

Medial foreset segment. The southern flank of the inner breccia prism is uplapped by planar foreset beds showing angular lower contacts (Fig. 8). This foreset-bed packet is *c.* 70 m wide along its base and shows increasing inclination from 15 to 25° outwards. It consists of a bedded pebble breccia (facies B) which becomes intercalated with lag-breccia horizons (facies C) towards lower reaches of the foresets.

Outer foreset segment. The medial segment is conformably enveloped, towards the W by a tangential foreset bundle, more than 500 m wide at the base (Figs 3, 8). The upper planar parts of the bundle dip at up to 25° and are built of the ill-stratified pebble hyaloclastite (facies F; Fig. 10). Bedding is poorly marked, although discontinuous truncations can be discernible (Fig. 13). This grades downdip into well-bedded toeset reaches composed of the bedded pebble breccia intercalated with breccia-lag horizons (facies B, C, Fig. 10). Oblique air photos suggest the presence of thick coarse-grained deposits within well-developed bottomset beds (Fig. 13); these, however, were not examined due to unfavourable outcrop conditions.

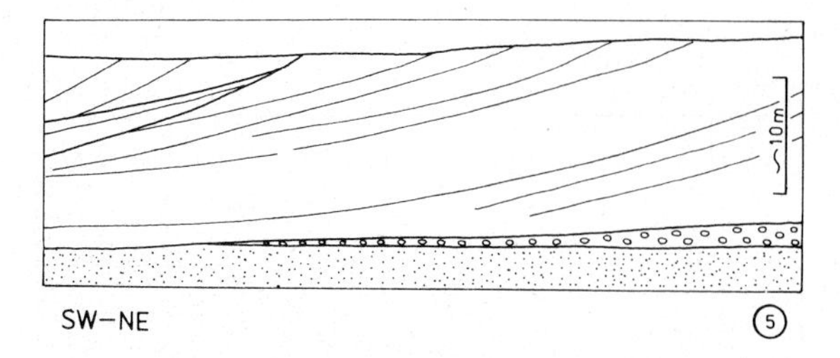

Fig. 13. Cliff wall sketched from perpendicular air-photo, showing the geometry of tangential foreset beds (outer foreset unit southwest of station 5 in Fig. 2). Note discontinuous truncations in upper planar foreset segments.

Southern deltaic body

Along its entire exposed width (800 m), the southern deltaic body reveals tangential foreset beds (Fig. 14). Their upper, relatively steep (up to 20°) segments are planar and composed of ill-stratified, bimodal hyaloclastite beds with rare intercalations of bedded pebble breccia (facies E, B), whereas the low-angle (less than 5°) toeset segments consist of amalgamated pebble-breccia beds. Further downdip, over a distance of 100 m, these latter become thinner and finer grading into the subhorizontal bottomset composed of the interbedded breccia and sandy hyaloclastite with rare lag-breccia intercalations (facies D, C). Further basinwards, the breccia beds tend to wedge out within the thin-bedded, graded-laminated hyaloclastites (facies F; Figs 12, 14). These hyaloclastic bottomsets share the basaltic composition and the high degree of halmyrolitic alteration with the underlying marine plane-bedded sandstones, but differ from the latter in having no thicker mudstone interbeds.

In these tangential foreset beds the coarsest debris tends to be concentrated in their low-angle toe segments. Upslope, there is a marked decrease in the mean grain size, sorting and bedding definition, whereas downslope there is a progressive tendency for sorting of gravel and sand into distinct beds, which goes hand in hand with the gradual development of internal grading and lamination. The toe- and proximal bottomsets display a crude fining-upwards cyclicity of an order of 3–5 m (Figs 11, 14). The cycles lose their identity up slope, where their boundaries probably merge with the discontinuous truncations locally discernible in the upper planar segments of the foreset beds (Figs 8, 13).

Topset unit

As revealed by helicopter photos, the northern deltaic body is capped by a flat-lying to gently westward-inclined lava-breccia unit which is more than 5 m thick and not less than 500 m in lateral extent (Figs 2, 8). In its northernmost outcrops the unit reveals pillow-shaped to strongly shatter-cracked masses of massive to weakly vesicular basalt, set in a sparse lithic matrix (Fig. 15).

The base of the topset unit, traced laterally southwards, becomes convex-down defining a local scoop-shaped depression, 80 m wide and up to 10 m deep, which cuts down into the inner foreset unit (Fig. 16). The depression has a west-dipping axis and is filled by a columnar-jointed to brecciated basalt. K. Birkenmajer (pers. comm., 1988) observed columnar-jointed to shattered sheets and fingers that detach from this lava-breccia topset downwards into the subjacent breccia foreset. Farther southwards, the lava-breccia unit wedges out and probably merges into a flat truncation that bounds at top the southern deltaic body (Fig. 2).

INTERPRETATION OF THE OBEREK DELTA

Origin of the delta bodies

The hydroclastic nature, Gilbert-type geometry and the stratigraphic context of the Oberek foreset bodies

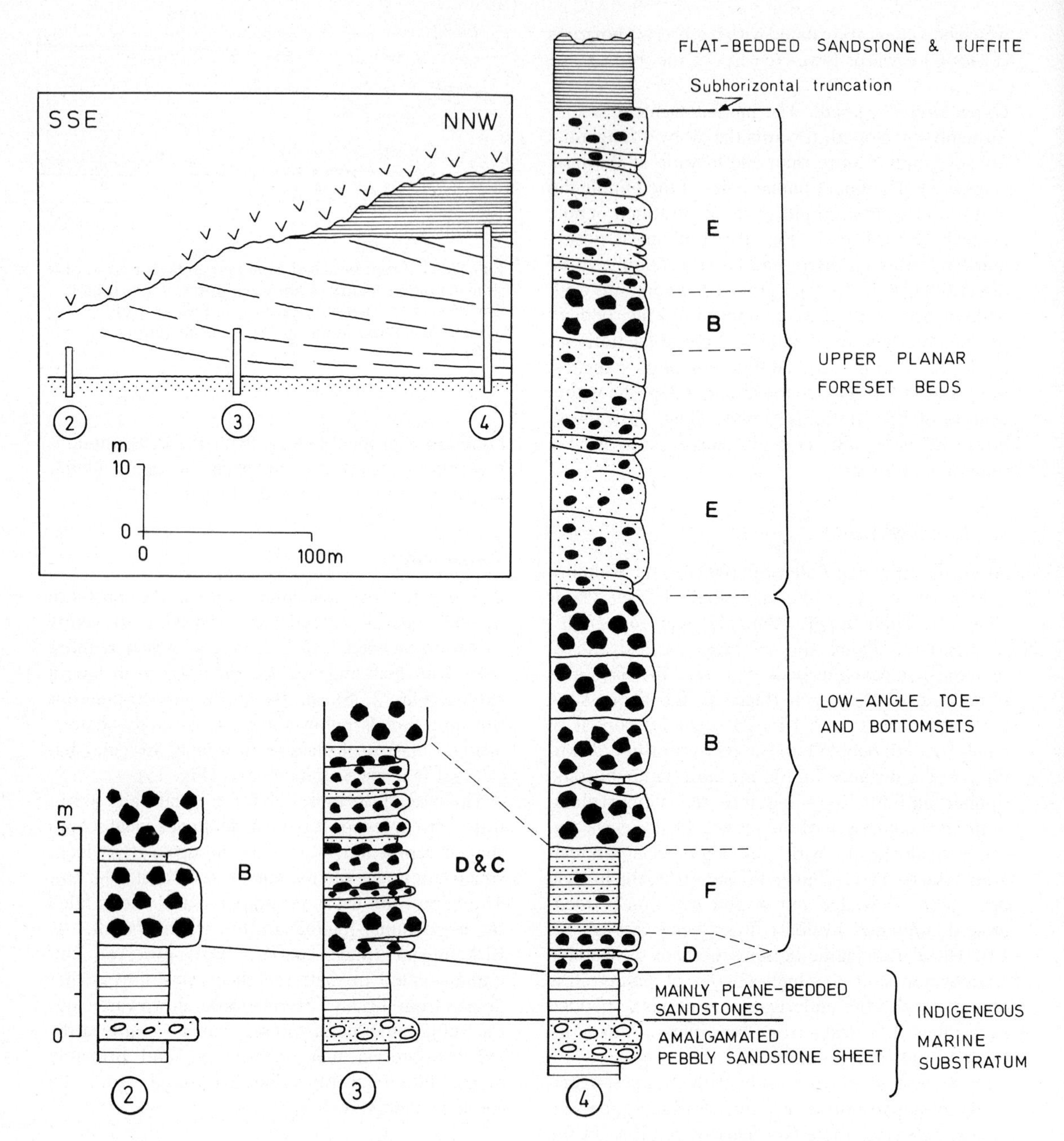

Fig. 14. Geometry and lithofacies logs of the southern body of the Oberek delta. Lithofacies code letters as in text; for location of the sections measured, see Fig. 2.

are best explained in terms of a flow of basaltic lava from land into the sea (Fig. 17). Other alternatives, such as deposition within a small volcano growing in shallow waters (1) or beneath grounded ice (2), are difficult to reconcile either with the closely associated laharic debris-flow deposition, or with the small height of the foreset bodies and the fossiliferous nature of the underlying deposits. Moreover,

Fig. 15. Close-up view of pillowed to irregular shattered basalt masses in the sheet-like lava-breccia topset (station 8 in Fig. 2).

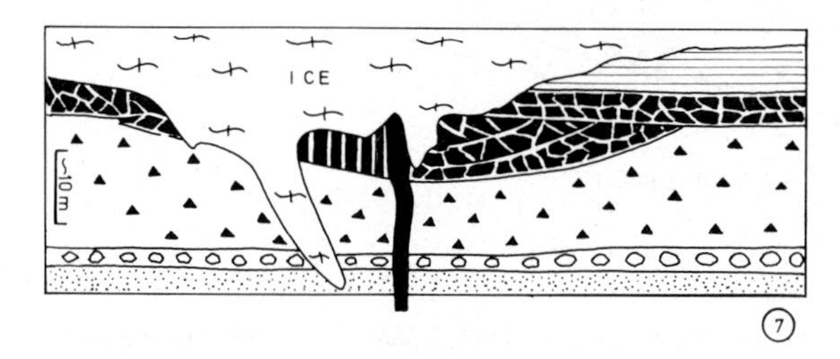

Fig. 16. Sketch modified from oblique air-photos, showing columnar-jointed to brecciated basalt which filled a scoop-shaped depression cut in the top of the inner indistinct-planar foreset segment and which merges upwards into the lava-breccia topset. Ornaments as in Fig. 2.

the non- to slightly vesicular character of the basaltic detritus suggests a hydroclastic fragmentation of volatile-free melts; their degassing must have taken place when the lava was flowing subaerially (cf. Nelson, 1966; Moore *et al.*, 1973).

The scarce admixtures of exotic clast types, though apparently supporting hypothesis 2 (see Walker & Blake, 1966; Jones, 1970), may have been derived either from the local diamictic substratum, through incorporation of unconsolidated gravels into the overriding lava (Furnes & Stuart, 1976), or from both stranded and floating growlers trapped in the lava or plumes of heated water. Similarly, the basaltic dyke piercing through the core zone of the northern deltaic body (Figs 2, 16), can hardly be taken as a vent to such a shoaling volcanoe (hypothesis 1), because the dyke follows closely the trend of other basaltic dykes in the area, that show consistently Miocene ages of intrusion (Birkenmajer *et al.*, 1986).

Local palaeogeography

Foreset-bed azimuths (Fig. 4) indicate that the lava flows which fed the Oberek delta system were issued from a volcanic centre located somewhere in the present-day Bransfield Strait area (Fig. 1B). They approached the coast from the east and poured into the sea at two main entry points at least 1.4 km apart, as indicated by the geometry of the coalescing zone. The resulting Gilbert-type hydroclastic bodies grew offshore to a depth of 23 m forming a deltaic prism that extended alongshore and seawards for at least 2.5 and 0.8 km, respectively. The lack of any clear signs of wave reworking throughout the foreset deposits suggest a very high rate of progradation, probably comparable to that of the Kelakomo lava delta which during 42 days of active growth reached about 1.5 km in length and 0.45 km in width (Moore *et al.*, 1973).

Feeding-flow phases and related delta-growth stages

The growth stages recognized in the Oberek deltaic bodies (Fig. 8) can be related to four distinct phases of the activity of a volcanic feeder system. The delta development was initiated by the emplacement of laharic debris flows, followed by high-viscosity lava streams that evolved, with time, towards lower vis-

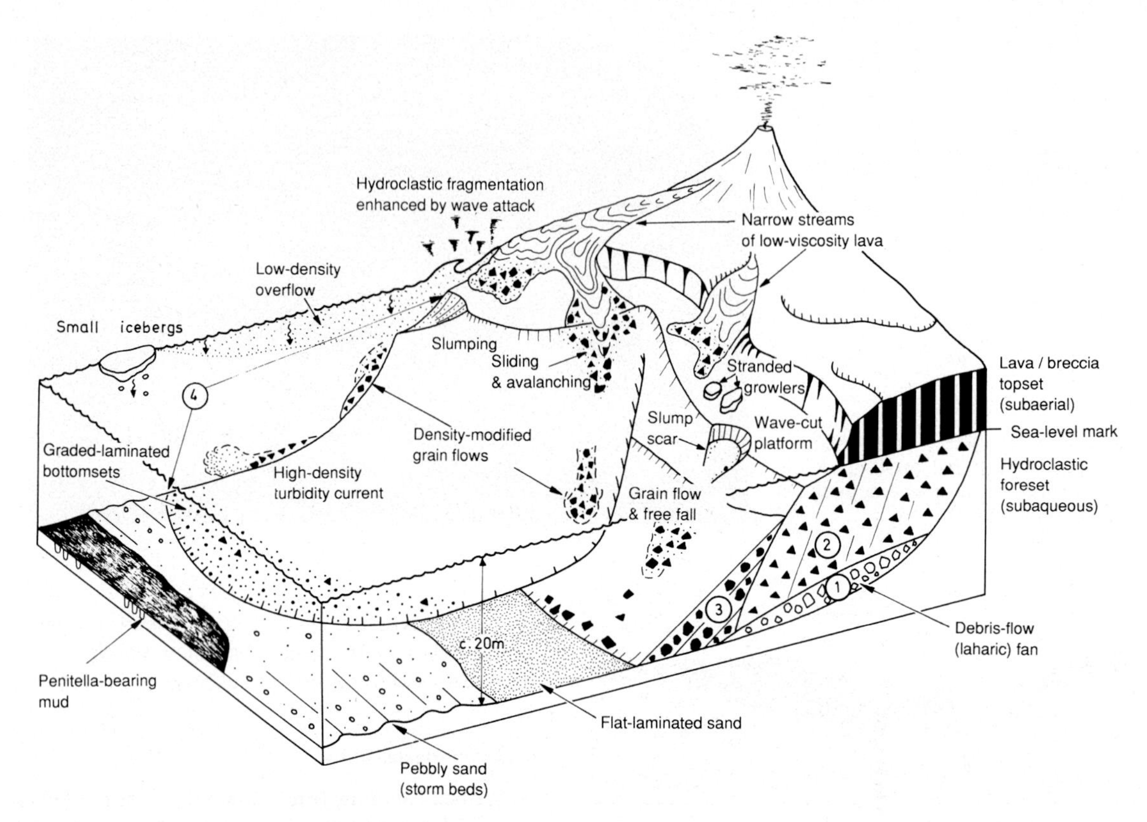

Fig. 17. Depositional model for the Oberek delta system. Numbers in circles indicate depositional units related to the four phases of delta growth described in the text.

cosities and smaller, possibly surging discharges (Fig. 17).

Phase of laharic flow (basal petromict-breccia unit)

The petromict breccia unit at the base of the northern deltaic body shows features consistent with a deposition from subaqueous cohesive debris flows (cf. Nemec & Steel, 1984) and reveals lateral textural trends (Figs 5, 7) which are compatible with those reported from small bodies of resedimented conglomerates (Porębski, 1984). The angular, petromict, volcaniclastic character of the breccia material suggests its derivation through collapse of a volcanic slope. This and the marine nearshore nature of the substratum (Siklawa Member) indicates that the breccia unit represents part of a small fan that was deposited on the lower shoreface from a series of laharic debris flows.

The lack of any signs of marine reworking in this fan strongly indicates that the successive flows closely followed each other, and that its emplacement was *immediately* followed by lava incursions which fed the overlying foreset body. This suggests, in turn, a close genetic link between the lahars and the lava incursions. A possible scenario to account for such a link could involve an emergent parent volcano having a snow/ice cover. A rapid melting of this cover due to the ascending magma, may have triggered the laharic debris flows which preceded lava effusions.

Phase of voluminous high-viscosity lava flow (inner foreset segment)

The coarse size and blocky non-vesicular lithic nature of the clasts, together with the scarcity of glassy sand matrix and absence of pillows, all indicate that the inner, indistinct planar-type foreset segment (Figs 3, 8) was fed by a relatively dense viscous lava flow which became fragmented on contact with the sea water due mainly to non-explosive thermal contraction. The considerable height and essentially unbedded core of this segment suggest that the initial flow phase involved rapid emplacement of voluminous highly coherent possibly *aa*-type flow. The hydroclastic fragmentation of the more rigid outer carapace of such a flow could have been aided through dynamic stress generated by the continued movement of the lower-viscosity lava forming the flow's interior (cf. Kokelaar, 1986; Busby-Spera & White, 1987).

The steep termini of the coherent brecciated prism provided neccessary slopes for launching foreset accretion, when new portions of hydroclastic debris became available for resedimentation at the top of the prism. This proceeded mainly through short-distance avalanching of grains and barely mobile grain masses, as indicated by the steep inclination and planar–angular geometry of the foreset beds. The absence of fine-grained fraction and the abundant supply of large angular clasts enhanced the stability of the slope and permitted it to maintain a its constant high dip (Colella *et al.*, 1987). Some lava tongues might overpass the prism's crest and break into coarse and fine hydroclasts locally further down on the slope. This could have prohibited gravity-sorting effects, as evidenced by the facies-A breccia. The scoop-shaped depression filled with the massive to brecciated basalt (Fig. 16) may be the result of an erosive action and freezing of a lava stream which remained coherent subaqueously, due to its exceptional thickness, and ploughed its way down through the talus slope.

Phase of ceased feeding flow (medial foreset segment)

The unconformable uplapping contact of the medial planar-type foreset segment against the distinct snout of the inner prism (Fig. 8) may reflect either a waning phase of the volcanic feeder system, or a temporal abandonment of the delta sector due to a shift of active lava streams. The planar–angular geometry and the distinct bedding created by regular alternation of the facies B and C breccia beds indicate that the medial foreset was emplaced by grain-flow-type avalanches and free-falling of larger clasts. The flows could have evolved from voluminous slumps initiated by wave attack on the crestal area which, if no longer supplied with new volumes of fresh hydroclastic debris, would become prone to retrogressive failure and smoothing. This resulted in a lengthening and lowering of the delta face, but the narrow coarse-size range of the detritus available apparently prevented the development of any mobile mass flows and creation of tangential foreset toes (Fig. 8).

Phase of small low-viscosity lava flows (outer foreset segment)

The considerable width and the sediment characteristics of the outer, tangential-type foreset segment, such as the wide grain-size spectrum, the abundance of basaltic glass (now largely converted into palagonite), and the blocky to pillow shapes of larger clasts, all suggest that this segment originated due to prolonged activity of small, discontinuous streams of low-viscosity lava.

Moore and others (1973) have noted that narrow streams of highly fluidal lava tend to be immediately quenched and fragmented into rubble and glassy sand on contact with the water, while thicker flows are occasionally able to maintain integrity for some distance down the delta face and produce lava tubes issuing pillow buds from cracks in a flow's solidified carapace. They noted also that: 'Fine material, including pebbles as much as 2 cm in diameter, were suspended in the water... and rained down to the old sea floor several hundred metres offshore' [beyond the talus slope] (Moore *et al.*, 1973, p. 542).

The alternation of the facies B, C and D beds in the toe parts of the tangential foreset beds, together with the occasional presence of inverse and normal grading in the breccia beds and the non-laminated nature of the sandy hyaloclastic interbeds, all suggest deposition from fluidal sediment-gravity flows, possibly density-modified grain flows transitional to small, high-density turbidity currents (cf. Lowe, 1982; Fig. 17). Such flows could have evolved from voluminous mass failures affecting the facies E hyaloclastite of the upper delta face. The high content of sand particles in this markedly bimodal sediment could significantly lower its frictional yield strength to promote large-scale slope failures along distinct slippage planes. The discontinuous truncations in the upper foreset parts (Figs 8, 13) may be traces of such slide/slump scars (cf. Ori & Roveri, 1987).

The large volumes of sediment set in motion provided a high initial momentum to the sliding gravel–sand mixtures which, further diluted by a water intake during flowage, increased in their mobility so that they could travel beyond the break of the delta slope (cf. Postma & Roep, 1985). Postma *et al.* (1988) have confirmed experimentally that such gravel–sand flows, while in motion, tend to split into a basal, coarse, inertia-flow layer and an overlying turbulent sandy layer, providing an effective mechanism for a downflow segregation of sand from gravel and for a transportation of outsized clasts along the rheological interface within a flow. Such mechanisms were probably responsible for the interbedding of facies D and C in the proximal bottomset beds (Figs 8, 14).

The thin-bedded graded–laminated distal bottomset (facies F) records chiefly a traction deposition from residual, turbulent sandy suspensions. Grainfall from a low-density 'overflow' which, as observed by Moore *et al.* (1973), may develop in the front of an advancing, sand-rich hydroclastic sheet (Fig. 17), probably contributed to the development and sandy nature of the distal bottomset beds.

SUMMARY AND CONCLUSIONS

The Oberek hydroclastic delta system of the Polonez Cove Formation resulted from coalescence of two, west-facing Gilbert-type bodies that descended down to a depth of 23 m on to a marine storm-wave dominated shoreface. A deltaic body has a basal unit (3.5 m thick) of flat-bedded petromict breccia, overlain by a foreset unit (16–23 m) of basaltic breccia and hyaloclastite, in turn, overlain by a topset (min. 5 m) of sheet-like lava breccia.

This tripartite stratigraphy is interpreted to reflect the initiation of a delta growth by subaerial laharic debris flows which came to rest on the lower shoreface to form a small subaqueous fan, and were followed immediately by incursions of basaltic lava streams. These latter, upon passing the shoreline, produced a subaqueous steep-flanked prism of hydroclastic debris and a subaerial platform of massive to brecciated lava on top of the prism. The fan and the lava-fed delta were supplied, probably during one eruptive cycle, from a volcanic centre located in the present-day Bransfield Strait area.

The evolution of the eruptive cycle, properties of the feeding lava flows and changes in the hydroclast-forming processes, can all be recognized from the geometry and lithofacies suite of the foreset-bedded deposits. A longitudinal section through the foreset unit reveals from the centre outwards: (1) a change from planar-angular to tangential wedge-shaped geometry of the foreset beds, accompanied by (2) improved bedding, (3) decreasing mean grain size and gravel/sand ratios and (4) a tendency for development of matrix-supported textures in the upper planar portions of otherwise tangential foresets, together with a downslope segregation of sand from gravel accompanying the development of bottomset beds. The (inner) planar–angular foreset beds accreted through grain avalanching and free-falling and by local emplacement of little fragmented lava tongues. The (outer) tangential foreset beds grew chiefly through deposition from fluidal, sediment gravity flows that evolved from large mass failures on the upper delta face. This change in the relative importance of particular progradation mechanisms is considered to reflect primarily a widening grain-size range of hydroclastic debris available for remobilization, in response to decreased viscosities and smaller, possibly surging discharges of the feeding lava streams.

This study shows a major similarity between the foreset developments of the studied lava-fed delta and coarse-grained alluvial deltas, such as described by Postma and Roep (1985), Postma and Cruickshank (1988), Colella *et al.* (1987), Ori and Roveri (1987). It is concluded that the mass-gravity transport plays the dominant role in the growth of such Gilbert-type deltaic systems, whether fed by lava streams or fluvial discharges.

ACKNOWLEDGEMENTS

Funding for this study was provided by the Polish Academy of Sciences research projects MR 1-29 and CPBP 03.03. Fieldwork was completed during V Polish Antarctic Expedition to Arctowski Station under leadership of Krzysztof Birkenmajer. C. Busby-Spera, F. Massari, W. Nemec and G. Postma are thanked for their criticism of the early drafts of the manuscript, and the editors for final improvement. SJP is grateful to A. Colella for her kind invitation to speak at the International Workshop on Fan Deltas, Calabria, 1988 and for financial support.

REFERENCES

Birkenmajer, K. (1982) Pliocene tillite-bearing succession of King George Island (South Shetland Islands, West Antarctica). *Studia geol. Polonica*, **74**, 7–72

Birkenmajer, K. (1988) Tertiary glacial and interglacial deposits, South Shetland Islands, Antarctica: geochronology versus biostratigraphy (a progress report). *Bull. Pol. Acad. Sci., Earth Sci.* (in press).

Birkenmajer, K., Delitala, M.B., Narebski, W., Nicoletti, M. & Petrucciani, C. (1986) Geochronology of Tertiary island-arc volcanics and glacigenic deposits,

King George Island, South Shetland Islands (West Antarctica). *Bull. Pol. Acad. Sci., Earth Sci.* **34**, (3), 257–273.

Birkenmajer, K. & Wieser, T. (1985) Petrology and provenance of magmatic and erratic blocks from Pliocene tillites of King George Island (South Shetland Islands, Antarctica). *Studia. geol. Polonica*, **81**, 53–97.

Busby-Spera, C.J. & White, J.D.L. (1987) Variation in peperite textures associated with differing host-sediment properties. *Bull. Volcanol.* **49**, 765–775.

Colella, A., De Boer, P.L. & Nio, S.D. (1987). Sedimentology of a marine intermontane Pleistocene Gilbert-type fan-delta complex in the Crati Basin, Calabria, southern Italy. *Sedimentology* **34**, 721–736.

Fisher, R.V. & Schmincke, H.U. (1984) *Pyroclastic Rocks*. Springer Verlag, Berlin, 472 pp.

Fuller, R.E. (1931) The aqueous chilling of basaltic lava on the Columbia River Plateau. *Am J. Sci.* **21**, 281–300.

Furnes, H. & Fridleifsson, I.B. (1974) Tidal effects on the formation of pillow lava/hyaloclastite deltas. *Geology*, **2**, 381–384.

Furnes, H. & Stuart, B.A. (1976) Beach/shallow-marine hyaloclastite deposits and their geological significance — an example from Gran Canaria. *J. geol.* **84**, 439–453.

Gazdzicka, E. & Gazdzicki, A. (1985) Oligocene coccoliths of the *Pecten* conglomerate, West Antarctica. *N. Jb. geol. Pal., Mh.* **12**, 727–735.

Gazdzicki, A., Gradziński, R., Porębski, S.J. & Wrona, R. (1982) Pholadid *Penitella* borings in glaciomarine sediments (Pliocene) of King George Island, Antarctica. *N. Jb. geol. Pal., Mh.* **12**, 723–735.

Gazdzicki, A. & Pugaczewska, H. (1984) Biota of the '*Pecten* conglomerate' (Polonez Cove Formation, Pliocene) of King George Island (South Shetland Islands, Antarctica). *Studia geol. Polonica*, **79**, 59–120.

Holmes, A. (1978) *Principles of Physical Geology. 3rd edn*. Thomas Nelson, London, 730 pp.

Jones, J.C. & Nelson, P.H.H. (1970) The flow of basalt lava from air into water – its structural expression and stratigraphic significance. *Geol. Mag.* **107**, 13–19.

Jones, J.G. (1970) Intraglacial volcanoes of the Laugarvatn region, southwest Iceland, II. *J. geol.* **78**, 127–140.

Kokelaar, B.P. (1986) Magma–water interactions in subaqueous and emergent basaltic volcanism. *Bull. Volcanol.* **48**, 275–289.

Lowe, D.R. (1982) Sediment-gravity flows: II. Depositional models with special reference to the deposits of high-density turbidity currents. *J. sedim. Petrol.* **52**, 279–297.

Moore, J.G., Phillips, R.L., Grigg, R.W., Peterson, D.W. & Swanson, D.A. (1973) Flow of lava into the sea 1969–1971, Kilauea Volcano, Hawaii. *Bull. geol. Soc. Am.* **84**, 537–546.

Nelson, P.H.H. (1966) The James Ross Island Volcanic Group of northeast Graham Land. *Br. Antarct. Surv. Sci. Rept.* **66**, 56 pp.

Nemec, W. & Steel, R.J. (1984) Alluvial and coastal conglomerates: their significant features and some comments on gravelly mass-flow deposits. In: *Sedimentology of Gravels and Conglomerates* (Ed. by E.H. Koster and R.J. Steel). Mem. Can. Soc. Petrol. Geol. 10, pp. 1–31.

Ori, G.G. & Roveri, M. (1987) Geometries of Gilbert-type deltas and large channels in the Meteora Conglomerate, Meso-Hellenic basin (Oligo-Miocene), central Greece. *Sedimentology* **34**, 845–859.

Porębski, S.J. (1984) Clast size and bed thickness trends in resedimented conglomerates; example from a Devonian fan-delta succession, southwest Poland. *Sedimentology of Gravels and Conglomerates* (Ed. by E.H. Koster and R.J. Steel). Mem. Can. Soc. Petrol. Geol. 10, pp. 399–411.

Porębski, S.J. & Gradziński, R. (1987) Depositional history of the Polonez Cove Formation (Oligocene), King George Island, West Antarctica: a record of continental glaciation, shallow-marine sedimentation and contemporaneous volcanism. *Studia geol. Polonica* **93**, 7–62.

Postma, G. & Cruickshank, C.H. (1988) Sedimentology of a late Weichselian to Holocene terraced fan delta, Varangerfjord, northern Norway. In: *Fan Deltas: Sedimentology and Tectonic Settings* (Ed. by W. Nemec and R.J. Steel), pp. 144–157. Blackie and Son, London.

Postma, G., Nemec, W. & Kleinspehn, K. (1988) Large floating clasts in turbidites: a mechanism for their emplacement. *Sedim. Geol.* **58**, 47–61.

Postma, G. & Roep, T.B (1985) Resedimented conglomerates in the bottomsets of Gilbert-type gravel deltas. *J. sedim. Petrol.* **55**, 874–885.

Snavely, P.D., Jr., Macleod, N.S. & Wagner, H.C. (1973) Miocene tholeiitic basalts of coastal Oregon and Washington and their relations to coeval basalts of the Columbia Plateau. *Bull. geol. Soc. Am.* **84**, 387–424.

Swanson, D.A. (1967) Yakima basalt of the Tieton river area, south central Washington. *Bull. geol. Soc. Am.* **78**, 1077–1110.

Walker, G.P.L. & Blake, D.H. (1966) Formation of a palagonite breccia mass beneath a valley glacier. *Q. J. geol. Soc. London* **122**, 45–61.

Waters, A.C. (1960) Determining the direction of flow in basalts. *Am J. Sci.* **258-A** (Bradley vol.), 350–366.

Weaver, S.D., Saunders, A.D. & Tarney, J. (1982) Mesozoic–Cenozoic volcanism in the South Shetland Islands and the Antarctic Peninsula: geochemical nature and plate tectonic significance. In: *Antarctic Geoscience* (Ed. by C. Craddock), pp. 263–273. Wisconsin Univ. Press, Madison.

Index

References to figures appear in *italic type*.
References to tables appear in **bold type**.